Study and Solutions Guide for
COLLEGE ALGEBRA
FOURTH EDITION
Larson / Hostetler

Dianna L. Zook

Indiana University
Purdue University at Fort Wayne, Indiana

HOUGHTON MIFFLIN COMPANY Boston New York

soring Editor: Christine B. Hoag
Senior Associate Editor: Maureen Brooks
Managing Editor: Catherine B. Cantin
Assistant Editor: Carolyn Johnson
Supervising Editor: Karen Carter
Associate Project Editor: Rachel D'Angelo Wimberly
Editorial Assistant: Caroline Lipscomb
Production Supervisor: Lisa Merrill
Art Supervisor: Gary Crespo
Marketing Manager: Charles Cavaliere
Marketing Associate: Ros Kane
Marketing Assistant: Kate Burden Thomas

Printed in the United States of America.

International Standard Book Number: 0-669-41753-X

23456789–VG 00 99 98 97 96

TO THE STUDENT

The *Study and Solutions Guide for College Algebra* is a supplement to the text by Roland E. Larson and Robert P. Hostetler.

As a mathematics instructor, I often have students come to me with questions about assigned homework. When I ask to see their work, the reply often is "I didn't know where to start." The purpose of the *Study Guide* is to provide brief summaries of the topics covered in the textbook and enough detailed solutions to problems so that you will be able to work the remaining exercises.

A special thanks to Larson Texts, Inc. for typing this guide. Also I would like to thank my husband Edward L. Schlindwein for his support during the several months I worked on this project.

If you have any corrections or suggestions for improving this *Study Guide*, I would appreciate hearing from you.

Good luck with your study of algebra.

Dianna L. Zook
Indiana University,
Purdue University at
Fort Wayne, Indiana 46805

CONTENTS

STUDY STRATEGIES

- Attend all classes and come prepared. Have your homework completed. Bring the text, paper, pen or pencil, and a calculator (scientific or graphing) to each class.

- Read the section in the text that is to be covered before class. Make notes about any questions that you have and, if not answered during the lecture, ask them at the appropriate time.

- Participate in class. As mentioned above, ask questions. Also, do not be afraid to answer questions.

- Take notes on all definitions, concepts, rules, formulas and examples. After class, read your notes and fill in any gaps, or make notations of any questions that you have.

- DO THE HOMEWORK!!! You learn mathematics by doing it yourself. Allow at least two hours outside of each class for homework. Do not fall behind.

- Seek help when needed. Visit your instructor during office hours and come prepared with specific questions; check with your school's tutoring service; find a study partner in class; check additional books in the library for more examples—just do something before the problem becomes insurmountable.

- Do not cram for exams. Each chapter in the text contains a chapter review and this study guide contains a practice test at the end of each chapter. (The answers are at the back of the study guide.) Work these problems a few days before the exam and review any areas of weakness.

PART I

CHAPTER P
Prerequisites

CHAPTER P
Prerequisites

Section P.1   Real Numbers

- You should know the following sets.
  - (a) The set of real numbers includes the rational numbers and the irrational numbers.
  - (b) The set of rational numbers includes all real numbers that can be written as the ratio p/q of two integers, where $q \neq 0$.
  - (c) The set of irrational numbers includes all real numbers which are not rational.
  - (d) The set of integers: $\{\ldots, -3, -2, -1, 0, 1, 2, 3, \ldots\}$
  - (e) The set of whole numbers: $\{0, 1, 2, 3, 4, \ldots\}$
  - (f) The set of natural numbers: $\{1, 2, 3, 4, \ldots\}$
- The real number line is used to represent the real numbers.
- Know the inequality symbols.
  - (a) $a < b$ means a is less than b.
  - (b) $a \leq b$ means a is less than or equal to b.
  - (c) $a > b$ means a is greater than b.
  - (d) $a \geq b$ means a is greater than or equal to b.
- You should know that
$$|a| = \begin{cases} a, & \text{if } a \geq 0 \\ -a, & \text{if } a < 0. \end{cases}$$
- Know the properties of absolute value.
  - (a) $|a| \geq 0$
  - (b) $|-a| = |a|$
  - (c) $|ab| = |a|\,|b|$
  - (d) $\left|\dfrac{a}{b}\right| = \dfrac{|a|}{|b|}$
- The distance between a and b on the real line is $|b - a| = |a - b|$.
- You should be able to identify the terms in an algebraic expression.
- You should know and be able to use the basic rules of algebra.
- Commutative Property
  - (a) Addition: $a + b = b + a$
  - (b) Multiplication: $a \cdot b = b \cdot a$
- Associative Property
  - (a) Addition: $(a + b) + c = a + (b + c)$
  - (b) Multiplication: $(ab)c = a(bc)$
- Identity Property
  - (a) Addition: 0 is the identity; $a + 0 = 0 + a = a$.
  - (b) Multiplication: 1 is the identity; $a \cdot 1 = 1 \cdot a = a$.
- Inverse Property
  - (a) Addition: $-a$ is the inverse of a; $a + (-a) = -a + a = 0$.
  - (b) Multiplication: $1/a$ is the inverse of a, $a \neq 0$; $a(1/a) = (1/a)a = 1$.
- Distributive Property
  - (a) Left: $a(b + c) = ab + ac$
  - (b) Right: $(a + b)c = ac + bc$

continued

■ Properties of Negatives

(a) $(-1)a = -a$

(b) $-(-a) = a$

(c) $(-a)b = a(-b) = -ab$

(d) $(-a)(-b) = ab$

(e) $-(a + b) = (-a) + (-b) = -a - b$

■ Properties of Zero

(a) $a \pm 0 = a$

(b) $a \cdot 0 = 0$

(c) $0 \div a = 0/a = 0, a \neq 0$

(d) If $ab = 0$, then $a = 0$ or $b = 0$.

(e) $a/0$ is undefined.

■ Properties of Fractions ($b \neq 0, d \neq 0$)

(a) Equivalent Fractions: $a/b = c/d$ if and only if $ad = bc$.

(b) Rule of Signs: $-a/b = a/-b = -(a/b)$ and $-a/-b = a/b$

(c) Equivalent Fractions: $a/b = ac/bc, c \neq 0$

(d) Addition and Subtraction

1. Like Denominators: $(a/b) \pm (c/b) = (a \pm c)/b$

2. Unlike Denominators: $(a/b) \pm (c/d) = (ad \pm bc)/bd$

(e) Multiplication: $(a/b) \cdot (c/d) = ac/bd$

(f) Division: $(a/b) \div (c/d) = (a/b) \cdot (d/c) = ad/bc$ if $c \neq 0$.

■ Properties of Equality

(a) If $a = b$, then $a + c = b + c$.

(b) If $a = b$, then $ac = bc$.

(c) If $a + c = b + c$, then $a = b$.

(d) If $ac = bc$ and $c \neq 0$, then $a = b$.

Solutions to Odd-Numbered Exercises

1. $-9, -\frac{7}{2}, 5, \frac{2}{3}, \sqrt{2}, 0, 1$

(a) Natural numbers: 5, 1

(b) Integers: $-9, 5, 0, 1$

(c) Rational numbers: $-9, -\frac{7}{2}, 5, \frac{2}{3}, 0, 1$

(d) Irrational numbers: $\sqrt{2}$

3. $2.01, 0.666\ldots, -13, 0.010110111\ldots$

(a) Natural numbers: none

(b) Integers: -13

(c) Rational numbers: $2.01, 0.666\ldots, -13$

(d) Irrational numbers: $0.010110111\ldots$

5. $-\pi, -\frac{1}{3}, \frac{6}{3}, \frac{1}{2}\sqrt{2}, -7.5$

(a) Natural numbers: $\frac{6}{3}$ (since it equals 2)

(b) Integers: $\frac{6}{3}$

(c) Rational numbers: $-\frac{1}{3}, \frac{6}{3}, -7.5$

(d) Irrational numbers: $-\pi, \frac{1}{2}\sqrt{2}$ *pg 2 = used as an terminating example of a rational number.*

7. $\frac{5}{8} = 0.625$

9. $\frac{41}{333} = 0.\overline{123}$

11. $-1 < 2.5$

13. $\frac{3}{2} < 7$

15. $-4 > -8$

17. $\frac{5}{6} > \frac{2}{3}$

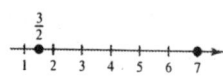

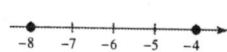

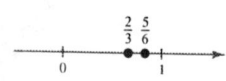

19. The inequality $x \le 5$ is the set of all real numbers less than or equal to 5. The interval is unbounded.

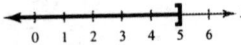

21. The inequality $x < 0$ is the set of all negative real numbers. The interval is unbounded.

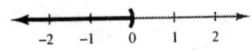

23. The inequality $x \ge 4$ is the set of all real numbers greater than or equal to 4. The interval is unbounded.

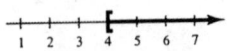

25. The inequality $-2 < x < 2$ is the set of all real numbers greater than -2 and less than 2. The interval is bounded.

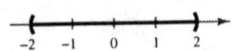

27. The inequality $-1 \le x < 0$ is the set of all negative real numbers greater than or equal to -1. The interval is bounded.

29. $\frac{127}{90} \approx 1.41111, \frac{584}{413} \approx 1.41404, \frac{7071}{5000} \approx 1.41420, \sqrt{2} \approx 1.41421, \frac{47}{33} \approx 1.42424$

31. $x < 0$

33. $y \ge 0$

35. $A \ge 30$

37. $|-10| = -(-10) = 10$

39. $|3 - \pi| = -(3 - \pi) = \pi - 3 \approx 0.1416$

41. $\dfrac{-5}{|-5|} = \dfrac{-5}{-(-5)} = \dfrac{-5}{5} = -1$

43. $-3|-3| = -3[-(-3)] = -9$

45. $-|16.25| + 20 = -16.25 + 20 = 3.75$

47. $|-3| > -|-3|$ since $3 > -3$.

49. $-5 = -|5|$ since $-5 = -5$.

51. $-|-2| = -|2|$ since $-2 = -2$.

53. $d(-1, 3) = |3 - (-1)| = |3 + 1| = 4$

55. $d\left(-\frac{5}{2}, 0\right) = \left|0 - \left(-\frac{5}{2}\right)\right| = \frac{5}{2}$

57. $d(126, 75) = |75 - 126| = 51$

59. $d\left(\frac{16}{5}, \frac{112}{75}\right) = \left|\frac{112}{75} - \frac{16}{5}\right| = \frac{128}{75}$

61. $d(x, 5) = |x - 5|$ and $d(x, 5) \le 3$, thus $|x - 5| \le 3$.

63. $d(7, 18) = |7 - 18| = 11$ miles

65. $d(y, 0) = |y - 0| = |y|$ and $d(y, 0) \ge 6$, thus $|y| \ge 6$.

67.

Budgeted Expense, b	Actual Expense, a	$|a - b|$	$0.05b$
$112,700	$113,356	$656	$5635

The actual expense difference is greater than $500 (but is less than 5% of the budget) so it does not pass the test.

69.

Budgeted Expense, b	Actual Expense, a	$|a - b|$	$0.05b$
$37,640	$37,335	$305	$1882

Since $305 < $500 and $305 < $1882, it passes the "budget variance test."

71. $|77.8 - 92.2| = \$14.4$ billion deficit for 1960

73. $|1031.3 - 1252.7| = \$221.4$ billion deficit for 1990

75. (a) $|u + v| \neq |u| + |v|$ if u is positive and v is negative or vice versa.

(b) $|u + v| \leq |u| + |v|$

They are equal when u and v have the same sign. If they differ in sign, $|u + v|$ is less than $|u| + |v|$.

77. $7x + 4$

Terms: $7x, 4$

79. $4x^3 + x - 5$

Terms: $4x^3, x, -5$

81. $4x - 6$

(a) $4(-1) - 6 = -4 - 6 = -10$

(b) $4(0) - 6 = 0 - 6 = -6$

83. $x^2 - 3x + 4$

(a) $(-2)^2 - 3(-2) + 4 = 4 + 6 + 4 = 14$

(b) $(2)^2 - 3(2) + 4 = 4 - 6 + 4 = 2$

85. $\dfrac{x + 1}{x - 1}$

(a) $\dfrac{1 + 1}{1 - 1} = \dfrac{2}{0}$ is undefined.

You cannot divide by zero.

(b) $\dfrac{-1 + 1}{-1 - 1} = \dfrac{0}{-2} = 0$

87. $x + 9 = 9 + x$

Commutative (addition)

89. $\dfrac{1}{(h + 6)}(h + 6) = 1, h \neq -6$

Inverse (multiplication)

91. $2(x + 3) = 2x + 6$

Distributive Property

93. $1 \cdot (1 + x) = 1 + x$

Identity (multiplication)

95. $x(3y) = (x \cdot 3)y$ Associative (multiplication)

$= (3x)y$ Commutative (multiplication)

97. $\dfrac{81 - (90 - 9)}{5} = \dfrac{81 - 81}{5} = \dfrac{0}{5} = 0$

99. $\dfrac{8 - 8}{-9 + (6 + 3)} = \dfrac{0}{-9 + 9} = \dfrac{0}{0}$ which is undefined.

101. $(4 - 7)(-2) = (-3)(-2) = 6$

103. $\frac{3}{16} + \frac{5}{16} = \frac{8}{16} = \frac{1}{2}$

105. $\dfrac{5}{8} - \dfrac{5}{12} + \dfrac{1}{6} = \dfrac{15}{24} - \dfrac{10}{24} + \dfrac{4}{24} = \dfrac{9}{24} = \dfrac{3}{8}$

107. $\dfrac{\cancel{4}}{5} \cdot \dfrac{1}{2} \cdot \dfrac{3}{\cancel{4}} \cdot = \dfrac{3}{10}$

109. $12 \div \frac{1}{4} = 12 \cdot \frac{4}{1} = 12 \cdot 4 = 48$

111. $-3 + \frac{3}{7} \approx -2.57$

113. $\dfrac{11.46 - 5.37}{3.91} \approx 1.56$

115.

n	1	0.5	0.01	0.0001	0.000001
$5/n$	5	10	500	50,000	5,000,000

117.

n	1	10	100	10,000	100,000
$5/n$	5	0.5	0.05	0.0005	0.00005

Section P.2   Exponents and Radicals

- ■ You should know the properties of exponents.

  (a) $a^1 = a$

  (b) $a^0 = 1, a \neq 0$

  (c) $a^m a^n = a^{m+n}$

  (d) $a^m/a^n = a^{m-n}, a \neq 0$

  (e) $a^{-n} = 1/a^n, a \neq 0$

  (f) $(a^m)^n = a^{mn}$

  (g) $(ab)^n = a^n b^n$

  (h) $(a/b)^n = a^n/b^n, b \neq 0$

  (i) $(a/b)^{-n} = (b/a)^n, a \neq 0, b \neq 0$

  (j) $|a^2| = |a|^2 = a^2$

- ■ You should be able to write numbers in scientific notation, $c \times 10^n$, where $1 \leq c < 10$ and n is an integer.

- ■ You should be able to use your calculator to evaluate expressions involving exponents.

- ■ You should know the properties of radicals.

  (a) $\sqrt[n]{a^m} = \left(\sqrt[n]{a}\right)^m$

  (b) $\sqrt[n]{a} \cdot \sqrt[n]{b} = \sqrt[n]{ab}$

  (c) $\dfrac{\sqrt[n]{a}}{\sqrt[n]{b}} = \sqrt[n]{\dfrac{a}{b}}$

  (d) $\sqrt[m]{\sqrt[n]{a}} = \sqrt[mn]{a}$

  (e) $\left(\sqrt[n]{a}\right)^n = a$

  (f) For n even, $\sqrt[n]{a^n} = |a|$.

  For n odd, $\sqrt[n]{a^n} = a$.

  (g) $a^{1/n} = \sqrt[n]{a}$

  (h) $a^{m/n} = \left(\sqrt[n]{a}\right)^m = \sqrt[n]{a^m}, a \geq 0$

- ■ You should be able to simplify radicals.

  (a) All possible factors have been removed from the radical sign.

  (b) All fractions have radical-free denominators.

  (c) The index for the radical has been reduced as far as possible

- ■ You should be able to use your calculator to evaluate radicals.

Solutions to Odd-Numbered Exercises

1. $-0.4^6 = -(0.4 \times 0.4 \times 0.4 \times 0.4 \times 0.4 \times 0.4)$

3. $(-10)(-10)(-10)(-10)(-10) = (-10)^5$

5. (a) $4^2 \cdot 3 = 16 \cdot 3 = 48$

    (b) $3 \cdot 3^3 = 3^4 = 81$

7. (a) $(3^3)^2 = 3^6 = 729$

    (b) $-3^2 = -9$

9. (a) $\dfrac{3}{3^{-4}} = 3^{1+4} = 3^5 = 243$

  (b) $24(-2)^{-5} = \dfrac{24}{(-2)^5} = \dfrac{24}{-32} = -\dfrac{3}{4}$

11. $(-4)^3(5^2) = (-64)(25) = -1600$

13. $\dfrac{3^6}{7^3} = \dfrac{729}{343} \approx 2.125$

15. When $x = 2$, $-3x^3 = -3(2)^3 = -24$.

17. When $x = 10$, $6x^0 - (6x)^0 = 6(10)^0 - (60)^0 = 6(1) - 1 = 5$.

19. (a) $(-5z)^3 = (-5)^3 z^3 = -125z^3$

  (b) $5x^4(x^2) = 5x^{4+2} = 5x^6$

21. (a) $6y^2(2y^4)^2 = 6y^2 2^2 y^8 = 6 \cdot 4y^{2+8} = 24y^{10}$

  (b) $\dfrac{3x^5}{x^3} = 3x^{5-3} = 3x^2$

23. (a) $\dfrac{7x^2}{x^3} = 7x^{2-3} = 7x^{-1} = \dfrac{7}{x}$

  (b) $\dfrac{12(x+y)^3}{9(x+y)} = \dfrac{4}{3}(x+y)^{3-1} = \dfrac{4}{3}(x+y)^2$

25. (a) $(x+5)^0 = 1$, $x \ne -5$

  (b) $(2x^2)^{-2} = \dfrac{1}{(2x^2)^2} = \dfrac{1}{4x^4}$

27. (a) $(-2x^2)^3 (4x^3)^{-1} = \dfrac{-8x^6}{4x^3} = -2x^3$

  (b) $\left(\dfrac{x}{10}\right)^{-1} = \dfrac{10}{x}$

29. (a) $(4a^{-2}b^3)^{-3} = 4^{-3}a^6 b^{-9} = \dfrac{a^6}{4^3 b^9} = \dfrac{a^6}{64b^9}$

  (b) $\left(\dfrac{5x^2}{y^{-2}}\right)^{-4} = (5x^2 y^2)^{-4} = \dfrac{1}{(5x^2 y^2)^4} = \dfrac{1}{625x^8 y^8}$

31. (a) $3^n \cdot 3^{2n} = 3^{n+2n} = 3^{3n}$

  (b) $\left(\dfrac{a^{-2}}{b^{-2}}\right)\left(\dfrac{b}{a}\right)^3 = \left(\dfrac{b^2}{a^2}\right)\left(\dfrac{b^3}{a^3}\right) = \dfrac{b^5}{a^5}$

Radical Form	*Rational Exponent Form*
33. $\sqrt{9} = 3$ Given	$9^{1/2} = 3$ Answer
35. $\sqrt[5]{32} = 2$ Answer	$32^{1/5} = 2$ Given
37. $\sqrt{196} = 14$ Answer	$196^{1/2} = 14$ Given
39. $\sqrt[3]{-216} = -6$ Given	$(-216)^{1/3} = -6$ Answer
41. $\sqrt[3]{27^2} = \left(\sqrt[3]{27}\right)^2 = 9$ Answer	$27^{2/3} = 9$ Given
43. $\sqrt[4]{81^3} = 27$ Given	$81^{3/4} = 27$ Answer

---

45. (a) $\sqrt{9} = 3$

  (b) $\sqrt[3]{8} = 2$

47. (a) $-\sqrt[3]{-27} = -(-3) = 3$

  (b) $\dfrac{4}{\sqrt{64}} = \dfrac{4}{8} = \dfrac{1}{2}$

49. (a) $\left(\sqrt[3]{-125}\right)^3 = -125$

  (b) $27^{1/3} = \sqrt[3]{27} = 3$

51. (a) $32^{-3/5} = \dfrac{1}{32^{3/5}} = \dfrac{1}{\left(\sqrt[5]{32}\right)^3} = \dfrac{1}{(2)^3} = \dfrac{1}{8}$

  (b) $\left(\dfrac{16}{81}\right)^{-3/4} = \left(\dfrac{81}{16}\right)^{3/4} = \left(\sqrt[4]{\dfrac{81}{16}}\right)^3 = \left(\dfrac{3}{2}\right)^3 = \dfrac{27}{8}$

53. (a) $\left(-\dfrac{1}{64}\right)^{-1/3} = (-64)^{1/3} = \sqrt[3]{-64} = -4$

(b) $\left(\dfrac{1}{\sqrt{32}}\right)^{-2/5} = \left(\sqrt{32}\right)^{2/5} = \sqrt[5]{\left(\sqrt{32}\right)^2} = \sqrt[5]{32} = 2$

55. (a) $\sqrt{57} \approx 7.550$

(b) $\sqrt[5]{-27^3} = (-27)^{3/5} \approx -7.225$

57. (a) $(1.2^{-2})\sqrt{75} + 3\sqrt{8} \approx 14.499$

(b) $\dfrac{-3 + \sqrt{21}}{3} \approx 0.528$

59. (a) $\sqrt{8} = \sqrt{4 \cdot 2} = \sqrt{4}\sqrt{2} = 2\sqrt{2}$

(b) $\sqrt[3]{24} = \sqrt[3]{8 \cdot 3} = \sqrt[3]{8}\sqrt[3]{3} = 2\sqrt[3]{3}$

61. (a) $\sqrt{72x^3} = \sqrt{36x^2 \cdot 2x} = 6x\sqrt{2x}$

(b) $\sqrt{\dfrac{18^2}{z^3}} = \dfrac{\sqrt{18^2}}{\sqrt{z^2 \cdot z}} = \dfrac{18}{z\sqrt{z}}$

63. (a) $\sqrt[3]{16x^5} = \sqrt[3]{8x^3 \cdot 2x^2} = 2x\sqrt[3]{2x^2}$

(b) $\sqrt{75x^2y^{-4}} = \sqrt{\dfrac{75x^2}{y^4}} = \dfrac{\sqrt{25x^2 \cdot 3}}{\sqrt{y^4}} = \dfrac{5|x|\sqrt{3}}{y^2}$

65. $5^{4/3} \cdot 5^{8/3} = 5^{12/3} = 5^4 = 625$

67. $\dfrac{(2x^2)^{3/2}}{2^{1/2}x^4} = \dfrac{2^{3/2}(x^2)^{3/2}}{2^{1/2}x^4} = \dfrac{2^{3/2}x^3}{2^{1/2}x^4} = 2^{3/2 - 1/2}x^{3-4} = 2^1 x^{-1} = \dfrac{2}{x}$

69. $\dfrac{x^{-3} \cdot x^{1/2}}{x^{3/2} \cdot x^{-1}} = \dfrac{x^{1/2} \cdot x^1}{x^{3/2} \cdot x^3} = x^{1/2 + 1 - 3/2 - 3} = x^{-3} = \dfrac{1}{x^3}\,, x > 0$

71. (a) $\dfrac{1}{\sqrt{3}} = \dfrac{1}{\sqrt{3}} \cdot \dfrac{\sqrt{3}}{\sqrt{3}} = \dfrac{\sqrt{3}}{3}$

(b) $\dfrac{8}{\sqrt[3]{2}} = \dfrac{8}{\sqrt[3]{2}} \cdot \dfrac{\sqrt[3]{4}}{\sqrt[3]{4}} = \dfrac{8\sqrt[3]{4}}{2} = 4\sqrt[3]{4}$

73. (a) $\dfrac{2x}{5 - \sqrt{3}} = \dfrac{2x}{5 - \sqrt{3}} \cdot \dfrac{5 + \sqrt{3}}{5 + \sqrt{3}}$

$$= \dfrac{2x(5 + \sqrt{3})}{25 - 3} = \dfrac{x(5 + \sqrt{3})}{11}$$

(b) $\dfrac{3}{\sqrt{5} + \sqrt{6}} \cdot \dfrac{\sqrt{5} - \sqrt{6}}{\sqrt{5} - \sqrt{6}} = \dfrac{3(\sqrt{5} - \sqrt{6})}{5 - 6}$

$$= -3(\sqrt{5} - \sqrt{6})$$

$$= 3(\sqrt{6} - \sqrt{5})$$

75. (a) $\dfrac{\sqrt{8}}{2} = \dfrac{\sqrt{4 \cdot 2}}{2} = \dfrac{2\sqrt{2}}{2} = \dfrac{\sqrt{2}}{1} \cdot \dfrac{\sqrt{2}}{\sqrt{2}} = \dfrac{2}{\sqrt{2}}$

(b) $\sqrt[3]{\dfrac{9}{25}} = \dfrac{\sqrt[3]{9}}{\sqrt[3]{25}} \cdot \dfrac{\sqrt[3]{3}}{\sqrt[3]{3}} = \dfrac{\sqrt[3]{27}}{\sqrt[3]{75}} = \dfrac{3}{\sqrt[3]{75}}$

77. (a) $\dfrac{\sqrt{5} + \sqrt{3}}{3} = \dfrac{\sqrt{5} + \sqrt{3}}{3} \cdot \dfrac{\sqrt{5} - \sqrt{3}}{\sqrt{5} - \sqrt{3}} = \dfrac{5 - 3}{3(\sqrt{5} - \sqrt{3})}$

$$= \dfrac{2}{3(\sqrt{5} - \sqrt{3})}$$

(b) $\dfrac{\sqrt{7} - 3}{4} = \dfrac{\sqrt{7} - 3}{4} \cdot \dfrac{\sqrt{7} + 3}{\sqrt{7} + 3} = \dfrac{7 - 9}{4(\sqrt{7} + 3)}$

$$= \dfrac{-2}{4(\sqrt{7} + 3)}$$

$$= -\dfrac{1}{2(\sqrt{7} + 3)}$$

79. (a) $\sqrt[4]{3^2} = 3^{2/4} = 3^{1/2} = \sqrt{3}$

(b) $\sqrt[6]{(x + 1)^4} = (x + 1)^{4/6} = (x + 1)^{2/3} = \sqrt[3]{(x + 1)^2}$

81. (a) $\sqrt{\sqrt{32}} = (32^{1/2})^{1/2} = 32^{1/4} = \sqrt[4]{32} = \sqrt[4]{16 \cdot 2} = 2\sqrt[4]{2}$

(b) $\sqrt{\sqrt[4]{2x}} = ((2x)^{1/4})^{1/2} = (2x)^{1/8} = \sqrt[8]{2x}$

83. (a) $2\sqrt{50} + 12\sqrt{8} = 2\sqrt{25 \cdot 2} + 12\sqrt{4 \cdot 2}$

$$= 2(5\sqrt{2}) + 12(2\sqrt{2}) = 10\sqrt{2} + 24\sqrt{2} = 34\sqrt{2}$$

(b) $10\sqrt{32} - 6\sqrt{18} = 10\sqrt{16 \cdot 2} - 6\sqrt{9 \cdot 2}$

$$= 10(4\sqrt{2}) - 6(3\sqrt{2}) = 40\sqrt{2} - 18\sqrt{2} = 22\sqrt{2}$$

85. (a) $5\sqrt{x} - 3\sqrt{x} = 2\sqrt{x}$

(b) $-2\sqrt{9y} + 10\sqrt{y} = -2(3\sqrt{y}) + 10\sqrt{y} = -6\sqrt{y} + 10\sqrt{y} = 4\sqrt{y}$

87. $\sqrt{5} + \sqrt{3} \approx 3.968$ and $\sqrt{5 + 3} = \sqrt{8} \approx 2.828$
Thus, $\sqrt{5} + \sqrt{3} > \sqrt{5 + 3}$.

89. $\sqrt{3^2 + 2^2} = \sqrt{9 + 4} = \sqrt{13} \approx 3.606$
Thus, $5 > \sqrt{3^2 + 2^2}$.

91. $57{,}500{,}000 = 5.75 \times 10^7$ square miles

93. $0.0000899 = 8.99 \times 10^{-5}$

95. $5.24 \times 10^8 = 524{,}000{,}000$ servings

97. $4.8 \times 10^{-10} = 0.00000000048$ electrostatic units

99. (a) $750\left(1 + \dfrac{0.11}{365}\right)^{800} \approx 954.448$

(b) $\dfrac{67,000,000 + 93,000,000}{0.0052} = 30,769,230,769.2 \approx 3.077 \times 10^{10}$

101. (a) $\sqrt{4.5 \times 10^9} \approx 67,082.039$

(b) $\sqrt[3]{6.3 \times 10^4} \approx 39.791$

103. When any positive integer is squared, the units digit is 0, 1, 4, 5, 6, or 9. Therefore, $\sqrt{5233}$ is not an integer.

105. $T = 2\pi\sqrt{\dfrac{2}{32}} = 2\pi\sqrt{\dfrac{1}{16}} = 2\pi\left(\dfrac{1}{4}\right) = \dfrac{\pi}{2} \approx 1.57$ seconds

107. $r = 1 - \left(\dfrac{3225}{12,000}\right)^{1/4} \approx 0.280$ or 28%

109. Time $= \dfrac{\text{Distance}}{\text{Rate}} = \dfrac{93,000,000 \text{ miles}}{11,160,000 \text{ miles per minute}} = \dfrac{25}{3}$ minutes

Section P.3   Polynomials and Special Products

- Given a polynomial in x, $a_n x^n + a_{n-1} x^{n-1} + \ldots + a_1 x + a_0$, where $a_n \neq 0$, and n is a nonnegative integer, you should be able to identify the following.
  - (a) Degree: n
  - (b) Terms: $a_n x^n, a_{n-1} x^{n-1}, \ldots, a_1 x, a_0$
  - (c) Coefficients: $a_n, a_{n-1}, \ldots, a_1, a_0$
  - (d) Leading coefficient: a_n
  - (e) Constant term: a
- You should be able to add and subtract polynomials.
- You should be able to multiply polynomials by either
  - (a) The Distributive Properties
  - (b) The Vertical Method.
- You should know the special binomial products.
  - (a) $(ax + b)(cx + d) = acx^2 + adx + bcx + bd$ FOIL
    $$= acx^2 + (ad + bc)x + bd$$
  - (b) $(u \pm v)^2 = u^2 \pm 2uv + v^2$
  - (c) $(u + v)(u - v) = u^2 - v^2$
  - (d) $(u \pm v)^3 = u^3 \pm 3u^2 v + 3uv^2 \pm v^3$

Solutions to Odd-Numbered Exercises

1. (d) 12 is a polynomial of degree zero.

3. (b) $1 - 2x^3 = -2x^3 + 1$ is a binomial with leading coefficient -2.

5. (f) $\frac{2}{3}x^4 + x^2 + 10$ is a trinomial with leading coefficient $\frac{2}{3}$.

7. Standard form: $2x^2 - x + 1$
   Degree: 2
   Leading coefficient: 2

9. Standard form: $x^5 - 1$
   Degree: 5
   Leading coefficient: 1

11. Standard form: $-4x^5 + 6x^4 - x + 1$
    Degree: 5
    Leading coefficient: -4

13. $2x - 3x^3 + 8$ *is* a polynomial.
    Standard form: $-3x^3 + 2x + 8$

15. $\dfrac{3x + 4}{x} = 3 + \dfrac{4}{x}$ is *not* a polynomial.

17. $y^2 - y^4 + y^3$ *is* a polynomial.
    Standard form: $-y^4 + y^3 + y^2$

19. $(6x + 5) - (8x + 15) = 6x + 5 - 8x - 15$
    $$= (6x - 8x) + (5 - 15)$$
    $$= -2x - 10$$

21. $-(x^3 - 2) + (4x^3 - 2x) = -x^3 + 2 + 4x^3 - 2x$
    $$= (4x^3 - x^3) - 2x + 2$$
    $$= 3x^3 - 2x + 2$$

23. $(15x^2 - 6) - (-8x^3 - 14x^2 - 17) = 15x^2 - 6 + 8x^3 + 14x^2 + 17$
    $$= 8x^3 + (15x^2 + 14x^2) + (-6 + 17)$$
    $$= 8x^3 + 29x^2 + 11$$

25. $5z - [3z - (10z + 8)] = 5z - (3z - 10z - 8)$
    $$= 5z - 3z + 10z + 8$$
    $$= (5z - 3z + 10z) + 8$$
    $$= 12z + 8$$

27. $3x(x^2 - 2x + 1) = 3x(x^2) + 3x(-2x) + 3x(1)$
    $$= 3x^3 - 6x^2 + 3x$$

29. $-5z(3z - 1) = -5z(3z) + (-5z)(-1)$
    $$= -15z^2 + 5z$$

31. $(1 - x^3)(4x) = 1(4x) - x^3(4x) = 4x - 4x^4 = -4x^4 + 4x$

33. Add: $7x^3 - 2x^2 + 8$
         $\underline{-3x^3 \qquad\quad - 4}$
         $4x^3 - 2x^2 + 4$

35. Subtract: $5x^2 - 3x + 8$
         $\underline{- \qquad\quad (x - 3)}$
         $5x^2 - 4x + 11$

37. Multiply: $-6x^2 + 15x - 4$
$$ 5x + 3$$
$$\overline{-30x^3 + 75x^2 - 20x}$$
$$ - 18x^2 + 45x - 12$$
$$\overline{-30x^3 + 57x^2 + 25x - 12}$$

39. Multiply: $x^2 - x - 4$
$$ x^2 + 9$$
$$\overline{x^4 - x^3 - 4x^2}$$
$$ 9x^2 - 9x - 36$$
$$\overline{x^4 - x^3 + 5x^2 - 9x - 36}$$

41. Multiply: $x^2 - x + 1$
$$ x^2 + x + 1$$
$$\overline{x^4 - x^3 + x^2}$$
$$ x^3 - x^2 + x$$
$$ x^2 - x + 1$$
$$\overline{x^4 - 0x^3 + x^2 + 0x + 1} = x^4 + x^2 + 1$$

43. $(x + 3)(x + 4) = x^2 + 4x + 3x + 12$ FOIL
$$= x^2 + 7x + 12$$

45. $(3x - 5)(2x + 1) = 6x^2 + 3x - 10x - 5$ FOIL
$$= 6x^2 - 7x - 5$$

47. $(2x + 3)^2 = (2x)^2 + 2(2x)(3) + 3^2$
$$= 4x^2 + 12x + 9$$

49. $(2x - 5y)^2 = 4x^2 - 2(5y)(2x) + 25y^2$
$$= 4x^2 - 20xy + 25y^2$$

51. $[(x - 3) + y]^2 = (x - 3)^2 + 2y(x - 3) + y^2$
$$= x^2 - 6x + 9 + 2xy - 6y + y^2$$
$$= x^2 + 2xy + y^2 - 6x - 6y + 9$$

53. $(x + 10)(x - 10) = x^2 - 100$

55. $(x + 2y)(x - 2y) = x^2 - (2y)^2 = x^2 - 4y^2$

57. $[(m - 3) + n][(m - 3) - n] = (m - 3)^2 - n^2$
$$= m^2 - 6m + 9 - n^2$$
$$= m^2 - n^2 - 6m + 9$$

59. $(2r^2 - 5)(2r^2 + 5) = (2r)^2 - 5^2 = 4r^4 - 25$

61. $(x + 1)^3 = x^3 + 3x^2(1) + 3x(1^2) + 1^3$
$$= x^3 + 3x^2 + 3x + 1$$

63. $(2x - y)^3 = (2x)^3 - 3(2x)^2y + 3(2x)y^2 - y^3$
$$= 8x^3 - 12x^2y + 6xy^2 - y^3$$

65. $(4x^3 - 3)^2 = (4x^3)^2 - 2(4x^3)(3) + (3)^2$
$$= 16x^6 - 24x^3 + 9$$

67. $5x(x + 1) - 3x(x + 1) = 2x(x + 1)$
$$= 2x^2 + 2x$$

69. $(u + 2)(u - 2)(u^2 + 4) = (u^2 - 4)(u^2 + 4)$
$$= u^4 - 16$$

71. $\left(\sqrt{x} + \sqrt{y}\right)\left(\sqrt{x} - \sqrt{y}\right) = \left(\sqrt{x}\right)^2 - \left(\sqrt{y}\right)^2 = x - y$

73. $\left(x + \sqrt{5}\right)\left(x - \sqrt{5}\right)(x + 4) = (x^2 - 5)(x + 4)$ FOIL
$$= x^3 + 4x^2 - 5x - 20$$

75. No. $(x^2 + 1) + (-x^2 + 3) = 4$ which is not a second degree polynomial.

77. (a) $(x - 1)(x + 1) = x^2 - 1$

(b) $(x - 1)(x^2 + x + 1) = x^3 + x^2 + x - x^2 - x - 1 = x^3 - 1$

(c) $(x - 1)(x^3 + x^2 + x + 1) = x^4 + x^3 + x^2 + x - x^3 - x^2 - x - 1$

$$= x^4 - 1$$

From this pattern we have $(x - 1)(x^4 + x^3 + x^2 + x + 1) = x^5 - 1$.

79. $(x + y)^2 \neq x^2 + y^2$

Let $x = 3$ and $y = 4$.

$(3 + 4)^2 = (7)^2 = 49$

$3^2 + 4^2 = 9 + 16 = 25$ ⟩ Not Equal

81. (a) $500(1 + r)^2 = 500(r + 1)^2 = 500(r^2 + 2r + 1)$

$$= 500r^2 + 1000r + 500$$

(b)

r	$5\frac{1}{2}\%$	7%	8%	$8\frac{1}{2}\%$	9%
$500(1 + r)^2$	$556.51	$572.45	$583.20	$588.61	$594.05

(c) As r increases, the amount increases.

83. $V = l \cdot w \cdot h = (26 - 2x)(18 - 2x)(x)$

$$= 2(13 - x)(2)(9 - x)(x)$$

$$= 4x(-1)(x - 13)(-1)(x - 9)$$

$$= 4x(x - 13)(x - 9)$$

When $x = 1$: $V = 4(1)(-12)(-8) = 384$ cubic inches

When $x = 2$: $V = 4(2)(-11)(-7) = 616$ cubic inches

When $x = 3$: $V = 4(3)(-10)(-6) = 720$ cubic inches

85. (a) Area of shaded region = Area of outer rectangle − Area of inner rectangle

$A = 2x(2x + 6) - x(x + 4)$

$$= 4x^2 + 12x - x^2 - 4x$$
$$= 3x^2 + 8x$$

(b) Area of shaded region = Area of outer triangle − Area of inner triangle

$A = \frac{1}{2}(9x)(12x) - \frac{1}{2}(6x)(8x)$

$$= 54x^2 - 24x^2$$
$$= 30x^2$$

87. (a) $T = R + B = 1.1x + (0.14x^2 - 4.43x + 58.40)$

$$= 0.14x^2 - 3.33x + 58.40$$

(b)

x mi/hr	30	40	55
T feet	84.50	149.20	298.75

(c) Stopping distance increases at an accelerating rate as speed increases.

89. $(x + 1)(x + 4) = x(x + 4) + 1(x + 4)$

$$= x^2 + 4x + x + 4$$

This illustrates the Distributive Property.

91. Since $x^m x^n = x^{m+n}$, the degree of the product is $m + n$.

93. $(x - 3)^2 = x^2 - 6x + 9 \neq x^2 + 9$
The student omitted the middle term.

Section P.4   Factoring

> ■ You should be able to factor out all common factors, the first step in factoring.
>
> ■ You should be able to factor the following special polynomial forms.
>
>   (a) $u^2 - v^2 = (u + v)(u - v)$
>
>   (b) $u^2 \pm 2uv + v^2 = (u \pm v)^2$
>
>   (c) $mx^2 + nx + r = (ax + b)(cx + d)$, where $m = ac, r = bd, n = ad + bc$
>
>   **Note:** Not all trinomials can be factored (using real coefficients).
>
>   (d) $u^3 \pm v^3 = (u \pm v)(u^2 \mp uv + v^2)$
>
> ■ You should be able to factor by grouping.
>
> ■ You should be able to factor some trinomials by grouping.

Solutions to Odd-Numbered Exercises

1. $90 = 2 \cdot 3 \cdot 3 \cdot 5$

$300 = 2 \cdot 2 \cdot 3 \cdot 5 \cdot 5$

Greatest common factor: $2 \cdot 3 \cdot 5 = 30$

3. $12x^2y^3 = 2 \cdot 2 \cdot 3 \cdot x \cdot x \cdot y \cdot y \cdot y$

$18x^2y = 2 \cdot 3 \cdot 3 \cdot x \cdot x \cdot y$

$24x^3y^2 = 2 \cdot 2 \cdot 2 \cdot 3 \cdot x \cdot x \cdot x \cdot y \cdot y$

Greatest common factor: $2 \cdot 3 \cdot x \cdot x \cdot y = 6x^2y$

5. $3x + 6 = 3(x + 2)$

7. $2x^3 - 6x = 2x(x^2 - 3)$

9. $(x - 1)^2 + 6(x - 1) = (x - 1)[(x - 1) + 6]$

$$= (x - 1)(x + 5)$$

11. $x^2 - 36 = x^2 - 6^2 = (x + 6)(x - 6)$

13. $16y^2 - 9 = (4y)^2 - 3^2 = (4y + 3)(4y - 3)$

15. $(x - 1)^2 - 4 = [(x - 1) + 2][(x - 1) - 2]$

$$= (x + 1)(x - 3)$$

17. $x^2 - 4x + 4 = x^2 - 2(2)x + 2^2 = (x - 2)^2$

19. $4t^2 + 4t + 1 = (2t)^2 + 2(2t)(1) + 1^2$

$$= (2t + 1)^2$$

21. $25y^2 - 10y + 1 = (5y)^2 - 2(5y)(1) + 1^2$

$$= (5y - 1)^2$$

23. $25 - 5x^2 = -5(-5 + x^2) = -5(x^2 - 5)$

25. $-2t^3 + 4t + 6 = -2(t^3 - 2t - 3)$

27. $x^2 + x - 2 = (x + 2)(x - 1)$

29. $s^2 - 5s + 6 = (s - 3)(s - 2)$

31. $20 - y - y^2 = -(y^2 + y - 20)$
$$= -(y + 5)(y - 4)$$

33. $x^2 - 30x + 200 = (x - 20)(x - 10)$

35. $3x^2 - 5x + 2 = (3x - 2)(x - 1)$

37. $-9z^2 + 3z + 2 = -(9z^2 - 3z - 2)$
$$= -(3z - 2)(3z + 1)$$

39. $5x^2 + 26x + 5 = (5x + 1)(x + 5)$

41. $x^3 - 8 = x^3 - 2^3 = (x - 2)(x^2 + 2x + 4)$

43. $y^3 + 64 = y^3 + 4^3 = (y + 4)(y^2 - 4y + 16)$

45. $8t^3 - 1 = (2t)^3 - 1^3 = (2t - 1)(4t^2 + 2t + 1)$

47. $x^3 - x^2 + 2x - 2 = x^2(x - 1) + 2(x - 1)$
$$= (x - 1)(x^2 + 2)$$

49. $2x^3 - x^2 - 6x + 3 = x^2(2x - 1) - 3(2x - 1)$
$$= (2x - 1)(x^2 - 3)$$

51. $6 + 2x - 3x^3 - x^4 = 2(3 + x) - x^3(3 + x)$
$$= (3 + x)(2 - x^3)$$

53. $a \cdot c = (3)(8) = 24$. Rewrite the middle term, $10x = 6x + 4x$, since $(6)(4) = 24$ and $6 + 4 = 10$.

$$3x^2 + 10x + 8 = 3x^2 + 6x + 4x + 8$$
$$= 3x(x + 2) + 4(x + 2)$$
$$= (x + 2)(3x + 4)$$

55. $a \cdot c = (6)(-2) = -12$. Rewrite the middle term, $x = 4x - 3x$, since $4(-3) = -12$ and $4 + (-3) = 1$.

$$6x^2 + x - 2 = 6x^2 + 4x - 3x - 2$$
$$= 2x(3x + 2) - 1(3x + 2)$$
$$= (3x + 2)(2x - 1)$$
$$= (2x - 1)(3x + 2)$$

57. $a \cdot c = (15)(2) = 30$. Rewrite the middle term, $-11x = -6x - 5x$, since $(-6)(-5) = 30$ and $(-6) + (-5) = -11$.

$$15x^2 - 11x + 2 = 15x^2 - 6x - 5x + 2$$
$$= 3x(5x - 2) - 1(5x - 2)$$
$$= (5x - 2)(3x - 1)$$
$$= (3x - 1)(5x - 2)$$

59. $x^3 - 9x = x(x^2 - 9) = x(x + 3)(x - 3)$

61. $x^3 - 4x^2 = x^2(x - 4)$

63. $x^2 - 2x + 1 = (x - 1)^2$

65. $1 - 4x + 4x^2 = (1 - 2x)^2$

67. $2x^2 + 4x - 2x^3 = -2x(-x - 2 + x^2)$
$$= -2x(x^2 - x - 2)$$
$$= -2x(x + 1)(x - 2)$$

69. $9x^2 + 10x + 1 = (9x + 1)(x + 1)$

71. $3x^3 + x^2 + 15x + 5 = x^2(3x + 1) + 5(3x + 1)$
$$= (3x + 1)(x^2 + 5)$$

73. $x^4 - 4x^3 + x^2 - 4x = x(x^3 - 4x^2 + x - 4)$
$$= x[x^2(x - 4) + (x - 4)]$$
$$= x(x - 4)(x^2 + 1)$$

75. $25 - (z + 5)^2 = [5 + (z + 5)][5 - (z + 5)]$
$$= -z(z + 10)$$

77. $(x^2 + 1)^2 - 4x^2 = [(x^2 + 1) + 2x][(x^2 + 1) - 2x]$
$$= (x^2 + 2x + 1)(x^2 - 2x + 1)$$
$$= (x + 1)^2(x - 1)^2$$

79. $2t^3 - 16 = 2(t^3 - 8) = 2(t - 2)(t^2 + 2t + 4)$

81. $4x(2x - 1) + (2x - 1)^2 = (2x - 1)[4x + (2x - 1)]$
$$= (2x - 1)(6x - 1)$$

83. $2(x + 1)(x - 3)^2 - 3(x + 1)^2(x - 3) = (x + 1)(x - 3)[2(x - 3) - 3(x + 1)]$
$$= (x + 1)(x - 3)[2x - 6 - 3x - 3]$$
$$= (x + 1)(x - 3)(-x - 9)$$
$$= -(x + 1)(x - 3)(x + 9)$$

85. $7x(2)(x^2 + 1)(2x) - (x^2 + 1)^2(7) = 7(x^2 + 1)[4x^2 - (x^2 + 1)]$
$$= 7(x^2 + 1)(3x^2 - 1)$$

87. $2x(x - 5)^4 - x^2(4)(x - 5)^3 = 2x(x - 5)^3[(x - 5) - 2x]$
$$= 2x(x - 5)^3(-x - 5)$$
$$= -2x(x - 5)^3(x + 5)$$

89. $\dfrac{x^2}{2}(x^2 + 1)^4 - (x^2 + 1)^5 = (x^2 + 1)^4\left[\dfrac{x^2}{2} - (x^2 + 1)\right]$
$$= (x^2 + 1)^4\left(-\dfrac{x^2}{2} - 1\right)$$
$$= -(x^2 + 1)^4\left(\dfrac{x^2}{2} + 1\right)$$

91. $a^2 - b^2 = (a + b)(a - b)$
Matches model (b).

93. $a^2 + 2a + 1 = (a + 1)^2$
Matches model (a).

95. $3x^2 + 7x + 2 = (3x + 1)(x + 2)$

97. $2x^2 + 7x + 3 = (2x + 1)(x + 3)$

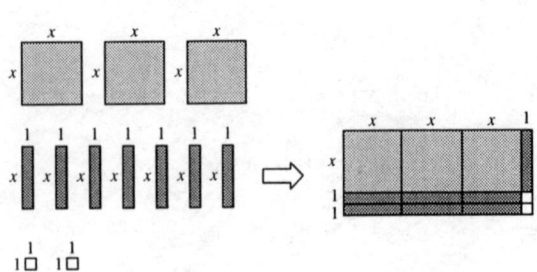

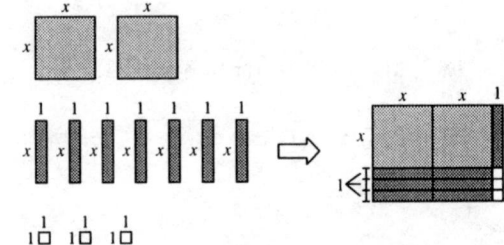

99. $A = \pi(r + 2)^2 - \pi r^2$

$\quad = \pi[(r + 2)^2 - r^2]$

$\quad = \pi[r^2 + 4r + 4 - r^2]$

$\quad = \pi(4r + 4)$

$\quad = 4\pi(r + 1)$

101. $A = 8(18) - 4x^2$

$\quad = 4(36 - x^2)$

$\quad = 4(6 - x)(6 + x)$

103. For $x^2 + bx - 15$ to be factorable, b must equal $m + n$ where $mn = -15$.

Factors of -15	Sum of factors
$(15)(-1)$	$15 + (-1) = 14$
$(-15)(1)$	$-15 + 1 = -14$
$(3)(-5)$	$3 + (-5) = -2$
$(-3)(5)$	$-3 + 5 = \ \ 2$

The possible b values are 14, -14, -2, or 2.

105. For $2x^2 + 5x + c$ to be factorable, the factors of $2c$ must add up to 5.

Possible c values	$2c$	Factors of $2c$ that add up to 5
2	4	$(1)(4) = 4$ and $1 + 4 = 5$
3	6	$(2)(3) = 6$ and $2 + 3 = 5$
-3	-6	$(6)(-1) = -6$ and $6 + (-1) = 5$
-7	-14	$(7)(-2) = -14$ and $7 + (-2) = 5$
-12	-24	$(8)(-3) = -24$ and $8 + (-3) = 5$

These are a few possible c values. There are <u>many</u> correct answers.

If $c = \ \ \ 2 : 2x^2 + 5x + 2 = (2x + 1)(x + 2)$

If $c = \ \ \ 3 : 2x^2 + 5x + 3 = (2x + 3)(x + 1)$

If $c = -3 : 2x^2 + 5x - 3 = (2x - 1)(x + 3)$

If $c = -7 : 2x^2 + 5x - 7 = (2x + 7)(x - 1)$

If $c = -12: 2x^2 + 5x - 12 = (2x - 3)(x + 4)$

107. $9x^2 - 9x - 54 = 9(x^2 - x - 6) = 9(x + 2)(x - 3)$

The error in the problem in the book was that 3 was factored out of the first binomial but not out of the second binomial.

$\quad (3x + 6)(3x - 9) = 3(x + 2)(3)(x - 3) = 9(x + 2)(x - 3)$

109. (a) $V = \pi R^2 h - \pi r^2 h$

$\qquad = \pi h(R^2 - r^2)$

$\qquad = \pi h(R - r)(R + r)$

(b) The average radius is $\dfrac{R + r}{2}$.

The thickness of the shell is $R - r$.

$V = \pi h(R - r)(R + r) = 2\pi\left(\dfrac{R + r}{2}\right)(R - r)h$

Section P.5   Fractional Expressions

- You should be able to find the domain of a fractional expression.
- You should know that a rational expression is the quotient of two polynomials.
- You should be able to simplify rational expressions by reducing them to lowest terms. This may involve factoring both the numerator and the denominator.
- You should be able to add, subtract, multiply, and divide rational expressions.
- You should be able to simplify compound fractions.

Solutions to Odd-Numbered Exercises

1. The domain of the polynomial $3x^2 - 4x + 7$ is the set of all real numbers.

3. The domain of the polynomial $4x^3 + 3, x \geq 0$ is the set of non-negative real numbers, since the polynomial is restricted to that set.

5. The domain of $\dfrac{1}{x-2}$ is the set of all real numbers x such that $x \neq 2$.

7. The domain of $\dfrac{x-1}{x(x-4)}$ is the set of all real numbers x such that $x \neq 0$ and $x \neq 4$.

9. The domain of $\sqrt{x+1}$ is the set of all real numbers x such that $x \geq -1$.

11. $\dfrac{5}{2x} = \dfrac{5(3x)}{(2x)(3x)} = \dfrac{5(3x)}{6x^2}, \quad x \neq 0$

The missing factor is $3x, x \neq 0$.

13. $\dfrac{x+1}{x} = \dfrac{(x+1)(x-2)}{x(x-2)}, \quad x \neq 2$

The missing factor is $x - 2, x \neq 2$.

15. $\dfrac{3x}{x-3} = \dfrac{3x(x)}{(x-3)(x)} = \dfrac{3x^2}{x^2 - 3x}, \quad x \neq 0$

The missing factor is $x, x \neq 0$.

17. $\dfrac{15x^2}{10x} = \dfrac{5x(3x)}{5x(2)} = \dfrac{3x}{2}, \quad x \neq 0$

19. $\dfrac{3xy}{xy+x} = \dfrac{x(3y)}{x(y+1)} = \dfrac{3y}{y+1}, \quad x \neq 0$

21. $\dfrac{x-5}{10-2x} = \dfrac{x-5}{-2(x-5)} = -\dfrac{1}{2}, \quad x \neq 5$

23. $\dfrac{x^3 + 5x^2 + 6x}{x^2 - 4} = \dfrac{x(x+2)(x+3)}{(x+2)(x-2)} = \dfrac{x(x+3)}{x-2}, \quad x \neq -2$

25. $\dfrac{y^2 - 7y + 12}{y^2 + 3y - 18} = \dfrac{(y-3)(y-4)}{(y+6)(y-3)} = \dfrac{y-4}{y+6}, \quad x \neq 3$

27. $\dfrac{2 - x + 2x^2 - x^3}{x-2} = \dfrac{(2-x) + x^2(2-x)}{-(2-x)} = \dfrac{(2-x)(1+x^2)}{-(2-x)} = -(1+x^2), \quad x \neq 2$

29. $\dfrac{z^3 - 8}{z^2 + 2z + 4} = \dfrac{(z-2)(z^2 + 2z + 4)}{z^2 + 2z + 4} = z - 2$

31.

x	0	1	2	3	4	5	6
$\dfrac{x^2 - 2x - 3}{x - 3}$	1	2	3	undef.	5	6	7
$x + 1$	1	2	3	4	5	6	7

The expressions are equivalent except at $x = 3$.

33. $\dfrac{5x^3}{2x^3 + 4} = \dfrac{5x^3}{2(x^3 + 2)}$. There are no common factors so this expression is in reduced form. In this case factors of terms were incorrectly cancelled.

35. $\dfrac{\pi r^2}{(2r)^2} = \dfrac{\pi r^2}{4r^2} = \dfrac{\pi}{4}, \quad r \neq 0$

37. $\dfrac{5}{x - 1} \cdot \dfrac{x - 1}{25(x - 2)} = \dfrac{1}{5(x - 2)}, \quad x \neq 1$

39. $\dfrac{(x + 5)(x - 3)}{x + 2} \cdot \dfrac{1}{(x + 5)(x + 2)} = \dfrac{x - 3}{(x + 2)^2}, \quad x \neq -5$

41. $\dfrac{r}{r - 1} \cdot \dfrac{r^2 - 1}{r^2} = \dfrac{r(r + 1)(r - 1)}{r^2(r - 1)} = \dfrac{r + 1}{r}, \quad r \neq 1$

43. $\dfrac{t^2 - t - 6}{t^2 + 6t + 9} \cdot \dfrac{t + 3}{t^2 - 4} = \dfrac{(t - 3)(t + 2)(t + 3)}{(t + 3)^2(t + 2)(t - 2)} = \dfrac{t - 3}{(t + 3)(t - 2)}, \quad t \neq -2$

45. $\dfrac{x^2 + xy - 2y^2}{x^3 + x^2y} \cdot \dfrac{x}{x^2 + 3xy + 2y^2} = \dfrac{(x + 2y)(x - y)}{x^2(x + y)} \cdot \dfrac{x}{(x + 2y)(x + y)}$

$$= \dfrac{x - y}{x(x + y)^2}, \quad x \neq -2y$$

47. $\dfrac{3(x + y)}{4} \div \dfrac{x + y}{2} = \dfrac{3(x + y)}{4} \cdot \dfrac{2}{x + y} = \dfrac{3}{2}, \quad x \neq -y$

49. $\dfrac{\left[\dfrac{x^2}{(x + 1)^2}\right]}{\left[\dfrac{x}{(x + 1)^3}\right]} = \dfrac{x^2}{(x + 1)^2} \cdot \dfrac{(x + 1)^3}{x} = x(x + 1), \quad x \neq -1, 0$

51. $\dfrac{5}{x - 1} + \dfrac{x}{x - 1} = \dfrac{5 + x}{x - 1} = \dfrac{x + 5}{x - 1}$

53. $6 - \dfrac{5}{x + 3} = \dfrac{6(x + 3)}{(x + 3)} - \dfrac{5}{x + 3} = \dfrac{6(x + 3) - 5}{x + 3} = \dfrac{6x + 13}{x + 3}$

55. $\dfrac{3}{x - 2} + \dfrac{5}{2 - x} = \dfrac{3}{x - 2} - \dfrac{5}{x - 2} = -\dfrac{2}{x - 2}$

57. $\dfrac{2}{x^2 - 4} - \dfrac{1}{x^2 - 3x + 2} = \dfrac{2}{(x + 2)(x - 2)} - \dfrac{1}{(x - 1)(x - 2)}$

$$= \dfrac{2(x - 1) - (x + 2)}{(x + 2)(x - 2)(x - 1)} = \dfrac{x - 4}{(x + 2)(x - 2)(x - 1)}$$

59. $\dfrac{1}{x^2 - x - 2} - \dfrac{x}{x^2 - 5x + 6} = \dfrac{1}{(x-2)(x+1)} - \dfrac{x}{(x-2)(x-3)}$

$$= \dfrac{(x-3) - x(x+1)}{(x+1)(x-2)(x-3)}$$

$$= \dfrac{-x^2 - 3}{(x+1)(x-2)(x-3)} = -\dfrac{x^2 + 3}{(x+1)(x-2)(x-3)}$$

61. $-\dfrac{1}{x} + \dfrac{2}{x^2 + 1} + \dfrac{1}{x^3 + x} = \dfrac{-(x^2+1)}{x(x^2+1)} + \dfrac{2x}{x(x^2+1)} + \dfrac{1}{x(x^2+1)}$

$$= \dfrac{-x^2 - 1 + 2x + 1}{x(x^2+1)} = \dfrac{-x^2 + 2x}{x(x^2+1)} = \dfrac{-x(x-2)}{x(x^2+1)}$$

$$= -\dfrac{x-2}{x^2+1} = \dfrac{2-x}{x^2+1}, \quad x \neq 0$$

63. $x^2(x^2+1)^{-5} - (x^2+1)^{-4} = (x^2+1)^{-5}[x^2 - (x^2+1)]$

$$= -\dfrac{1}{(x^2+1)^5}$$

65. $\dfrac{x+4}{x+2} - \dfrac{3x-8}{x+2} = \dfrac{(x+4) - (3x-8)}{x-2}$

$$= \dfrac{x+4-3x+8}{x-2}$$

$$= \dfrac{-2x+12}{x-2}$$

$$= \dfrac{-2(x-6)}{x-2}$$

67. $\dfrac{\left(\dfrac{x}{2} - 1\right)}{(x-2)} = \dfrac{\left(\dfrac{x}{2} - \dfrac{2}{2}\right)}{\left(\dfrac{x-2}{1}\right)}$

$$= \dfrac{x-2}{2} \cdot \dfrac{1}{x-2}$$

$$= \dfrac{1}{2}, \quad x \neq 2$$

The error was an incorrect subtraction in the numerator.

69. $\dfrac{\left(\dfrac{1}{x} - \dfrac{1}{x+1}\right)}{\left(\dfrac{1}{x+1}\right)} = \dfrac{\dfrac{(x+1) - x}{x(x+1)}}{\dfrac{1}{x+1}} = \dfrac{1}{x(x+1)} \cdot \dfrac{x+1}{1} = \dfrac{1}{x}, \quad x \neq -1$

71. $\dfrac{\left(\dfrac{x+3}{x-3}\right)^2}{\dfrac{1}{x+3} + \dfrac{1}{x-3}} = \dfrac{\dfrac{(x+3)^2}{(x-3)^2}}{\dfrac{(x-3) + (x+3)}{(x+3)(x-3)}}$

$$= \dfrac{(x+3)^2}{(x-3)^2} \cdot \dfrac{(x+3)(x-3)}{2x} = \dfrac{(x+3)^3}{2x(x-3)}, \quad x \neq -3$$

73. $\dfrac{\left[\dfrac{1}{(x+h)^2}-\dfrac{1}{x^2}\right]}{h}=\dfrac{\left[\dfrac{1}{(x+h)^2}-\dfrac{1}{x^2}\right]}{h}\cdot\dfrac{x^2(x+h)^2}{x^2(x+h)^2}$

$\qquad\qquad =\dfrac{x^2-(x+h)^2}{hx^2(x+h)^2}$

$\qquad\qquad =\dfrac{x^2-(x^2+2xh+h^2)}{hx^2(x+h)^2}$

$\qquad\qquad =\dfrac{-h(2x+h)}{hx^2(x+h)^2}$

$\qquad\qquad =-\dfrac{2x+h}{x^2(x+h)^2},\quad h\neq 0$

75. $\dfrac{\left(\sqrt{x}-\dfrac{1}{2\sqrt{x}}\right)}{\sqrt{x}}=\dfrac{\left(\sqrt{x}-\dfrac{1}{2\sqrt{x}}\right)}{\sqrt{x}}\cdot\dfrac{2\sqrt{x}}{2\sqrt{x}}=\dfrac{2x-1}{2x},\quad x>0$

77. $\dfrac{\dfrac{t^2}{\sqrt{t^2+1}}-\sqrt{t^2+1}}{t^2}=\dfrac{\left[\dfrac{t^2}{\sqrt{t^2+1}}-\sqrt{t^2+1}\right]}{t^2}\cdot\dfrac{\sqrt{t^2+1}}{\sqrt{t^2+1}}$

$\qquad\qquad =\dfrac{t^2-(t^2+1)}{t^2\sqrt{t^2+1}}=-\dfrac{1}{t^2\sqrt{t^2+1}}$

79. $\dfrac{x(x+1)^{-3/4}-(x+1)^{1/4}}{x^2}=\dfrac{x(x+1)^{-3/4}-(x+1)^{1/4}}{x^2}\cdot\dfrac{(x+1)^{3/4}}{(x+1)^{3/4}}$

$\qquad\qquad =\dfrac{x(x+1)^0-(x+1)^1}{x^2(x+1)^{3/4}}$

$\qquad\qquad =\dfrac{x-x-1}{x^2(x+1)^{3/4}}$

$\qquad\qquad =-\dfrac{1}{x^2(x+1)^{3/4}}$

81. $\dfrac{\sqrt{x+2}-\sqrt{x}}{2}=\dfrac{\sqrt{x+2}-\sqrt{x}}{2}\cdot\dfrac{\sqrt{x+2}+\sqrt{x}}{\sqrt{x+2}+\sqrt{x}}$

$\qquad\qquad =\dfrac{(x+2)-x}{2(\sqrt{x+2}+\sqrt{x})}=\dfrac{2}{2(\sqrt{x+2}+\sqrt{x})}$

$\qquad\qquad =\dfrac{1}{\sqrt{x+2}+\sqrt{x}}$

83. (a) $\dfrac{1}{16}$ minute

(b) $x\left(\dfrac{1}{16}\right)=\dfrac{x}{16}$ minutes

(c) $\dfrac{60}{16}=\dfrac{15}{4}$ minutes

85. $\text{Average} = \dfrac{\left(\dfrac{x}{3} + \dfrac{2x}{5}\right)}{2} = \dfrac{\left(\dfrac{x}{3} + \dfrac{2x}{5}\right)}{2} \cdot \dfrac{15}{15} = \dfrac{5x + 6x}{30} = \dfrac{11x}{30}$

87. (a) $r = \dfrac{\left(\dfrac{24[48(400) - 15,000]}{48}\right)}{\left[15,000 + \dfrac{48(400)}{12}\right]} \approx 0.1265 = 12.65\%$

(b) $r = \dfrac{\left[\dfrac{24(NM - P)}{N}\right]}{\left(P + \dfrac{NM}{12}\right)} = \dfrac{24(NM - P)}{N} \cdot \dfrac{12}{12P + NM} = \dfrac{288(NM - P)}{N(12P + NM)}$

$r = \dfrac{288[48(400) - 15,000]}{48[12(15,000) + 48(400)]} \approx 0.1265 = 12.65\%$

89. $T = 10\left(\dfrac{4t^2 + 16t + 75}{t^2 + 4t + 10}\right)$

(a)

t	0	1	2	3	4	5
T	75°	63.3°	55.9°	51.3°	48.3°	46.4°

(b)

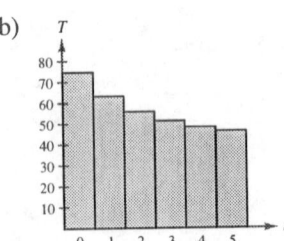

91. $\dfrac{x\left(\dfrac{x}{2}\right)}{x(2x + 1)} = \dfrac{\dfrac{x}{2}}{2x + 1} \cdot \dfrac{2}{2} = \dfrac{x}{2(2x + 1)}$

Section P.6   Errors and the Algebra of Calculus

■ You should be able to recognize and avoid the common algebraic errors involving parentheses, fractions, exponents, radicals, and cancellation.

■ You should be able to "unsimplify" algebraic expressions by the following methods.
   (a) Unusual Factoring
   (b) Inserting Factors or Terms
   (c) Rewriting with Negative Exponents
   (d) Writing a Fraction as a Sum of Terms

Solutions to Odd-Numbered Exercises

1. $2x - (3y + 4) = 2x - 3y - 4$

Distribute the minus sign.

3. $5z + 3(x - 2) = 5z + 3x - 6$

Use the Distributive Property.

5. $-\dfrac{x - 3}{x - 1} = \dfrac{-(x - 3)}{x - 1}$

$\qquad = \dfrac{3 - x}{x - 1}$

Only the numerator is multiplied by (-1).

7. $a\left(\dfrac{x}{y}\right) = \dfrac{a}{1} \cdot \dfrac{x}{y} = \dfrac{ax}{y}$

Only the numerator is multiplied by a.

9. $(4x)^2 = 4^2x^2 = 16x^2$

Square both factors.

11. $\sqrt{x + 9}$ does not simplify.

Do not apply the radical to the terms.

13. $\dfrac{6x + y}{6x - y}$ does not simplify.

Reduce common factors, not common factors of terms.

15. $\dfrac{1}{x + y^{-1}} = \dfrac{1}{x + (1/y)} \cdot \dfrac{y}{y} = \dfrac{y}{xy + 1}$

The negative exponent is on a term of the denominator, not a factor.

17. $x(2x + 1)^2 = x(4x^2 + 4x + 1)$

Exponents are applied before multiplying.

19. $\sqrt[3]{x^3 + 7x^2} = \sqrt[3]{x^2(x + 7)} = \sqrt[3]{x^2}\,\sqrt[3]{x + 7}$

Radicals apply to every factor of the radicand.

21. $\dfrac{3}{x} + \dfrac{4}{y} = \dfrac{3}{x} \cdot \dfrac{y}{y} + \dfrac{4}{y} \cdot \dfrac{x}{x} = \dfrac{3y + 4x}{xy}$

To add fractions, they must have a common denominator.

23. $\dfrac{1}{2y} = \dfrac{1}{2} \cdot \dfrac{1}{y}$

Use the definition for multiplying fractions.

25. $\dfrac{3x + 2}{5} = \dfrac{1}{5}(3x + 2)$

The required factor is $3x + 2$.

27. $\dfrac{2}{3}x^2 + \dfrac{1}{3}x + 5 = \dfrac{2}{3}x^2 + \dfrac{1}{3}x + \dfrac{15}{3} = \dfrac{1}{3}(2x^2 + x + 15)$

The required factor is $2x^2 + x + 15$.

29. $\frac{1}{3}x^3 + 5 = \frac{1}{3}x^3 + \frac{15}{3} = \frac{1}{3}(x^3 + 15)$

The required factor is $\frac{1}{3}$.

31. $x(2x^2 + 15) = \dfrac{2x}{2}(2x^2 + 15) = \left(\dfrac{1}{2}\right)(2x)(2x^2 + 15)$

$\qquad = \left(\dfrac{1}{2}\right)(2x^2 + 15)(2x)$

The required factor is $\frac{1}{2}$.

33. $x(1 - 2x^2)^3 = \dfrac{-4x}{-4}(1 - 2x^2)^3 = \left(-\dfrac{1}{4}\right)(-4x)(1 - 2x^2)^3$

$\qquad = \left(-\dfrac{1}{4}\right)(1 - 2x^2)^3(-4x)$

The required factor is $-\frac{1}{4}$.

35. $\dfrac{1}{\sqrt{x}\left(1 + \sqrt{x}\right)^2} = \dfrac{1}{\sqrt{x}} \cdot \dfrac{1}{\left(1 + \sqrt{x}\right)^2} = (2)\left(\dfrac{1}{2\sqrt{x}}\right)\dfrac{1}{\left(1 + \sqrt{x}\right)^2}$

$$= (2)\dfrac{1}{\left(1 + \sqrt{x}\right)^2}\left(\dfrac{1}{2\sqrt{x}}\right)$$

The required factor is 2.

37. $\dfrac{x + 1}{(x^2 + 2x - 3)^2} = \dfrac{1}{2} \cdot \dfrac{2(x + 1)}{(x^2 + 2x - 3)^2} = \left(\dfrac{1}{2}\right)\left(\dfrac{1}{(x^2 + 2x - 3)^2}\right)(2x + 2)$

The required factor is $\dfrac{1}{2}$.

39. $\dfrac{3}{x} + \dfrac{5}{2x^2} - \dfrac{3}{2}x = \dfrac{6x}{2x^2} + \dfrac{5}{2x^2} - \dfrac{3x^3}{2x^2} = \left(\dfrac{1}{2x^2}\right)(6x + 5 - 3x^3)$

The required factor is $\dfrac{1}{2x^2}$.

41. $\dfrac{9x^2}{25} + \dfrac{16y^2}{49} = \dfrac{9}{25} \cdot \dfrac{x^2}{1} + \dfrac{16}{49} \cdot \dfrac{y^2}{1}$

$$= \dfrac{1}{25/9} \cdot \dfrac{x^2}{1} + \dfrac{1}{49/16} \cdot \dfrac{y^2}{1}$$

$$= \dfrac{x^2}{(25/9)} + \dfrac{y^2}{(49/16)}$$

The required factors are $\dfrac{25}{9}$ and $\dfrac{49}{16}$.

43. $\dfrac{x^2}{1/12} - \dfrac{y^2}{2/3} = x^2\left(\dfrac{12}{1}\right) - y^2\left(\dfrac{3}{2}\right) = \dfrac{12x^2}{1} - \dfrac{3y^2}{2}$

The required factors are 1 and 2.

45. $\sqrt{x} + \left(\sqrt{x}\right)^3 = \sqrt{x}\left(1 + \left(\sqrt{x}\right)^2\right) = \sqrt{x}\left(1 + x\right)$

The required factor is $1 + x$.

47. $3(2x + 1)x^{1/2} + 4x^{3/2} = x^{1/2}[3(2x + 1) + 4x]$

$$= x^{1/2}(6x + 3 + 4x)$$

$$= x^{1/2}(10x + 3)$$

The required factor is $10x + 3$.

49. $\dfrac{x^2}{\sqrt{x^2 + 1}} - \sqrt{x^2 + 1} = \dfrac{x^2}{\sqrt{x^2 + 1}} - \dfrac{\sqrt{x^2 + 1}}{1} \cdot \dfrac{\sqrt{x^2 + 1}}{\sqrt{x^2 + 1}}$

$$= \dfrac{x^2 - (x^2 + 1)}{\sqrt{x^2 + 1}} = \dfrac{-1}{\sqrt{x^2 + 1}}$$

$$= \dfrac{1}{\sqrt{x^2 + 1}}(-1)$$

The required factor is -1.

51. $\frac{1}{10}(2x + 1)^{5/2} - \frac{1}{6}(2x + 1)^{3/2} = \frac{3}{30}(2x + 1)^{3/2}(2x + 1)^1 - \frac{5}{30}(2x + 1)^{3/2}$

$$= \frac{1}{30}(2x + 1)^{3/2}[3(2x + 1) - 5]$$

$$= \frac{1}{30}(2x + 1)^{3/2}(6x - 2)$$

$$= \frac{1}{30}(2x + 1)^{3/2}2(3x - 1)$$

$$= \frac{1}{15}(2x + 1)^{3/2}(3x - 1)$$

The required factor is $3x - 1$.

53. $\dfrac{16 - 5x - x^2}{x} = \dfrac{16}{x} - \dfrac{5x}{x} - \dfrac{x^2}{x} = \dfrac{16}{x} - 5 - x$

55. $\dfrac{4x^3 - 7x^2 + 1}{x^{1/3}} = \dfrac{4x^3}{x^{1/3}} - \dfrac{7x^2}{x^{1/3}} + \dfrac{1}{x^{1/3}}$

$$= 4x^{3-1/3} - 7x^{2-1/3} + \frac{1}{x^{1/3}}$$

$$= 4x^{8/3} - 7x^{5/3} + \frac{1}{x^{1/3}}$$

57. $\dfrac{3 - 5x^2 - x^4}{\sqrt{x}} = \dfrac{3}{\sqrt{x}} - \dfrac{5x^2}{\sqrt{x}} - \dfrac{x^4}{\sqrt{x}}$

$$= \frac{3}{\sqrt{x}} - 5x^{2-1/2} - x^{4-1/2}$$

$$= \frac{3}{\sqrt{x}} - 5x^{3/2} - x^{7/2}$$

59. $\dfrac{-2(x^2 - 3)^{-3}(2x)(x + 1)^3 - 3(x + 1)^2(x^2 - 3)^{-2}}{[(x + 1)^3]^2} = \dfrac{(x^2 - 3)^{-3}(x + 1)^2[-4x(x + 1) - 3(x^2 - 3)]}{(x + 1)^6}$

$$= \frac{-4x^2 - 4x - 3x^2 + 9}{(x^2 - 3)^3(x + 1)^4}$$

$$= \frac{-7x^2 - 4x + 9}{(x^2 - 3)^3(x + 1)^4}$$

61. $\dfrac{(6x + 1)^3(27x^2 + 2) - (9x^3 + 2x)(3)(6x + 1)^2(6)}{[(6x + 1)^3]^2} = \dfrac{(6x + 1)^2[(6x + 1)(27x^2 + 2) - 18(9x^3 + 2x)]}{(6x + 1)^6}$

$$= \frac{162x^3 + 12x + 27x^2 + 2 - 162x^3 - 36x}{(6x + 1)^4}$$

$$= \frac{27x^2 - 24x + 2}{(6x + 1)^4}$$

63. $\dfrac{(x + 2)^{3/4}(x + 3)^{-2/3} - (x + 3)^{1/3}(x + 2)^{-1/4}}{[(x + 2)^{3/4}]^2} = \dfrac{(x + 2)^{-1/4}(x + 3)^{-2/3}[(x + 2) - (x + 3)]}{(x + 2)^{6/4}}$

$$= \frac{x + 2 - x - 3}{(x + 2)^{1/4}(x + 3)^{2/3}(x + 2)^{6/4}}$$

$$= -\frac{1}{(x + 3)^{2/3}(x + 2)^{7/4}}$$

65. $\dfrac{2(3x-1)^{1/3} - (2x+1)(\frac{1}{3})(3x-1)^{-2/3}(3)}{(3x-1)^{2/3}} = \dfrac{(3x-1)^{-2/3}[2(3x-1)-(2x+1)]}{(3x-1)^{2/3}}$

$$= \dfrac{6x-2-2x-1}{(3x-1)^{2/3}(3x-1)^{2/3}}$$

$$= \dfrac{4x-3}{(3x-1)^{4/3}}$$

67. (a) $y_1 = x^2\left(\dfrac{1}{3}\right)(x^2+1)^{-2/3}(2x) + (x^2+1)^{1/3}(2x)$

$$= 2x(x^2+1)^{-2/3}\left[\dfrac{x^2}{3} + (x^2+1)\right]$$

$$= 2x(x^2+1)^{-2/3}\left(\dfrac{4x^2}{3} + \dfrac{3}{3}\right)$$

$$= \dfrac{2x}{(x^2+1)^{2/3}} \cdot \dfrac{4x^2+3}{3}$$

$$= \dfrac{2x(4x^2+3)}{3(x^2+1)^{2/3}}$$

$$= y_2$$

(b)

x	-2	-1	$-\frac{1}{2}$	0	1	2	$\frac{5}{2}$
y_1	-8.7	-2.9	-1.1	0	2.9	8.7	12.5
y_2	-8.7	-2.9	-1.1	0	2.9	8.7	12.5

69. $y_1 = 2x\sqrt{1-x^2} - \dfrac{x^3}{\sqrt{1-x^2}}$ $\qquad y_2 = \dfrac{2-3x^2}{\sqrt{1-x^2}}$

When $x = 0$, $y_1 = 0$. $\qquad\qquad$ When $x = 0$, $y_2 = 1$.

Thus, $y_1 \neq y_2$.

$y_1 = \dfrac{2x\sqrt{1-x^2}}{1} - \dfrac{x^3}{\sqrt{1-x^2}} = \dfrac{2x\sqrt{1-x^2}}{1} \cdot \dfrac{\sqrt{1-x^2}}{\sqrt{1-x^2}} - \dfrac{x^3}{\sqrt{1-x^2}}$

$\qquad = \dfrac{2x(1-x^2)-x^3}{\sqrt{1-x^2}} = \dfrac{2x-2x^3-x^3}{\sqrt{1-x^2}}$

$\qquad = \dfrac{2x-3x^3}{\sqrt{1-x^2}}$

Let $y_2 = \dfrac{2x-3x^3}{\sqrt{1-x^2}}$. Then $y_1 = y_2$.

Section P.7   Graphical Representation of Data

- You should be able to plot points.
- You should know that the distance between (x_1, y_1) and (x_2, y_2) in the plane is

    $$d = \sqrt{(x_2 - x_1)^2 + (y_2 - y_1)^2}.$$

- You should know that the midpoint of the line segment joining (x_1, y_1) and (x_2, y_2) is

    $$\left(\frac{x_1 + x_2}{2}, \frac{y_1 + y_2}{2} \right).$$

Solutions to Odd-Numbered Exercises

1.

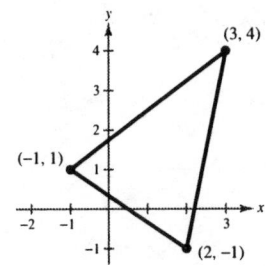

3.

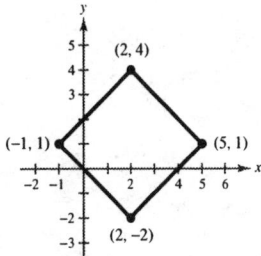

5. A: $(2, 6)$, B: $(-6, -2)$, C: $(4, -4)$, D: $(-3, 2)$

7. $(-3, 4)$

9. $(-5, -5)$

11. On the x-axis, $y = 0$.
On the y-axis, $x = 0$.

13. $x > 0$ and $y < 0$ in Quadrant IV.

15. $x = -4$ and $y > 0$ in Quadrant II.

17. $y < -5$ in Quadrants III and IV.

19. $(x, -y)$ is in the second Quadrant means that (x, y) is in Quadrant III.

21. (x, y), $xy > 0$ means x and y have the same signs. This occurs in Quadrants I and III.

23. $(-2 + 2, -4 + 5) = (0, 1)$

$(2 + 2, -3 + 5) = (4, 2)$

$(-1 + 2, -1 + 5) = (1, 4)$

25.

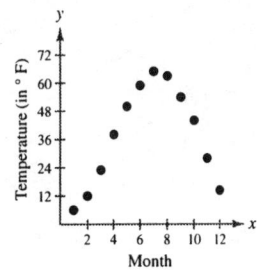

27. $y = 2 - \frac{1}{2}x$

x	-2	-1	$-\frac{1}{2}$	0	$\frac{1}{2}$	1	2
y	3	$\frac{5}{2}$	$\frac{9}{4}$	2	$\frac{7}{4}$	$\frac{3}{2}$	1

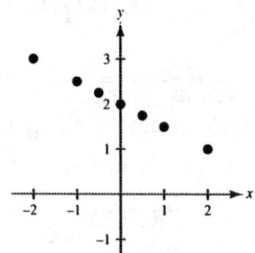

29. The highest price of milk is approximately \$13.70 per 100 lb. This occurred in 1990.

31. $\left(\dfrac{1200}{1995} \div \dfrac{51}{1967}\right)100 \approx 2300\%$

33. The minimum wage increased most rapidly in the 1970s.

35. The point $(65, 83)$ represents an entrance exam score of 65.

37. $d = |5 - (-3)| = 8$ **39.** $d = |2 - (-3)| = 5$

41. (a) The distance between $(0, 2)$ and $(4, 2)$ is 4.
The distance between $(4, 2)$ and $(4, 5)$ is 3.
The distance between $(0, 2)$ and $(4, 5)$ is
$$\sqrt{(4 - 0)^2 + (5 - 2)^2} = \sqrt{16 + 9} = \sqrt{25} = 5.$$
(b) $4^2 + 3^2 = 16 + 9 = 25 = 5^2$

43. (a) The distance between $(-1, 1)$ and $(9, 1)$ is 10.
The distance between $(9, 1)$ and $(9, 4)$ is 3.
The distance between $(-1, 1)$ and $(9, 4)$ is
$$\sqrt{(9 - (-1))^2 + (4 - 1)^2} = \sqrt{100 + 9} = \sqrt{109}.$$
(b) $10^2 + 3^2 = 109 = \left(\sqrt{109}\right)^2$

45. (a)

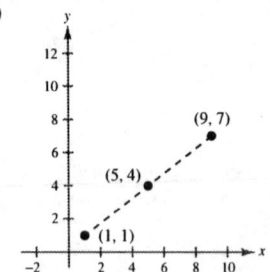

(b) $d = \sqrt{(9 - 1)^2 + (7 - 1)^2}$
$= \sqrt{64 + 36} = 10$

(c) $\left(\dfrac{9 + 1}{2}, \dfrac{7 + 1}{2}\right) = (5, 4)$

47. (a)

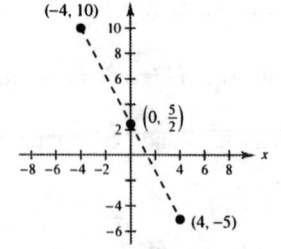

(b) $d = \sqrt{(4 + 4)^2 + (-5 - 10)^2}$
$= \sqrt{64 + 225} = 17$

(c) $\left(\dfrac{4 - 4}{2}, \dfrac{-5 + 10}{2}\right) = \left(0, \dfrac{5}{2}\right)$

49. (a)

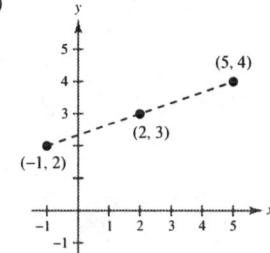

(b) $d = \sqrt{(5 + 1)^2 + (4 - 2)^2}$

$= \sqrt{36 + 4} = 2\sqrt{10}$

(c) $\left(\dfrac{-1 + 5}{2}, \dfrac{2 + 4}{2}\right) = (2, 3)$

51. (a)

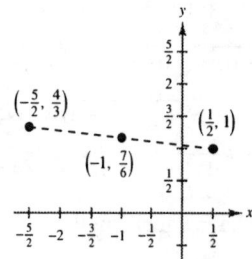

(b) $d = \sqrt{\left(\dfrac{1}{2} + \dfrac{5}{2}\right)^2 + \left(1 - \dfrac{4}{3}\right)^2}$

$d = \sqrt{9 + \dfrac{1}{9}} = \dfrac{\sqrt{82}}{3}$

(c) $\left(\dfrac{-\frac{5}{2} + \frac{1}{2}}{2}, \dfrac{\frac{4}{3} + 1}{2}\right) = \left(-1, \dfrac{7}{6}\right)$

53. (a)

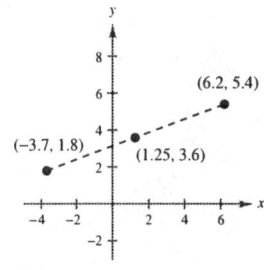

(b) $d = \sqrt{(6.2 + 3.7)^2 + (5.4 - 1.8)^2}$

$= \sqrt{98.01 + 12.96}$

$= \sqrt{110.97}$

(c) $\left(\dfrac{6.2 - 3.7}{2}, \dfrac{5.4 + 1.8}{2}\right) = (1.25, 3.6)$

55. (a)

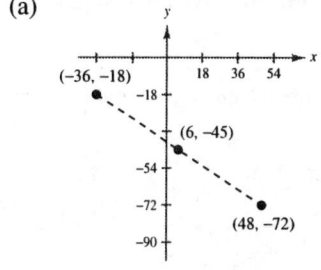

(b) $d = \sqrt{(48 + 36)^2 + (-72 + 18)^2}$

$= \sqrt{7056 + 2916}$

$= \sqrt{9972} = 6\sqrt{277}$

(c) $\left(\dfrac{-36 + 48}{2}, \dfrac{-18 - 72}{2}\right) = (6, -45)$

57. $\left(\dfrac{1991 + 1995}{2}, \dfrac{\$520{,}000 + \$740{,}000}{2}\right) = (1993, \$630{,}000)$

In 1993 the sales were \$630,000.

59. $d_1 = \sqrt{(4 - 2)^2 + (0 - 1)^2} = \sqrt{5}$

$d_2 = \sqrt{(4 + 1)^2 + (0 + 5)^2} = \sqrt{50}$

$d_3 = \sqrt{(2 + 1)^2 + (1 + 5)^2} = \sqrt{45}$

$\left(\sqrt{5}\right)^2 + \left(\sqrt{45}\right)^2 = \left(\sqrt{50}\right)^2$

61. $d_1 = \sqrt{(0 - 1)^2 + (0 - 2)^2} = \sqrt{5}$

$d_2 = \sqrt{(0 - 2)^2 + (0 - 1)^2} = \sqrt{5}$

$d_3 = \sqrt{(3 - 1)^2 + (3 - 2)^2} = \sqrt{5}$

$d_4 = \sqrt{(3 - 2)^2 + (3 - 1)^2} = \sqrt{5}$

$d_1 = d_2 = d_3 = d_4$

63. $d_1 = \sqrt{(0-2)^2 + (9-5)^2} = \sqrt{4+16} = \sqrt{20} = 2\sqrt{5}$

$d_2 = \sqrt{(-2-0)^2 + (0-9)^2} = \sqrt{4+81} = \sqrt{85}$

$d_3 = \sqrt{(0-(-2))^2 + (-4-0)^2} = \sqrt{4+16} = \sqrt{20} = 2\sqrt{5}$

$d_4 = \sqrt{(0-2)^2 + (-4-5)^2} = \sqrt{4+81} = \sqrt{85}$

Opposite sides have equal lengths of $2\sqrt{5}$ and $\sqrt{85}$.

65. Since $x_m = \dfrac{x_1 + x_2}{2}$ and $y_m = \dfrac{y_1 + y_2}{2}$ we have:

$$2x_m = x_1 + x_2 \qquad\qquad 2y_m = y_1 + y_2$$

$$2x_m - x_1 = x_2 \qquad\qquad 2y_m - y_1 = y_2$$

Thus, $(x_2, y_2) = (2x_m - x_1, 2y_m - y_1)$.

67. The midpoint of the given line segment is $\left(\dfrac{x_1 + x_2}{2}, \dfrac{y_1 + y_2}{2}\right)$.

The midpoint between (x_1, y_1) and $\left(\dfrac{x_1 + x_2}{2}, \dfrac{y_1 + y_2}{2}\right)$ is

$$\left(\dfrac{x_1 + \dfrac{x_1 + x_2}{2}}{2}, \dfrac{y_1 + \dfrac{y_1 + y_2}{2}}{2}\right) = \left(\dfrac{3x_1 + x_2}{4}, \dfrac{3y_1 + y_2}{4}\right).$$

The midpoint between $\left(\dfrac{x_1 + x_2}{2}, \dfrac{y_1 + y_2}{2}\right)$ and (x_2, y_2) is

$$\left(\dfrac{\dfrac{x_1 + x_2}{2} + x_2}{2}, \dfrac{\dfrac{y_1 + y_2}{2} + y_2}{2}\right) = \left(\dfrac{x_1 + 3x_2}{4}, \dfrac{y_1 + 3y_2}{4}\right).$$

Thus, the three points are

$$\left(\dfrac{3x_1 + x_2}{4}, \dfrac{3y_1 + y_2}{4}\right), \left(\dfrac{x_1 + x_2}{2}, \dfrac{y_1 + y_2}{2}\right), \text{ and } \left(\dfrac{x_1 + 3x_2}{4}, \dfrac{y_1 + 3y_2}{4}\right).$$

69. $d = \sqrt{(45-10)^2 + (40-15)^2} = \sqrt{35^2 + 25^2} = \sqrt{1850} = 5\sqrt{74} \approx \textbf{43 yards}$

71.   **The points are reflected through the *y*-axis.**

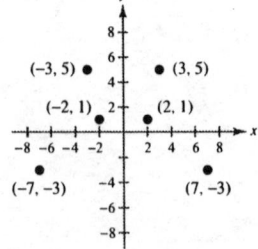

73. (a) It appears that the number of artists elected alternates between 7 and 8 per year in the 1990s. If this pattern continues, 8 would be elected in 1996.

(b) Since 1986 and 1987 were the first two years that artists were elected, there was a larger number of artists chosen.

❑ Review Exercises for Chapter P

Solutions to Odd-Numbered Exercises

1. $\{11, -14, -\frac{8}{9}, \frac{5}{2}, \sqrt{6}, 0.4\}$

  (a) Natural numbers: 11

  (b) Integers: $11, -14$

  (c) Rational numbers: $11, -14, -\frac{8}{9}, \frac{5}{2}, 0.4$

  (d) Irrational numbers: $\sqrt{6}$

3. (a) $\frac{5}{6} = 0.8\overline{3}$

  (b) $\frac{7}{8} = 0.875$

5. $x \le 7$ The set consists of all real numbers less than or equal to 7.

7. $d(x, 7) = |x - 7|$ and $d(x, 7) \ge 4$, thus $|x - 7| \ge 4$.

9. $d(y, -30) = |y - (-30)| = |y + 30|$ and $d(y, -30) < 5$, thus $|y + 30| < 5$.

11. $|-3| + 4(-2) - 6 = 3 - 8 - 6 = -11$

13. $\dfrac{5}{18} \div \dfrac{10}{3} = \dfrac{\cancel{5}}{\cancel{18}_6} \cdot \dfrac{\cancel{3}}{\cancel{10}_2} = \dfrac{1}{12}$

15. $6[4 - 2(6 + 8)] = 6[4 - 2(14)] = 6[4 - 28] = 6(-24) = -144$

17. $2x + (3x - 10) = (2x + 3x) - 10$

Illustrates the Associative Property of Addition.

19. $\dfrac{2}{y + 4} \cdot \dfrac{y + 4}{2} = 1, \quad y \ne -4$

Illustrates the Multiplicative Inverse Property.

21. (a) $\dfrac{6^2 u^3 v^{-3}}{12 u^{-2} v} = \dfrac{36 u^{3 - (-2)} v^{-3 - 1}}{12} = 3u^5 v^{-4} = \dfrac{3u^5}{v^4}$

  (b) $\dfrac{3^{-4} m^{-1} n^{-3}}{9^{-2} mn^{-3}} = \dfrac{9^2 n^3}{3^4 mmn^3} = \dfrac{81}{81 m^2} = \dfrac{1}{m^2} = m^{-2}$

23. $30{,}296{,}000{,}000 = 3.0296 \times 10^{10}$

25. $4.833 \times 10^8 = 483{,}300{,}000$

27. (a) $1800(1 + 0.08)^{24} \approx 11,414.125$

    (b) $0.0024\,(7,658,400) = 18,380.160$

29. Radical Form: $\sqrt{16} = 4$

    Rational Exponent Form: $16^{1/2} = 4$

31. (a) $\sqrt{4x^4} = 2x^2$

    (b) $\sqrt{\dfrac{18u^2}{b^3}} = \sqrt{\dfrac{9u^2}{b^2} \cdot \dfrac{2}{b}} = \dfrac{3|u|}{b}\sqrt{\dfrac{2}{b}}$

33. $\dfrac{1}{2 - \sqrt{3}} = \dfrac{1}{2 - \sqrt{3}} \cdot \dfrac{2 + \sqrt{3}}{2 + \sqrt{3}} = \dfrac{2 + \sqrt{3}}{4 - 3} = \dfrac{2 + \sqrt{3}}{1} = 2 + \sqrt{3}$

35. $\sqrt{50} - \sqrt{18} = \sqrt{25 \cdot 2} - \sqrt{9 \cdot 2} = 5\sqrt{2} - 3\sqrt{2} = 2\sqrt{2}$

37. $A = wh = 8\sqrt{3}\;\sqrt{24^2 - \left(8\sqrt{3}\right)^2} = 8\sqrt{3}\;\sqrt{384}$

    $= 8\sqrt{3}\left(8\sqrt{6}\right) = 64\sqrt{18} = 64\left(3\sqrt{2}\right) = 192\sqrt{2}$

39. $10(4 \cdot 7) = 10(28) = 280$

    The error in $40 \cdot 70$ is an improper use of the Distributive Property.

41. $4\left(\dfrac{3}{7}\right) = \dfrac{4}{1}\left(\dfrac{3}{7}\right) = \dfrac{12}{7}$

    Only the numerator is multiplied by 4.

43. $\dfrac{x - 1}{1 - x} = \dfrac{x - 1}{-(x - 1)} = -1,\, x \neq 1$

    The error is an improper cancellation.

45. $(-x)^6 = x^6$

    The exponent is to be applied to the whole quantity inside the parentheses.

47. $-x^2(-x^2 + 3) = x^4 - 3x^2$

    The minus sign is to be distributed to both terms inside the parentheses.

49. $(5 + 8)^2 = 5^2 + 2(5)(8) + 8^2 = 25 + 80 + 64 = 169$

    The middle term was omitted.

51. $\sqrt{7x}\;\sqrt[3]{2} = (7x)^{1/2}(2)^{1/3} = (7x)^{3/6}(2)^{2/6}$

    $= \sqrt[6]{(7x)^3}\;\sqrt[6]{(2)^2} = \sqrt[6]{343x^3}\;\sqrt[6]{4}$

    $= \sqrt[6]{1372x^3}$

    The indices must be the same to multiply the radicands.

53. $-(3x^2 + 2x) + (1 - 5x) = -3x^2 - 2x + 1 - 5x$

    $= -3x^2 - 7x + 1$

55. $(2x - 3)^2 = (2x)^2 - 2(2x)(3) + 3^2$

    $= 4x^2 - 12x + 9$

57. Multiply: $2x^2 + 3x + 5$

$$\underline{ x^3 - 3x}$$
$$2x^5 + 3x^4 + 5x^3$$
$$\underline{ -6x^3 - 9x^2 - 15x}$$
$$2x^5 + 3x^4 - x^3 - 9x^2 - 15x$$

59. (a)

The surface is the sum of the area of the side, $2\pi rh$, and the areas of the top and bottom which are each πr^2.

$$S = 2\pi rh + \pi r^2 + \pi r^2 = 2\pi rh + 2\pi r^2$$

(b) $S = 2\pi rh + 2\pi r^2 = 2\pi r(r + h)$

61. $x^3 - x = x(x^2 - 1) = x(x + 1)(x - 1)$

63. $2x^2 + 21x + 10 = (2x + 1)(x + 10)$

65. $x^3 - x^2 + 2x - 2 = x^2(x - 1) + 2(x - 1)$
$$= (x - 1)(x^2 + 2)$$

67. $\frac{2}{3}x^4 - \frac{3}{8}x^3 + \frac{5}{6}x^2 = \frac{16}{24}x^4 - \frac{9}{24}x^3 + \frac{20}{24}x^2$
$$= \frac{1}{24}x^2(16x^2 - 9x + 20)$$

The missing factor is $16x^2 - 9x + 20$.

69. $2x(x^2 - 3)^{1/3} - 5(x^2 - 3)^{4/3} = (x^2 - 3)^{1/3}[2x - 5(x^2 - 3)]$
$$= (x^2 - 3)^{1/3}(-5x^2 + 2x + 15)$$

The missing factor is $-5x^2 + 2x + 15$.

71. $\dfrac{x^2 - 4}{x^4 - 2x^2 - 8} \cdot \dfrac{x^2 + 2}{x^2} = \dfrac{(x^2 - 4)(x^2 + 2)}{(x^2 - 4)(x^2 + 2)x^2} = \dfrac{1}{x^2}, \, x \neq \pm 2$

73. $2x + \dfrac{3}{2(x - 4)} - \dfrac{1}{2(x + 2)} = \dfrac{2x(2)(x - 4)(x + 2) + 3(x + 2) - (x - 4)}{2(x - 4)(x + 2)}$

$$= \dfrac{4x(x^2 - 2x - 8) + 3x + 6 - x + 4}{2(x - 4)(x + 2)} = \dfrac{4x^3 - 8x^2 - 32x + 2x + 10}{2(x - 4)(x + 2)}$$

$$= \dfrac{4x^3 - 8x^2 - 30x + 10}{2(x - 4)(x + 2)} = \dfrac{2x^3 - 4x^2 - 15x + 5}{(x - 4)(x + 2)}$$

75. $\dfrac{1}{x - 1} + \dfrac{1 - x}{x^2 + x + 1} = \dfrac{x^2 + x + 1 + (1 - x)(x - 1)}{(x - 1)(x^2 + x + 1)}$

$$= \dfrac{x^2 + x + 1 + x - 1 - x^2 + x}{(x - 1)(x^2 + x + 1)} = \dfrac{3x}{(x - 1)(x^2 + x + 1)}$$

77. $\dfrac{\left(\dfrac{\frac{3a}{a^2}-1}{x}\right)}{\left(\dfrac{a}{x}-1\right)} = \dfrac{\left(\dfrac{\frac{3a}{a^2-x}}{x}\right)}{\left(\dfrac{a-x}{x}\right)} = \dfrac{3a}{1} \cdot \dfrac{x}{a^2-x} \cdot \dfrac{x}{a-x} = \dfrac{3ax^2}{(a^2-x)(a-x)}$

79.

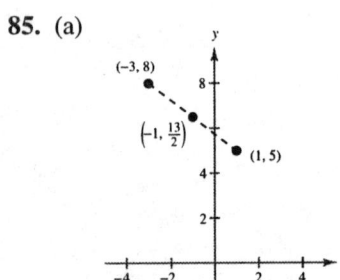

$d_1 = \sqrt{(13-5)^2 + (11-22)^2} = \sqrt{64+121} = \sqrt{185}$

$d_2 = \sqrt{(2-13)^2 + (3-11)^2} = \sqrt{121+64} = \sqrt{185}$

$d_3 = \sqrt{(2-5)^2 + (3-22)^2} = \sqrt{9+361} = \sqrt{370}$

$d_1^2 + d_2^2 = 185 + 185 = 370 = d_3^2$

Thus, the triangle is a right triangle.

81. $x > 0$ and $y = -2$ in Quadrant IV.

83. $(-x, y)$ is in the third Quadrant means that (x, y) is in Quadrant IV.

85. (a)

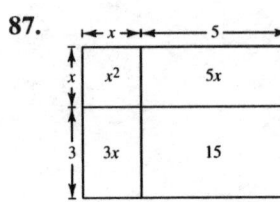

(b) $d = \sqrt{(-3-1)^2 + (8-5)^2} = \sqrt{16+9} = 5$

(c) $\left(\dfrac{-3+1}{2}, \dfrac{8+5}{2}\right) = \left(-1, \dfrac{13}{2}\right)$

87.

Using $A = l \cdot w$ we have $(x+5)(x+3)$. We could also add up the areas of the four inner rectangles, $x^2 + 3x + 5x + 15$. Notice that $(x+5)(x+3) = x^2 + 3x + 5x + 15$ by the Distributive Property (FOIL).

89.

n	1	10	10^2	10^4	10^6	10^{10}
$\dfrac{5}{\sqrt{n}}$	5	1.5811	0.5	0.05	0.0005	0.00005

$\dfrac{5}{\sqrt{n}}$ approaches 0 as n increases without bound.

❑ Practice Test for Chapter P

1. Evaluate $\dfrac{|-42| - 20}{15 - |-4|}$.

2. Simplify $\dfrac{x}{z} - \dfrac{z}{y}$.

3. The distance between x and 7 is no more than 4. Use absolute value notation to describe this expression.

4. Evaluate $10(-x)^3$ for $x = 5$.

5. Simplify $(-4x^3)(-2x^{-5})\left(\frac{1}{16}x\right)$.

6. Change 0.0000412 to scientific notation.

7. Evaluate $125^{2/3}$.

8. Simplify $\sqrt[4]{64x^7y^9}$.

9. Rationalize the denominator and simplify $\dfrac{6}{\sqrt{12}}$.

10. Simplify $3\sqrt{80} - 7\sqrt{500}$.

11. Simplify $(8x^4 - 9x^2 + 2x - 1) - (3x^3 + 5x + 4)$.

12. Multiply $(x - 3)(x^2 + x - 7)$.

13. Multiply $[(x - 2) - y]^2$.

14. Factor $16x^4 - 1$.

15. Factor $6x^2 + 5x - 4$.

16. Factor $x^3 - 64$.

17. Combine and simplify $-\dfrac{3}{x} + \dfrac{x}{x^2 + 2}$.

18. Combine and simplify $\dfrac{x - 3}{4x} \div \dfrac{x^2 - 9}{x^2}$.

19. Simplify $\dfrac{1 - (1/x)}{1 - \dfrac{1}{1 - (1/x)}}$.

20. (a) Plot the points $(-3, 7)$ and $(5, -1)$,

     (b) find the distance between the points, and

     (c) find the midpoint of the line segment joining the points.

CHAPTER 1
Equations and Inequalities

CHAPTER 1
Equations and Inequalities

Section 1.1   Graphs and Graphing Utilities

- You should be able to use the point-plotting method of graphing.
- You should be able to find x- and y-intercepts.
  - (a) To find the x-intercepts, let $y = 0$ and solve for x.
  - (b) To find the y-intercepts, let $x = 0$ and solve for y.
- You should be able to test for symmetry.
  - (a) To test for x-axis symmetry, replace y with $-y$.
  - (b) To test for y-axis symmetry, replace x with $-x$.
  - (c) To test for origin symmetry, replace x with $-x$ and y with $-y$.
- You should know the standard equation of a circle with center (h, k) and radius r:
  $$(x - h)^2 + (y - k)^2 = r^2$$

Solutions to Odd-Numbered Exercises

1. $y = \sqrt{x + 4}$

  (a) $(0, 2)$: $2 \overset{?}{=} \sqrt{0 + 4}$

   $2 = 2$ ✓

   Yes, the point *is* on the graph.

  (b) $(5, 3)$: $3 \overset{?}{=} \sqrt{5 + 4}$

   $3 = \sqrt{9}$ ✓

   Yes, the point *is* on the graph.

3. $y = 4 - |x - 2|$

  (a) $(1, 5)$: $5 \overset{?}{=} 4 - |1 - 2|$

   $5 \neq 4 - 1$

   No, the point *is not* on the graph.

  (b) $(6, 0)$: $0 \overset{?}{=} 4 - |6 - 2|$

   $0 = 4 - 4$ ✓

   Yes, the point *is* on the graph.

5. $2x - y - 3 = 0$

  (a) $(1, 2)$: $2(1) - (2) - 3 \overset{?}{=} 0$

   $-3 \neq 0$

   No, the point *is not* on the graph.

  (b) $(1, -1)$: $2(1) - (-1) - 3 \overset{?}{=} 0$

   $2 + 1 - 3 = 0$ ✓

   Yes, the point *is* on the graph.

7. $x^2y - x^2 + 4y = 0$

  (a) $\left(1, \frac{1}{5}\right)$: $(1)^2\left(\frac{1}{5}\right) - (1)^2 + 4\left(\frac{1}{5}\right) \overset{?}{=} 0$

   $\frac{1}{5} - 1 + \frac{4}{5} = 0$ ✓

   Yes, the point *is* on the graph.

  (b) $\left(2, \frac{1}{2}\right)$: $(2)^2\left(\frac{1}{2}\right) - (2)^2 + 4\left(\frac{1}{2}\right) \overset{?}{=} 0$

   $2 - 4 + 2 = 0$ ✓

   Yes, the point *is* on the graph.

9. $y = -2x + 3$

x	-1	0	1	$\frac{3}{2}$	2
y	5	3	1	0	-1

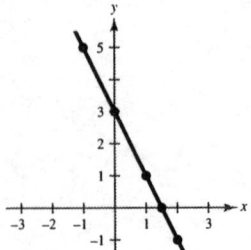

11. $y = x^2 - 2x$

x	-1	0	1	2	3
y	3	0	-1	0	3

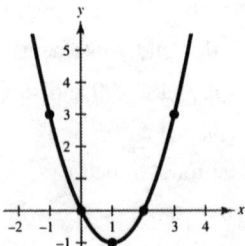

13.

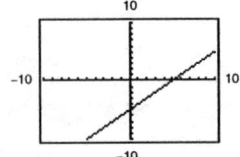

$y = x - 5$

Intercepts: $(5, 0), (0, -5)$

15.

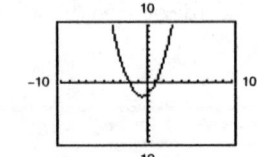

$y = x^2 + x - 2$

Intercepts: $(1, 0), (-2, 0), (0, -2)$

17.

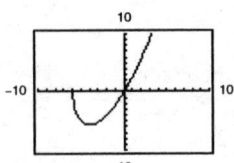

$y = x\sqrt{x + 6}$

Intercepts: $(0, 0), (-6, 0)$

19.

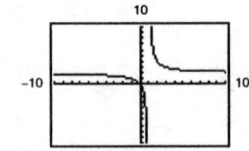

$y = \dfrac{2x}{x - 1}$

Intercept: $(0, 0)$

21. $(-x)^2 - y = 0 \implies x^2 - y = 0$

y-axis symmetry

23. $x - (-y)^2 = 0 \implies x - y^2 = 0$

x-axis symmetry

25. $-y = (-x)^3 \implies y = x^3$

Origin symmetry

27. $-y = \dfrac{-x}{(-x)^2 + 1} \implies y = \dfrac{x}{x^2 + 1}$

Origin symmetry

29. y-axis symmetry

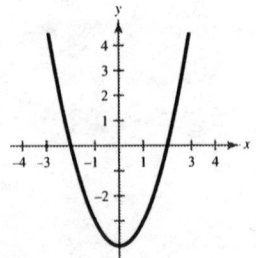

31. Origin symmetry

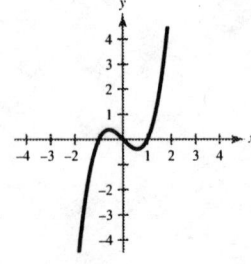

33. $y = 1 - x$ has intercepts $(1, 0)$ and $(0, 1)$. Matches graph (c).

35. $y = \sqrt{9 - x^2}$ has intercepts $(\pm 3, 0)$ and $(0, 3)$. Matches graph (f).

37. $y = x^3 - x + 1$ has a y-intercept of $(0, 1)$ and the points $(1, 1)$ and $(-2, -5)$ are on the graph. Matches graph (b).

39. $y = -3x + 2$
No symmetry

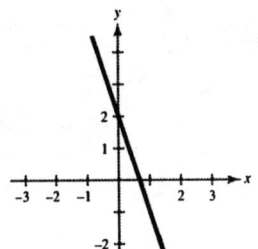

41. $y = 1 - x^2$
y-axis symmetry

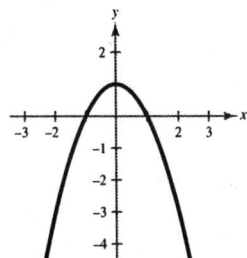

43. $y = x^2 - 3x$
No symmetry

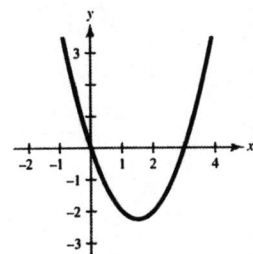

45. $y = x^3 + 2$
No symmetry

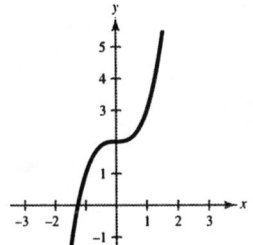

47. $y = \sqrt{x - 3}$
No symmetry
Domain: $x \geq 3$

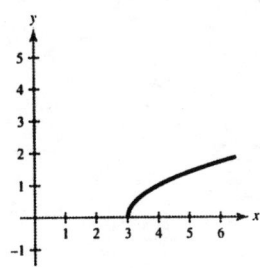

49. $y = |x - 2|$
No symmetry

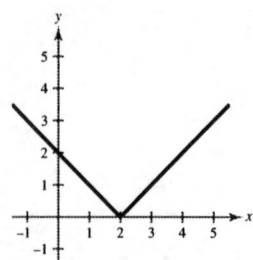

51. $x = y^2 - 1$
x-axis symmetry

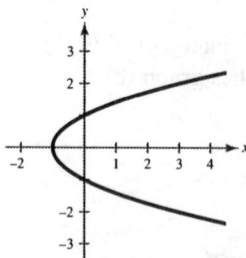

53. $y = 3 - \frac{1}{2}x$

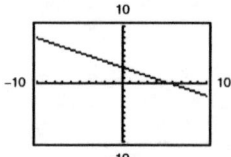

Intercepts: (6,0), (0, 3)

55. $y = x^2 - 4x + 3$

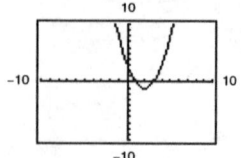

Intercepts: (3, 0), (1, 0), (0, 3)

57. $y = x(x - 2)^2$

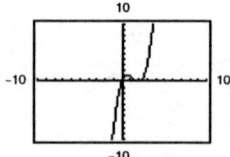

Intercepts: (0, 0), (2, 0)

59. $y = \sqrt[3]{x}$

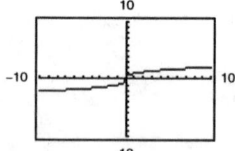

Intercepts: (0, 0)

61. $y = \frac{5}{2}x + 5$

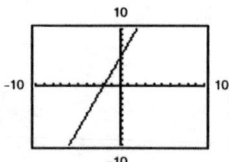

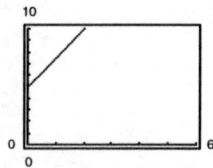

The standard setting gives a more complete graph.

63. $y = -x^2 + 10x - 5$

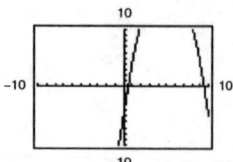

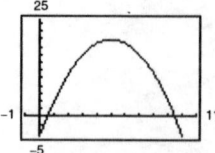

The specified setting gives a more complete graph.

65. $y = 4x^2 - 25$

Range/Window

Xmin = -5
Xmax = 5
Xscl = 1
Ymin = -30
Ymax = 30
Yscl = 10

67. $y = |x| + |x + 10|$

Range/Window

Xmin = -10
Xmax = 20
Xscl = 5
Ymin = -5
Ymax = 30
Yscl = 5

69. $x^2 + y^2 = 3^2$

$x^2 + y^2 = 9$

71. $(x - 2)^2 + [y - (-1)]^2 = 4^2$

$(x - 2)^2 + (y + 1)^2 = 16$

73. $r = \sqrt{(0 - (-1))^2 + (0 - 2)^2} = \sqrt{1 + 4} = \sqrt{5}$

$[x - (-1)]^2 + (y - 2)^2 = (\sqrt{5})^2$

$(x + 1)^2 + (y - 2)^2 = 5$

75. $r = \dfrac{1}{2}\sqrt{(6 - 0)^2 + (8 - 0)^2} = 5$

$\text{Center} = \left(\dfrac{0 + 6}{2}, \dfrac{0 + 8}{2}\right) = (3, 4)$

$(x - 3)^2 + (y - 4)^2 = 25$

77. Center: $(-2, -3)$; radius: 2

$[x - (-2)]^2 + [y - (-3)]^2 = 2^2$

$(x + 2)^2 + (y + 3)^2 = 4$

79. $x^2 + y^2 = 4$

Center: $(0, 0)$; radius: 2

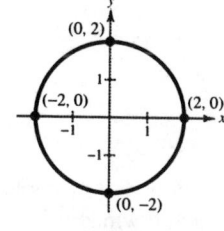

81. $(x - 1)^2 + (y + 3)^2 = 4$

$(x - 1)^2 + [y - (-3)]^2 = 2^2$

Center: $(1, -3)$; radius: 2

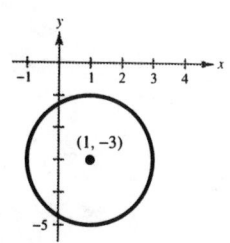

83. $\left(x - \frac{1}{2}\right)^2 + \left(y - \frac{1}{2}\right)^2 = \frac{9}{4}$

Center: $\left(\frac{1}{2}, \frac{1}{2}\right)$; radius: $\frac{3}{2}$

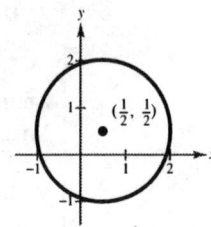

85. $y_1 = \sqrt{9 - x^2}$

$y_2 = -\sqrt{9 - x^2}$

A circle is bounded by their graphs.

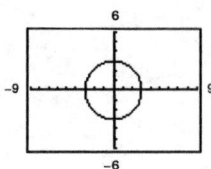

87. $y_1 = \frac{1}{4}(x^2 - 8)$

$y_2 = \frac{1}{4}x^2 - 2$

The graphs are identical. The Distributive Property is illustrated.

89. $y_1 = \frac{1}{5}[10(x^2 - 1)]$

$y_2 = 2(x^2 - 1)$

The graphs are identical. The Associative Property of Multiplication is illustrated.

91. $y = 225,000 - 20,000t, \ 0 \le t \le 8$

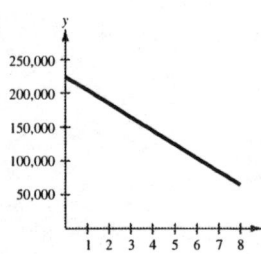

93. Most likely you would need to change the viewing window. For example, let $y_1 = x^2 + 12$. This graph would not show up on the standard window. Change the range/window to the following setting and try again.

> Xmin = -5
> Xmax = 5
> Xscl = 1
> Ymin = -5
> Ymax = 40
> Yscl = 5

95. (a)

Year	1920	1930	1940	1950	1960	1970	1980	1990
Life Expectancy	54.1	59.7	62.9	68.2	69.7	70.8	73.7	75.4
Model	52.8	58.7	63.3	66.9	69.9	72.4	74.6	76.4

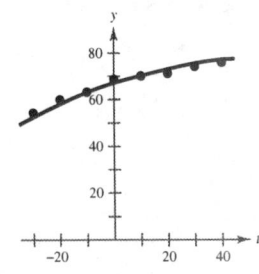

(b) When $t = 48$, $y \approx 77.7$ years.

(c) When $t = 50$, $y \approx 78.0$ years.

97. $y = 0.086t + 0.872, 0 \le t \le 4$

Year	1990	1991	1992	1993	1994
t	0	1	2	3	4
y	0.872	0.958	1.044	1.130	1.216

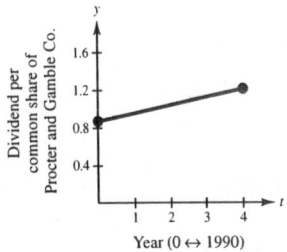

99. $9x^5 + 4x^3 - 7$

Terms: $9x^5, 4x^3, -7$

101. $\dfrac{1}{3 \cdot 4^{-1}} = \dfrac{4}{3} \neq 3 \cdot 4$ False

103. $\sqrt{18x} - \sqrt{2x} = 3\sqrt{2x} - \sqrt{2x} = 2\sqrt{2x}$

105. $\dfrac{70}{\sqrt{7x}} = \dfrac{70}{\sqrt{7x}} \cdot \dfrac{\sqrt{7x}}{\sqrt{7x}} = \dfrac{70\sqrt{7x}}{7x} = \dfrac{10\sqrt{7x}}{x}$

107. $\sqrt[6]{t^2} = t^{2/6} = |t|^{1/3} = \sqrt[3]{|t|}$

Section 1.2   Linear Equations

- You should know how to solve linear equations.
  $ax + b = 0$
- An identity is an equation whose solution consists of every real number in its domain.
- To solve an equation you can:
  (a) Add or subtract the same quantity from both sides.
  (b) Multiply or divide both sides by the same nonzero quantity.
- To solve an equation that can be simplified to a linear equation:
  (a) Remove all symbols of grouping and all fractions.
  (b) Combine like terms.
  (c) Solve by algebra.
  (d) Check the answer.
- A "solution" that does not satisfy the original equation is called an extraneous solution.

Solutions to Odd-Numbered Exercises

1. $5x - 3 = 3x + 5$

  (a) $5(0) - 3 \stackrel{?}{=} 3(0) + 5$

         $-3 \neq 5$

     $x = 0$ *is not* a solution.

  (c) $5(4) - 3 \stackrel{?}{=} 3(4) + 5$

         $17 = 17$

     $x = 4$ *is* a solution.

  (b) $5(-5) - 3 \stackrel{?}{=} 3(-5) + 5$

         $-28 \neq -10$

     $x = -5$ *is not* a solution.

  (d) $5(10) - 3 \stackrel{?}{=} 3(10) + 5$

         $47 \neq 35$

     $x = 10$ *is not* a solution.

3. $3x^2 + 2x - 5 = 2x^2 - 2$

  (a) $3(-3) + 2(-3) - 5 \stackrel{?}{=} 2(-3)^2 - 2$

         $16 = 16$

     $x = -3$ *is* a solution.

  (c) $3(4)^2 + 2(4) - 5 \stackrel{?}{=} 2(4)^2 - 2$

         $51 \neq 30$

     $x = 4$ *is not* a solution.

  (b) $3(1)^2 + 2(1) - 5 \stackrel{?}{=} 2(1)^2 - 2$

         $0 = 0$

     $x = 1$ *is* a solution.

  (d) $3(-5)^2 + 2(-5) - 5 \stackrel{?}{=} 2(-5)^2 - 2$

         $60 \neq 48$

     $x = -5$ *is not* a solution.

5. $\dfrac{5}{2x} - \dfrac{4}{x} = 3$

  (a) $\dfrac{5}{2(-1/2)} - \dfrac{5}{(-1/2)} \stackrel{?}{=} 3$

                 $3 = 3$

     $x = -\frac{1}{2}$ *is* a solution.

  (c) $\dfrac{5}{2(0)} - \dfrac{4}{0}$ is undefined.

     $x = 0$ *is not* a solution.

  (b) $\dfrac{5}{2(4)} - \dfrac{4}{4} \stackrel{?}{=} 3$

          $-\dfrac{3}{8} \neq 3$

     $x = 4$ *is not* a solution.

  (d) $\dfrac{5}{2(1/4)} - \dfrac{4}{1/4} \stackrel{?}{=} 3$

            $-6 \neq 3$

     $x = \frac{1}{4}$ *is not* a solution.

7. $(x + 5)(x - 3) = 20$

  (a) $(3 + 5)(3 - 3) \stackrel{?}{=} 20$

           $0 \neq 20$

     $x = 3$ *is not* a solution.

  (c) $(0 + 5)(0 - 3) \stackrel{?}{=} 20$

         $-15 \neq 20$

     $x = 0$ *is not* a solution.

  (b) $(-2 + 5)(-2 - 3) \stackrel{?}{=} 20$

          $-15 \neq 20$

     $x = -2$ *is not* a solution.

  (d) $(-7 + 5)(-7 - 3) \stackrel{?}{=} 20$

          $20 = 20$

     $x = -7$ *is* a solution.

9. $2(x - 1) = 2x - 2$ is an *identity* by the Distributive Property. It is true for all real values of x.

11. $-6(x - 3) + 5 = -2x + 10$ is *conditional*. There are real values of x for which the equation is not true.

13. $4(x + 1) - 2x = 4x + 4 - 2x = 2x + 4 = 2(x + 2)$
This is an *identity* by simplification. It is true for all real values of x.

15. $x^2 - 8x + 5 = (x - 4)^2 - 11$ is an *identity* since $(x - 4)^2 - 11 = x^2 - 8x + 16 - 11 = x^2 - 8x + 5$.

17. $3 + \dfrac{1}{x + 1} = \dfrac{4x}{x + 1}$ is *conditional*. There are real values of x for which the equation is not true.

19. Equivalent equations are derived from the substitution principle and simplification techniques. They have the same solution(s).

$2x + 3 = 8$ and $2x = 5$ are equivalent equations.

21.

$$4x + 32 = 83 \qquad \text{Original Equation}$$

$$4x + 32 - 32 = 83 - 32 \qquad \text{Subtract 32 from both sides}$$

$$4x = 51 \qquad \text{Simplify}$$

$$\frac{4x}{4} = \frac{51}{4} \qquad \text{Divide both sides by 4}$$

$$x = \frac{51}{4} \qquad \text{Simplify}$$

23. $3x = 15$

$\quad\ x = 5$

25. $s + 12 = 18$

$\qquad\quad s = 6$

27. $\quad 3(x - 1) = 4 \qquad$ or $\qquad 3(x - 1) = 4$

$\qquad\ x - 1 = \frac{4}{3} \qquad\qquad\qquad 3x - 3 = 4$

$\qquad\qquad x = \frac{4}{3} + 1 \qquad\qquad\qquad 3x = 7$

$\qquad\qquad x = \frac{7}{3} \qquad\qquad\qquad\quad\ x = \frac{7}{3}$

The second way is easier since you are not working with fractions until the end of the solution.

29. $\quad \frac{1}{3}(x + 2) = 5 \qquad$ or $\qquad \frac{1}{3}(x + 2) = 5$

$\quad 3\left(\frac{1}{3}\right)(x + 2) = 3(5) \qquad\qquad \frac{1}{3}x + \frac{2}{3} = 5$

$\qquad\quad x + 2 = 15 \qquad\qquad\qquad \frac{1}{3}x = 5 - \frac{2}{3}$

$\qquad\qquad\ x = 13 \qquad\qquad\qquad\ \frac{1}{3}x = \frac{13}{3}$

$\qquad\qquad\qquad\qquad\qquad\qquad\qquad x = 3\left(\frac{13}{3}\right)$

$\qquad\qquad\qquad\qquad\qquad\qquad\qquad x = 13$

The first way is easier here. The fraction is eliminated in the first step.

31.
$$x + 10 = 15$$
$$x + 10 - 10 = 15 - 10$$
$$x = 5$$

33.
$$7 - 2x = 15$$
$$7 - 2x - 7 = 15 - 7$$
$$-2x = 8$$
$$x = -4$$

35.
$$8x - 5 = 3x + 10$$
$$8x - 3x - 5 + 5 = 3x - 3x + 10 + 5$$
$$5x = 15$$
$$x = 3$$

37.
$$2(x + 5) - 7 = 3(x - 2)$$
$$2x + 10 - 7 = 3x - 6$$
$$2x + 3 = 3x - 6$$
$$-x = -9$$
$$x = 9$$

39.
$$6[x - (2x + 3)] = 8 - 5x$$
$$6[-x - 3] = 8 - 5x$$
$$-6x - 18 = 8 - 5x$$
$$-x = 26$$
$$x = -26$$

41.
$$\frac{5x}{4} + \frac{1}{2} = x - \frac{1}{2}$$
$$4\left(\frac{5x}{4}\right) + 4\left(\frac{1}{2}\right) = 4(x) - 4\left(\frac{1}{2}\right)$$
$$5x + 2 = 4x - 2$$
$$x = -4$$

43.
$$\frac{3}{2}(z + 5) - \frac{1}{4}(z + 24) = 0$$
$$4\left(\frac{3}{2}\right)(z + 5) - 4\left(\frac{1}{4}\right)(z + 24) = 4(0)$$
$$6(z + 5) - (z + 24) = 0$$
$$6z + 30 - z - 24 = 0$$
$$5z = -6$$
$$z = -\frac{6}{5}$$

45.
$$0.25x + 0.75(10 - x) = 3$$
$$4(0.25x) + 4(0.75)(10 - x) = 4(3)$$
$$x + 3(10 - x) = 12$$
$$x + 30 - 3x = 12$$
$$-2x = -18$$
$$x = 9$$

47. $y = 2(x - 1) - 4$ $0 = 2(x - 1) - 4$

$$0 = 2x - 2 - 4$$
$$0 = 2x - 6$$
$$6 = 2x$$
$$3 = x$$
$$x = 3$$

The *x*-intercept is at 3.

The solution to $0 = 2(x - 1) - 4$ and the *x*-intercept of $y = 2(x - 1) - 4$ are the same. They are both $x = 3$.

49. $y = 20 - (3x - 10)$ $0 = 20 - (3x - 10)$

$$0 = 20 - 3x + 10$$
$$0 = 30 - 3x$$
$$3x = 30$$
$$x = 10$$

The *x*-intercept is at 10.

The solution to $0 = 20 - (3x - 10)$ and the *x*-intercept of $y = 20 - (3x - 10)$ are the same. They are both $x = 10$.

51. $x + 8 = 2(x - 2) - x$

$$x + 8 = 2x - 4 - x$$
$$x + 8 = x - 4$$
$$8 = -4$$

Contradiction: no solution

53. $$\frac{100 - 4u}{3} = \frac{5u + 6}{4} + 6$$

$$12\left(\frac{100 - 4u}{3}\right) = 12\left(\frac{5u + 6}{4}\right) + 12(6)$$
$$4(100 - 4u) = 3(5u + 6) + 72$$
$$400 - 16u = 15u + 18 + 72$$
$$-31u = -310$$
$$u = 10$$

55. $$\frac{5x - 4}{5x + 4} = \frac{2}{3}$$

$$3(5x - 4) = 2(5x + 4)$$
$$15x - 12 = 10x + 8$$
$$5x = 20$$
$$x = 4$$

57. $$10 - \frac{13}{x} = 4 + \frac{5}{x}$$

$$\frac{10x - 13}{x} = \frac{4x + 5}{x}$$
$$10x - 13 = 4x + 5$$
$$6x = 18$$
$$x = 3$$

59. $$\frac{1}{x - 3} + \frac{1}{x + 3} = \frac{10}{x^2 - 9}$$

$$\frac{(x + 3) + (x - 3)}{x^2 - 9} = \frac{10}{x^2 - 9}$$
$$2x = 10$$
$$x = 5$$

61. $$\frac{x}{x + 4} + \frac{4}{x + 4} + 2 = 0$$

$$\frac{x + 4}{x + 4} + 2 = 0$$
$$1 + 2 = 0$$
$$3 = 0$$

Contradiction : no solution

63. $$\frac{7}{2x + 1} - \frac{8x}{2x - 1} = -4$$

$$7(2x - 1) - 8x(2x + 1) = -4(2x + 1)(2x - 1)$$
$$14x - 7 - 16x^2 - 8x = -16x^2 + 4$$
$$6x = 11$$
$$x = \frac{11}{6}$$

65. $$\frac{1}{x} + \frac{2}{x - 5} = 0$$

$$1(x - 5) + 2x = 0$$
$$3x - 5 = 0$$
$$3x = 5$$
$$x = \frac{5}{3}$$

67. $\dfrac{3}{x(x-3)} + \dfrac{4}{x} = \dfrac{1}{x-3}$

$3 + 4(x - 3) = x$

$3 + 4x - 12 = x$

$3x = 9$

$x = 3$

A check reveals that $x = 3$ is an extraneous solution, so there is no solution.

69. $(x + 2)^2 + 5 = (x + 3)^2$

$x^2 + 4x + 4 + 5 = x^2 + 6x + 9$

$4x + 9 = 6x + 9$

$-2x = 0$

$x = 0$

71. $(x + 2)^2 - x^2 = 4(x + 1)$

$x^2 + 4x + 4 - x^2 = 4x + 4$

$4 = 4$

The equation is an identity; every real number is a solution.

73. $4(x + 1) - ax = x + 5$

$4x + 4 - ax = x + 5$

$3x - ax = 1$

$x(3 - a) = 1$

$x = \dfrac{1}{3 - a}, a \neq 3$

75. $6x + ax = 2x + 5$

$4x + ax = 5$

$x(4 + a) = 5$

$x = \dfrac{5}{4 + a}, a \neq -4$

77. $0.275x + 0.725(500 - x) = 300$

$0.275x + 362.5 - 0.725x = 300$

$-0.45x = -62.5$

$x = \dfrac{62.5}{0.45}$

≈ 138.889

79. $\dfrac{x}{0.6821} + \dfrac{x}{0.0692} = 1000$

$0.0692x + 0.6321x = 1000(0.6321)(0.0692)$

$0.7013x = 43.74132$

$x = \dfrac{43.74132}{0.7013}$

≈ 62.372

81. $\dfrac{2}{7.398} - \dfrac{4.405}{x} = \dfrac{1}{x}$

$2x - (4.405)(7.398) = 7.398$

$2x = (4.405)(7.398) + 7.398$

$2x = (5.405)(7.398)$

$x = \dfrac{(5.405)(7.398)}{2}$

≈ 19.993

83. $\dfrac{1 + 0.73205}{1 - 0.73205}$

(a) 6.46

(b) $\dfrac{1.73}{0.27} \approx 6.41$

The second method introduced an additional round-off error.

85. $\dfrac{3.33 + \dfrac{1.98}{0.74}}{4 + \dfrac{6.25}{3.15}}$

(a) 1.00

(b) $\dfrac{6.01}{5.98} \approx 1.01$

The second method introduced an additional round-off error.

87. (a)

x	-1	0	1	2	3	4
$3.2x - 5.8$	-9	-5.8	-2.6	0.6	3.8	7

(b) Since the sign changes from negative at 1 to positive at 2, the root is somewhere between 1 and 2. $1 < x < 2$

(c)

x	1.5	1.6	1.7	1.8	1.9	2
$3.2x - 5.8$	-1	-0.68	-0.36	-0.04	0.28	0.6

(d) Since the sign changes from negative at 1.8 to positive at 1.9, the root is somewhere between 1.8 and 1.9. $1.8 < x < 1.9$.

To improve accuracy, evaluate the expression in this interval and determine where the sign changes.

89. $16 = 0.432x - 10.44$

$26.44 = 0.432x$

$\dfrac{26.44}{0.432} = x$

$x \approx 61.2$ inches

91. $T = I + S = x + (10{,}000 - \tfrac{1}{2}x) = 10{,}000 + \tfrac{1}{2}x,\ \ 0 \le x \le 20{,}000$

93. $13{,}800 = 10{,}000 + \tfrac{1}{2}x$

$3{,}800 = \tfrac{1}{2}x$

$7{,}600 = x$

Earned income: $7,600

95. $248 = 2(24) + 2(4x) + 2(6x)$

$248 = 48 + 8x + 12x$

$200 = 20x$

$x = 10$ centimeters

97. $10{,}000 = 0.32m + 2500$

$7{,}500 = 0.32m$

$\dfrac{7{,}500}{0.32} = m$

$m = 23{,}437.5$ miles

Section 1.3   Modeling with Linear Equations

- ■ You should be able to set up mathematical models to solve problems.
- ■ You should be able to translate key words and phrases.

  (a) Equality:

  Equals, equal to, is, are, was, will be, represents

  (b) Addition:

  Sum, plus, greater, increased by, more than, exceeds, total of

  (c) Subtraction:

  Difference, minus, less than, decreased by, subtracted from, reduced by, the remainder

  (d) Multiplication:

  Product, multiplied by, twice, times, percent of

  (e) Division:

  Quotient, divided by, ratio, per

  (f) Consecutive:

  Next, subsequent

- ■ You should know the following formulas:

  (a) Perimeter:

  1. Square: $P = 4s$
  2. Rectangle: $P = 2L + 2W$
  3. Circle: $C = 2\pi r$

  (b) Area:

  1. Square: $A = s^2$
  2. Rectangle: $A = LW$
  3. Circle: $A = \pi r^2$
  4. Triangle: $A = \left(\dfrac{1}{2}\right)bh$

  (c) Volume

  1. Cube: $V = s^3$
  2. Rectangular solid: $V = LWH$
  3. Cylinder: $V = \pi r^2 h$
  4. Sphere: $V = \left(\dfrac{4}{3}\right)\pi r^3$

  (d) Simple Interest: $I = Prt$

  (e) Compound Interest: $A = P\left(1 + \dfrac{r}{n}\right)^{nt}$

  (f) Distance: $D = r \cdot t$

  (g) Temperature: $F = \dfrac{9}{5}C + 32$

- ■ You should be able to solve word problems. Study the examples in the text carefully.

Solutions to Odd-Numbered Exercises

1. $x + 4$

The sum of a number and 4. A number increased by 4.

3. $\dfrac{u}{5}$

The ratio of u and 5. The quotient of u and 5. A number divided by 5.

5. $\dfrac{y - 4}{5}$

The difference of a number and 4 is divided by 5. A number decreased by 4 is divided by 5.

7. *Verbal Model:* (Sum) = (first number) + (second number)
   Labels: Sum = S, first number = n, second number = $n + 1$
   Expression: $S = n + (n + 1) = 2n + 1$

 9. *Verbal Model:* Product = (first odd integer)(second odd integer)
   Labels: Product = P, first odd integer = $2n - 1$, second odd integer = $2n - 1 + 2 = 2n + 1$
   Expression: $P = (2n - 1)(2n + 1) = 4n^2 - 1.$

11. *Verbal Model:* (Distance) = (rate) × (time)
   Labels: Distance = d, rate = 50 mph, time = t
   Expression: $d = 50t$

13. *Verbal Model:* (Amount of acid) = 20% × (amount of solution)
   Labels: Amount of acid (in gallons) = A, amount of solution (in gallons) = x
   Expression: $A = 0.20x$

15. *Verbal Model:* Perimeter = 2(width) + 2(length)
   Labels: Perimeter = P, width = x, length = 2(width) = $2x$
   Expression: $P = 2x + 2(2x) = 6x$

17. *Verbal Model:* (Total cost) = (unit cost)(number of units) + (fixed cost)
   Labels: Total cost = C, fixed cost = \$1200, unit cost = \$25, number of units = x
   Expression: $C = 25x + 1200$

19.

Area = Area of top rectangle + Area of bottom rectangle

$A = 4x + 8x = 12x$

21. *Verbal Model:* Sum = (first number) + (second number)
   Labels: Sum = 525, first number = n, second number = $n + 1$
   Equation: $525 = n + (n + 1)$
   $n = 262$

   Answer: First number = $n = 262$, second number = $n + 1 = 263$

23. *Verbal Model:* Difference = (one number) − (another number)
   Labels: Difference = 148, one number = $5x$, another number = x
   Equation: $148 = 5x - x$
   $x = 37$
   $5x = 185$

   Answer: The two numbers are 37 and 185.

25. *Verbal Model:* Product = (first number) × (second number) = (first number)^2 − 5

Labels: First number = n, second number = $n + 1$

Equation: $n(n + 1) = n^2 - 5$

$\qquad\qquad\; n^2 + n = n^2 - 5$

$\qquad\qquad\qquad\; n = -5$

Answer: First number = $n = -5$, second number = $n + 1 = -4$

27. 30% of L

0.30L

29. N is what percent of 500?

$N = p(500)$

31. x = percent • number

$\quad = (30\%)(45)$

$\quad = 0.30(45)$

$\quad = 13.5$

33. x = percent • number

$x = 0.045\% \times 2{,}650{,}000$

$x = (0.00045)(2{,}650{,}000)$

$x = 1192.5$

35. $459 = \text{percent} \cdot 340$

$459 = p(340)$

$\dfrac{459}{340} = p$

$p = 1.35 = 135\%$

37. $\qquad 70 = 40\% \cdot \text{number}$

$\text{number} = \dfrac{70}{40\%} = \dfrac{70}{0.40} = 175$

39. *Verbal Model:* Loan payments = 58.6% • Annual income

Labels: Loan payments = 13,077.75

$\qquad\quad$ Annual income = I

Equation: $13{,}077.75 = 0.586I$

$\qquad\qquad\quad I \approx 22{,}316.98$

The family's annual income is 22,316.98.

41. Income Tax: 45% of 1,147,588,000,000

$\qquad\qquad\quad = 0.45\,(1{,}147{,}588{,}000{,}000)$

$\qquad\qquad\quad \approx \$516 \text{ billion}$

Corporation Taxes: 9.3% of 1,147,588,000,000

$\qquad\qquad\qquad = 0.093\,(1{,}147{,}588{,}000{,}000)$

$\qquad\qquad\qquad \approx \$107 \text{ billion}$

Social Security: 37.3% of 1,147,588,000,000

$\qquad\qquad\quad = 0.373\,(1{,}147{,}588{,}000{,}000)$

$\qquad\qquad\quad \approx \$428 \text{ billion}$

Other: 8.4% of 1,147,588,000,000

$\qquad\quad = 0.084\,(1{,}147{,}588{,}000{,}000)$

$\qquad\quad \approx \$96 \text{ billion}$

43. *Verbal Model:* (Total profit) = (January profit) + (February profit)

Labels: Total profit = \$157,498, January profit = x, February profit = $x + 20\%$ of $x = x + 0.2x$

Equation: $157{,}498 = x + (x + 0.2x) = 2.2x$

$$x = \frac{157{,}498}{2.2} = 71{,}590$$

Answer: January profit = x = \$71,590.00, February profit = $x + 0.2x$ = \$85,908.00

45. *Verbal Model:* (1992 weekly earnings) = (percentage increase)(1980 weekly earnings) + (1980 weekly earnings)

Labels: 1992 weekly earnings = \$688, percentage increase = p, 1980 weekly earnings = \$400

Equation: $688 = 400p + 400$

$$\frac{288}{400} = p$$

$$p = 0.72 = 72\%$$

Answer: percentage increase $= p = 72\%$

47. *Verbal Model:* (1992 price of gold) = (percentage decrease)(1980 price of gold) + (1980 price of gold)

Labels: 1992 price of gold = \$350, percentage decrease = p, 1980 price of gold = \$613

Equation: $350 = 613 - 613p$

$$-263 = -613p$$

$$\frac{-263}{-613} = p$$

$$p \approx 0.43 = 43\%$$

Answer: percentage decrease $= 43\%$

49. (a)

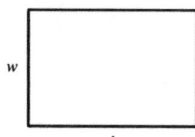

(b) $l = 1.5w$

$P = 2l + 2w$

$ = 2(1.5w) + 2w$

$ = 5w$

(c) $25 = 5w$

$ 5 = w$

Width: $w = 5$ meters

Length: $l = 1.5w = 7.5$ meters

Dimensions: 7.5 m $\times$ 5 m

51. Model: $\text{Average} = \dfrac{(\text{test \#1}) + (\text{test \#2}) + (\text{test \#3}) + (\text{test \#4})}{4}$

Labels: Average = 90, test \#1 = 87, test \#2 = 92, test \#3 = 84, test \#4 = x

Equation: $90 = \dfrac{87 + 92 + 84 + x}{4}$

$$360 = 263 + x$$

$$x = 97$$

Answer: test \#4 $= x = 97$ (or greater)

53. $\text{Rate} = \dfrac{\text{distance}}{\text{time}} = \dfrac{50 \text{ kilometers}}{\frac{1}{2} \text{ hours}} = 100 \text{ kilometers/hours}$

$\text{Total time} = \dfrac{\text{total distance}}{\text{rate}} = \dfrac{300 \text{ kilometers}}{100 \text{ kilometers/hours}} = 3 \text{ hours}$

55. $\text{distance} = \text{rate} \times \text{time}$

$d_1 = 40 \text{ mph} \times t$

$d_2 = 55 \text{ mph} \times t$

$(\text{distance between cars}) = (\text{second distance}) - (\text{first distance})$

$5 = d_2 - d_1$

$5 = 55t - 40t = 15t$

$t = \tfrac{1}{3} \text{ hour}$

57. (a) Time for the first family: $t_1 = \dfrac{d}{r_1} = \dfrac{160}{42} \approx 3.8 \text{ hours}$

Time for the other family: $t_2 = \dfrac{d}{r_2} = \dfrac{160}{50} = 3.2 \text{ hours}$

(b) $t = \dfrac{d}{r} = \dfrac{100}{42 + 50} = \dfrac{100}{92} \approx 1.1 \text{ hours}$

(c) $d = rt = 42\left(\dfrac{160}{42} - \dfrac{160}{50}\right) = 25.6 \text{ miles}$

59. Let x = wind speed, then the rate to the city = $600 + x$, the rate from the city = $600 - x$, the distance to the city = 1500 kilometers, the distance traveled so far in the return trip = $1500 - 300 = 1200$ kilometers.

$$\text{time} = \dfrac{\text{distance}}{\text{rate}}$$

$$\dfrac{1500}{600 + x} = \dfrac{1200}{600 - x}$$

$$1500(600 - x) = 1200(600 + x)$$

$$900{,}000 - 1500x = 720{,}000 + 1200x$$

$$180{,}000 = 2700x$$

$$66\tfrac{2}{3} = x$$

Wind speed: $66\tfrac{2}{3}$ kilometers per hour

61. $\text{time} = \dfrac{\text{distance}}{\text{rate}}$

$t = \dfrac{3.86 \times 10^8 \text{ meters}}{3.0 \times 10^8 \text{ meters per second}}$

$t \approx 1.29 \text{ seconds}$

63. Let h = height of the building in feet.

$$\frac{h \text{ feet}}{80 \text{ feet}} = \frac{4 \text{ feet}}{3.5 \text{ feet}}$$

$$\frac{h}{80} = \frac{4}{3.5}$$

$$3.5h = 320$$

$$h \approx 91.4 \text{ feet}$$

65.

$$\frac{50}{32 + x} = \frac{6}{x}$$

$$50x = 6(32 + x)$$

$$50x = 192 + 6x$$

$$44x = 192$$

$$x \approx 4.36 \text{ feet}$$

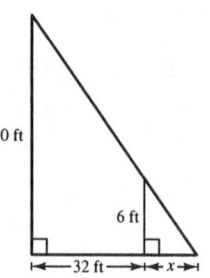

67. Let x = amount in the $12\frac{1}{2}\%$ fund. Then $25{,}000 - x$ = amount in the 11% fund.

$$3000 = 0.125x + 0.11(25{,}000 - x)$$

$$3000 = 0.125x + 2750 - 0.11x$$

$$250 = 0.015x$$

$$\frac{250}{0.015} = x$$

$$x \approx \$16{,}666.67$$

69.   interest = (interest rate) × principal × (portion of year)

$$i_1 = x \times \$10{,}000 \times \tfrac{3}{4} \text{ year}$$

$$i_2 = (x - 1.5\%) \times \$10{,}000 \times \tfrac{1}{4} \text{ year}$$

total interest = (first three quarter's interest) + (last quarter's interest)

$$1112.50 = i_1 + i_2$$

$$1112.50 = 7500x + 2500(x - 0.015)$$

$$x = \frac{1150}{10{,}000} = 0.115 = 11.5\%$$

The first three quarter's rate was 11.5% and the last quarter's rate was 10%.

71. (Final concentration)(Amount) = (Solution 1 concentration)(Amount) +

(Solution 2 concentration)(Amount)

$$(75\%)(55 \text{ gal}) = (40\%)(55 - x) + (100\%)x$$

$$41.25 = 0.60x + 22$$

$$x \approx 32.1 \text{ gallons}$$

73. Let x = number of pounds of \$2.49 nuts. Then $100 - x$ = number of pounds of \$3.89 nuts.

$$2.49x + 3.89(100 - x) = 3.19(100)$$

$$2.49x + 389 - 3.89x = 319$$

$$-1.40x = -70$$

$$x = \frac{-70}{-1.40}$$

$$x = 50 \text{ pounds of } \$2.49 \text{ nuts}$$

$$100 - x = 50 \text{ pounds of } \$3.89 \text{ nuts}$$

Use 50 pounds of each kind of nuts.

75. Cost = Fixed cost + Variable cost · Number of units

$$\$85{,}000 = \$10{,}000 + \$9.30x$$

$$x = \frac{75{,}000}{9.3} \approx 8064.52 \text{ units}$$

At most the company can manufacture 8064 units.

77. $W_1x = W_2(L - x)$

$50x = 75(10 - x)$

$50x = 750 - 75x$

$125x = 750$

$x = 6$ feet from 50 pound child.

79. $A = \dfrac{1}{2}bh$

$2A = bh$

$\dfrac{2A}{b} = h$

81. $S = C + RC$

$S = C(1 + R)$

$\dfrac{S}{1 + R} = C$

83. $A = P + Prt$

$A - P = Prt$

$\dfrac{A - P}{Pt} = r$

85. $A = \dfrac{1}{2}(a + b)h$

$\dfrac{2A}{h} = a + b$

$\dfrac{2A - ah}{h} = b$

87. $V = \dfrac{1}{3}\pi h^2(3r - h)$

$3V = \pi h^2(3r - h)$

$3V = 3\pi rh^2 - \pi h^3$

$3V + \pi h^3 = 3\pi rh^2$

$\dfrac{3V + \pi h^3}{3\pi h^2} = r$

89. $L = L_0[1 + \alpha(\Delta t)]$

$L = L_0 + L_0\alpha(\Delta t)$

$L - L_0 = L_0\alpha(\Delta t)$

$\dfrac{L - L_0}{L_0(\Delta t)} = \alpha$

91. $F = \alpha\dfrac{m_1 m_2}{r^2}$

$Fr^2 = \alpha m_1 m_2$

$\dfrac{Fr^2}{\alpha m_1} = m_2$

93. $\dfrac{1}{f} = (n - 1)\left(\dfrac{1}{R_1} - \dfrac{1}{R_2}\right)$

$\dfrac{1}{(n - 1)f} = \dfrac{1}{R_1} - \dfrac{1}{R_2}$

$\dfrac{1}{(n - 1)f} + \dfrac{1}{R_2} = \dfrac{1}{R_1}$

$\dfrac{R_2 + (n - 1)f}{(n - 1)f R_2} = \dfrac{1}{R_1}$

$R_1 = \dfrac{(n - 1)f R_2}{R_2 + (n - 1)f}$

$\left(R_1 \text{ is the reciprocal of } \dfrac{1}{R_1}.\right)$

95.
$$L = a + (n - 1)d$$
$$L - a = nd - d$$
$$L - a + d = nd$$
$$\frac{L - a + d}{d} = n$$

97.
$$S = \frac{rL - a}{r - 1}$$
$$S(r - 1) = rL - a$$
$$Sr - S = rL - a$$
$$Sr - rL = S - a$$
$$r(S - L) = S - a$$
$$r = \frac{S - a}{S - L}$$

99. $(5x^4)(25x^2)^{-1} = \dfrac{5x^4}{25x^2} = \dfrac{x^2}{5},\ x \neq 0$

101. $\dfrac{6}{\sqrt{10} - 2} = \dfrac{6}{\sqrt{10} - 2} \cdot \dfrac{\sqrt{10} + 2}{\sqrt{10} + 2} = \dfrac{6(\sqrt{10} + 2)}{(\sqrt{10})^2 - (2)^2}$

$$= \frac{6(\sqrt{10} + 2)}{10 - 4} = \frac{6(\sqrt{10} + 2)}{6} = \sqrt{10} + 2$$

Section 1.4   Quadratic Equations and Applications

- You should be able to solve a quadratic equation by factoring, if possible.
- You should be able to solve a quadratic equation of the form $u^2 = d$ by extracting square roots.
- You should be able to solve a quadratic equation by completing the square.
- You should know and be able to use the Quadratic Formula: For $ax^2 + bx + c = 0, a \neq 0$,
$$x = \frac{-b \pm \sqrt{b^2 - 4ac}}{2a}.$$
- You should be able to determine the types of solutions of a quadratic equation by checking the discriminant $b^2 - 4ac$.
  (a) If $b^2 - 4ac > 0$, there are two distinct real solutions.
  (b) If $b^2 - 4ac = 0$, there is one repeating real solution.
  (c) If $b^2 - 4ac < 0$, there is no real solution.
- You should be able to use your calculator to solve quadratic equations.
- You should be able to solve word problems involving quadratic equations. Study the examples in the text carefully.

Solutions to Odd-Numbered Exercises

1. $2x^2 = 3 - 5x$

Standard form: $2x^2 + 5x - 3 = 0$

3. $(x - 3)^2 = 2$

$x^2 - 6x + 9 = 2$

Standard form: $x^2 - 6x + 7 = 0$

5. $\frac{1}{5}(3x^2 - 10) = 12x$

$3x^2 - 10 = 60x$

Standard form: $3x^2 - 60x - 10 = 0$

7. $6x^2 + 3x = 0$

$3x(2x + 1) = 0$

$3x = 0$ or $2x + 1 = 0$

$x = 0$ or $x = -\frac{1}{2}$

9. $x^2 - 2x - 8 = 0$

$(x - 4)(x + 2) = 0$

$x - 4 = 0$ or $x + 2 = 0$

$x = 4$ or $x = -2$

11. $x^2 + 10x + 25 = 0$

$(x + 5)^2 = 0$

$x + 5 = 0$

$x = -5$

13. $3 + 5x - 2x^2 = 0$

$(3 - x)(1 + 2x) = 0$

$3 - x = 0$ or $1 + 2x = 0$

$x = 3$ or $x = -\frac{1}{2}$

15. $x^2 + 4x = 12$

$x^2 + 4x - 12 = 0$

$(x + 6)(x - 2) = 0$

$x + 6 = 0$ or $x - 2 = 0$

$x = -6$ or $x = 2$

17. $x^2 + 2ax + a^2 = 0$

$(x + a)^2 = 0$

$x + a = 0$

$x = -a$

19. $x^2 = 16$

$x = \pm 4 = \pm 4.00$

21. $x^2 = 7$

$x = \pm \sqrt{7} \approx 2.65$

23. $3x^2 = 36$

$x^2 = 12$

$x = \pm 2\sqrt{3}$

$\approx \pm 3.46$

25. $(x - 12)^2 = 18$

$x - 12 = \pm 3\sqrt{2}$

$x = 12 \pm 3\sqrt{2}$

$x \approx 16.24$ or $x \approx 7.76$

27. $(x + 2)^2 = 12$

$x + 2 = \pm 2\sqrt{3}$

$x = -2 \pm 2\sqrt{3}$

$x \approx 1.46$ or $x \approx -5.46$

29. $(x - 7)^2 = (x + 3)^2$

$x - 7 = \pm(x + 3)$

$x - 7 = x + 3$ or $x - 7 = -x - 3$

$-7 \neq 3$ or $2x = 4$

No solution $x = 2 = 2.00$

31. $x^2 - 2x = 0$

$x^2 - 2x + 1^2 = 0 + 1$

$x^2 - 2x + 1 = 1$

$(x - 1)^2 = 1$

$x - 1 = \pm\sqrt{1}$

$x = 1 \pm 1$

$x = 0$ or $x = 2$

33. $x^2 + 4x - 32 = 0$

$x^2 + 4x = 32$

$x^2 + 4x + 2^2 = 32 + 2^2$

$(x + 2)^2 = 36$

$x + 2 = \pm 6$

$x = -2 \pm 6$

$x = 4$ or $x = -8$

35. $x^2 + 6x + 2 = 0$

$x^2 + 6x = -2$

$x^2 + 6x + 3^2 = -2 + 3^2$

$(x + 3)^2 = 7$

$x + 3 = \pm\sqrt{7}$

$x = -3 \pm \sqrt{7}$

37. $9x^2 - 18x + 3 = 0$

$x^2 - 2x + \frac{1}{3} = 0$

$x^2 - 2x = -\frac{1}{3}$

$x^2 - 2x + 1^2 = -\frac{1}{3} + 1^2$

$(x - 1)^2 = \frac{2}{3}$

$x - 1 = \pm\sqrt{\frac{2}{3}}$

$x = 1 \pm \sqrt{\frac{2}{3}}$

$x = 1 \pm \frac{\sqrt{6}}{3}$

39. $8 + 4x - x^2 = 0$

$-x^2 + 4x + 8 = 0$

$x^2 - 4x - 8 = 0$

$x^2 - 4x = 8$

$x^2 - 4x + 2^2 = 8 + 2^2$

$(x - 2)^2 = 12$

$x - 2 = \pm\sqrt{12}$

$x = 2 \pm 2\sqrt{3}$

41. $\dfrac{1}{x^2 + 2x + 5} = \dfrac{1}{x^2 + 2x + 1^2 - 1^2 + 5} = \dfrac{1}{(x + 1)^2 + 4}$

43. $\dfrac{1}{\sqrt{6x - x^2}} = \dfrac{1}{\sqrt{-1(x^2 - 6x + 3^2 - 3^2)}} = \dfrac{1}{\sqrt{-1[(x - 3)^2 - 9]}}$

$= \dfrac{1}{\sqrt{-(x - 3)^2 + 9}} = \dfrac{1}{\sqrt{9 - (x - 3)^2}}$

45. $y = (x + 3)^2 - 4$

$0 = (x + 3)^2 - 4$

$4 = (x + 3)^2$

$\pm\sqrt{4} = x + 3$

$-3 \pm 2 = x$

$x = -1 \text{ or } x = -5$

The x-intercepts are $(-1, 0)$ and $(-5, 0)$.
The x-intercepts of the graph are solutions to the equation $0 = (x + 3)^2 - 4$.

47. $y = -4x^2 + 4x + 3$

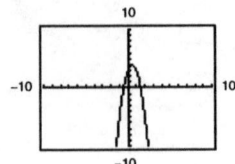

The x-intercepts are $\left(-\frac{1}{2}, 0\right)$ and $\left(\frac{3}{2}, 0\right)$.
The x-intercepts of the graph are solutions
to the equation $0 = -4x^2 + 4x + 3$.

$$0 = -4x^2 + 4x + 3$$
$$4x^2 - 4x = 3$$
$$4(x^2 - x) = 3$$
$$x^2 - x = \tfrac{3}{4}$$
$$x^2 - x + \left(\tfrac{1}{2}\right)^2 = \tfrac{3}{4} + \left(\tfrac{1}{2}\right)^2$$
$$\left(x - \tfrac{1}{2}\right)^2 = 1$$
$$x - \tfrac{1}{2} = \pm\sqrt{1}$$
$$x = \tfrac{1}{2} \pm 1$$
$$x = \tfrac{3}{2} \text{ or } x = -\tfrac{1}{2}$$

49. $2x^2 - 5x + 5 = 0$

$b^2 - 4ac = (-5)^2 - 4(2)(5) = -15 < 0$

No real solution

51. $\frac{1}{5}x^2 + \frac{6}{5}x - 8 = 0$

$b^2 - 4ac = \left(\frac{6}{5}\right)^2 - 4\left(\frac{1}{5}\right)(-8) = \frac{196}{25} > 0$

Two real solutions

53. $2x^2 + x - 1 = 0$

$$x = \frac{-b \pm \sqrt{b^2 - 4ac}}{2a}$$
$$= \frac{-1 \pm \sqrt{1^2 - 4(2)(-1)}}{2(2)}$$
$$= \frac{-1 \pm 3}{4} = \frac{1}{2}, -1$$

55. $16x^2 + 8x - 3 = 0$

$$x = \frac{-b \pm \sqrt{b^2 - 4ac}}{2a}$$
$$= \frac{-8 \pm \sqrt{8^2 - 4(16)(-3)}}{2(16)}$$
$$= \frac{-8 \pm 16}{32} = \frac{1}{4}, -\frac{3}{4}$$

57. $2 + 2x - x^2 = 0$

$$x = \frac{-b \pm \sqrt{b^2 - 4ac}}{2a}$$
$$= \frac{-2 \pm \sqrt{2^2 - 4(-1)(2)}}{2(-1)}$$
$$= \frac{-2 \pm 2\sqrt{3}}{-2} = 1 \pm \sqrt{3}$$

59. $x^2 + 14x + 44 = 0$

$$x = \frac{-b \pm \sqrt{b^2 - 4ac}}{2a}$$
$$= \frac{-14 \pm \sqrt{14^2 - 4(1)(44)}}{2(1)}$$
$$= \frac{-14 \pm 2\sqrt{5}}{2} = -7 \pm \sqrt{5}$$

61. $x^2 + 8x - 4 = 0$

$$x = \frac{-b \pm \sqrt{b^2 - 4ac}}{2a}$$
$$= \frac{-8 \pm \sqrt{8^2 - 4(1)(-4)}}{2(1)}$$
$$= \frac{-8 \pm 4\sqrt{5}}{2} = -4 \pm 2\sqrt{5}$$

63.

$$12x - 9x^2 = -3$$

$$-9x^2 + 12x + 3 = 0$$

$$x = \frac{-b \pm \sqrt{b^2 - 4ac}}{2a}$$

$$= \frac{-12 \pm \sqrt{12^2 - 4(-9)(3)}}{2(-9)}$$

$$= \frac{-12 \pm 6\sqrt{7}}{-18} = \frac{2}{3} \pm \frac{\sqrt{7}}{3}$$

65. $3x + x^2 - 1 = 0$

$$x^2 + 3x - 1 = 0$$

$$x = \frac{-b \pm \sqrt{b^2 - 4ac}}{2a}$$

$$= \frac{-3 \pm \sqrt{3^2 - 4(1)(-1)}}{2(1)}$$

$$= \frac{-3 \pm \sqrt{13}}{2} = -\frac{3}{2} \pm \frac{\sqrt{13}}{2}$$

67.

$$4x^2 + 4x = 7$$

$$4x^2 + 4x - 7 = 0$$

$$x = \frac{-b \pm \sqrt{b^2 - 4ac}}{2a}$$

$$= \frac{-4 \pm \sqrt{4^2 - 4(4)(-7)}}{2(4)}$$

$$= \frac{-4 \pm 8\sqrt{2}}{8} = -\frac{1}{2} \pm \sqrt{2}$$

69.

$$28x - 49x^2 = 4$$

$$-49x^2 + 28x - 4 = 0$$

$$x = \frac{-b \pm \sqrt{b^2 - 4ac}}{2a}$$

$$= \frac{-28 \pm \sqrt{28^2 - 4(-49)(-4)}}{2(-49)}$$

$$= \frac{-28 \pm 0}{-98} = \frac{2}{7}$$

71.

$$8t = 5 + 2t^2$$

$$-2t^2 + 8t - 5 = 0$$

$$t = \frac{-b \pm \sqrt{b^2 - 4ac}}{2a}$$

$$= \frac{-8 \pm \sqrt{8^2 - 4(-2)(-5)}}{2(-2)}$$

$$= \frac{-8 \pm 2\sqrt{6}}{-4} = 2 \pm \frac{\sqrt{6}}{2}$$

73.

$$(y - 5)^2 = 2y$$

$$y^2 - 12y + 25 = 0$$

$$y = \frac{-b \pm \sqrt{b^2 - 4ac}}{2a}$$

$$= \frac{-(-12) \pm \sqrt{(-12)^2 - 4(1)(25)}}{2(1)}$$

$$= \frac{12 \pm 2\sqrt{11}}{2} = 6 \pm \sqrt{11}$$

75. $5.1x^2 - 1.7x - 3.2 = 0$

$$x = \frac{1.7 \pm \sqrt{(-1.7)^2 - 4(5.1)(-3.2)}}{2(5.1)}$$

$$x \approx 0.976, \ -0.643$$

77. $422x^2 - 506x - 347 = 0$

$$x = \frac{506 \pm \sqrt{(-506)^2 - 4(422)(-347)}}{2(422)}$$

$$x \approx 1.687, \ -0.488$$

79. $x^2 - 2x - 1 = 0$

$$x^2 - 2x = 1$$

$$x^2 - 2x + 1^2 = 1 + 1^2$$

$$(x - 1)^2 = 2$$

$$x - 1 = \pm\sqrt{2}$$

$$x = 1 \pm \sqrt{2}$$

81. $(x + 3)^2 = 81$

$$x + 3 = \pm 9$$

$$x + 3 = 9 \quad \text{or} \quad x + 3 = -9$$

$$x = 6 \quad \text{or} \qquad x = -12$$

83. $x^2 - x - \dfrac{11}{4} = 0$

$$x^2 - x = \dfrac{11}{4}$$

$$x^2 - x + \left(\dfrac{1}{2}\right)^2 = \dfrac{11}{4} + \left(\dfrac{1}{2}\right)^2$$

$$\left(x - \dfrac{1}{2}\right)^2 = \dfrac{12}{4}$$

$$x - \dfrac{1}{2} = \pm\sqrt{\dfrac{12}{4}}$$

$$x = \dfrac{1}{2} \pm \sqrt{3}$$

85. $(x + 1)^2 = x^2$

$$x^2 = (x + 1)^2$$

$$x = \pm(x + 1)$$

For $x = +(x + 1)$:

$$0 = 1 \quad \text{No solution}$$

For $x = -(x + 1)$:

$$2x = -1$$

$$x = -\dfrac{1}{2}$$

87. $3x + 4 = 2x - 7$

$$x + 4 = -7$$

$$x = -11$$

89. False. The product must equal zero in order for the Zero-Factor Property to be used.

91. $(x - (-4))(x - 6) = 0$

$$(x + 4)(x - 6) = 0$$

$$x^2 - 2x - 24 = 0$$

93. (a)

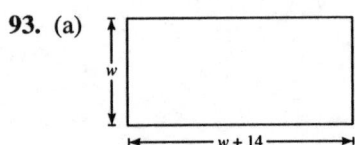

(b) $w(w + 14) = 1632$

(c) $w^2 + 14w - 1632 = 0$

$$(w + 48)(w - 34) = 0$$

$$w = -48 \quad \text{or} \quad w = 34$$

Since w must be greater than zero, we have
$w = 34$ feet and the length is $w + 14 = 48$ feet.

95. $S = x^2 + 4xh$

$$84 = x^2 + 4x(2)$$

$$0 = x^2 + 8x - 84$$

$$0 = (x + 14)(x - 6)$$

$$x = -14 \quad \text{or} \quad x = 6$$

Since x must be positive, we have $x = 6$ inches. The dimensions of the base are
6 inches $\times$ 6 inches.

97. $(200 - 2x)(100 - 2x) = \dfrac{1}{2}(100)(200)$

$20{,}000 - 600x + 4x^2 = 10{,}000$

$4x^2 - 600x + 10{,}000 = 0$

$4(x^2 - 150x + 2500) = 0$

Thus, $a = 1$, $b = -150$, $c = 2500$.

$$x = \frac{150 \pm \sqrt{(-150)^2 - 4(1)(2500)}}{2(1)}$$

$$\approx \frac{150 \pm 111.8034}{2}$$

$$x \approx \frac{150 + 111.8034}{2} \approx 130.902 \text{ feet}$$ Not possible since the lot is only 100 feet wide.

$$x \approx \frac{150 - 111.8034}{2} \approx 19.098 \text{ feet}$$

The person must go around the lot

$$\frac{19.098 \text{ feet}}{24 \text{ inches}} = \frac{19.098 \text{ feet}}{2 \text{ feet}} = 9.5 \text{ times.}$$

99. (a) $s = -16t^2 + v_0 t + s_0$

Since the object was dropped, $v_0 = 0$, and the initial height is $s_0 = 1821$.
Thus, $s = -16t^2 + 1821$.

(b)

t	0	2	4	6	8	10
s	1821	1757	1565	1245	797	221

(c) When $t = 10$ seconds, the object is still 221 feet above ground, so we know
$t > 10$ seconds.

$$0 = -16t^2 + 1821$$

$$16t^2 = 1821$$

$$4t = \pm\sqrt{1821}$$

Since $t > 0$, we have $t = \dfrac{\sqrt{1821}}{4} \approx 10.67$ seconds.

101. $x^2 + x^2 = 5^2$ Pythagorean Theorem

$2x^2 = 25$

$x^2 = \dfrac{25}{2}$

$x = \sqrt{\dfrac{25}{2}}$

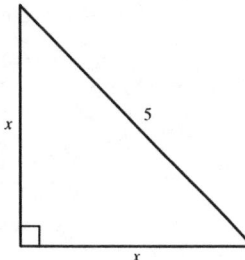

$= \dfrac{5}{\sqrt{2}}$

$= \dfrac{5\sqrt{2}}{2} \approx 3.54$ centimeters

103.

$$d_N = (3 \text{ hours})(r + 50 \text{ mph})$$

$$d_E = (3 \text{ hours})(r \text{ mph})$$

$$d_N{}^2 + d_E{}^2 = 2440^2$$

$$9(r + 50)^2 + 9r^2 = 2440^2$$

$$18r^2 + 900r - 5{,}931{,}100 = 0$$

$$r = \frac{-900 \pm \sqrt{900^2 - 4(18)(-5{,}931{,}100)}}{2(18)} = \frac{-900 \pm 60\sqrt{118{,}847}}{36}$$

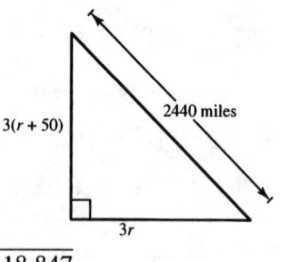

Using the positive value for r, we have the plane moving northbound at $r + 50 \approx 600$ miles per hour and one plane moving eastbound at $r \approx 550$ miles per hour.

105. $x(20 - 0.0002x) = 500{,}000$

$$0 = 0.0002x^2 - 20x + 500{,}000$$

$$0 = x^2 - 100{,}000x + 2{,}500{,}000{,}000$$

$$0 = (x - 50{,}000)^2$$

$$x = 50{,}000 \text{ units}$$

107. $0.125x^2 + 20x + 500 = 14{,}000$

$$0.125x^2 + 20x - 13{,}500 = 0$$

$$x = \frac{-20 \pm \sqrt{20^2 - 4(0.125)(-13{,}500)}}{2(0.125)}$$

Using the positive value for x, we have

$$x = \frac{-20 + \sqrt{7150}}{0.25} \approx 258 \text{ units.}$$

109. $800 + 0.04x + 0.002x^2 = 1680$

$$0.002x^2 + 0.04 - 880 = 0$$

$$x = \frac{-0.04 \pm \sqrt{(0.04)^2 - 4(0.002x)(-880)}}{2(0.002)}$$

$$= \frac{-0.04 \pm \sqrt{7.0416}}{0.004}$$

Choosing the positive value for x, we have

$$x = \frac{-0.04 + \sqrt{7.0416}}{0.004} \approx 653 \text{ units.}$$

111. $250 = 0.6942t^2 + 6.183$

$$243.817 = 0.6942t^2$$

$$t = \sqrt{\frac{243.817}{0.6942}} \approx 19$$

Since $t = 0$ represents 1800 and $t = 1$ represents 1810, $t = 19$ represents $1800 + 10(19) = 1990$. The model was a good representation through 1890, but not through 1990.

113. $S = 0.157t^2 - 1.041t + 7.385$

(a)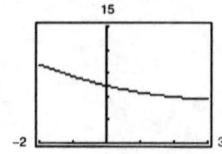

(b) Yes, the sales will exceed 7 billion dollars.

$$7 = 0.157t^2 - 1.041t + 7.385$$

$$0 = 0.157t^2 - 1.041t + 0.385$$

By the Quadratic Formula, we have $t \approx 0.393$ and $t \approx 6.237$. Since ' to forecast future sales,

we use $t \approx 6.237$ which corresponds to the year 1996.

115. $(10x)y = 10(xy)$ by the Associative Property of Multiplication.

117. $\left(\dfrac{6u^2}{5v^{-3}}\right)^{-1} = \left(\dfrac{6u^2v^3}{5}\right)^{-1} = \dfrac{5}{6u^2v^3}$

119. $x^5 - 27x^2 = x^2(x^3 - 27) = x^2(x - 3)(x^2 + 3x + 9)$

121. $\dfrac{x^2 - 100}{10 - x} = \dfrac{(x + 10)(x - 10)}{-(x - 10)} = \dfrac{x + 10}{-1} = -(x + 10),\ x \neq 10$

123.
Xmin = -2
Xmax = 5
Xscl = 1
Ymin = -2
Ymax = 12
Yscl = 2

Section 1.5   Complex Numbers

- You should know how to work with complex numbers.
- Operations on complex numbers
  - (a) Addition: $(a + bi) + (c + di) = (a + c) + (b + d)i$
  - (b) Subtraction: $(a + bi) - (c + di) = (a - c) + (b - d)i$
  - (c) Multiplication: $(a + bi)(c + di) = (ac - bd) + (ad + bc)i$
  - (d) Division: $\dfrac{a + bi}{c + di} = \dfrac{a + bi}{c + di} \cdot \dfrac{c - di}{c - di} = \dfrac{ac + bd}{c^2 + d^2} + \dfrac{bc - ad}{c^2 + d^2}i$
- The complex conjugate of $a + bi$ is $a - bi$:
  $$(a + bi)(a - bi) = a^2 + b^2$$
- The additive inverse of $a + bi$ is $-a - bi$.
- The multiplicative inverse of $a + bi$ is
  $$\dfrac{a - bi}{a^2 + b^2}.$$
- $\sqrt{-a} = \sqrt{a}\,i$ for $a > 0$.

Solutions to Odd-Numbered Exercises

1. $a + bi = -10 + 6i$
   $\quad\quad a = -10$
   $\quad\quad b = 6$

3. $(a - 1) + (b + 3)i = 5 + 8i$
   $\quad a - 1 = 5 \quad \Rightarrow \quad a = 6$
   $\quad b + 3 = 8 \quad \Rightarrow \quad b = 5$

5. $4 + \sqrt{-9} = 4 + 3i$

7. $2 - \sqrt{-27} = 2 - \sqrt{27}\,i = 2 - 3\sqrt{3}\,i$

9. $\sqrt{-75} = \sqrt{75}\,i = 5\sqrt{3}\,i$

11. $-6i + i^2 = -6i - 1 = -1 - 6i$

13. $8 = 8 + 0i = 8$

15. $\sqrt{-0.09} = \sqrt{0.09}\,i = 0.3i$

17. $(5 + i) + (6 - 2i) = 11 - i$

19. $(8 - i) - (4 - i) = 8 - i - 4 + i = 4$

21. $\left(-2 + \sqrt{-8}\right)\left(5 - \sqrt{50}\right) = -2 + 2\sqrt{2}\,i + 5 - 5\sqrt{2}\,i$
   $\quad\quad\quad\quad\quad\quad\quad\quad\quad = 3 - 3\sqrt{2}\,i$

23. $13i - (14 - 7i) = 13i - 14 + 7i = -14 + 20i$

25. $-\left(\frac{3}{2} + \frac{5}{2}i\right) + \left(\frac{5}{3} + \frac{11}{3}i\right) = -\frac{3}{2} - \frac{5}{2}i + \frac{5}{3} + \frac{11}{3}i$
   $\quad\quad\quad\quad\quad\quad\quad\quad\quad\quad = -\frac{9}{6} - \frac{15}{6}i + \frac{10}{6} + \frac{22}{6}i$
   $\quad\quad\quad\quad\quad\quad\quad\quad\quad\quad = \frac{1}{6} + \frac{7}{6}i$

27. $\sqrt{-6} \cdot \sqrt{-2} = \left(\sqrt{6}\,i\right)\left(\sqrt{2}\,i\right) = \sqrt{12}\,i^2 = \left(2\sqrt{3}\right)(-1) = -2\sqrt{3}\,i$

29. $\left(\sqrt{-10}\right)^2 = \left(\sqrt{10}\,i\right)^2 = 10i^2 = -10$

31. $(1 + i)(3 - 2i) = 3 - 2i + 3i - 2i^2$
   $\quad\quad\quad\quad\quad\quad = 3 + i + 2$
   $\quad\quad\quad\quad\quad\quad = 5 + i$

33. $6i(5 - 2i) = 30i - 12i^2 = 30i + 12 = 12 + 30i$

35. $\left(\sqrt{14} + \sqrt{10}\,i\right)\left(\sqrt{14} - \sqrt{10}\,i\right) = 14 - 10i^2 = 14 + 10 = 24$

37. $(4 + 5i)^2 = 16 + 40i + 25i^2 = 16 + 40i - 25$
   $\quad\quad\quad\quad\quad = -9 + 40i$

39. $(2 + 3i)^2 + (2 - 3i)^2 = 4 + 12i + 9i^2 + 4 - 12i + 9i^2$
   $\quad\quad\quad\quad\quad\quad\quad\quad\quad = 4 + 12i - 9 + 4 - 12i - 9$
   $\quad\quad\quad\quad\quad\quad\quad\quad\quad = -10$

41. $\sqrt{-6}\sqrt{-6} = \sqrt{6}\,i\sqrt{6}\,i = 6i^2 = -6$

43. The complex conjugate of $5 + 3i$ is $5 - 3i$.

$(5 + 3i)(5 - 3i) = 25 - 9i^2 = 25 + 9 = 34$

45. The complex conjugate of $-2 - \sqrt{5}\,i$ is $-2 + \sqrt{5}\,i$.

$\left(-2 - \sqrt{5}\,i\right)\left(-2 + \sqrt{5}\,i\right) = 4 - 5t^2 = 4 + 5 = 9$

47. The complex conjugate of $20i$ is $-20i$.

$(20i)(-20i) = -400i^2 = 400$

49. The complex conjugate of $\sqrt{8}$ is $\sqrt{8}$.

$\left(\sqrt{8}\right)\left(\sqrt{8}\right) = 8$

51. $\dfrac{6}{i} = \dfrac{6}{i} \cdot \dfrac{-i}{-i} = \dfrac{-6i}{-i^2} = \dfrac{-6i}{1} = -6i$

53. $\dfrac{4}{4 - 5i} = \dfrac{4}{4 - 5i} \cdot \dfrac{4 + 5i}{4 + 5i} = \dfrac{4(4 + 5i)}{16 + 25} = \dfrac{16 + 20i}{41} = \dfrac{16}{41} + \dfrac{20}{41}i$

55. $\dfrac{2 + i}{2 - i} = \dfrac{2 + i}{2 - i} \cdot \dfrac{2 + i}{2 + i} = \dfrac{4 + 4i + i^2}{4 + 1} = \dfrac{3 + 4i}{5} = \dfrac{3}{5} + \dfrac{4}{5}i$

57. $\dfrac{6 - 7i}{i} = \dfrac{6 - 7i}{i} \cdot \dfrac{-i}{-i} = \dfrac{-6i + 7i^2}{1} = -7 - 6i$

59. $\dfrac{1}{(4 - 5i)^2} = \dfrac{1}{16 - 40i + 25i^2} = \dfrac{1}{-9 - 40i} \cdot \dfrac{-9 + 40i}{-9 + 40i}$

$= \dfrac{-9 + 40i}{81 + 1600} = \dfrac{-9 + 40i}{1681} = -\dfrac{9}{1681} + \dfrac{40}{1681}i$

61. $\dfrac{2}{1 + i} - \dfrac{3}{1 - i} = \dfrac{2(1 - i) - 3(1 + i)}{(1 + i)(1 - i)}$

$= \dfrac{2 - 2i - 3 - 3i}{1 + 1}$

$= \dfrac{-1 - 5i}{2}$

$= -\dfrac{1}{2} - \dfrac{5}{2}i$

63. $\dfrac{i}{3-2i} + \dfrac{2i}{3+8i} = \dfrac{i(3+8i) + 2i(3-2i)}{(3-2i)(3+8i)}$

$= \dfrac{3i + 8i^2 + 6i - 4i^2}{9 + 24i - 6i - 16i^2}$

$= \dfrac{4i^2 + 9i}{9 + 18i + 16}$

$= \dfrac{-4 + 9i}{25 + 18i} \cdot \dfrac{25 - 18i}{25 - 18i}$

$= \dfrac{-100 + 72i + 225i - 162i^2}{625 + 324}$

$= \dfrac{-100 + 297i + 162}{949}$

$= \dfrac{62 + 297i}{949}$

$= \dfrac{62}{949} + \dfrac{297}{949}i$

65. $x^2 - 2x + 2 = 0;\ a = 1,\ b = -2,\ c = 2$

$x = \dfrac{-(-2) \pm \sqrt{(-2)^2 - 4(1)(2)}}{2(1)} = \dfrac{2 \pm \sqrt{-4}}{2} = \dfrac{2 \pm 2i}{2} = 1 \pm i$

67. $4x^2 + 16x + 17 = 0;\ a = 4,\ b = 16,\ c = 17$

$x = \dfrac{-16 \pm \sqrt{(16)^2 - 4(4)(17)}}{2(4)}$

$= \dfrac{-16 \pm \sqrt{-16}}{8} = \dfrac{-16 + 4i}{8}$

$= -2 \pm \dfrac{1}{2}i$

69. $4x^2 + 16x + 15 = 0;\ a = 4,\ b = 16,\ c = 15$

$x = \dfrac{-16 \pm \sqrt{(16)^2 - 4(4)(15)}}{2(4)} = \dfrac{-16 \pm \sqrt{16}}{8} = \dfrac{-16 \pm 4}{8}$

$x = -\dfrac{12}{8} = -\dfrac{3}{2}$ or $x = \dfrac{-20}{8} = -\dfrac{5}{2}$

71. $16t^2 - 4t + 3 = 0;\ a = 16,\ b = -4,\ c = 3$

$t = \dfrac{-(-4) \pm \sqrt{(-4)^2 - 4(16)(3)}}{2(16)}$

$= \dfrac{4 \pm \sqrt{-176}}{32} = \dfrac{4 \pm 4\sqrt{11}i}{32}$

$= \dfrac{1}{8} \pm \dfrac{\sqrt{11}}{8}i$

73. $y = \frac{1}{4}(4x^2 - 20x + 25)$

x-intercept: $\left(\frac{5}{2}, 0\right)$

$$0 = \frac{1}{4}(4x^2 - 20x + 25)$$
$$0 = 4x^2 - 20x + 25$$
$$0 = (2x - 5)^2$$
$$0 = 2x - 5$$
$$-2x = -5$$
$$x = \frac{5}{2}$$

75. $y = -(x^2 - 4x + 5)$

No x-intercepts

$$0 = -(x^2 - 4x + 5)$$
$$0 = x^2 - 4x + 5$$
$$x = \frac{-(-4) \pm \sqrt{(-4)^2 - 4(1)(5)}}{2(1)}$$
$$= \frac{4 \pm 2i}{2} = 2 \pm i$$

No real solutions

77. The number of x-intercepts of the graph of $y = ax^2 + bx + c$ corresponds to the number of real solutions of the equation $0 = ax^2 + bx + c$. If there are no x-intercepts, the quadratic equation has two complex solutions.

79. $-6i^3 + i^2 = -6i^2i + i^2$

$= -6(-1)i + (-1)$

$= 6i - 1$

$= -1 + 6i$

81. $-5i^5 = -5i^2i^2i$

$= -5(-1)(-1)i$

$= -5i$

83. $\left(\sqrt{-75}\right)^3 \left(5\sqrt{3}\,i\right)^3 = 5^3\left(\sqrt{3}\right)^3 i^3$

$125\left(3\sqrt{3}\right)(-1)$

$= -375\sqrt{3}\,i$

85. $\dfrac{1}{i^3} = \dfrac{1}{-i} = \dfrac{1}{-i} \cdot \dfrac{i}{i} = \dfrac{i}{-i^2} = \dfrac{i}{1} = i$

87. $(2)^3 = 8$

$(1 + \sqrt{3}i)^3 = (-1)^3 + 3(-1)^2(\sqrt{3}i) + 3(-1)(\sqrt{3}i)^2 + (\sqrt{3}i)^3$

$\qquad = -1 + 3\sqrt{3}i - 9i^2 + 3\sqrt{3}i^3$

$\qquad = -1 + 3\sqrt{3}\,i + 9 - 3\sqrt{3}\,i$

$\qquad = 8$

$(-1 - \sqrt{3}i)^3 = (-1)^3 + 3(-1)^2\left(-\sqrt{3}i\right) + 3(-1)\left(-\sqrt{3}i\right)^2 + \left(-\sqrt{3}i\right)^3$

$\qquad = -1 - 3\sqrt{3}i - 9i^2 - 3\sqrt{3}i^3$

$\qquad = -1 - 3\sqrt{3}\,i + 9 + 3\sqrt{3}\,i$

$\qquad = 8$

89. $(a + bi) + (a - bi) = 2a$ which is a real number.

91. $(a + bi)(a - bi) = a^2 - (bi)^2 = a^2 - b^2i^2$

$\qquad\qquad\qquad\qquad = a^2 + b^2$ which is a real number.

93. $(a_1 + b_1i) + (a_2 + b_2i) = (a_1 + a_2) + (b_1 + b_2)i$

The complex conjugate of the sum is $(a_1 + a_2) - (b_1 + b_2)i$, and the sum of the conjugates is

$$(a_1 - b_1i) + (a_2 - b_2i) = (a_1 + a_2) + (-b_1 - b_2)i$$
$$= (a_1 + a_2) - (b_1 + b_2)i.$$

Thus, the conjugate of the sum is the sum of the conjugates.

95. $(4 + 3x) + (8 - 6x - x^2) = -x^2 - 3x + 12$

97. $(2x - 5)^2 = (2x)^2 - 2(2x)(5) + (5)^2$
$$= 4x^2 - 20x + 25$$

99.
$$V = \frac{4}{3}\pi a^2 b$$
$$3V = 4\pi a^2 b$$
$$\frac{3V}{4\pi b} = a^2$$
$$\sqrt{\frac{3V}{4\pi b}} = a$$
$$a = \frac{1}{2}\sqrt{\frac{3V}{\pi b}}$$

101. Let x = # liters withdrawn and replaced.
$$0.50(5 - x) + 1.00x = 0.60(5)$$
$$2.50 - 0.50x + 1.00x = 3.00$$
$$0.50x = 0.50$$
$$x = 1 \text{ liter}$$

Section 1.6   Other Types of Equations

- ■ You should be able to solve certain types of nonlinear or nonquadratic equations.
- ■ For equations involving radicals or fractional powers, raise both sides to the same power.
- ■ For equations that are of the quadratic type, $au^2 + bu + c = 0, a \neq 0$, use either factoring or the quadratic formula.
- ■ For equations with fractions, multiply both sides by the least common denominator to clear the fractions.
- ■ For equations involving absolute value, remember that the expression inside the absolute value can be positive or negative.
- ■ Always check for extraneous solutions.

Solutions to Odd-Numbered Exercises

1. $4x^4 - 18x^2 = 0$

$2x^2(2x^2 - 9) = 0$

$2x^2 = 0 \implies x = 0$

$2x^2 - 9 = 0 \implies x = \pm\dfrac{3\sqrt{2}}{2}$

3. $x^4 - 81 = 0$

$(x^2 + 9)(x + 3)(x - 3) = 0$

$x^2 + 9 = 0 \implies x = \pm 3i$

$x + 3 = 0 \implies x = -3$

$x - 3 = 0 \implies x = 3$

5. $5x^3 + 30x^2 + 45x = 0$

$5x(x^2 + 6x + 9) = 0$

$5x(x + 3)^2 = 0$

$5x = 0 \implies x = 0$

$x + 3 = 0 \implies x = -3$

7. $x^3 - 3x^2 - x + 3 = 0$

$x^2(x - 3) - (x - 3) = 0$

$(x - 3)(x^2 - 1) = 0$

$(x - 3)(x + 1)(x - 1) = 0$

$x - 3 = 0 \implies x = 3$

$x + 1 = 0 \implies x = -1$

$x - 1 = 0 \implies x = 1$

9. $x^4 - x^3 + x - 1 = 0$

$x^3(x - 1) + (x - 1) = 0$

$(x - 1)(x^3 + 1) = 0$

$(x - 1)(x + 1)(x^2 - x + 1) = 0$

$x - 1 = 0 \implies x = 1$

$x + 1 = 0 \implies x = -1$

$x^2 - x + 1 = 0 \implies x = \dfrac{1}{2} \pm \dfrac{\sqrt{3}}{2}i$ (By the Quadratic Formula)

11. $x^4 - 4x^2 + 3 = 0$

$(x^2 - 3)(x^2 - 1) = 0$

$\left(x + \sqrt{3}\right)\left(x - \sqrt{3}\right)(x + 1)(x - 1) = 0$

$x + \sqrt{3} = 0 \implies x = -\sqrt{3}$

$x - \sqrt{3} = 0 \implies x = \sqrt{3}$

$x + 1 = 0 \implies x = -1$

$x - 1 = 0 \implies x = 1$

13. $4x^4 - 65x^2 + 16 = 0$

$(4x^2 - 1)(x^2 - 16) = 0$

$(2x + 1)(2x - 1)(x + 4)(x - 4) = 0$

$2x + 1 = 0 \implies x = -\dfrac{1}{2}$

$2x - 1 = 0 \implies x = \dfrac{1}{2}$

$x + 4 = 0 \implies x = -4$

$x - 4 = 0 \implies x = 4$

15.
$$x^6 + 7x^3 - 8 = 0$$
$$(x^3 + 8)(x^3 - 1) = 0$$
$$(x + 2)(x^2 - 2x + 4)(x - 1)(x^2 + x + 1) = 0$$
$$x + 2 = 0 \implies x = -2$$
$$x^2 - 2x + 4 = 0 \implies x = 1 \pm \sqrt{3} \text{ (By the Quadratic Formula)}$$
$$x - 1 = 0 \implies x = 1$$
$$x^2 + x + 1 = 0 \implies x = -\frac{1}{2} \pm \frac{\sqrt{3}}{2}i \text{ (By the Quadratic Formula)}$$

17.
$$\frac{1}{t^2} + \frac{8}{t} + 15 = 0$$
$$1 + 8t + 15t^2 = 0$$
$$(1 + 3t)(1 + 5t) = 0$$
$$1 + 3t = 0 \implies t = -\frac{1}{3}$$
$$1 + 5t = 0 \implies t = -\frac{1}{5}$$

19.
$$2x + 9\sqrt{x} - 5 = 0$$
$$\left(2\sqrt{x} - 1\right)\left(\sqrt{x} + 5\right) = 0$$
$$\sqrt{x} = \frac{1}{2} \implies x = \frac{1}{4}$$
$$\left(\sqrt{x} = -5 \text{ is not a solution.}\right)$$

21.
$$3x^{1/3} + 2x^{2/3} = 5$$
$$2x^{2/3} + 3x^{1/3} - 5 = 0$$
$$(2x^{1/3} + 5)(x^{1/3} - 1) = 0$$
$$2x^{1/3} + 5 = 0 \implies x^{1/3} = -\frac{5}{2} \implies x = \left(-\frac{5}{2}\right)^3 = -\frac{125}{8}$$
$$x^{1/3} - 1 = 0 \implies x^{1/3} = 1 \implies x = (1)^3 = 1$$

23. $y = x^3 - 2x^2 - 3x$

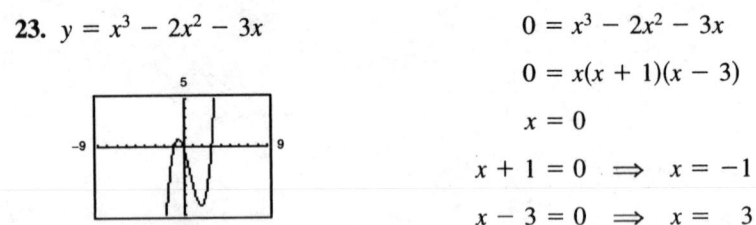

$$0 = x^3 - 2x^2 - 3x$$
$$0 = x(x + 1)(x - 3)$$
$$x = 0$$
$$x + 1 = 0 \implies x = -1$$
$$x - 3 = 0 \implies x = 3$$

x-intercepts: $(-1, 0), (0, 0), (3, 0)$

25. $y = x^4 - 10x^2 + 9$

$$0 = x^4 - 10x^2 + 9$$
$$0 = (x^2 - 1)(x^2 - 9)$$
$$0 = (x + 1)(x - 1)(x + 3)(x - 3)$$
$$x + 1 = 0 \implies x = -1$$
$$x - 1 = 0 \implies x = 1$$
$$x + 3 = 0 \implies x = -3$$
$$x - 3 = 0 \implies x = 3$$

x-intercepts: $(\pm 1, 0), \ (\pm 3, 0)$

27. $\sqrt{2x} - 10 = 0$

$\sqrt{2x} = 10$

$2x = 100$

$x = 50$

29. $\sqrt{x - 10} - 4 = 0$

$\sqrt{x - 10} = 4$

$x - 10 = 16$

$x = 26$

31. $\sqrt[3]{2x + 5} + 3 = 0$

$\sqrt[3]{2x + 5} = -3$

$2x + 5 = -27$

$2x = -32$

$x = -16$

33. $-\sqrt{26 - 11x} + 4 = x$

$4 - x = \sqrt{26 - 11x}$

$16 - 8x + x^2 = 26 - 11x$

$x^2 + 3x - 10 = 0$

$(x + 5)(x - 2) = 0$

$x + 5 = 0 \implies x = -5$

$x - 2 = 0 \implies x = 2$

35. $\sqrt{x + 1} - 3x = 1$

$\sqrt{x + 1} = 3x + 1$

$x + 1 = 9x^2 + 6x + 1$

$0 = 9x^2 + 5x$

$0 = x(9x + 5)$

$x = 0$

$9x + 5 = 0 \implies x = -\frac{5}{9}$, extraneous

37. $\sqrt{x} - \sqrt{x - 5} = 1$

$\sqrt{x} = 1 + \sqrt{x - 5}$

$\left(\sqrt{x}\right)^2 = \left(1 + \sqrt{x - 5}\right)^2$

$x = 1 + 2\sqrt{x - 5} + x - 5$

$4 = 2\sqrt{x - 5}$

$2 = \sqrt{x - 5}$

$4 = x - 5$

$9 = x$

39. $\sqrt{x + 5} + \sqrt{x - 5} = 10$

$\sqrt{x + 5} = 10 - \sqrt{x - 5}$

$\left(\sqrt{x + 5}\right)^2 = \left(10 - \sqrt{x - 5}\right)^2$

$x + 5 = 100 - 20\sqrt{x - 5} + x - 5$

$-90 = -20\sqrt{x - 5}$

$9 = 2\sqrt{x - 5}$

$81 = 4(x - 5)$

$81 = 4x - 20$

$101 = 4x$

$\frac{101}{4} = x$

41. $(x - 5)^{2/3} = 16$

$x - 5 = \pm 16^{3/2}$

$x - 5 = \pm 64$

$x = 69, -59$

43. $(x + 3)^{2/3} = 8$

$x + 3 = \pm 8^{3/2}$

$x + 3 = \pm\sqrt{512}$

$x = -3 \pm 16\sqrt{2}$

45. $(x^2 - 5)^{2/3} = 16$

$x^2 - 5 = \pm 16^{3/2}$

$x^2 - 5 = \pm 64$

$x^2 = 69 \implies x = \pm\sqrt{69}$

$x^2 = -59 \implies x = \pm\sqrt{59}i$

47. $3x(x - 1)^{1/2} + 2(x - 1)^{3/2} = 0$

$(x - 1)^{1/2}[3x + 2(x - 1)] = 0$

$(x - 1)^{1/2}(5x - 2) = 0$

$(x - 1)^{1/2} = 0 \implies x - 1 = 0 \implies x = 1$

$5x - 2 = 0 \implies x = \frac{2}{5}$ which is extraneous.

49. $y = \sqrt{11x - 30} - x$

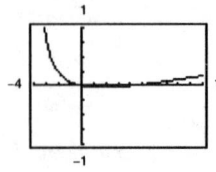

x-intercepts: (5, 0), (6, 0)

$$0 = \sqrt{11x - 30} - x$$
$$x = \sqrt{11x - 30}$$
$$x^2 = 11x - 30$$
$$x^2 - 11x + 30 = 0$$
$$(x - 5)(x - 6) = 0$$
$$x - 5 = 0 \implies x = 5$$
$$x - 6 = 0 \implies x = 6$$

51. $y = \sqrt{7x + 36} - \sqrt{5x + 16} - 2$

x-intercepts: (0, 0), (4, 0)

53. $\dfrac{20 - x}{x} = x$
$$20 - x = x^2$$
$$0 = x^2 + x - 20$$
$$0 = (x + 5)(x - 4)$$
$$x + 5 = 0 \implies x = -5$$
$$x - 4 = 0 \implies x = 4$$

$$0 = \sqrt{7x + 36} - \sqrt{5x + 16} - 2$$
$$-\sqrt{7x + 36} = -\sqrt{5x - 16} - 2$$
$$\sqrt{7x + 36} = 2 + \sqrt{5x + 16}$$
$$\left(\sqrt{7x + 36}\right)^2 = \left(2 + \sqrt{5x + 16}\right)^2$$
$$7x + 36 = 4 + 4\sqrt{5x + 16} + 5x + 16$$
$$7x + 36 = 5x + 20 + 4\sqrt{5x + 16}$$
$$2x + 16 = 4\sqrt{5x + 16}$$
$$x + 8 = 2\sqrt{5x + 16}$$
$$(x + 8)^2 = \left(2\sqrt{5x + 16}\right)^2$$
$$x^2 + 16x + 64 = 4(5x + 16)$$
$$x^2 + 16x + 64 = 20x + 64$$
$$x^2 - 4x = 0$$
$$x(x - 4) = 0$$
$$x = 0$$
$$x - 4 = 0 \implies x = 4$$

55.
$$\frac{1}{x} - \frac{1}{x + 1} = 3$$
$$x(x + 1)\frac{1}{x} - x(x + 1)\frac{1}{x + 1} = x(x + 1)(3)$$
$$x + 1 - x = 3x(x + 1)$$
$$1 = 3x^2 + 3x$$
$$0 = 3x^2 + 3x - 1; \ a = 3, \ b = 3, \ c = -1$$
$$x = \frac{-3 \pm \sqrt{(3)^2 - 4(3)(-1)}}{2(3)} = \frac{-3 \pm \sqrt{21}}{6}$$

57.
$$x = \frac{3}{x} + \frac{1}{2}$$

$$(2x)(x) = (2x)\left(\frac{3}{x}\right) + (2x)\left(\frac{1}{2}\right)$$

$$2x^2 = 6 + x$$

$$2x^2 - x - 6 = 0$$

$$(2x + 3)(x - 2) = 0$$

$$2x + 3 = 0 \implies x = -\frac{3}{2}$$

$$x - 2 = 0 \implies x = 2$$

59.
$$\frac{4}{x + 1} - \frac{3}{x + 2} = 1$$

$$4(x + 2) - 3(x + 1) = (x + 1)(x + 2), x \neq -2, -1$$

$$4x + 8 - 3x - 3 = x^2 + 3x + 2$$

$$x^2 + 2x - 3 = 0$$

$$(x - 1)(x + 3) = 0$$

$$x - 1 = 0 \implies x = 1$$

$$x + 3 = 0 \implies x = -3$$

61. $|2x - 1| = 5$

$$2x - 1 = 5 \implies x = 3$$

$$-(2x - 1) = 5 \implies x = -2$$

63. $|x| = x^2 + x - 3$

$$x = x^2 + x - 3 \quad \text{OR} \qquad\qquad -x = x^2 + x - 3$$

$$x^2 - 3 = 0 \qquad\qquad\qquad x^2 + 2x - 3 = 0$$

$$x = \pm\sqrt{3} \qquad\qquad\qquad (x - 1)(x + 3) = 0$$

$$x - 1 = 0 \implies x = 1$$

$$x + 3 = 0 \implies x = -3$$

Only $x = \sqrt{3}$, and $x = -3$ are solutions to the original equation. $x = -\sqrt{3}$ and $x = 1$ are extraneous.

65. $|x + 1| = x^2 - 5$

$$x + 1 = x^2 - 5 \qquad\qquad \text{OR} \qquad -(x + 1) = x^2 - 5$$

$$x^2 - x - 6 = 0 \qquad\qquad\qquad\qquad -x - 1 = x^2 - 5$$

$$(x - 3)(x + 2) = 0 \qquad\qquad\qquad\qquad x^2 + x - 4 = 0$$

$$x - 3 = 0 \implies x = 3 \qquad\qquad\qquad x = \frac{-1 \pm \sqrt{17}}{2}$$

$$x + 2 = 0 \implies x = -2$$

Only $x = 3$ and $x = \dfrac{-1 - \sqrt{17}}{2}$ are solutions to the original equation. $x = -2$ and

$x = \dfrac{-1 + \sqrt{17}}{2}$ are extraneous.

67. $y = \dfrac{1}{x} - \dfrac{4}{x-1} - 1$

x-intercept: $(-1, 0)$

$0 = \dfrac{1}{x} - \dfrac{4}{x-1} - 1$

$0 = (x-1) - 4x - x(x-1)$

$0 = x - 1 - 4x - x^2 + x$

$0 = -x^2 - 2x - 1$

$0 = x^2 + 2x + 1$

$0 = (x+1)^2$

$x + 1 = 0 \implies x = -1$

69. $y = |x+1| - 2$

x-intercepts: $(1, 0), (-3, 0)$

$0 = |x+1| - 2$

$2 = |x+1|$

$x + 1 = 2 \quad$ or $\quad -(x+1) = 2$

$x = 1 \quad$ or $\quad -x - 1 = 2$

$-x = 3$

$x = -3$

71. $3.2x^4 - 1.5x^2 - 2.1 = 0$

$x^2 = \dfrac{1.5 \pm \sqrt{1.5^2 - 4(3.2)(-2.1)}}{2(3.2)}$

Using the positive value for x^2, we have $x = \pm\sqrt{\dfrac{1.5 + \sqrt{29.13}}{6.4}} \approx \pm 1.038.$

73. $\qquad 1.8x - 6\sqrt{x} - 5.6 = 0 \qquad$ Given equation

$\quad 1.8\left(\sqrt{x}\right)^2 - 6\sqrt{x} - 5.6 = 0 \qquad$ Quadratic form with $u = \sqrt{x}$

Use the Quadratic Formula with $a = 1.8$, $b = -6$, and $c = -5.6$.

$$\sqrt{x} = \dfrac{6 \pm \sqrt{36 - 4(1.8)(-5.6)}}{2(1.8)} \approx \dfrac{6 \pm 8.7361}{3.6}$$

Considering only the positive value for $\sqrt{x}$, we have

$\sqrt{x} \approx 4.0934$

$x \approx 16.756.$

75. $(x - (-3))(x - 5) = 0$

$(x + 3)(x - 5) = 0$

$x^2 - 2x - 15 = 0$

77. $\left(x - \sqrt{2}\right)\left(x + \sqrt{2}\right)(x - 4) = 0$

$(x^2 - 2)(x - 4) = 0$

$x^3 - 4x^2 - 2x + 8 = 0$

79.   Number of students $= x$

Cost per students $= x$

$$fx = 1700 \quad \Rightarrow \quad f = \frac{1700}{x}$$

$$(f - 7.50)(x + 6) = 1700$$

$$\left(\frac{1700}{x} - 7.5\right)(x + 6) = 1700$$

$$(3400 - 15x)(x + 6) = 3400x \quad \text{Multiply both sides by } 2x \text{ to clear fractions.}$$

$$-15x^2 - 90x + 20,400 = 0$$

$$x = \frac{90 \pm \sqrt{(-90)^2 - 4(-15)(20,400)}}{2(-15)} = \frac{90 \pm 1110}{-30}$$

Using the positive value for x we conclude that the original number was $x = 34$ students.

81. Formula: Time $= \dfrac{\text{Distance}}{\text{Rate}}$

Let $x =$ average speed of the plane. Then we have a travel time of $t = \dfrac{720}{x}$.

If the average speed is increased by 40 mph, then

$$t - \frac{12}{60} = \frac{720}{x + 40}$$

$$t = \frac{720}{x + 40} + \frac{1}{5}.$$

Now, we equate these two equations and solve for x.

$$\frac{720}{x} = \frac{720}{x + 40} + \frac{1}{5}$$

$$720(5)(x + 40) = 720(5)x + x(x + 40)$$

$$3600x + 144,000 = 3600x + x^2 + 40x$$

$$0 = x^2 + 40x - 144,000$$

$$0 = (x + 400)(x - 360)$$

Using the positive value for x we have $x = 360$ mph and $x + 40 = 400$ mph.
The airspeed required to obtain the decrease in travel time is 400 miles per hour.

83.

$$A = P\left(1 + \frac{r}{n}\right)^{nt}$$

$$3544.06 = 2500\left(1 + \frac{r}{12}\right)^{(12)(5)}$$

$$1.417636 = \left(1 + \frac{r}{12}\right)^{60}$$

$$(1.417636)^{1/60} = 1 + \frac{r}{12}$$

$$[(1.417636)^{1/60} - 1](12) = r$$

$$r \approx 0.07 = 7\%$$

85. The distance between $(1, 2)$ and $(x, -10)$ is 13.

$$\sqrt{(x - 1)^2 + (-10 - 2)^2} = 13$$
$$(x - 1)^2 + (-12)^2 = 13^2$$
$$x^2 - 2x + 1 + 144 = 169$$
$$x^2 - 2x - 24 = 0$$
$$(x + 4)(x - 6) = 0$$
$$x + 4 = 0 \implies x = -4$$
$$x - 6 = 0 \implies x = 6$$

87. The distance between $(0, 0)$ and $(8, y)$ is 17.

$$\sqrt{(8 - 0)^2 + (y - 0)^2} = 17$$
$$(8)^2 + (y)^2 = 17^2$$
$$64 + y^2 = 289$$
$$y^2 = 225$$
$$y = \pm\sqrt{225}$$
$$= \pm 15$$

89. When $C = 2.5$, we have:

$$2.5 = \sqrt{0.2x + 1}$$
$$6.25 = 0.2x + 1$$
$$5.25 = 0.2x$$
$$x = 26.25 = 26{,}250 \text{ passengers}$$

91.

$$37.55 = 40 - \sqrt{0.01x + 1}$$
$$\sqrt{0.01x + 1} = 2.45$$
$$0.01x + 1 = 6.0025$$
$$0.01x = 5.0025$$
$$x = 500.25$$

Rounding x to the nearest whole unit yields $x \approx 500$ units.

93. *Verbal Model:*   Cost = Cost underwater · Distance underwater + Cost overland · Distance overland

Labels: Cost = 1,098,662.40

Cost underwater = \$30 per foot

Distance underwater in feet $= 5280\sqrt{x^2 + (3/4)^2} = 1320\sqrt{16x^2 + 9}$

Cost overland = \$24 per foot

Distance overland in feet $= 5280(8 - x)$

Equation: $1,098,662.40 = 30\left(1320\sqrt{16x^2 + 9}\right) + 24[5280(8 - x)]$

$1,098,622.40 = 6(1320)[5\sqrt{16x^2 + 9} + 16(8 - x)]$

$138.72 = 5\sqrt{16x^2 + 9} + 128 - 16x$

$16x + 10.72 = 5\sqrt{16x^2 + 9}$

$(16x + 10.72)^2 = \left(5\sqrt{16x^2 + 9}\right)^2$

$256x^2 + 343.04x + 114.9184 = 25(16x^2 + 9)$

$256x^2 + 343.04x + 114.9184 = 400x^2 + 225$

$0 = 144x^2 - 343.04x + 110.0816$

By the Quadratic Formula, we have $x \approx 2$ miles or $x \approx 0.382$ mile.

95. $d = \sqrt{100^2 + h^2}$

(a)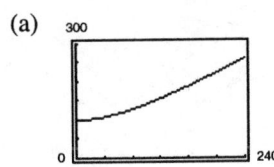

$d = 200$ when $h \approx 173$ feet.

(b)

h	160	165	170	175	180	185
d	188.7	192.9	197.2	201.6	205.9	210.3

$d = 200$ when h is between 170 and 175 feet.

(c) $200 = \sqrt{100^2 + h^2}$
    $40,000 = 10,000 + h^2$
    $30,000 = h^2$
    $h = \sqrt{30,000}$
    $h \approx 173.2$ feet

(d) Solving graphically or numerically yields an approximation. An exact solution is obtained algebraically.

97. $S = \pi r\sqrt{r^2 + h^2}$

$S^2 = \pi^2 r^2(r^2 + h^2)$

$S^2 = \pi^2 r^4 + \pi^2 r^2 h^2$

$\dfrac{S^2 - \pi^2 r^4}{\pi^2 r^2} = h^2$

$h = \dfrac{\sqrt{S^2 - \pi^2 r^4}}{\pi r} = \dfrac{1}{\pi r}\sqrt{S^2 - \pi^2 r^4}$

99. $20 + \sqrt{20 - a} = b$

$$\sqrt{20 - a} = b - 20$$

$$20 - a = b^2 - 40b + 400$$

$$-a = b^2 - 40b + 380$$

$$a = -b^2 + 40b - 380$$

This formula gives the relationship between a and b. From the original equation we know that $a \leq 20$ and $b \geq 20$. Choose a b value, where $b \geq 20$ and then solve for a, keeping in mind that $a \leq 20$.

Some possibilities are: $b = 20,\ \ a = 20$

$$b = 21,\ a = 19$$

$$b = 22,\ a = 16$$

$$b = 23,\ a = 11$$

$$b = 24,\ a = \ \ 4 \quad \leftarrow \left(\begin{array}{l}\text{This is the one given}\\ \text{in your textbook.}\end{array}\right)$$

$$b = 25,\ a = -5$$

101. $25y^2 / \dfrac{xy}{5} = \dfrac{25y^2}{1} \cdot \dfrac{5}{xy} = \dfrac{125y}{x}, y \neq 0$

103. $x^2 \cdot \dfrac{x + 1}{x^2 - x} \cdot \dfrac{(5x - 5)^2}{x^2 + 6x + 5} = \dfrac{x^2}{1} \cdot \dfrac{x + 1}{x(x - 1)} \cdot \dfrac{5^2(x - 1)^2}{(x + 1)(x + 5)}$

$$= \dfrac{25x(x - 1)}{x + 5}, x \neq 0, \pm 1$$

105. $\dfrac{8}{3t} + \dfrac{3}{2t} = \dfrac{16}{6t} + \dfrac{9}{6t} = \dfrac{25}{6t}$

107. $\dfrac{2}{x^2 - 4} - \dfrac{1}{x^2 - 3x + 2} = \dfrac{2}{(x + 2)(x - 2)} - \dfrac{1}{(x - 1)(x - 2)}$

$$= \dfrac{2(x - 1) - (x + 2)}{(x - 1)(x + 2)(x - 2)}$$

$$= \dfrac{2x - 2 - x - 2}{(x - 1)(x^2 - 4)}$$

$$= \dfrac{x - 4}{(x - 1)(x^2 - 4)}$$

Section 1.7   Linear Inequalities

- You should know the properties of inequalities.

  (a) Transitive: $a < b$ and $b < c$ implies $a < c$.

  (b) Addition: $a < b$ and $c < d$ implies $a + c < b + d$.

  (c) Adding or Subtracting a Constant: $a \pm c < b \pm c$ if $a < b$.

  (d) Multiplying or Dividing a Constant: For $a < b$,

   1. If $c > 0$, then $ac < bc$ and $\dfrac{a}{c} < \dfrac{b}{c}$.

   2. If $c < 0$, then $ac > bc$ and $\dfrac{a}{c} > \dfrac{b}{c}$.

- You should know that

$$|x| = \begin{cases} x & \text{if } x \geq 0 \\ -x & \text{if } x < 0 \end{cases}.$$

- You should be able to solve absolute value inequalities.

  (a) $|x| < a$ if and only if $-a < x < a$.

  (b) $|x| > a$ if and only if $x < -a$ or $x > a$.

Solutions to Odd-Numbered Problems

1. $x \leq 25$

   x is less than or equal to 25.

   x is no more than 25.

3. Interval: $[-1, 3]$

   Inequality: $-1 \leq x \leq 3$

   The interval is bounded.

5. Interval: $(10, \infty)$

   Inequality: $x > 10$ or $10 < x < \infty$

   The interval is unbounded.

7. $x < 3$

   Matches (c).

9. $-3 < x \leq 4$

   Matches (f).

11. $|x| < 3 \implies -3 < x < 3$

   Matches (g).

13. $|x - 4| > 2 \implies x - 4 < -2$ or $x - 4 > 2$

   $x < 2$ or $x > 6$

   Matches (b).

15. (a) $x = 3$

   $5(3) - 12 \overset{?}{>} 0$

   $3 > 0$

   Yes, $x = 3$ is a solution.

   (b) $x = -3$

   $5(-3) - 12 \overset{?}{>} 0$

   $-27 \not> 0$

   No, $x = -3$ is not a solution.

   (c) $x = \frac{5}{2}$

   $5\left(\frac{5}{2}\right) - 12 \overset{?}{>} 0$

   $\frac{1}{2} > 0$

   Yes, $x = \frac{5}{2}$ is a solution.

   (d) $x = \frac{3}{2}$

   $5\left(\frac{3}{2}\right) - 12 \overset{?}{>} 0$

   $-\frac{9}{2} \not> 0$

   No, $x = \frac{3}{2}$ is not a solution.

17. (a) $x = 4$

$$0 \overset{?}{<} \frac{4 - 2}{4} \overset{?}{<} 2$$

$$0 < \frac{1}{2} < 2$$

Yes, $x = 4$ is a solution.

(b) $x = 10$

$$0 \overset{?}{<} \frac{10 - 2}{4} \overset{?}{<} 2$$

$$0 < 2 \not< 2$$

No, $x = 10$ is not a solution.

(c) $x = 0$

$$0 \overset{?}{<} \frac{0 - 2}{4} \overset{?}{<} 2$$

$$0 \not< -\frac{1}{2} < 2$$

No, $x = 0$ is not a solution.

(d) $x = \frac{7}{2}$

$$0 \overset{?}{<} \frac{(7/2) - 2}{4} \overset{?}{<} 2$$

$$0 < \frac{3}{8} < 2$$

Yes, $x = \frac{7}{2}$ is a solution.

19. (a) $x = 13$

$$|13 - 10| \overset{?}{\geq} 3$$

$$3 \geq 3$$

Yes, $x = 13$ is a solution.

(b) $x = -1$

$$|-1 - 10| \overset{?}{\geq} 3$$

$$11 \geq 3$$

Yes, $x = -1$ is a solution.

(c) $x = 14$

$$|14 - 10| \overset{?}{\geq} 3$$

$$4 \geq 3$$

Yes, $x = 14$ is a solution.

(d) $x = 9$

$$|9 - 10| \overset{?}{\geq} 3$$

$$1 \not\geq 3$$

No, $x = 9$ is not a solution.

21. $4x < 12$

$$\frac{1}{4}(4x) < \frac{1}{4}(12)$$

$$x < 3$$

23. $-10x < 40$

$$-\frac{1}{10}(-10) > -\frac{1}{10}(40)$$

$$x > -4$$

25. $x - 5 \geq 7$

$$x \geq 12$$

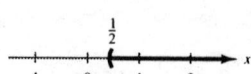

27. $2x + 7 < 3$

$$2x < -4$$

$$x < -2$$

29. $2x - 1 \geq 0$

$$2x \geq 1$$

$$x \geq \frac{1}{2}$$

31. $4 - 2x < 3$

$$-2x < -1$$

$$x > \frac{1}{2}$$

33. $1 < 2x + 3 < 9$

$-2 < 2x < 6$

$-1 < x < 3$

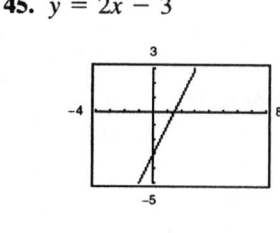

35. $-4 < \dfrac{2x - 3}{3} < 4$

$-12 < 2x - 3 < 12$

$-9 < 2x < 15$

$-\dfrac{9}{2} < x < \dfrac{15}{2}$

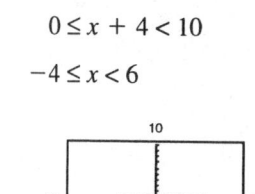

37. $\dfrac{3}{4} > x + 1 > \dfrac{1}{4}$

$-\dfrac{1}{4} > x > -\dfrac{3}{4}$

$-\dfrac{3}{4} < x < -\dfrac{1}{4}$

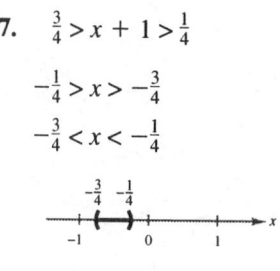

39. $6x > 12$

$x > 2$

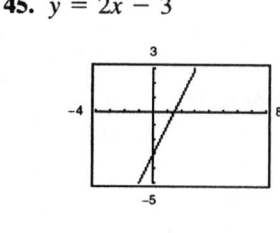

41. $5 - 2x \geq 1$

$-2x \geq -4$

$x \leq 2$

43. $0 \leq 2(x + 4) < 20$

$0 \leq x + 4 < 10$

$-4 \leq x < 6$

45. $y = 2x - 3$ (a) $y \geq 1$

$2x - 3 \geq 1$

$2x \geq 4$

$x \geq 2$

(b) $y \leq 0$

$2x - 3 \leq 0$

$2x \leq 3$

$x \leq \dfrac{3}{2}$

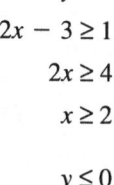
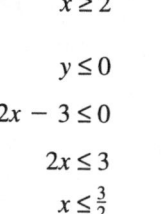

47. $y = -\dfrac{1}{2}x + 2$ (a) $0 \leq y \leq 3$

$0 \leq -\dfrac{1}{2}x + 2 \leq 3$

$-2 \leq -\dfrac{1}{2}x \leq 1$

$4 \geq x \geq -2$

(b) $y \geq 0$

$-\dfrac{1}{2}x + 2 \geq 0$

$-\dfrac{1}{2}x \geq -2$

$x \leq 4$

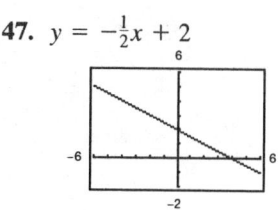

49. $x - 5 \geq 0$

$x \geq 5$

$[5, \infty)$

51. $x + 3 \geq 0$

$x \geq -3$

$[-3, \infty)$

53. $7 - 2x \geq 0$

$-2x \geq -7$

$x \leq \dfrac{7}{2}$

$\left(-\infty, \dfrac{7}{2}\right]$

55. $|x| < 5$

$-5 < x < 5$

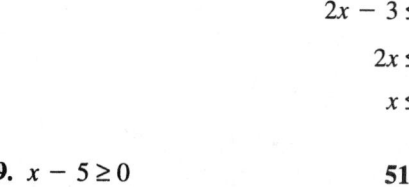

57. $\left|\dfrac{x}{2}\right| > 3$

$\dfrac{x}{2} < -3$ or $\dfrac{x}{2} > 3$

$x < -6$ $x > 6$

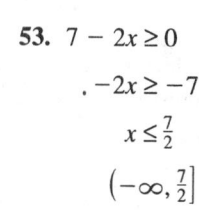

59. $|x - 20| \leq 4$

$-4 \leq x - 20 \leq 4$

$16 \leq x \leq 24$

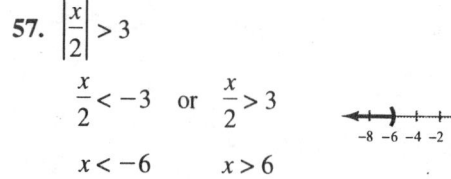

61. $|x - 20| \geq 4$

$x - 20 \leq -4$ or $x - 20 \geq 4$

$x \leq 16$ $x \geq 24$

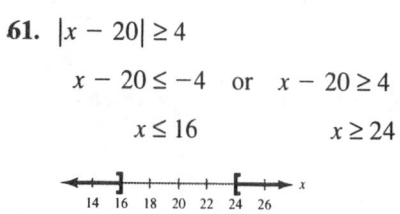

63. $\left|\dfrac{x-3}{2}\right| \geq 5$

$\dfrac{x-3}{2} \leq -5$ or $\dfrac{x-3}{2} \geq 5$

$x - 3 \leq -10$ $x - 3 \geq 10$

$x \leq -7$ $x \geq 13$

65. $|9 - 2x| - 2 < -1$

$|9 - 2x| < 1$

$-1 < 9 - 2x < 1$

$-10 < -2x < -8$

$5 > x > 4$

$4 < x < 5$

67. $2|x + 10| \geq 9$

$|x + 10| \geq \frac{9}{2}$

$x + 10 \leq -\frac{9}{2}$ or $x + 10 \geq \frac{9}{2}$

$x \leq -\frac{29}{2}$ $x \geq -\frac{11}{2}$

69. $|x - 5| < 0$

$|x + 10| \geq \frac{9}{2}$

No solution. The absolute value of a number can never be less than zero.

71. $y = |x - 3|$

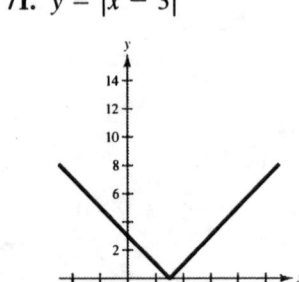

(a) $y \leq 2$

$|x - 3| \leq 2$

$-2 \leq x - 3 \leq 2$

$1 \leq x \leq 5$

(b) $y \geq 4$

$|x - 3| \geq 4$

$x - 3 \leq -4$ or $x - 3 \geq 4$

$x \leq -1$ $x \geq 7$

73. The midpoint of the interval $[-3, 3]$ is 0. The interval represents all real numbers x no more than 3 units from 0.

$|x - 0| \leq 3$

$|x| \leq 3$

75. The graph shows all real numbers at least 3 units from 7.

$|x - 7| \geq 3$

77. All real numbers within 10 units of 12.

$|x - 12| < 10$

79. All real numbers more than 5 units from -3.

$|x - (-3)| > 5$

$|x + 3| > 5$

81. $|x - 10| < 8$ represents all real numbers within 8 units of 10.

83. $150 + 0.25x > 250$

$\qquad 0.25x > 100$

$\qquad\quad x > 400$

If you drive more than 400 miles in a week, the rental fee for Company B is greater than the rental fee for Company A.

85. $1250 < 1000[1 + r(2)]$

$\qquad 1.250 < 1 + 2r$

$\qquad 0.250 < 2r$

$\qquad 0.125 < r$

$\qquad\quad r > 12.5\%$

87. $\qquad\quad R > C$

$\qquad 115.95x > 95x + 750$

$\qquad\quad 20.95x > 750$

$\qquad\qquad\quad x > 35.7995$

$\qquad\qquad\quad x \geq 36 \text{ units}$

89. (a) and (b) $y = 0.067x - 5.638$

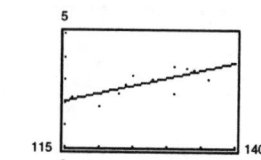

(c) $\qquad 3 \geq 0.067x - 5.638$

$\qquad 8.638 \geq 0.067x$

$\qquad\quad x \geq 129$

(d) IQ scores are not a good predictor of GPAs. Other factors include study habits, class attendance, and attitude.

91. $|s - 10.4| \leq \frac{1}{16}$

$\qquad -\frac{1}{16} \leq s - 10.4 \leq \frac{1}{16}$

$\qquad -0.0625 \leq s - 10.4 \leq 0.0625$

$\qquad 10.3375 \leq s \leq 10.4625$

Since $A = s^2$, we have

$\qquad (10.3375)^2 \leq \text{area} \leq (10.4625)^2$

$\qquad 106.864 \quad \leq \text{area} \leq \ 109.464.$

93. $\left| \dfrac{h - 68.5}{2.7} \right| \leq 1$

$\qquad -1 \leq \dfrac{h - 68.5}{2.7} \leq 1$

$\qquad -2.7 \leq h - 68.5 \leq 2.7$

$\qquad 65.8 \text{ inches} \leq h \leq 71.2 \text{ inches}$

95. $|ax - b| \leq c \implies c$ must be greater than or equal to zero.

$\qquad -c \leq ax - b \leq c$

$\qquad b - c \leq ax \leq b + c$

Let $a = 1$, then $b - c = 0$ and $b + c = 10$. This is true when $b = c = 5$. One set of values is : $a = 1,\ b = 5,\ c = 5$

(Note: This solution is not unique. Any positive multiple of these
   values will also work, such as:

$\qquad a = 2,\ b = c = 10$

$\qquad a = 3,\ b = c = 15$).

97. $V = \dfrac{500}{3}(2) \approx 333$ vibrations per second

99. $200 < \dfrac{500}{3}t < 400$

$600 < 500t < 1200$

$1.2 < t < 2.4$ millimeters

101. $(-3, 10)$

103. $d = \sqrt{[1 - (-4)]^2 + (12 - 2)^2} = \sqrt{(5)^2 + (10)^2} = \sqrt{25 + 100} = \sqrt{125} = 5\sqrt{5}$

Section 1.8   Other Types of Inequalities

■ You should be able to solve inequalities.

(a) Find the critical number.

1. Values that make the expression zero

2. Values that make the expression undefined

(b) Test one value in each interval on the real number line resulting from the critical numbers.

(c) Determine the solution intervals.

Solutions to Odd Numbered Exercises

1. $x^2 - 3 < 0$

(a) $x = 3$

$(3)^2 - 3 \overset{?}{<} 0$

$6 \not< 0$

No, $x = 3$ is not a solution.

(b) $x = 0$

$(0)^2 - 3 \overset{?}{<} 0$

$-3 < 0$

Yes, $x = 0$ is a solution.

(c) $x = \frac{3}{2}$

$\left(\frac{3}{2}\right)^2 - 3 \overset{?}{<} 0$

$-\frac{3}{4} < 0$

Yes, $x = \frac{3}{2}$ is a solution.

(d) $x = -5$

$(-5)^2 - 3 \overset{?}{<} 0$

$22 \not< 0$

No, $x = -5$ is not a solution.

3. $\dfrac{x+2}{x-4} \geq 3$

(a) $x = 5$

$$\dfrac{5+2}{5-4} \overset{?}{\geq} 3$$

$$7 \geq 3$$

Yes, $x = 5$ is a solution.

(c) $x = -\dfrac{9}{2}$

$$\dfrac{-\frac{9}{2}+2}{-\frac{9}{2}-4} \overset{?}{\geq} 3$$

$$\dfrac{5}{17} \not\geq 3$$

No, $x = -\dfrac{9}{2}$ is not a solution.

(b) $x = 4$

$$\dfrac{4+2}{4-4} \overset{?}{\geq} 3$$

$$\dfrac{6}{0} \text{ is undefined.}$$

No, $x = 4$ is not a solution.

(d) $x = \dfrac{9}{2}$

$$\dfrac{\frac{9}{2}+2}{\frac{9}{2}-4} \overset{?}{\geq} 3$$

$$13 \geq 3$$

Yes, $x = \dfrac{9}{2}$ is a solution.

5. $2x^2 - x - 6 = (2x + 3)(x - 2)$

The critical numbers are $x = -\dfrac{3}{2}$ and $x = 2$.

7. $2 + \dfrac{3}{x-5} = \dfrac{2(x-5)+3}{x-5} = \dfrac{2x-7}{x-5}$

The critical numbers are $x = \dfrac{7}{2}$ and $x = 5$.

9. $x^2 \leq 9$

$x^2 - 9 \leq 0$

$(x+3)(x-3) \leq 0$

Critical numbers: $x = \pm 3$

Test intervals: $(-\infty, -3), (-3, 3), (3, \infty)$

Test: Is $(x+3)(x-3) \leq 0$?

Solution set: $[-3, 3]$

11. $x^2 > 4$

$x^2 - 4 > 0$

$(x+2)(x-2) > 0$

Critical numbers: $x = \pm 2$

Test intervals: $(-\infty, -2), (-2, 2), (2, \infty)$

Test: Is $x^2 - 4 > 0$?

Solution set: $(-\infty, -2) \cup (2, \infty)$

13. $(x+2)^2 < 25$

$x^2 + 4x + 4 < 25$

$x^2 + 4x - 21 < 0$

$(x+7)(x-3) < 0$

Critical numbers: $x = -7, x = 3$

Test intervals: $(-\infty, -7), (-7, 3), (3, \infty)$

Test: Is $(x+7)(x-3) < 0$?

Solution set: $(-7, 3)$

15. $x^2 + 4x + 4 \geq 9$

$x^2 + 4x - 5 \geq 0$

$(x+5)(x-1) \geq 0$

Critical numbers: $x = -5, x = 1$

Test intervals: $(-\infty, -5), (-5, 1), (1, \infty)$

Test: Is $(x+5)(x-1) \geq 0$?

Solution set: $(-\infty, -5] \cup [1, \infty)$

17.
$$x^2 + x < 6$$
$$x^2 + x - 6 < 0$$
$$(x + 3)(x - 2) < 0$$
Critical numbers: $x = -3, x = 2$
Test intervals: $(-\infty, -3), (-3, 2), (2, \infty)$
Test: Is $(x + 3)(x - 2) < 0$?
Solution set: $(-3, 2)$

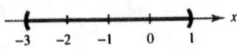

19. $3(x - 1)(x + 1) > 0$
Critical numbers: $x = \pm 1$
Test intervals: $(-\infty, -1), (-1, 1), (1, \infty)$
Test: Is $3(x - 1)(x + 1) > 0$?
Solution set: $(-\infty, -1) \cup (1, \infty)$

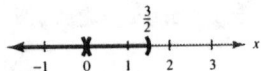

21. $x^2 + 2x - 3 < 0$
$$(x + 3)(x - 1) < 0$$
Critical numbers: $x = -3, x = 1$
Test intervals: $(-\infty, -3), (-3, 1), (1, \infty)$
Test: Is $(x + 3)(x - 1) < 0$?
Solution set: $(-3, 1)$

[Note: Compare this problem to #18.]

23. $4x^3 - 6x^2 < 0$
$$2x^2(2x - 3) < 0$$
Critical numbers: $x = 0, x = \frac{3}{2}$
Test intervals: $(-\infty, 0), \left(0, \frac{3}{2}\right), \left(\frac{3}{2}, \infty\right)$
Test: Is $2x^2(2x - 3) < 0$?
Solution set: $(-\infty, 0) \cup \left(0, \frac{3}{2}\right)$

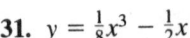

25.
$$x^3 - 4x \geq 0$$
$$x(x + 2)(x - 2) \geq 0$$
Critical numbers: $x = 0, x = \pm 2$
Test intervals: $(-\infty, -2), (-2, 0), (0, 2), (2, \infty)$
Test: Is $x(x + 2)(x - 2) \geq 0$?
Solution set: $[-2, 0] \cup [2, \infty)$

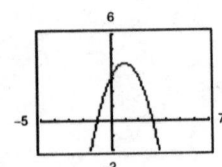

27. $(x - 1)^2(x + 2)^3 \geq 0$
Critical numbers: $x = 1, x = -2$
Test intervals: $(-\infty, -2), (-2, 1), (1, \infty)$
Test: Is $(x - 1)^2(x + 3)^3 \geq 0$?
Solution set: $[-2, \infty)$

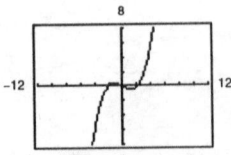

29. $y = -x^2 + 2x + 3$

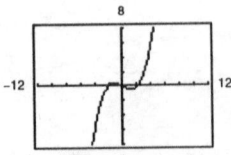

(a) $y \leq 0$ when $x \leq -1$ or $x \geq 3$.

(b) $y \geq 3$ when $0 \leq x \leq 2$.

31. $y = \frac{1}{8}x^3 - \frac{1}{2}x$

(a) $y \geq 0$ when $-2 \leq x \leq 0, 2 \leq x < \infty$.

(b) $y \leq 6$ when $x \leq 4$.

33. $\dfrac{1}{x} - x > 0$

$\dfrac{1 - x^2}{x} > 0$

Critical numbers: $x = 0, x = \pm 1$

Test intervals: $(-\infty, -1), (-1, 0), (0, 1), (1, \infty)$

Test: Is $\dfrac{1 - x^2}{x} > 0$?

Solution set: $(-\infty, -1) \cup (0, 1)$

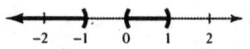

35. $\dfrac{x + 6}{x + 1} - 2 < 0$

$\dfrac{x + 6 - 2(x + 1)}{x + 1} < 0$

$\dfrac{4 - x}{x + 1} < 0$

Critical numbers: $x = -1, x = 4$

Test intervals: $(-\infty, -1), (-1, 4), (4, \infty)$

Test: Is $\dfrac{4 - x}{x + 1} < 0$?

Solution set: $(-\infty, -1) \cup (4, \infty)$

37. $\dfrac{3x - 5}{x - 5} > 4$

$\dfrac{3x - 5}{x - 5} - 4 > 0$

$\dfrac{3x - 5 - 4(x - 5)}{x - 5} > 0$

$\dfrac{15 - x}{x - 5} > 0$

Critical numbers: $x = 5, x = 15$

Test intervals: $(-\infty, 5), (5, 15), (15, \infty)$

Test: Is $\dfrac{15 - x}{x - 5} > 0$?

Solution set: $(5, 15)$

39.
$$\frac{4}{x+5} > \frac{1}{2x+3}$$

$$\frac{4}{x+5} - \frac{1}{2x+3} > 0$$

$$\frac{4(2x+3) - (x+5)}{(x+5)(2x+3)} > 0$$

$$\frac{7x+7}{(x+5)(2x+3)} > 0$$

Critical numbers: $x = -1, x = -5, x = -\frac{3}{2}$

Test intervals: $(-\infty, -5), \left(-5, -\frac{3}{2}\right),$

$$\left(-\frac{3}{2}, -1\right), (-1, \infty)$$

Test: Is $\dfrac{7(x+1)}{(x+5)(2x+3)} > 0$?

Solution set: $\left(-5, -\frac{3}{2}\right) \cup (-1, \infty)$

41.
$$\frac{1}{x-3} \le \frac{9}{4x+3}$$

$$\frac{1}{x-3} - \frac{9}{4x+3} \le 0$$

$$\frac{4x+3 - 9(x-3)}{(x-3)(4x+3)} \le 0$$

$$\frac{30 - 5x}{(x-3)(4x+3)} \le 0$$

Critical numbers: $x = 3, x = -\frac{3}{4}, x = 6$

Test intervals: $\left(-\infty, -\frac{3}{4}\right), \left(-\frac{3}{4}, 3\right), (3, 6), (6, \infty)$

Test: Is $\dfrac{5(6-x)}{(x-3)(4x+3)} \le 0$?

Solution set: $\left(-\frac{3}{4}, 3\right) \cup [6, \infty)$

43.
$$\frac{x^2 + 2x}{x^2 - 9} \le 0$$

$$\frac{x(x+2)}{(x+3)(x-3)} \le 0$$

Critical numbers: $x = 0, x = -2, x = \pm 3$

Test intervals: $(-\infty, -3), (-3, -2), (-2, 0), (0, 3), (3, \infty)$

Test: Is $\dfrac{x(x+2)}{(x+3)(x-3)} \le 0$?

Solution set: $(-3, -2] \cup [0, 3)$

45.
$$\frac{3}{x-1} - \frac{2}{x+1} < 1$$

$$\frac{3}{x-1} - \frac{2}{x+1} - 1 < 0$$

$$\frac{3(x+1) - 2(x-1) - (x-1)(x+1)}{(x-1)(x+1)} < 0$$

$$\frac{-x^2 + x + 6}{(x-1)(x+1)} < 0$$

$$\frac{-(x-3)(x+2)}{(x-1)(x+1)} < 0$$

Critical numbers: $x = 3, x = -2, x = \pm 1$

Test intervals: $(-\infty, -2), (-2, -1), (-1, 1), (1, 3), (3, \infty)$

Test: Is $\dfrac{-(x-3)(x+2)}{(x-1)(x+1)} < 0$?

Solution set: $(-\infty, -2) \cup (-1, 1) \cup (3, \infty)$

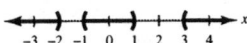

47. $y = \dfrac{3x}{x-2}$

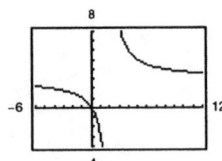

(a) $y \le 0$ when $0 \le x < 2$.

(b) $y \ge 6$ when $2 < x \le 4$.

49. $y = \dfrac{2x^2}{x^2 + 4}$

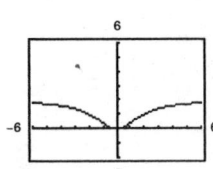

(a) $y \ge 1$ when $x \le -2$ or $x \ge 2$.

　　This can also be expressed as $|x| \ge 2$.

(b) $y \le 2$ for all real numbers x.

　　This can also be expressed as $-\infty < x < \infty$.

51.
$$4 - x^2 \ge 0$$

$$(2 + x)(2 - x) \ge 0$$

Critical numbers: $x = \pm 2$

Test intervals: $(-\infty, -2), (-2, 2), (2, \infty)$

Test: Is $4 - x^2 \ge 0$?

Domain: $[-2, 2]$

53. $x^2 - 7x + 12 \ge 0$

$$(x - 3)(x - 4) \ge 0$$

Critical numbers: $x = 3, x = 4$

Test intervals: $(-\infty, 3), (3, 4), (4, \infty)$

Test: Is $(x - 3)(x - 4) \ge 0$?

Domain: $(-\infty, 3] \cup [4, \infty)$

55. $12 - x - x^2 \geq 0$

$(4 + x)(3 - x) \geq 0$

Critical numbers: $x = -4, x = 3$

Test intervals: $(-\infty, -4), (-4, 3), (3, \infty)$

Test: Is $(4 + x)(3 - x) \geq 0$?

Domain: $[-4, 3]$

57. $0.4x^2 + 5.26 < 10.2$

$0.4x^2 - 4.94 < 0$

$0.4(x^2 - 12.35) < 0$

Critical numbers: $x \approx \pm 3.51$

Test intervals: $(-\infty, -3.51), (-3.51, 3.51), (3.51, \infty)$

Solution set: $(-3.51, 3.51)$

59. $-0.5x^2 + 12.5x + 1.6 > 0$

The zeros are $x = \dfrac{-12.5 \pm \sqrt{(12.5)^2 - 4(-0.5)(1.6)}}{2(-0.5)}$.

Critical numbers: $x \approx -0.13, x \approx 25.13$

Test intervals: $(-\infty, -0.13), (-0.13, 25.13),$
$$(25.13, \infty)$$

Solution set: $(-0.13, 25.13)$

61. $\dfrac{1}{2.3x - 5.2} > 3.4$

$\dfrac{1}{2.3x - 5.2} - 3.4 > 0$

$\dfrac{-7.82x + 18.68}{2.3x - 5.2} > 0$

Critical numbers: $x \approx 2.39, x = 2.26$

Test intervals: $(-\infty, 2.26), (2.26, 2.39), (2.39, \infty)$

Solution set: $(2.26, 2.39)$

63. $x^2 + bx + 4 = 0$

To have at least one real solution, $b^2 - 16 \geq 0$.
This occurs when $b \leq -4$ or $b \geq 4$.
This can be written as $(-\infty, -4] \cup [4, \infty)$.

65. $3x^2 + bx + 10 = 0$

To have at least one real solution, $b^2 - 4(3)(10) \geq 0$.

$$b^2 - 120 \geq 0$$

$$\left(b + \sqrt{120}\right)\left(b - \sqrt{120}\right) \geq 0$$

Critical numbers: $b = \pm\sqrt{120} = \pm 2\sqrt{30}$

Test intervals: $\left(-\infty, -2\sqrt{30}\right), \left(-2\sqrt{30}, 2\sqrt{30}\right), \left(2\sqrt{30}, \infty\right)$

Test: Is $b^2 - 120 \geq 0$?

Solution set: $\left(-\infty, -2\sqrt{30}\right] \cup \left[2\sqrt{30}, \infty\right)$

67. If $a > 0$ and $c \leq 0$, then b can be any real number. If $a > 0$ and $c > 0$, then for $b^2 - 4ac$
to be greater than or equal to zero, b is restricted to $b < -2\sqrt{ac}$ or $b > 2\sqrt{ac}$.

69. $s = -16t^2 + v_0t + s_0 = -16t^2 + 160t$

(a) $-16t^2 + 160t = 0$

$-16t(t - 10) = 0$

$t = 0, t = 10$

It will be back on the ground in 10 seconds.

(b) $-16t^2 + 160t > 384$

$-16t^2 + 160t - 384 > 0$

$-16(t^2 - 10t + 24) > 0$

$t^2 - 10t + 24 < 0$

$(t - 4)(t - 6) < 0$

$4 < t < 6$ seconds

71. $2L + 2W = 100 \implies W = 50 - L$

$LW \geq 500$

$L(50 - L) \geq 500$

$-L^2 + 50L - 500 \geq 0$

By the Quadratic Formula we have:

Critical numbers: $L = 25 \pm 5\sqrt{5}$

Test: Is $-L^2 + 50L - 500 \geq 0$?

Solution set: $25 - 5\sqrt{5} \leq L \leq 25 + 5\sqrt{5}$

13.8 meters $\leq L \leq$ 36.2 meters

73. $\dfrac{1}{R} = \dfrac{1}{R_1} + \dfrac{1}{2}$

$2R_1 = 2R + RR_1$

$2R_1 = R(2 + R_1)$

$\dfrac{2R_1}{2 + R_1} = R$

Since $R \geq 1$, we have

$\dfrac{2R_1}{2 + R_1} \geq 1$

$\dfrac{2R_1}{2 + R_1} - 1 \geq 0$

$\dfrac{R_1 - 2}{2 + R_1} \geq 0.$

Since $R_1 > 0$, the only critical number is $R_1 = 2$.

The inequality is satisfied when $R_1 \geq 2$ ohms.

75. $7.34 + 0.41t + 0.002t^2 \geq 25$

$0.002t^2 + 0.41t - 17.66 \geq 0$

By the Quadratic Formula, the critical numbers are $t \approx -241.55$ and $t \approx 36.55$.

The only one that makes sense here is $t \approx 36.55$.

This corresponds to the year 1996.

77.

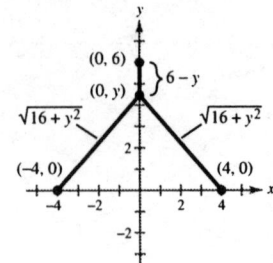

(a) $L = (6 - y) + 2\sqrt{16 + y^2}$

(b) $0 \le y \le 6$

When $y = 0$: $L = 14$

When $y = 6$: $L = 2\sqrt{52} = 4\sqrt{13} \approx 14.4$

(c)

(d) $6 - y + 2\sqrt{16 + y^2} < 13$

$2\sqrt{16 + y^2} - y - 7 < 0$

To find the critical numbers, set

$2\sqrt{16 + y^2} - y - 7 = 0$

$2\sqrt{16 + y^2} = y + 7$

$4(16 + y^2) = y^2 + 14y + 49$

$64 + 4y^2 = y^2 + 14y + 49$

$3y^2 - 14y + 15 = 0$

$(3y - 5)(y - 3) = 0$

$y = \frac{5}{3}, y = 3.$

By testing, the solution set is $\frac{5}{3} < y < 3$.

79. $(x + 3)^2 - 16 = [(x + 3) + 4][(x + 3) - 4]$

$= (x + 7)(x - 1)$

81. $2x^4 - 54x = 2x(x^3 - 27)$

$= 2x(x - 3)(x^2 + 3x + 9)$

❑ Review Exercises for Chapter 1

Solutions to Odd-Numbered Exercises

1. $y = -\frac{1}{2}x + 2$

x	-4	-2	0	2	4
y	4	3	2	1	0

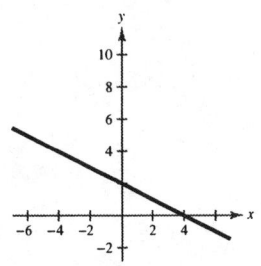

3. $y - 2x - 3 = 0$

$y = 2x + 3$

Line with x-intercept $\left(-\frac{3}{2}, 0\right)$ and y-intercept $(0, 3)$

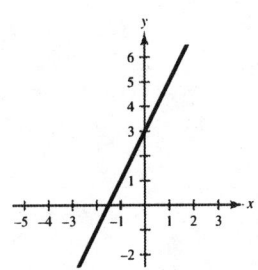

5. $x - 5 = 0$

$x = 5$ is a vertical line through $(5, 0)$.

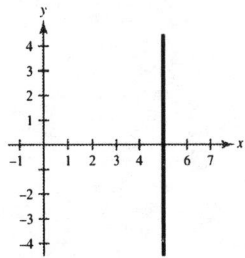

7. $y = \sqrt{5 - x}$

Domain: $(-\infty, 5]$

x	5	4	1	-4
y	0	1	2	3

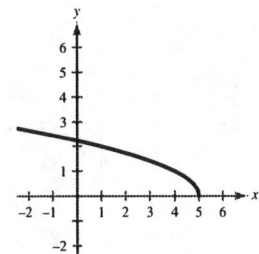

9. $y + 2x^2 = 0$

$y = -2x^2$ is a parabola.

x	0	± 1	± 2
y	0	-2	-8

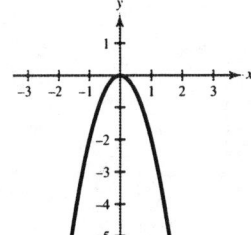

11. $y = \sqrt{25 - x^2}$

Domain: $-5 \le x \le 5$

x	0	± 3	± 4	± 5
y	5	4	3	0

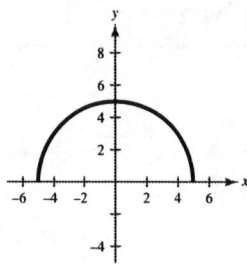

13. $y = \frac{1}{4}(x + 1)^3$

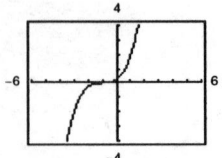

Intercepts: $(-1, 0), \left(0, \frac{1}{4}\right)$

15. $y = \frac{1}{4}x^4 - 2x^2$

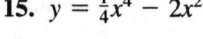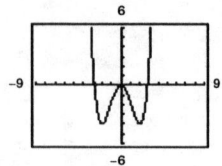

Intercepts: $(0, 0), \left(\pm 2\sqrt{2}, 0\right)$

17. $y = x\sqrt{9 - x^2}$

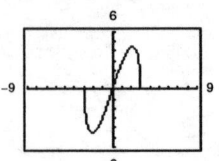

Intercepts: $(0, 0), (\pm 3, 0)$

19. $y = |x - 4| - 4$

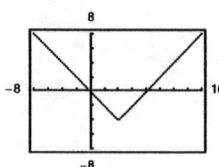

Intercepts: $(0, 0), (8, 0)$

21. $y^2 = 25 - x^2$

$y = \pm\sqrt{25 - x^2}$

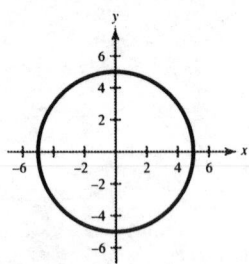

23. $y = 10x^3 - 21x^2$

```
Xmin = -2
Xmax = 3
Xscl = 1
Ymin = -20
Ymax = 15
Yscl = 5
```

25. $(x - 3)^2 + (y + 1)^2 = 9$

$(x - 3)^2 + [y - (-1)]^2 = 3^2$

Center: $(3, -1)$
Radius: $r = 3$

27.
$$6 - (x - 2)^2 = 2 + 4x - x^2$$
$$6 - (x^2 - 4x + 4) = 2 + 4x - x^2$$
$$2 + 4x - x^2 = 2 + 4x - x^2$$
$$0 = 0 \quad \text{Identity}$$

All real numbers are solutions.

29. $3x^2 + 7x = x^2 + 4$

(a) $x = 0$
$$3(0)^2 + 7(0) \stackrel{?}{=} (0)^2 + 4$$
$$0 \neq 4$$

No, $x = 0$ is not a solution.

(b) $x = -4$
$$3(-4)^2 + 7(-4) \stackrel{?}{=} (-4)^2 + 4$$
$$20 = 20$$

Yes, $x = -4$ is a solution.

(c) $x = \frac{1}{2}$
$$3\left(\tfrac{1}{2}\right)^2 + 7\left(\tfrac{1}{2}\right) \stackrel{?}{=} \left(\tfrac{1}{2}\right)^2 + 4$$
$$\tfrac{17}{4} = \tfrac{17}{4}$$

Yes, $x = \frac{1}{2}$ is a solution.

(d) $x = -1$
$$3(-1)^2 + 7(-1) \stackrel{?}{=} (-1)^2 + 4$$
$$-4 \neq 5$$

No, $x = -1$ is not a solution.

31. $3x - 2(x + 5) = 10$
$$3x - 2x - 10 = 10$$
$$x = 20$$

33. $4(x + 3) - 3 = 2(4 - 3x) - 4$
$$4x + 12 - 3 = 8 - 6x - 4$$
$$4x + 9 = -6x + 4$$
$$10x = -5$$
$$x = -\frac{1}{2}$$

35. $3\left(1 - \dfrac{1}{5t}\right) = 0$
$$1 - \frac{1}{5t} = 0$$
$$1 = \frac{1}{5t}$$
$$5t = 1$$
$$t = \frac{1}{5}$$

37. $6x = 3x^2$
$$0 = 3x^2 - 6x$$
$$0 = 3x(x - 2)$$
$$3x = 0 \implies x = 0$$
$$x - 2 = 0 \implies x = 2$$

39. $(x + 4)^2 = 18$
$$x + 4 = \pm\sqrt{18}$$
$$x = -4 \pm 3\sqrt{2}$$

41. $x^2 - 12x + 30 = 0$
$$x^2 - 12x = -30$$
$$x^2 - 12x + 36 = -30 + 36$$
$$(x - 6)^2 = 6$$
$$x - 6 = \pm\sqrt{6}$$
$$x = 6 \pm \sqrt{6}$$

43.
$$5x^4 - 12x^3 = 0$$
$$x^3(5x - 12) = 0$$
$$x^3 = 0 \quad \text{or} \quad 5x - 12 = 0$$
$$x = 0 \quad \text{or} \quad x = \tfrac{12}{5}$$

45. $\dfrac{4}{(x-4)^2} = 1$

$4 = (x-4)^2$

$\pm 2 = x - 4$

$4 \pm 2 = x$

$x = 6$ or $x = 2$

47. $\sqrt{x+4} = 3$

$\left(\sqrt{x+4}\right)^2 = (3)^2$

$x + 4 = 9$

$x = 5$

49. $2\sqrt{x} - 5 = 0$

$2\sqrt{x} = 5$

$4x = 25$

$x = \dfrac{25}{4}$

51. $\sqrt{2x+3} + \sqrt{x-2} = 2$

$\left(\sqrt{2x+3}\right)^2 = \left(2 - \sqrt{x-2}\right)^2$

$2x + 3 = 4 - 4\sqrt{x-2} + x - 2$

$x + 1 = -4\sqrt{x-2}$

$(x+1)^2 = \left(-4\sqrt{x-2}\right)^2$

$x^2 + 2x + 1 = 16(x-2)$

$x^2 - 14x + 33 = 0$

$(x-3)(x-11) = 0$

$x = 3$, extraneous   or $x = 11$, extraneous

No solution

53. $(x-1)^{2/3} - 25 = 0$

$(x-1)^{2/3} = 25$

$(x-1)^2 = 25^3$

$x - 1 = \pm\sqrt{25^3}$

$x = 1 \pm 125$

$x = 126$ or $x = -124$

55. $(x+4)^{1/2} + 5x(x+4)^{3/2} = 0$

$(x+4)^{1/2}[1 + 5x(x+4)] = 0$

$(x+4)^{1/2}(5x^2 + 20x + 1) = 0$

$(x+4)^{1/2} = 0$

$x = -4$ OR $5x^2 + 20x + 1 = 0$

$x = \dfrac{-20 \pm \sqrt{400 - 20}}{10}$

$x = \dfrac{-20 \pm 2\sqrt{95}}{10}$

$x = -2 \pm \dfrac{\sqrt{95}}{5}$

57. $|x - 5| = 10$

$x - 5 = -10$ or $x - 5 = 10$

$x = -5$ $x = 15$

59. $|x^2 - 3| = 2x$

$$x^2 - 3 = 2x \quad \text{or} \quad x^2 - 3 = -2x$$

$$x^2 - 2x - 3 = 0 \qquad x^2 + 2x - 3 = 0$$

$$(x - 3)(x + 1) = 0 \qquad (x + 3)(x - 1) = 0$$

$$x = 3 \quad \text{or} \quad x = -1 \qquad x = -3 \quad \text{or} \quad x = 1$$

The only solutions to the original equation are $x = 3$ or $x = 1$.
($x = -3$ and $x = -1$ are extraneous.)

61. $y = 4x^3 - 12x^2 + 8x$

$$0 = 4x^3 - 12x^2 + 8x$$

$$0 = 4x(x^2 - 3x + 2)$$

$$0 = 4x(x - 1)(x - 2)$$

$$x = 0, \ x = 1, \ \text{or} \ x = 2$$

x-intercepts: $(0, 0)$, $(1, 0)$, $(2, 0)$

63. $y = \dfrac{1}{x} + \dfrac{1}{x + 1} - 2$

$$0 = \frac{1}{x} + \frac{1}{x + 1} - 2$$

$$2 = \frac{1}{x} + \frac{1}{x + 1}$$

$$2x(x + 1) = (x + 1) + x$$

$$2x^2 + 2x = 2x + 1$$

$$2x^2 = 1$$

x-intercepts: $\left(\pm \dfrac{\sqrt{2}}{2}, 0 \right)$

$$x^2 = \frac{1}{2}$$

$$x = \pm \sqrt{\frac{1}{2}} = \pm \frac{\sqrt{2}}{2}$$

65.
$$V = \frac{1}{3}\pi r^2 h$$

$$3V = \pi r^2 h$$

$$\frac{3V}{\pi h} = r^2$$

$$r = \sqrt{\frac{3V}{\pi h}}$$

Since r represents the radius of a cone, r is positive only.

67.
$$L = \frac{k}{3\pi r^2 p}$$

$$3\pi r^2 p L = k$$

$$p = \frac{k}{3\pi r^2 L}$$

69. $y = C\sqrt{x + 1}$

$8 = C\sqrt{3 + 1}$

$8 = C(2)$

$4 = C$

71. $x^2 - 2x \geq 3$

$x^2 - 2x - 3 \geq 0$

$(x + 1)(x - 3) \geq 0$

Critical numbers: $x = -1, x = 3$

Test intervals: $(-\infty, -1), (-1, 3), (3, \infty)$

Test: Is $(x + 1)(x - 3) \geq 0$?

Solution set: $(-\infty, -1] \cup [3, \infty)$

73. $\dfrac{x - 5}{3 - x} < 0$

Critical numbers: $x = 5, x = 3$

Test intervals: $(-\infty, 3), (3, 5), (5, \infty)$

Test: Is $\dfrac{x - 5}{3 - x} < 0$?

Solution set: $(-\infty, 3) \cup (5, \infty)$

75. $|x - 2| < 1$

$-1 < x - 2, < 1$

$1 < x < 3,$

which can be written as $(1, 3)$.

77. $\left| x - \dfrac{3}{2} \right| \geq \dfrac{3}{2}$

$x - \dfrac{3}{2} \leq -\dfrac{3}{2}$ or $x - \dfrac{3}{2} \geq \dfrac{3}{2}$

$x \leq 0$ or $x \geq 3,$

which can be written as $(-\infty, 0] \cup [3, \infty)$.

79. $\dfrac{x}{5} - 6 \leq -\dfrac{x}{2} + 6$

$10\left(\dfrac{x}{5} - 6\right) \leq 10\left(-\dfrac{x}{2} + 6\right)$

$2x - 60 \leq -5x + 60$

$7x \leq 120$

$x \leq \dfrac{120}{7}$ or $\left(-\infty, \dfrac{120}{7}\right]$

81. $(x - 4)|x| > 0$

Critical numbers: $x = 4, x = 0$

Test intervals: $(-\infty, 0), (0, 4), (4, \infty)$

Test: Is $(x - 4)|x| > 0$?

Solution set: $(4, \infty)$

83. $2x - 10 \geq 0$

$2x \geq 10$

$x \geq 5$

Domain: $[5, \infty)$

85. $(7 + 5i) + (-4 + 2i) = (7 - 4) + (5i + 2i) = 3 + 7i$

87. $5i(13 - 8i) = 65i - 40i^2 = 40 + 65i$

89. $(10 - 8i)(2 - 3i) = 20 - 30i - 16i + 24i^2 = -4 - 46i$

91. $\dfrac{6 + i}{i} = \dfrac{6 + i}{i} \cdot \dfrac{-i}{-i} = \dfrac{-6i - i^2}{-i^2} = \dfrac{-6i + 1}{1} = 1 - 6i$

93. $\dfrac{4}{-3i} = \dfrac{4}{-3i} \cdot \dfrac{3i}{3i} = \dfrac{12i}{9} = \dfrac{4i}{3} = \dfrac{4}{3}i$

95. $3x^2 + 1 = 0$

$$3x^2 = -1$$

$$x^2 = -\frac{1}{3}$$

$$x = \pm\sqrt{-\frac{1}{3}}$$

$$= \pm\sqrt{\frac{1}{3}}\, i$$

97. September's profit + October's profit = 689,000

Let x = September's profit.

Then $x + 0.12x$ = October's profit.

$$x + (x + 0.12x) = 689,000$$

$$2.12x = 689,000$$

$$x = 325,000$$

$$x + 0.12x = 364,000$$

September: \$325,000

October: \$364,000

99. Let x = the number of liters of pure antifreeze.

$$30\% \text{ of } (10 - x) + 100\% \text{ of } x = 50\% \text{ of } 10$$

$$0.30(10 - x) + 1.00x = 0.50(10)$$

$$3 - 0.30x + 1.00x = 5$$

$$0.70x = 2$$

$$x = \frac{2}{0.70} = \frac{20}{7} = 2\frac{6}{7} \text{ liters}$$

101. Let x = the number of farmers in the group.

$$\text{Cost per farmer} = \frac{48,000}{x}$$

If two more farmers join the group, the cost per farmer will be $\dfrac{48,000}{x + 2}$.

Since this new cost is \$4000 less than the original cost,

$$\frac{48,000}{x} - 4000 = \frac{48,000}{x + 2}$$

$$48,000(x + 2) - 4000x(x + 2) = 48,000x$$

$$12(x + 2) - x(x + 2) = 12x \qquad \text{Divide both sides by 4000.}$$

$$12x + 24 - x^2 - 2x = 12x$$

$$0 = x^2 + 2x - 24$$

$$0 = (x + 6)(x - 4)$$

$$x = -6, \text{extraneous} \quad \text{or} \quad x = 4$$

$$x = 4 \text{ farmers.}$$

103.

	Rate	Time	Distance
To work	r	$\dfrac{56}{r}$	56
From work	$r + 8$	$\dfrac{56}{r + 8}$	56

$$\text{time} = \frac{\text{distance}}{\text{rate}}$$

time to work = time from work + 10 minutes

$$\frac{56}{r} = \frac{56}{r + 8} + \frac{1}{6} \qquad \text{Convert minutes to portion of an hour.}$$

$$6(r + 8)(56) = 6r(56) + r(r + 8)$$

$$336r + 2688 = 336r + r^2 + 8r$$

$$0 = r^2 + 8r - 2688$$

$$0 = (r - 48)(r + 56)$$

Using the positive value for r, we have $r = 48$ miles per hour. The average speed on the trip home was $r + 8 = 56$ miles per hour.

105. $M = 500x(20 - x)$

(a)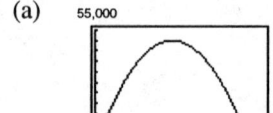

(b) $M = 0$ when $x = 0$ or $x = 20$

(c) The highest point on the graph occurs when $x = 10$.

(d) $500x(20 - x) < 40,000$

$\quad x(20 - x) < 80 \qquad$ Divide both sides by 500.

$\quad 20x - x^2 < 80$

$\qquad 0 < x^2 - 20x + 80$

Use the Quadratic Formula to find the critical numbers of $x \approx 5.53$ and $x \approx 14.47$.

Test intervals: $(0, 5.53), (5.53, 14.47), (14.47, 20)$

Solution set: $(0, 5.53) \cup (14.47, 20)$

Thus, $M < 40,000$ when $x < 5.53$ or $x > 14.47$.

107. $T = 2\pi\sqrt{\dfrac{L}{32}}$

Since $T \geq 2$ and $L > 0$, we have

$$2\pi\sqrt{\frac{L}{32}} \geq 2$$

$$\sqrt{\frac{L}{32}} \geq \frac{1}{\pi}$$

$$\frac{L}{32} \geq \frac{1}{\pi^2}$$

$$L \geq \frac{32}{\pi^2}.$$

❑ Practice Test for Chapter 1

1. Graph $3x - 5y = 15$.

2. Graph $y = \sqrt{9 - x}$.

3. Solve $5x + 4 = 7x - 8$.

4. Solve $\dfrac{x}{3} - 5 = \dfrac{x}{5} + 1$.

5. Solve $\dfrac{3x + 1}{6x - 7} = \dfrac{2}{5}$.

6. Solve $(x - 3)^2 + 4 = (x + 1)^2$.

7. Solve $A = \frac{1}{2}(a + b)h$ for a.

8. 301 is what percent of 4300?

9. Cindy has \$6.05 in quarters and nickels. How many of each coin does she have if there are 53 coins in all?

10. Ed has \$15,000 invested in two funds paying $9\frac{1}{2}\%$ and 11% simple interest, respectively. How much is invested in each if the yearly interest is \$1582.50?

11. Solve $28 + 5x - 3x^2 = 0$ by factoring.

12. Solve $(x - 2)^2 = 24$ by taking the square root of both sides.

13. Solve $x^2 - 4x - 9 = 0$ by completing the square.

14. Solve $x^2 + 5x - 1 = 0$ by the Quadratic Formula.

15. Solve $3x^2 - 2x + 4 = 0$ by the Quadratic Formula.

16. The perimeter of a rectangle is 1100 feet. Find the dimension so that the enclosed area will be 60,000 square feet.

17. Find two consecutive even positive integers whose product is 624.

18. Solve $x^3 - 10x^2 + 24x = 0$ by factoring.

19. Solve $\sqrt[3]{6 - x} = 4$.

20. Solve $(x^2 - 8)^{2/5} = 4$.

21. Solve $x^4 - x^2 - 12 = 0$.

22. Solve $4 - 3x > 16$.

23. Solve $\left| \dfrac{x - 3}{2} \right| < 5$.

24. Solve $\dfrac{x + 1}{x - 3} < 2$.

25. Solve $|3x - 4| \geq 9$.

C H A P T E R 2
Functions and Their Graphs

C H A P T E R   2
Functions and Their Graphs

Section 2.1   Lines in the Plane and Slope

You should know the following important facts about lines.

- ■ The graph of $y = mx + b$ is a straight line. It is called a linear equation.
- ■ The slope of the line through (x_1, y_1) and (x_2, y_2) is

$$m = \frac{y_2 - y_1}{x_2 - x_1}.$$

- ■ (a) If $m > 0$, the line rises from left to right.

  (b) If $m = 0$, the line is horizontal.

  (c) If $m < 0$, the line falls from left to right.

  (d) If m is undefined, the line is vertical.

- ■ Equations of Lines

  (a) Slope-Intercept: $y = mx + b$

  (b) Point-Slope: $y - y_1 = m(x - x_1)$

  (c) Two-Point: $y - y_1 = \dfrac{y_2 - y_1}{x_2 - x_1}(x - x_1)$

  (d) General: $Ax + By + C = 0$

  (e) Vertical: $x = a$

  (f) Horizontal: $y = b$

- ■ Given two distinct nonvertical lines

$$L_1: y = m_1 x + b_1 \quad \text{and} \quad L_2: y = m_2 x + b_2$$

  (a) L_1 is parallel to L_2 if and only if $m_1 = m_2$ and $b_1 \neq b_2$.

  (b) L_1 is perpendicular to L_2 if and only if $m_1 = -1/m_2$.

Solutions to Odd-Numbered Exercises

1. (a) $m = \frac{2}{3}$. Since the slope is positive, the line rises. Matches L_2.

   (b) m is undefined. The line is vertical. Matches L_3.

   (c) $m = -2$. The line falls. Matches L_1.

5. Slope $= \dfrac{\text{rise}}{\text{run}} = \dfrac{8}{5}$

7. Slope $= \dfrac{\text{rise}}{\text{run}} = \dfrac{0}{1} = 0$

3.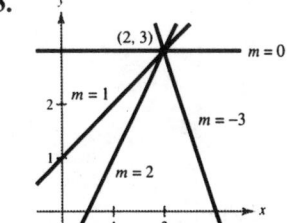

9. Slope $= \dfrac{\text{rise}}{\text{run}} = \dfrac{-8}{2} = -4$

11.

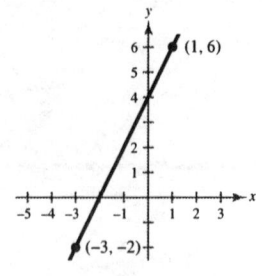

$$\text{slope} = \frac{6 + 2}{1 + 3} = 2$$

13.

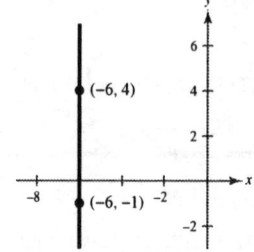

Slope is undefined.

15.

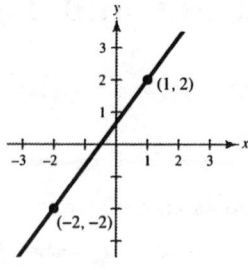

$$\text{slope} = \frac{2 + 2}{1 + 2} = \frac{4}{3}$$

17. Since $m = 0$, y does not change. Three points are $(0, 1)$, $(3, 1)$, and $(-1, 1)$.

19. Since $m = 1$, y increases by 1 for every one unit increase in x. Three points are $(6, -5)$, $(7, -4)$, and $(8, -3)$.

21. Since m is undefined, x does not change. Three points are $(-8, 0)$, $(-8, 2)$, and $(-8, 3)$.

23. Slope of L_1: $m = \dfrac{9 + 1}{5 - 0} = 2$

Slope of L_2: $m = \dfrac{1 - 3}{4 - 0} = -\dfrac{1}{2}$

L_1 and L_2 are perpendicular.

25. Slope of L_1: $m = \dfrac{0 - 6}{-6 - 3} = \dfrac{2}{3}$

Slope of L_2: $m = \dfrac{\frac{7}{3} + 1}{5 - 0} = \dfrac{2}{3}$

L_1 and L_2 are parallel.

27. Yes, any pair of points on a line can be used to calculate the slope of the line. The rate of change remains the same on a line.

29. (a) $m = 135$. The sales are increasing 135 units per year.

(b) $m = 0$. There is no change in sales.

(c) $m = -40$. The sales are decreasing 40 units per year.

31. (a) The slope is negative and steep in 1989 to 1990.

(b) The slope is positive and steep in 1988 to 1989.

33. Slope $= \dfrac{\text{rise}}{\text{run}}$

$-\dfrac{12}{100} = -\dfrac{2000}{y}$

$-12y = -200{,}000$

$y = 16{,}666\frac{2}{3}$ feet ≈ 3.16 miles

35. $5x - y + 3 = 0$

$y = 5x + 3$

Slope: $m = 5$

y-intercept: $(0, 3)$

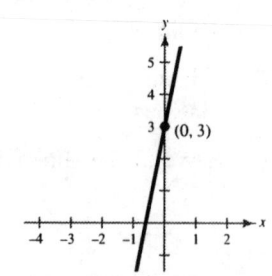

37. $5x - 2 = 0$

$$x = \frac{2}{5}$$

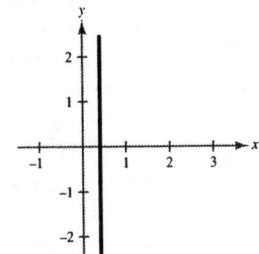

Slope: undefined

No y-intercept

39. $7x + 6y - 30 = 0$

$$y = -\frac{7}{6}x + 5$$

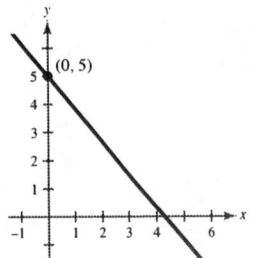

Slope: $m = -\frac{7}{6}$

y-intercept: $(0, 5)$

41. $y + 1 = \dfrac{5 + 1}{-5 - 5}(x - 5)$

$$y = -\frac{3}{5}(x - 5) - 1$$

$$y = -\frac{3}{5}x + 2 \implies 3x + 5y - 10 = 0$$

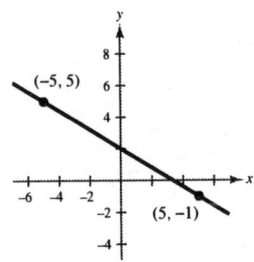

43. $y - \dfrac{1}{2} = \dfrac{\frac{5}{4} - \frac{1}{2}}{\frac{1}{2} - 2}(x - 2)$

$$y = -\frac{1}{2}(x - 2) + \frac{1}{2}$$

$$y = -\frac{1}{2}x + \frac{3}{2} \implies x + 2y - 3 = 0$$

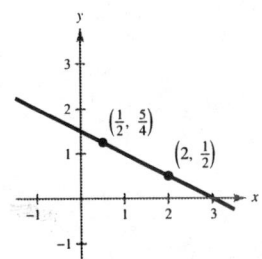

45. Since both points have $x = -8$, the slope is undefined.

$x = -8 \implies x + 8 = 0$

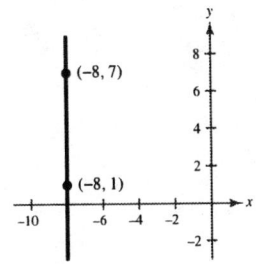

47. $y - 0.6 = \dfrac{-0.6 - 0.6}{-2 - 1}(x - 1)$

$$y = 0.4(x - 1) + 0.6$$

$$y = 0.4x + 0.2 \implies 2x - 5y + 1 = 0$$

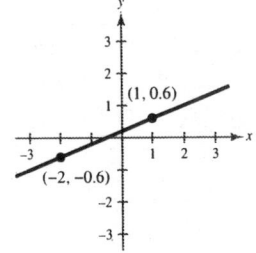

49. $y + 2 = 3(x - 0)$

$y = 3x - 2 \implies 3x - y - 2 = 0$

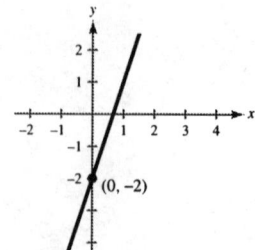

51. $y - 6 = -2(x + 3)$

$y = -2x \implies 2x + y = 0$

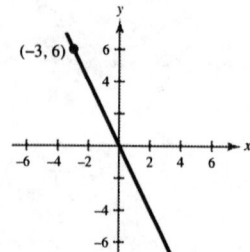

53. $y - 0 = -\frac{1}{3}(x - 4)$

$y = -\frac{1}{3}x + \frac{4}{3} \implies x + 3y - 4 = 0$

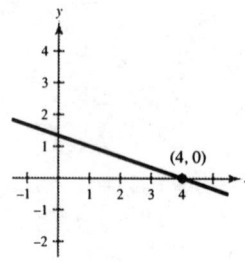

55. $x = 6$

$x - 6 = 0$

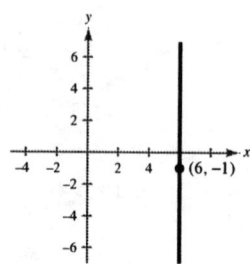

57. $y - \frac{5}{2} = \frac{4}{3}(x - 4)$

$y = \frac{4}{3}x - \frac{17}{6} \implies 8x - 6y - 17 = 0$

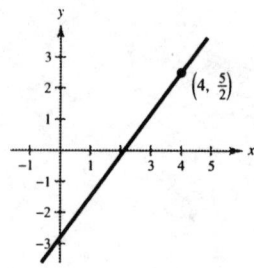

59. $\dfrac{x}{2} + \dfrac{y}{3} = 1$

$3x + 2y - 6 = 0$

61. $\dfrac{x}{-1/6} + \dfrac{y}{-2/3} = 1$

$6x + \dfrac{3}{2}y = -1$

$12x + 3y + 2 = 0$

63. $\dfrac{x}{a} + \dfrac{y}{a} = 1,\ a \neq 0$

$x + y = a$

$1 + 2 = a$

$3 = a$

$x + y = 3$

$x + y - 3 = 0$

65. $4x - 2y = 3$

$$y = 2x - \tfrac{3}{2}$$

slope: $m = 2$

(a) $y - 1 = 2(x - 2)$

$$y = 2x - 3 \implies 2x - y - 3 = 0$$

(b) $y - 1 = -\tfrac{1}{2}(x - 2)$

$$y = -\tfrac{1}{2}x + 2 \implies x + 2y - 4 = 0$$

67. $3x + 4y = 7$

$$y = -\tfrac{3}{4}x + \tfrac{7}{4}$$

slope: $m = -\tfrac{3}{4}$

(a) $y - 4 = -\tfrac{3}{4}(x + 6)$

$$y = -\tfrac{3}{4}x - \tfrac{1}{2} \implies 3x + 4y + 2 = 0$$

(b) $y - 4 = \tfrac{4}{3}(x + 6)$

$$y = \tfrac{4}{3}x + 12 \implies 4x - 3y + 36 = 0$$

69. $y = -3$

slope: $m = 0$

(a) $y = 0$

(b) $x = -1 \implies x + 1 = 0$

71. $L_1: y = \tfrac{1}{3}x - 2$

$L_2: y = \tfrac{1}{3}x + 3$

The lines are parallel.

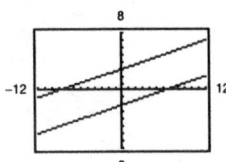

73. $L_1: y = \tfrac{1}{2}x - 3$

$L_2: y = -\tfrac{1}{2}x + 1$

Neither parallel nor perpendicular

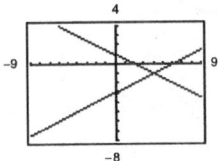

75. $L_1: y = \tfrac{2}{3}x - 3$

$L_2: y = -\tfrac{3}{2}x + 2$

The lines are perpendicular.

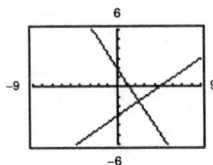

77. $y = 0.5x - 3$

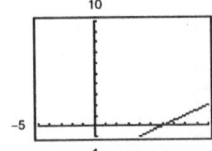 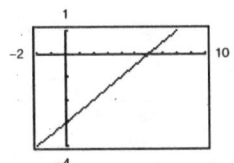

The second setting shows the x- and y-intercepts more clearly.

79. (a) $y = 2x$ (b) $y = -2x$ (c) $y = \tfrac{1}{2}x$

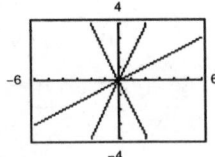

(b) and (c) are perpendicular.

81. (a) $y = -\tfrac{1}{2}x$ (b) $y = -\tfrac{1}{2}x + 3$ (c) $y = 2x - 4$

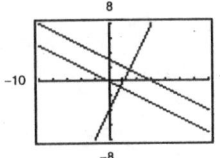

(a) and (b) are parallel.

(c) is perpendicular to (a) and (b).

83. $(6, 2540), m = 125$

$$V - 2540 = 125(t - 6)$$

$$V - 2540 = 125t - 750$$

$$V = 125t + 1790$$

85. The slope is $m = -20$. This represents the decrease in the amount of the loan each week. Matches graph (b).

87. The slope is $m = 0.32$. This represents the increase in travel cost for each mile driven. Matches graph (a).

89. Set the distance between $(4, -1)$ and (x, y) equal to the distance between $(-2, 3)$ and (x, y).

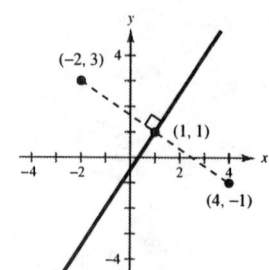

$$\sqrt{(x - 4)^2 + [y - (-1)]^2} = \sqrt{[x - (-2)]^2 + (y - 3)^2}$$

$$(x - 4)^2 + (y + 1)^2 = (x + 2)^2 + (y - 3)^2$$

$$x^2 - 8x + 16 + y^2 + 2y + 1 = x^2 + 4x + 4 + y^2 - 6y + 9$$

$$-8x + 2y + 17 = 4x - 6y + 13$$

$$0 = 12x - 8y - 4$$

$$0 = 4(3x - 2y - 1)$$

$$0 = 3x - 2y - 1$$

This line is the perpendicular bisector of the line segment connecting $(4, -1)$ and $(-2, 3)$.

91. Using the points $(0, 32)$ and $(100, 212)$, we have

$$m = \frac{212 - 32}{100 - 0} = \frac{180}{100} = \frac{9}{5}$$

$$F - 32 = \frac{9}{5}(C - 0)$$

$$F = \frac{9}{5}C + 32.$$

93. Using the points $(1995, 28{,}500)$ and $(1997, 32{,}900)$, we have

$$m = \frac{32{,}900 - 28{,}500}{1997 - 1995} = \frac{4400}{2} = 2200$$

$$S - 28{,}500 = 2200(t - 1995)$$

$$S = 2200t - 4{,}360{,}500$$

When $t = 2000$ we have $S = 2200(2000) - 4{,}360{,}500$ or \$39,500.

95. Using the points $(0, 875)$ and $(5, 0)$, where the first coordinate represents the year t and the second coordinate represents the value V, we have

$$m = \frac{0 - 875}{5 - 0} = -175$$

$$V = -175t + 875, \ 0 \le t \le 5.$$

97. Sale price = List price $- 15\%$ of the list price

$$S = L - 0.15L$$

$$S = 0.85L$$

99. (a) $C = 36{,}500 + 5.25t + 11.50t$

$\qquad = 16.75t + 36{,}500$

(b) $R = 27t$

(c) $P = R - C$

$\qquad = 27t - (16.75t + 36{,}500)$

$\qquad = 10.25t - 36{,}500$

(d) $\qquad 0 = 10.25t - 36{,}500$

$\quad 36{,}500 = 10.25t$

$\qquad\quad t \approx 3561$ hours

101. (a)

(b) $y = 2(15 + 2x) + 2(10 + 2x)$

$\qquad = 8x + 50$

(c)

(d) Since $m = 8$, each 1 meter increase in x will increase y by 8 meters.

103. $C = 120 + 0.26x$

105. Two approximate points on this line are $(6, 710)$ and $(10, 1075)$.

$$m = \frac{1075 - 710}{10 - 6} \approx 91$$

$$y - 710 = 91(x - 6)$$

$$y = 91x + 164$$

This answer may vary depending on the points used.

107. $y = 8 - 3x$ is a linear equation with slope $m = -3$. Matches graph (d).

109. $y = \frac{1}{2}x^2 + 2x + 1$ is a quadratic equation. Its graph is a parabola. Matches graph (a).

Section 2.2   Functions

■ Given a set or an equation, you should be able to determine if it represents a function.

■ Given a function, you should be able to do the following.

(a) Find the domain.

(b) Evaluate it at specific values.

Solutions to Odd-Numbered Exercises

1. Yes, it does represent a function. Each domain value is matched with only one range value.

3. No, it does not represent a function. The domain values are each matched with three range values.

5. Yes, it does represent a function. Each input value is matched with only one output value.

7. No, it does not represent a function. The input values of 10 and 7 are each matched with two output values.

9. (a) Each element of A is matched with exactly one element of B, so it does represent a function.

    (b) The element 1 in A is matched with two elements, -2 and 1 of B, so it does not represent a function.

    (c) Each element of A is matched with exactly one element of B, so it does represent a function.

    (d) The element 2 in A is not matched with an element of B, so it does not represent a function.

11. Each is a function. For each year there corresponds one and only one circulation.

13. $x^2 + y^2 = 4 \implies y = \pm\sqrt{4 - x^2}$

    No, y *is not* a function of x.

15. $x^2 + y = 4 \implies y = 4 - x^2$

    Yes, y *is* a function of x.

17. $2x + 3y = 4 \implies y = \frac{1}{3}(4 - 2x)$

    Yes, y *is* a function of x.

19. $y^2 = x^2 - 1 \implies y = \pm\sqrt{x^2 - 1}$

    No, y *is not* a function of x.

21. $y = |4 - x|$

    Yes, y is a function of x.

23. $f(s) = \dfrac{1}{s + 1}$

    (a) $f(4) = \dfrac{1}{(4) + 1} = \dfrac{1}{5}$

    (b) $f(0) = \dfrac{1}{(0) + 1} = 1$

    (c) $f(4x) = \dfrac{1}{(4x) + 1} = \dfrac{1}{4x + 1}$

    (d) $f(x + c) = \dfrac{1}{(x + c) + 1} = \dfrac{1}{x + c + 1}$

25. $f(x) = 2x - 3$

    (a) $f(1) = 2(1) - 3 = -1$

    (b) $f(-3) = 2(-3) - 3 = -9$

    (c) $f(x - 1) = 2(x - 1) - 3 = 2x - 5$

27. $h(t) = t^2 - 2t$

    (a) $h(2) = 2^2 - 2(2) = 0$

    (b) $h(1.5) = (1.5)^2 - 2(1.5) = -0.75$

    (c) $h(x + 2) = (x + 2)^2 - 2(x + 2) = x^2 + 2x$

29. $f(y) = 3 - \sqrt{y}$

    (a) $f(4) = 3 - \sqrt{4} = 1$

    (b) $f(0.25) = 3 - \sqrt{0.25} = 2.5$

    (c) $f(4x^2) = 3 - \sqrt{4x^2} = 3 - 2|x|$

31. $q(x) = \dfrac{1}{x^2 - 9}$

(a) $q(0) = \dfrac{1}{0^2 - 9} = -\dfrac{1}{9}$

(b) $q(3) = \dfrac{1}{3^2 - 9}$ is undefined.

(c) $q(y + 3) = \dfrac{1}{(y + 3)^2 - 9} = \dfrac{1}{y^2 + 6y}$

33. $f(x) = \dfrac{|x|}{x}$

(a) $f(2) = \dfrac{|2|}{2} = 1$

(b) $f(-2) = \dfrac{|-2|}{-2} = -1$

(c) $f(x - 1) = \dfrac{|x - 1|}{x - 1}$

35. $f(x) = \begin{cases} 2x + 1, & x < 0 \\ 2x + 2, & x \geq 0 \end{cases}$

(a) $f(-1) = 2(-1) + 1 = -1$

(b) $f(0) = 2(0) + 2 = 2$

(c) $f(2) = 2(2) + 2 = 6$

37. $f(x) = x^2 - 3$

x	-2	-1	0	1	2
$f(x)$	1	-2	-3	-2	1

39. $h(t) = \frac{1}{2}|t + 3|$

t	-5	-4	-3	-2	-1
$h(t)$	1	$\frac{1}{2}$	0	$\frac{1}{2}$	1

41. $f(x) = \begin{cases} -\frac{1}{2}x + 4, & x \leq 0 \\ (x - 2)^2, & x > 0 \end{cases}$

x	-2	-1	0	1	2
$f(x)$	5	$\frac{9}{2}$	4	1	0

43. $15 - 3x = 0$

$3x = 15$

$x = 5$

45. $x^2 - 9 = 0$

$x^2 = 9$

$x = \pm 3$

47. $\quad f(x) = g(x)$

$x^2 = x + 2$

$x^2 - x - 2 = 0$

$(x + 1)(x - 2) = 0$

$x = -1 \quad \text{or} \quad x = 2$

49. $\quad f(x) = g(x)$

$\sqrt{3x} + 1 = x + 1$

$\sqrt{3x} = x$

$3x = x^2$

$0 = x^2 - 3x$

$0 = x(x - 3)$

$x = 0 \quad \text{or} \quad x = 3$

51. $f(x) = 5x^2 + 2x - 1$

Since $f(x)$ is a polynomial, the domain is all real numbers x.

53. $h(t) = \dfrac{4}{t}$

Domain: All real numbers except $t = 0$

55. $g(y) = \sqrt{y - 10}$

Domain: $y - 10 \geq 0$

$y \geq 10$

57. $f(x) = \sqrt[4]{1 - x^2}$

Domain: $1 - x^2 \geq 0$

$-x^2 \geq -1$

$x^2 \leq 1$

$x^2 - 1 \leq 0$

$-1 \leq x \leq 1$ (See Section 1.8.)

59. $g(x) = \dfrac{1}{x} - \dfrac{1}{x + 2}$

Domain: All real numbers except
$x = 0, \ x = -2$

61. $f(x) = x^2$

$\{(-2, 4), (-1, 1), (0, 0), (1, 1), (2, 4)\}$

63. $f(x) = \sqrt{x + 2}$

$\{(-2, 0), (-1, 1), (0, \sqrt{2}), (1, \sqrt{3}), (2, 2)\}$

65. The domain is the set of inputs of the function and the range is the set of corresponding outputs.

67. By plotting the points, we have a parabola, so $g(x) = cx^2$. Since $(-4, -32)$ is on the graph, we have $-32 = c(-4)^2 \implies c = -2$. Thus, $g(x) = -2x^2$.

69. Since the function is undefined at 0, we have $r(x) = c/x$. Since $(-8, -4)$ is on the graph, we have $-4 = c/-8 \implies c = 32$. Thus, $r(x) = 32/x$.

71.
$$f(x) = x^2 - x + 1$$
$$f(2 + h) = (2 + h)^2 - (2 + h) + 1$$
$$= 4 + 4h + h^2 - 2 - h + 1$$
$$= h^2 + 3h + 3$$
$$f(2) = (2)^2 - 2 + 1 = 3$$
$$f(2 + h) - f(2) = h^2 + 3h$$
$$\frac{f(2 + h) - f(2)}{h} = h + 3, \ h \neq 0$$

73. $f(x) = x^3$
$$f(x + c) = (x + c)^3 = x^3 + 3x^2c + 3xc^2 + c^3$$
$$\frac{f(x + c) - f(x)}{c} = \frac{(x^3 + 3x^2c + 3xc^2 + c^3) - x^3}{c}$$
$$= \frac{c(3x^2 + 3xc + c^2)}{c}$$
$$= 3x^2 + 3xc + c^2, \ c \neq 0$$

75. $g(x) = 3x - 1$

$$\frac{g(x) - g(3)}{x - 3} = \frac{(3x - 1) - 8}{x - 3} = \frac{3x - 9}{x - 3} = \frac{3(x - 3)}{x - 3} = 3, \ x \neq 3$$

77. $A = \pi r^2, \ C = 2\pi r$

$$r = \frac{C}{2\pi}$$

$$A = \pi \left(\frac{C}{2\pi} \right)^2 = \frac{C^2}{4\pi}$$

79. (a)

Height, x	Width	Volume, V
1	24 − 2(1)	1[24 − 2(1)]² = 484
2	24 − 2(2)	2[24 − 2(2)]² = 800
3	24 − 2(3)	3[24 − 2(3)]² = 972
4	24 − 2(4)	4[24 − 2(4)]² = 1024
5	24 − 2(5)	5[24 − 2(5)]² = 980
6	24 − 2(6)	6[24 − 2(6)]² = 864

The volume is maximum when $x = 4$.

(b)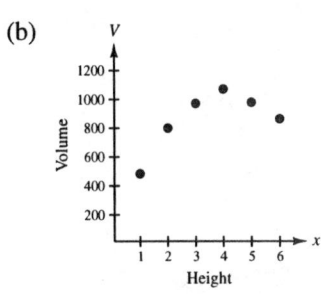

(c) $V = x(24 - 2x)^2$

Domain: $0 < x < 12$

V is a function of x.

81. $A = \dfrac{1}{2}bh = \dfrac{1}{2}xy$

Since $(0, y)$, $(2, 1)$, and $(x, 0)$ all lie on the same line, the slopes between any pair are equal.

$$\frac{1 - y}{2 - 0} = \frac{0 - 1}{x - 2}$$

$$\frac{1 - y}{2} = \frac{-1}{x - 2}$$

$$y = \frac{2}{x - 2} + 1$$

$$y = \frac{x}{x - 2}$$

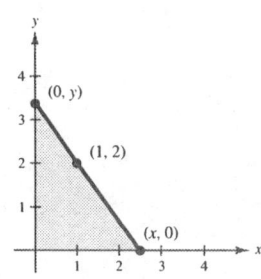

Therefore,

$$A = \frac{1}{2}x\left(\frac{x}{x - 2}\right) = \frac{x^2}{2(x - 2)}.$$

The domain of A includes x-values such that $x^2/[2(x - 2)] > 0$. Using methods of Section 1.8 we find that the domain is $x > 2$.

83. $V = l \cdot w \cdot h = x \cdot y \cdot x = x^2y$ where $4x + y = 108$.

Thus, $y = 108 - 4x$ and $V = x^2(108 - 4x) = 108x^2 - 4x^3$ where $0 < x < 27$.

85. (a) Cost = variable costs + fixed costs

$$C = 12.30x + 98{,}000$$

(b) Revenue = price per unit × number of units

$$R = 17.98x$$

(c) Profit = Revenue − Cost

$$P = 17.98x - (12.30x + 98{,}000)$$

$$P = 5.68x - 98{,}000$$

87. (a) $R = n(\text{rate}) = n[8.00 - 0.05(n - 80)], \ n \geq 80$

$$R = 12.00n - 0.05n^2 = 12n - \frac{n^2}{20} = \frac{240n - n^2}{20}, \ n \geq 80$$

(b)

n	90	100	110	120	130	140	150
$R(n)$	\$675	\$700	\$715	\$720	\$715	\$700	\$675

The revenue is maximum when 120 people take the trip.

89. (a)

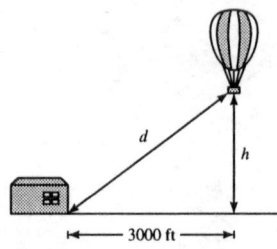

(b) $(3000)^2 + h^2 = d^2$

$$h = \sqrt{d^2 - (3000)^2}$$

Domain: $[3000, \infty)$

(since both $d \geq 0$ and $d^2 - (3000)^2 \geq 0$)

91. $\dfrac{t}{3} + \dfrac{t}{5} = 1$

$$15\left(\frac{t}{3} + \frac{t}{5}\right) = 15(1)$$

$$5t + 3t = 15$$

$$8t = 15$$

$$t = \frac{15}{8}$$

93. $\dfrac{3}{x(x + 1)} - \dfrac{4}{x} = \dfrac{1}{x + 1}$

$$x(x + 1)\left[\frac{3}{x(x + 1)} - \frac{4}{x}\right] = x(x + 1)\left(\frac{1}{x + 1}\right)$$

$$3 - 4(x + 1) = x$$

$$3 - 4x - 4 = x$$

$$-1 = 5x$$

$$-\frac{1}{5} = x$$

Section 2.3   Analyzing Graphs of Functions

- ■ You should be able to determine the domain and range of a function from its graph.
- ■ You should be able to use the vertical line test for functions.
- ■ You should be able to determine when a function is constant, increasing, or decreasing.
- ■ You should know that f is
  - (a) odd if $f(-x) = -f(x)$.
  - (b) even if $f(-x) = f(x)$.

Solutions to Odd-Numbered Exercises

1. $f(x) = 1 - x^2$

Domain: All real numbers

Range: $(-\infty, 1]$

3. $f(x) = \sqrt{x^2 - 1}$

Domain: $(-\infty, -1] \cup [1, \infty)$

Range: $[0, \infty)$

5. $h(x) = \sqrt{16 - x^2}$

Domain: $[-4, 4]$

Range: $[0, 4]$

7. $y = \frac{1}{2}x^2$

A vertical line intersects the graph just once, so y is a function of x.

9. $x - y^2 = 1 \implies y = \pm\sqrt{x - 1}$

y is not a function of x.

11. $x^2 = 2xy - 1$

A vertical line intersects the graph just once, so y is a function of x.

13. Yes, the graph in Exercise 9 does represent x as a function of y. For each y-value there corresponds one and only one x-value.

15. $f(x) = -0.2x^2 + 3x + 32$

The second setting shows the most complete graph.

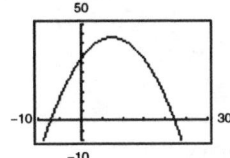

17. $f(x) = 4x^3 - x^4$

The first setting shows the most complete graph.

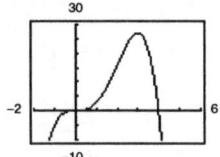

19. $f(x) = \frac{3}{2}x$

(a) f is increasing on $(-\infty, \infty)$.

(b) Since $f(-x) = -f(x)$, f is odd.

21. $f(x) = x^3 - 3x^2 + 2$

(a) f is increasing on $(-\infty, 0)$ and $(2, \infty)$.

f is decreasing on $(0, 2)$.

(b) $f(-x) \neq -f(x)$

$f(-x) \neq f(x)$

f is neither odd nor even.

23. $f(x) = 3x^4 - 6x^2$

(a)

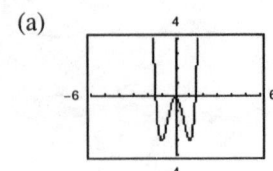

(b) Increasing on $(-1, 0)$ and $(1, \infty)$

Decreasing on $(-\infty, -1)$ and $(0, 1)$

(c) Since $f(-x) = f(x)$, f is even.

25. $f(x) = x\sqrt{x + 3}$

(a)

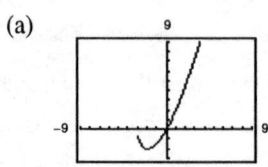

(b) Increasing on $(-2, \infty)$

Decreasing on $(-3, -2)$

(c) $f(-x) \neq -f(x)$

$f(-x) \neq f(x)$

f is neither odd nor even.

27. $f(-x) = (-x)^6 - 2(-x)^2 + 3$

$\qquad = x^6 - 2x^2 + 3$

$\qquad = f(x)$

f is even.

29. $g(-x) = (-x)^3 - 5(-x)$

$\qquad = -x^3 + 5x$

$\qquad = -g(x)$

g is odd.

31. $f(-t) = (-t)^2 + 2(-t) - 3$

$\qquad = t^2 - 2t - 3$

$\qquad \neq f(t) \neq -f(t)$

f is neither even nor odd.

33. $\left(-\frac{3}{2}, 4\right)$

(a) If f is even, another point is $\left(\frac{3}{2}, 4\right)$.

(b) If f is odd, another point is $\left(\frac{3}{2}, -4\right)$.

35. $f(x) = 3$, even

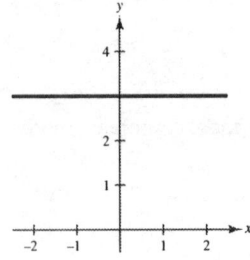

37. $f(x) = 5 - 3x$, neither even nor odd

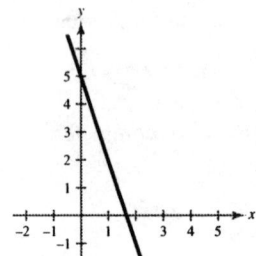

39. $g(s) = \dfrac{s^2}{4}$, even

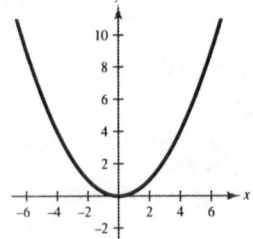

41. $f(x) = \sqrt{1 - x}$, neither even nor odd

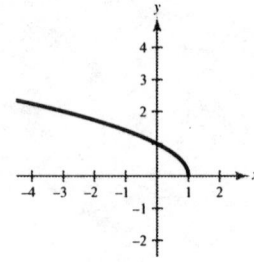

43. $g(t) = \sqrt[3]{t} - 1$, neither even nor odd

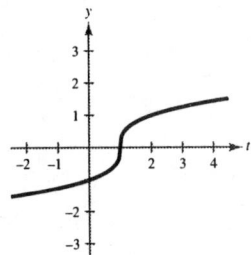

45. $f(x) = \begin{cases} x + 3, & x \le 0 \\ 3, & 0 < x \le 2 \\ 2x - 1, & x > 2 \end{cases}$

Neither even nor odd

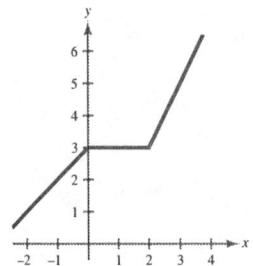

47. $f(x) = 4 - x$

$f(x) \ge 0$ on $(-\infty, 4]$.

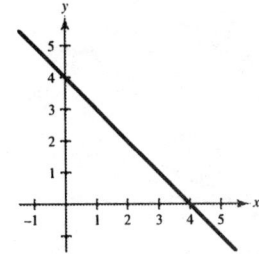

49. $f(x) = x^2 - 9$

$f(x) \ge 0$ on $(-\infty, -3]$ and $[3, \infty)$.

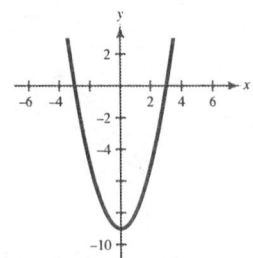

51. $f(x) = 1 - x^4$

$f(x) \ge 0$ on $[-1, 1]$.

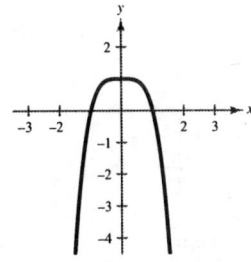

53. $f(x) = x^2 + 1$

$f(x) \ge 0$ on $(-\infty, \infty)$.

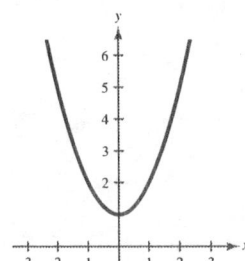

55. $f(x) = -5$, $f(x) < 0$ for all x.

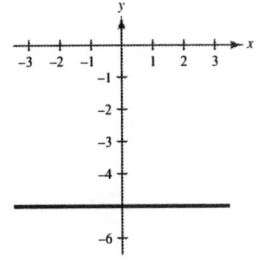

57. $f(x) = \begin{cases} 2x + 3, & x < 0 \\ 3 - x, & x \geq 0 \end{cases}$

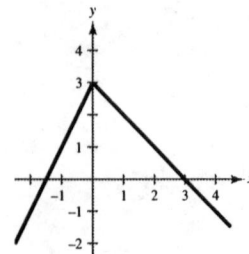

59. $f(x) = \begin{cases} x^2 + 5, & x \leq 1 \\ -x^2 + 4x + 3, & x > 1 \end{cases}$

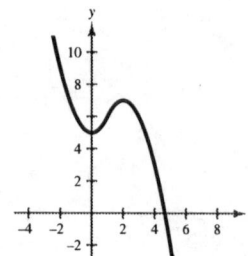

61. $f(x) = |x + 3|$

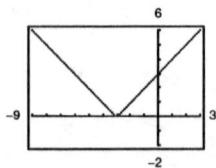

Domain: All real numbers or $(-\infty, \infty)$
Range: $[0, \infty)$

63. $s(x) = 2\left(\frac{1}{4}x - \left[\!\left[\frac{1}{4}x\right]\!\right]\right)$

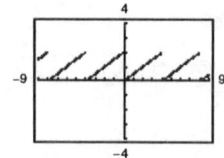

Domain: $(-\infty, \infty)$
Range: $[0, 2)$
Sawtooth pattern

65. (a) $y = x$

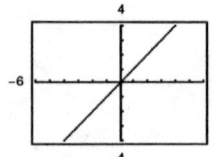

(b) $y = x^2$

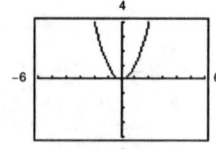

(c) $y = x^3$

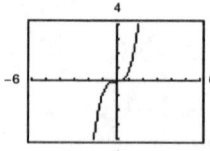

(d) $y = x^4$

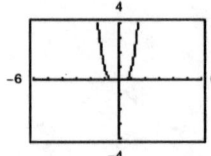

(e) $y = x^5$

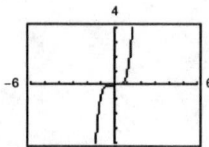

(f) $y = x^6$

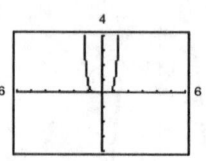

All the graphs pass through the origin. The graphs of the odd powers of x are symmetric to the origin and the graphs of the even powers are symmetric to the y-axis. As the powers increase, the graphs become narrower in the interval $-1 < x < 1$.

67. (a) $C_2(t) = 0.65 - 0.4\left[\!\left[-(t - 1)\right]\!\right]$ is the appropriate model since the cost does not increase until after the next minute of conversation has started.

(b)

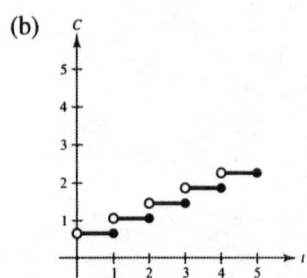

$C = 0.65 + 0.40(18) = \$7.85$

69. $P = R - C = xp - C = x(100 - 0.0001x) - (350,000 + 30x)$

$\quad\quad = -0.0001x^2 + 70x - 350,000, \ 0 \le x$

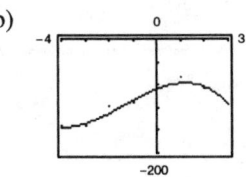

This function is maximized when $x = 350,000$ units.

71. $h = \text{top} - \text{bottom}$

$\quad = (-x^2 + 4x - 1) - 2$

$\quad = -x^2 + 4x - 3$

73. $h = \text{top} - \text{bottom}$

$\quad = (4x - x^2) - 2x$

$\quad = 2x - x^2$

75. $L = \text{right} - \text{left}$

$\quad = \frac{1}{2}y^2 - 0$

$\quad = \frac{1}{2}y^2$

77. $L = \text{right} - \text{left}$

$\quad = 4 - y^2$

79. $y = -87.49 + 16.28t - 4.82t^2 - 1.20t^3$

(a) Domain: $-4 \le t \le 3$

(b)

(c) Most accurate in 1986

Least accurate in 1990

(d) The balance would continue to decrease.

81. (a) For average salaries of college professors, a scale of \$10,000 would be appropriate.

(b) For the population of the United States, use a scale of 50,000,000.

(c) For the percent of the civilian workforce that is unemployed, use a scale of 1%.

83. $f(x) = a_{2n+1}x^{2n+1} + a_{2n-1}x^{2n-1} + \cdots + a_3x^3 + a_1x$

$f(-x) = a_{2n+1}(-x)^{2n+1} + a_{2n-1}(-x)^{2n-1} + \cdots + a_3(-x)^3 + a_1(-x)$

$\quad\quad = -a_{2n+1}x^{2n+1} - a_{2n-1}x^{2n-1} - \cdots - a_3x^3 - a_1x = -f(x)$

Therefore, $f(x)$ is odd.

85. $x^2 - 10x = 0$

$x(x - 10) = 0$

$x = 0 \ \ \text{or} \ \ x = 10$

87. $x^3 + x = 0$

$x(x^2 + 1) = 0$

$x = 0 \ \ \text{or} \ \ x^2 + 1 = 0$

$\quad\quad\quad\quad\quad x^2 = -1$

$\quad\quad\quad\quad\quad x = \pm\sqrt{-1} = \pm i$

Section 2.4   Translations and Combinations

■   You should know the basic types of transformations.

Let $y = f(x)$ and let c be a positive real number.

   1. $h(x) = f(x) + c$ Vertical shift c units upward

   2. $h(x) = f(x) - c$ Vertical shift c units downward

   3. $h(x) = f(x - c)$ Horizontal shift c units to the right

   4. $h(x) = f(x + c)$ Horizontal shift c units to the left

   5. $h(x) = -f(x)$ Reflection in the x-axis

   6. $h(x) = f(-x)$ Reflection in the y-axis

   7. $h(x) = cf(x), c > 1$ Vertical stretch

   8. $h(x) = cf(x), 0 < c < 1$ Vertical shrink

■   Given two functions, f and g, you should be able to form the following functions (if defined):

   1. Sum: $(f + g)(x) = f(x) + g(x)$

   2. Difference: $(f - g)(x) = f(x) - g(x)$

   3. Product: $(fg)(x) = f(x)g(x)$

   4. Quotient: $(f/g)(x) = f(x)/g(x), g(x) \neq 0$

   5. Composition of f with g: $(f \circ g)(x) = f(g(x))$

   6. Composition of g with f: $(g \circ f)(x) = g(f(x))$

Solutions to Odd-Numbered Exercises

1. (a) $f(x) = x^3 + c$

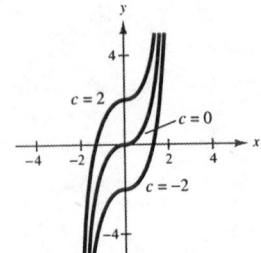

                              (b) $f(x) = (x - c)^3$

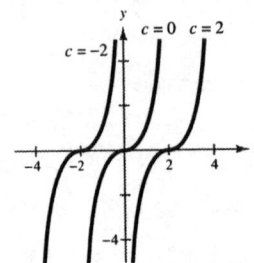

3. (a) $f(x) = |x + c|$

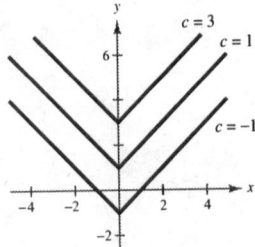

     (b) $f(x) = |x - c|$

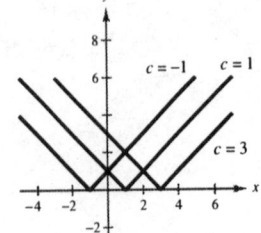

     (c) $f(x) = |x + 4| + c$

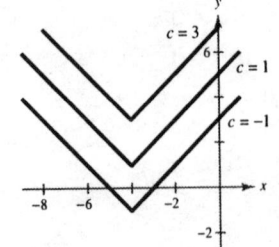

5. (a) $y = f(x) + 2$

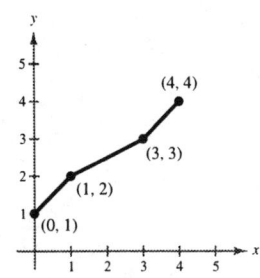

(b) $y = -f(x)$

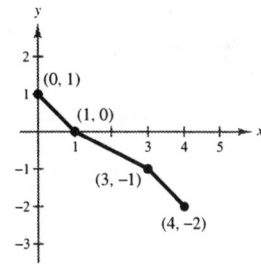

(c) $y = f(x - 2)$

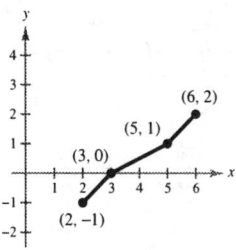

(d) $y = f(x + 3)$

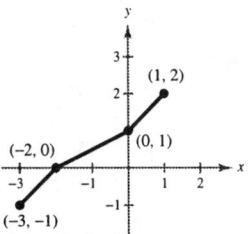

(e) $y = f(2x)$

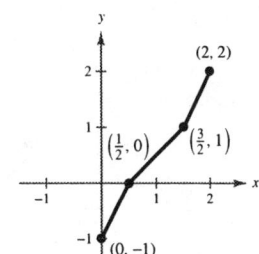

(f) $y = f(-x)$

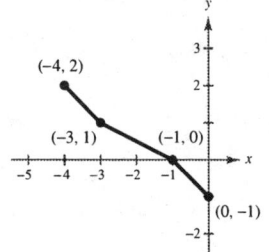

7. (a) Vertical shift one unit downward

$$y = x^2 - 1$$

(b) Vertical shift one unit upward, horizontal shift one unit to the left, and a reflection in the *x*-axis

$$y = 1 - (x + 1)^2$$

9. Horizontal shift two units to the right of $y = x^3$

$$y = (x - 2)^3$$

11. Reflection in the *x*-axis of $y = x^2$

$$y = -x^2$$

13. Reflection in the *x*-axis and a vertical shift one unit upward of $y = \sqrt{x}$

$$y = 1 - \sqrt{x}$$

15.

x	0	1	2	3
$f(x)$	2	3	1	2
$g(x)$	-1	0	$\frac{1}{2}$	0
$h(x) = f(x) + g(x)$	1	3	$\frac{3}{2}$	2

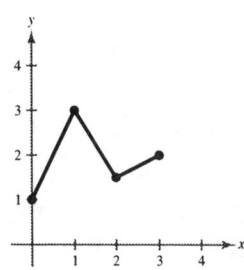

17.

x	-2	-1	0	1	2	3
$f(x)$	2	1	0	1	2	3
$g(x)$	4	3	2	1	0	1
$h(x) = f(x) + g(x)$	6	4	2	2	2	4

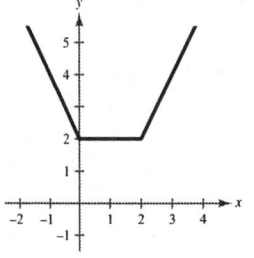

19. $f(x) = x + 1, g(x) = x - 1$

$(f + g)(x) = f(x) + g(x) = (x + 1) + (x - 1) = 2x$

$(f - g)(x) = f(x) - g(x) = (x + 1) - (x - 1) = 2$

$(fg)(x) = f(x) \cdot g(x) = (x + 1)(x - 1) = x^2 - 1$

$\left(\dfrac{f}{g}\right)(x) = \dfrac{f(x)}{g(x)} = \dfrac{x + 1}{x - 1}, \ x \neq 1$

21. $f(x) = x^2, g(x) = 1 - x$

$(f + g)(x) = f(x) + g(x) = x^2 + (1 - x) = x^2 - x + 1$

$(f - g)(x) = f(x) - g(x) = x^2 - (1 - x) = x^2 + x - 1$

$(fg)(x) = f(x) \cdot g(x) = x^2(1 - x) = x^2 - x^3$

$\left(\dfrac{f}{g}\right)(x) = \dfrac{f(x)}{g(x)} = \dfrac{x^2}{1 - x}, x \neq 1$

23. $f(x) = x^2 + 5, g(x) = \sqrt{1 - x}$

$(f + g)(x) = f(x) + g(x) = (x^2 + 5) + \sqrt{1 - x}$

$(f - g)(x) = f(x) - g(x) = (x^2 + 5) - \sqrt{1 - x}$

$(fg)(x) = f(x) \cdot g(x) = (x^2 + 5)\sqrt{1 - x}$

$\left(\dfrac{f}{g}\right)(x) = \dfrac{f(x)}{g(x)} = \dfrac{x^2 + 5}{\sqrt{1 - x}}, \ x < 1$

25. $f(x) = \dfrac{1}{x}, g(x) = \dfrac{1}{x^2}$

$(f + g)(x) = f(x) + g(x) = \dfrac{1}{x} + \dfrac{1}{x^2} = \dfrac{x + 1}{x^2}$

$(f - g)(x) = f(x) - g(x) = \dfrac{1}{x} - \dfrac{1}{x^2} = \dfrac{x - 1}{x^2}$

$(fg)(x) = f(x) \cdot g(x) = \dfrac{1}{x}\left(\dfrac{1}{x^2}\right) = \dfrac{1}{x^3}$

$\left(\dfrac{f}{g}\right)(x) = \dfrac{f(x)}{g(x)} = \dfrac{1/x}{1/x^2} = \dfrac{x^2}{x} = x, \ x \neq 0$

27. $(f + g)(3) = f(3) + g(3) = (3^2 + 1) + (3 - 4) = 9$

29. $(f - g)(0) = f(0) - g(0) = [0^2 + 1] - (0 - 4) = 5$

31. $(f - g)(2t) = f(2t) - g(2t) = [(2t)^2 + 1] - (2t - 4) = 4t^2 - 2t + 5$

33. $(fg)(4) = f(4)g(4) = (4^2 + 1)(4 - 4) = 0$

35. $\left(\dfrac{f}{g}\right)(5) = \dfrac{f(5)}{g(5)} = \dfrac{5^2 + 1}{5 - 4} = 26$

37. $\left(\dfrac{f}{g}\right)(-1) - g(3) = \dfrac{f(-1)}{g(-1)} - g(3)$

$$= \dfrac{(-1)^2 + 1}{-1 - 4} - (3 - 4)$$

$$= -\dfrac{2}{5} + 1 = \dfrac{3}{5}$$

39. $f(x) = \frac{1}{2}x$, $g(x) = x - 1$, $(f + g)(x) = \frac{3}{2}x - 1$

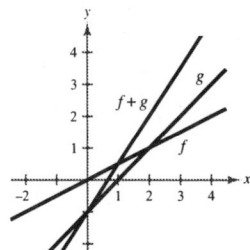

41. $f(x) = x^2$, $g(x) = -2x$, $(f + g)(x) = x^2 - 2x$

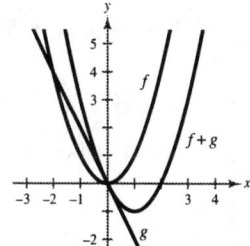

43. $f(x) = 3x$, $g(x) = -\dfrac{x^3}{10}$, $(f + g)(x) = 3x - \dfrac{x^3}{10}$

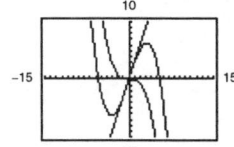

For $0 \le x \le 2$, $f(x)$ contributes most to the magnitude.
For $x > 6$, $g(x)$ contributes most to the magnitude.

45. $T(x) = R(x) + B(x) = \frac{3}{4}x + \frac{1}{15}x^2$

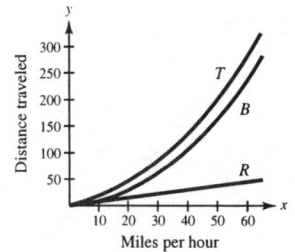

47.

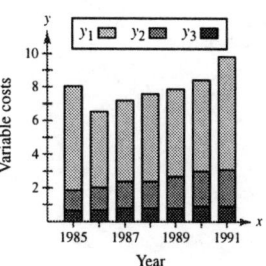

49. (a) T is a function of t since for each time t there corresponds one and only one temperature T.

(b) $T(4) = 60°$
$T(15) = 72°$

(c) $H(t) = T(t - 1)$; All the temperature changes would be one hour later.

(d) $H(t) = T(t) - 1$; The temperature would be decreased by one degree.

51. $f(x) = x^2, g(x) = x - 1$

    (a) $(f \circ g)(x) = f(g(x)) = f(x - 1) = (x - 1)^2$

    (b) $(g \circ f)(x) = g(f(x)) = g(x^2) = x^2 - 1$

    (c) $(f \circ f)(x) = f(f(x)) = f(x^2) = (x^2)^2 = x^4$

53. $f(x) = 3x + 5, g(x) = 5 - x$

    (a) $(f \circ g)(x) = f(g(x)) = f(5 - x) = 3(5 - x) + 5 = 20 - 3x$

    (b) $(g \circ f)(x) = g(f(x)) = g(3x + 5) = 5 - (3x + 5) = -3x$

    (c) $(f \circ f)(x) = f(f(x)) = f(3x + 5) = 3(3x + 5) + 5 = 9x + 20$

55. (a) $(f \circ g)(x) = f(g(x)) = f(x^2) = \sqrt{x^2 + 4}$

    (b) $(g \circ f)(x) = g(f(x)) = g\left(\sqrt{x + 4}\right) = \left(\sqrt{x + 4}\right)^2 = x + 4, \ x \geq 4$

57. (a) $(f \circ g)(x) = f(g(x)) = f(3x + 1) = \frac{1}{3}(3x + 1) - 3 = x - \frac{8}{3}$

    (b) $(g \circ f)(x) = g(f(x)) = g\left(\frac{1}{3}x - 3\right) = 3\left(\frac{1}{3}x - 3\right) + 1 = x - 8$

59. (a) $(f \circ g)(x) = f(g(x)) = f\left(\sqrt{x}\right) = (x^{1/2})^{1/2} = x^{1/4} = \sqrt[4]{x}$

    (b) Since $f(x) = g(x), (g \circ f)(x) = (f \circ g)(x) = \sqrt[4]{x}$.

61. (a) $(f \circ g)(x) = f(g(x)) = f(x + 6) = |x + 6|$

    (b) $(g \circ f)(x) = g(f(x)) = g(|x|) = |x| + 6$

63. (a) $(f + g)(3) = f(3) + g(3) = 2 + 1 = 3$

    (b) $\left(\dfrac{f}{g}\right)(2) = \dfrac{f(2)}{g(2)} = \dfrac{0}{2} = 0$

65. (a) $(f \circ g)(2) = f(g(2)) = f(2) = 0$

    (b) $(g \circ f)(2) = g(f(2)) = g(0) = 4$

67. $g(x) = f(x) + 2$

Vertical shift 2 units upward

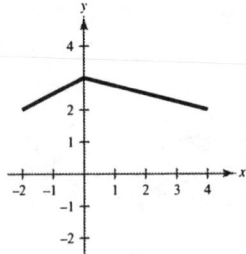

69. $g(x) = f(-x)$

Reflection in the y-axis

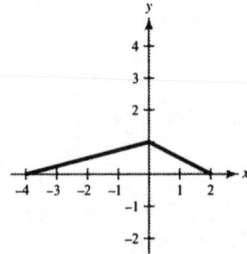

71. Let $f(x) = x^2$ and $g(x) = 2x + 1$, then $(f \circ g)(x) = h(x)$. This is not a unique solution. For example, if $f(x) = (x + 1)^2$ and $g(x) = 2x$, then $(f \circ g)(x) = h(x)$ as well.

73. Let $f(x) = \sqrt[3]{x}$ and $g(x) = x^2 - 4$, then $(f \circ g)(x) = h(x)$.
This answer is not unique. Other possibilities may be:

$$f(x) = \sqrt[3]{x - 4} \text{ and } g(x) = x^2$$
$$\text{or } f(x) = \sqrt[3]{-x} \text{ and } g(x) = 4 - x^2$$
$$\text{or } f(x) = \sqrt[9]{x} \text{ and } g(x) = (4 - x^2)^3$$

75. Let $f(x) = 1/x$ and $g(x) = x + 2$, then $(f \circ g)(x) = h(x)$. Again, this is not a unique solution. Other possibilities may be:

$$f(x) = \frac{1}{x + 2} \text{ and } g(x) = x$$

$$\text{or } f(x) = \frac{1}{x + 1} \text{ and } g(x) = x + 1$$

$$\text{or } f(x) = \frac{1}{x^2 + 2} \text{ and } g(x) = \sqrt{x}$$

77. (a) The domain of $f(x) = \sqrt{x}$ is $x \geq 0$.

(b) The domain of $g(x) = x^2 + 1$ is all real numbers.

(c) $(f \circ g)(x) = f(g(x)) = f(x^2 + 1) = \sqrt{x^2 + 1}$

The domain of $f \circ g$ is all real numbers.

79. (a) The domain of $f(x) = \dfrac{3}{x^2 - 1}$ is all real numbers except $x = \pm 1$.

(b) The domain of $g(x) = x + 1$ is all real numbers.

(c) $(f \circ g)(x) = f(g(x)) = f(x + 1) = \dfrac{3}{(x + 1)^2 - 1} = \dfrac{3}{x^2 + 2x} = \dfrac{3}{x(x + 2)}$

This domain of $f \circ g$ is all real numbers except $x = 0$ and $x = -2$.

81. $f(x) = 3x - 4$

$$\frac{f(x + h) - f(x)}{h} = \frac{[3(x + h) - 4] - (3x - 4)}{h}$$

$$= \frac{3x + 3h - 4 - 3x + 4}{h}$$

$$= \frac{3h}{h}$$

$$= 3$$

83. $f(x) = \dfrac{4}{x}$

$$\frac{f(x+h) - f(x)}{h} = \frac{\dfrac{4}{x+h} - \dfrac{4}{x}}{h} = \frac{\dfrac{4x - 4(x+h)}{x(x+h)}}{\dfrac{h}{1}}$$

$$= \frac{4x - 4x - 4h}{x(x+h)} \cdot \frac{1}{h}$$

$$= \frac{-4h}{x(x+h)} \cdot \frac{1}{h}$$

$$= \frac{-4}{x(x+h)}$$

85. (a) $r(x) = \dfrac{x}{2}$

    (b) $A(r) = \pi r^2$

    (c) $(A \circ r)(x) = A(r(x)) = A\left(\dfrac{x}{2}\right) = \pi\left(\dfrac{x}{2}\right)^2$

    $(A \circ r)(x)$ represents the area of the circular base of the tank on the square foundation with side length y.

87. $(C \circ x)(t) = C(x(t))$

$$= 60(50t) + 750$$

$$= 3000t + 750$$

$(C \circ x)(t)$ represents the cost after t production hours.

89. (a) $R = p - 1200$

    (b) $S = p - 0.08p = 0.92p$

    (c) $(R \circ S)(p) = R(S(p)) = R(0.92p) = 0.92p - 1200$

    $(S \circ R)(p) = S(R(p)) = S(p - 1200) = 0.92(p - 1200)$

    $R \circ S$ represents taking a discount of 8% of the retail price and then receiving a $1200 rebate. $S \circ R$ represents taking the $1200 rebate first and then receiving an 8% discount on the difference.

    (d) $(R \circ S)(18,400) = 0.92(18,400) - 1200 = \$15,728$

    $(S \circ R)(18,400) = 0.92(18,400 - 1200) = \$15,824$

    $R \circ S$ is a better deal. $S \circ R$ takes an 8% discount on a smaller amount.

91. Let $f(x)$ be an odd function, $g(x)$ be an even function and define $h(x) = f(x)g(x)$. Then

$$h(-x) = f(-x)g(-x)$$

$$= [-f(x)]g(x) \qquad \text{Since } f \text{ is odd and } g \text{ is even.}$$

$$= -f(x)g(x)$$

$$= -h(x)$$

Thus, h is odd.

Section 2.5   Inverse Functions

■ Two functions f and g are inverses of each other if $f(g(x)) = x$ for every x in the domain of g and $g(f(x)) = x$ for every x in the domain of f.

■ Be able to find the inverse of a function, if it exists.

    1. Replace $f(x)$ with y.

    2. Interchange x and y.

    3. Solve for y. If this equation represents y as a function of x, then you have found $f^{-1}(x)$. If this equation does not represent y as a function of x, then f does not have an inverse function.

■ A function f has an inverse function if and only if no **horizontal** line crosses the graph of f at more than one point.

Solutions to Odd-Numbered Exercises

1. The inverse is a line through $(-1, 0)$.
Matches graph (c).

3. The inverse is half a parabola starting at $(1, 0)$.
Matches graph (a).

5. $f^{-1}(x) = \dfrac{x}{8} = \dfrac{1}{8}x$

$f(f^{-1}(x)) = f\left(\dfrac{x}{8}\right) = 8\left(\dfrac{x}{8}\right) = x$

$f^{-1}(f(x)) = f^{-1}(8x) = \dfrac{8x}{8} = x$

7. $f^{-1}(x) = x - 10$

$f(f^{-1}(x)) = f(x - 10) = (x - 10) + 10 = x$

$f^{-1}(f(x)) = f^{-1}(x + 10) = (x + 10) - 10 = x$

9. $f^{-1}(x) = x^3$

$f(f^{-1}(x)) = f(x^3) = \sqrt[3]{x^3} = x$

$f^{-1}(f(x)) = f^{-1}(\sqrt[3]{x}) = (\sqrt[3]{x})^3 = x$

11. (a) $f(g(x)) = f\left(\dfrac{x}{2}\right) = 2\left(\dfrac{x}{2}\right) = x$

    $g(f(x)) = g(2x) = \dfrac{2x}{2} = x$

    (b)

13. (a) $f(g(x)) = f\left(\dfrac{x - 1}{5}\right) = 5\left(\dfrac{x - 1}{5}\right) + 1 = x$

    $g(f(x)) = g(5x + 1) = \dfrac{(5x + 1) - 1}{5} = x$

    (b)

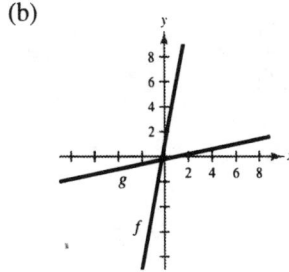

15. (a) $f(g(x)) = f(\sqrt[3]{x}) = (\sqrt[3]{x})^3 = x$

$g(f(x)) = g(x^3) = \sqrt[3]{x^3} = x$

(b)

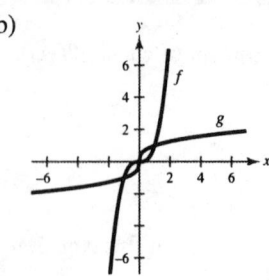

17. (a) $f(g(x)) = f(x^2 + 4),\ x \geq 0$

$= \sqrt{(x^2 + 4) - 4} = x$

$g(f(x)) = g(\sqrt{x - 4})$

$= (\sqrt{x - 4})^2 + 4 = x$

(b)

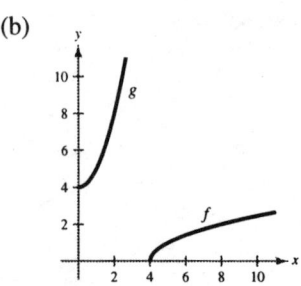

19. (a) $f(g(x)) = f(\sqrt{9 - x}),\ x \leq 9$

$= 9 - (\sqrt{9 - x})^2 = x$

$g(f(x)) = g(9 - x^2),\ x \geq 0$

$= \sqrt{9 - (9 - x^2)} = x$

(b)

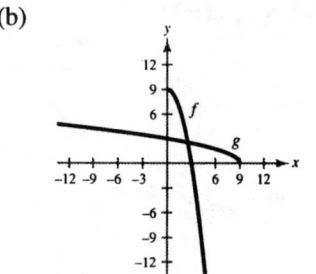

21. No, $\{(-2, -1), (1, 0), (2, 1), (1, 2), (-2, 3), (-6, 4)\}$ does not represent a function.

23. Since no horizontal line crosses the graph of f at more than one point, f **has** an inverse.

25. Since some horizontal lines cross the graph of f twice, f does **not** have an inverse.

27. $g(x) = \dfrac{4 - x}{6}$

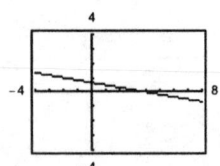

g passes the horizontal line test, so g
has an inverse.

29. $h(x) = |x + 4| - |x - 4|$

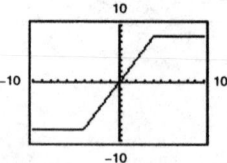

h does not pass the horizontal line test, so h
does **not** have an inverse.

31. $f(x) = -2x\sqrt{16 - x^2}$

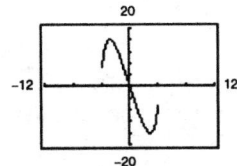

f does not pass the horizontal line test, so *f* does **not** have an inverse.

33. $f(x) = 2x - 3$

$y = 2x - 3$

$x = 2y - 3$

$y = \dfrac{x + 3}{2}$

$f^{-1}(x) = \dfrac{x + 3}{2}$

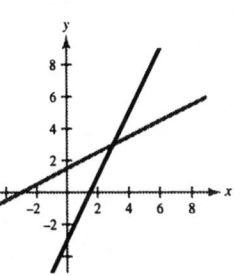

35. $f(x) = x^5$

$y = x^5$

$x = y^5$

$y = \sqrt[5]{x}$

$f^{-1}(x) = \sqrt[5]{x}$

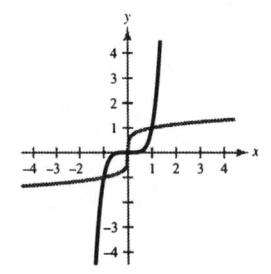

37. $f(x) = \sqrt{x}$

$y = \sqrt{x}$

$x = \sqrt{y}$

$y = x^2$

$f^{-1}(x) = x^2, \ x \geq 0$

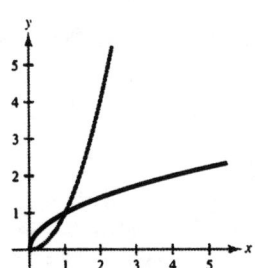

39. $f(x) = \sqrt{4 - x^2}, \ 0 \leq x \leq 2$

$y = \sqrt{4 - x^2}$

$x = \sqrt{4 - y^2}$

$f^{-1}(x) = \sqrt{4 - x^2}, \ 0 \leq x \leq 2$

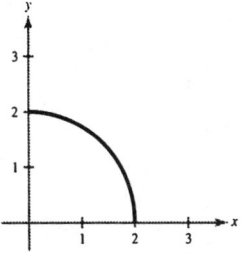

41. $f(.\ = \sqrt[3]{x - 1}$

$y = \sqrt[3]{x - 1}$

$x = \sqrt[3]{y - 1}$

$x^3 = y - 1$

$y = x^3 + 1$

$f^{-1}(x) = x^3 + 1$

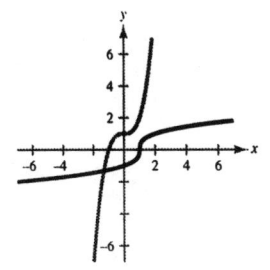

43. $f(x) = x^4$

$y = x^4$

$x = y^4$

$y = \pm \sqrt[4]{x}$

This does not represent *y* as a function of *x*.

f does not have an inverse.

45. $g(x) = \dfrac{x}{8}$

$y = \dfrac{x}{8}$

$x = \dfrac{y}{8}$

$y = 8x$

This is a function of x, so g has an inverse.

$g^{-1}(x) = 8x$

49. $f(x) = (x + 3)^2,\ x \geq -3\ \Rightarrow\ y \geq 0$

$y = (x + 3)^2,\ x \geq -3,\ y \geq 0$

$x = (y + 3)^2,\ y \geq -3,\ x \geq 0$

$\sqrt{x} = y + 3,\ y \geq -3,\ x \geq 0$

$y = \sqrt{x} - 3,\ x \geq 0,\ y \geq -3$

This is a function of x, so f has an inverse.

$f^{-1}(x) = \sqrt{x} - 3,\ x \geq 0$

53. $f(x) = \sqrt{2x + 3}\ \Rightarrow\ x \geq -\dfrac{3}{2},\ y \geq 0$

$y = \sqrt{2x + 3},\ x \geq -\dfrac{3}{2},\ y \geq 0$

$x = \sqrt{2y + 3},\ y \geq -\dfrac{3}{2},\ x \geq 0$

$x^2 = 2y + 3,\ x \geq 0,\ y \geq -\dfrac{3}{2}$

$y = \dfrac{x^2 - 3}{2},\ x \geq 0,\ y \geq -\dfrac{3}{2}$

This is a function of x, so f has an inverse.

$f^{-1}(x) = \dfrac{x^2 - 3}{2},\ x \geq 0$

47. $p(x) = -4$

$y = -4$

Since $y = -4$ for all x, the graph is a horizontal line and fails the horizontal line test. p does not have an inverse.

51. $h(x) = \dfrac{1}{x}$

$y = \dfrac{1}{x}$

$xy = 1$

$y = \dfrac{1}{x}$

This is a function of x, so h has an inverse.

$h^{-1}(x) = \dfrac{1}{x}$

55. $g(x) = x^2 - x^4$

The graph fails the horizontal line test, so g does not have an inverse.

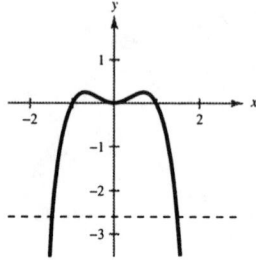

57. $f(x) = 25 - x^2,\ x \leq 0\ \Rightarrow\ y \leq 25$

$y = 25 - x^2,\ x \leq 0,\ y \leq 25$

$x = 25 - y^2,\ y \leq 0,\ x \leq 25$

$y^2 = 25 - x,\ x \leq 25,\ y \leq 0$

$y = -\sqrt{25 - x},\ x \leq 25,\ y \leq 0$

This is a function of x, so f has an inverse.

$f^{-1}(x) = -\sqrt{25 - x},\ x \leq 25$

59. If we let $f(x) = (x - 2)^2$, $x \geq 2$, then f has an inverse. [Note: we could also let $x \leq 2$.]

$$f(x) = (x - 2)^2, \ x \geq 2 \implies y \geq 0$$
$$y = (x - 2)^2, \ x \geq 2, \ y \geq 0$$
$$x = (y - 2)^2, \ x \geq 0, \ y \geq 2$$
$$\sqrt{x} = y - 2, \ x \geq 0, \ y \geq 2$$
$$\sqrt{x} + 2 = y, \ x \geq 0, \ y \geq 2$$
Thus, $f^{-1}(x) = \sqrt{x} + 2$, $x \geq 0$.

61. If we let $f(x) = |x + 2|$, $x \geq -2$, then f has an inverse. [Note: we could also let $x \leq -2$.]

$$f(x) = |x + 2|, \ x \geq -2$$
$$f(x) = x + 2 \ \text{ when } \ x \geq -2.$$
$$y = x + 2, \ x \geq -2, \ y \geq 0$$
$$x = y + 2, \ x \geq 0, \ y \geq -2$$
$$x - 2 = y, \ x \geq 0, \ y \geq -2$$
Thus, $f^{-1}(x) = x - 2$, $x \geq 0$.

63.

x	$f(x)$
-2	-4
-1	-2
1	2
3	3

x	$f^{-1}(x)$
-4	-2
-2	-1
2	1
3	3

65. False, $f(x) = x^2$ is even and does not have an inverse.

67. True

In Exercises 69, 71, and 73, $f(x) = \frac{1}{8}x - 3$, $f^{-1}(x) = 8(x + 3)$, $g(x) = x^3$, $g^{-1}(x) = \sqrt[3]{x}$.

69. $(f^{-1} \circ g^{-1})(1) = f^{-1}(g^{-1}(1)) = f^{-1}(\sqrt[3]{1}) = 8(\sqrt[3]{1} + 3) = 32$

71. $(f^{-1} \circ f^{-1})(6) = f^{-1}(f^{-1}(6)) = f^{-1}(8[6 + 3]) = 8[8(6 + 3) + 3] = 600$

73. $(f \circ g)(x) = f(g(x)) = f(x^3) = \frac{1}{8}x^3 - 3$

$$y = \frac{1}{8}x^3 - 3$$
$$x = \frac{1}{8}y^3 - 3$$
$$x + 3 = \frac{1}{8}y^3$$
$$8(x + 3) = y^3$$
$$\sqrt[3]{8(x + 3)} = y$$
$$(f \circ g)^{-1}(x) = 2\sqrt[3]{x + 3}$$

In Exercises 75 and 77, $f(x) = x + 4$, $f^{-1}(x) = x - 4$, $g(x) = 2x - 5$, $g^{-1}(x) = \dfrac{x + 5}{2}$.

75. $(g^{-1} \circ f^{-1})(x) = g^{-1}(f^{-1}(x)) = g^{-1}(x - 4) = \dfrac{(x - 4) + 5}{2} = \dfrac{x + 1}{2}$

77. $(f \circ g)(x) = f(g(x)) = f(2x - 5) = (2x - 5) + 4 = 2x - 1$

$$(f \circ g)^{-1}(x) = \dfrac{x + 1}{2}$$

Note: Comparing Exercises 75 and 77, we see that $(f \circ g)^{-1}(x) = (g^{-1} \circ f^{-1})(x)$.

79. (a) $y = 8 + 0.75x$

$x = 8 + 0.75y$

$x - 8 = 0.75y$

$\dfrac{x - 8}{0.75} = y$

$f^{-1}(x) = \dfrac{x - 8}{0.75}$

(b) x = hourly wage

y = number of units produced

(c) $y = \dfrac{22.25 - 8}{0.75} = 19$ units

81. (a) $y = 0.03x^2 + 254.50, \ 0 < x < 100$

$x = 0.03y^2 + 254.50$

$x - 254.50 = 0.03y^2$

$\dfrac{x - 254.50}{0.03} = y^2$

$\sqrt{\dfrac{x - 254.50}{0.03}} = y, \ 254.5 < x < 545.5$

$f^{-1}(x) = \sqrt{\dfrac{x - 254.50}{0.03}}$

x = temperature in degrees Fahrenheit

y = percent load for a diesel engine

(b)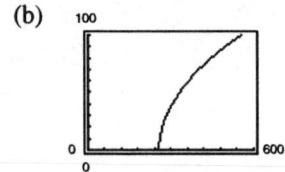

(c) $0.03x^2 + 254.50 < 500$

$0.03x^2 < 245.5$

$x^2 < 8183\tfrac{1}{3}$

$x < 90.46$

Thus, $0 < x < 90.46$.

83. (a) Yes, since no y-value is paired with two different x-values, f^{-1} does exist.

(b) f^{-1} yields the year for a given average fuel consumption.

(c) $f^{-1}(19.95) = 8$

85. $x^2 = 64$

$x = \pm\sqrt{64} = \pm 8$

87. $4x^2 - 12x + 9 = 0$

$(2x - 3)^2 = 0$

$2x - 3 = 0$

$x = \tfrac{3}{2}$

89. $x^2 - 6x + 4 = 0$

$\quad\quad x^2 - 6x = -4$

$x^2 - 6x + 9 = -4 + 9$

$\quad\quad (x - 3)^2 = 5$

$\quad\quad x - 3 = \pm\sqrt{5}$

$\quad\quad\quad x = 3 \pm \sqrt{5}$

91. $50 + 5x = 3x^2$

$\quad\quad 0 = 3x^2 - 5x - 50$

$\quad\quad 0 = (3x + 10)(x - 5)$

$3x + 10 = 0 \implies x = -\frac{10}{3}$

$x - 5 = 0 \implies x = 5$

93. Let $2n$ = first positive even integer. Then $2n + 2$ = next positive even integer.

$\quad\quad 2n(2n + 2) = 288$

$4n^2 + 4n - 288 = 0$

$\quad 4(n^2 + n - 72) = 0$

$4(n + 9)(n - 8) = 0$

$n + 9 = 0 \implies n = -9$ Not a solution since the integers are positive.

$n - 8 = 0 \implies n = 8$

Thus, $2n = 16$ and $2n + 2 = 18$.

95.

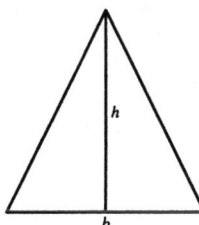

Given $b = h$ and $A = 10$ sq ft:

$A = \frac{1}{2}bh$

$10 = \frac{1}{2}bb$

$20 = b^2$

$\sqrt{20} = b$

$2\sqrt{5} = b$

Thus, $b = h = 2\sqrt{5}$ feet.

❑ **Review Exercises for Chapter 2**

Solutions to Odd-Numbered Exercises

1. (a) $m = \frac{3}{2} > 0 \implies$ The line rises. Matches L_2.

(b) $m = 0 \implies$ The line is horizontal. Matches L_3.

(c) $m = -3 < 0 \implies$ The line falls. Matches L_1.

3. $(-4.5, 6), (2.1, 3)$

$$m = \frac{3 - 6}{2.1 - (-4.5)} = \frac{-3}{6.6} = -\frac{30}{66} = -\frac{5}{11}$$

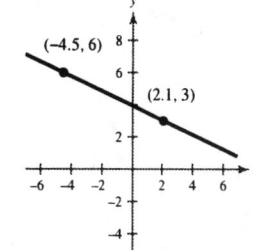

5. $(-2, 5)$, $(0, t)$, $(1, 1)$ are collinear.

$$\frac{t - 5}{0 - (-2)} = \frac{1 - 5}{1 - (-2)}$$

$$\frac{t - 5}{2} = \frac{-4}{3}$$

$$3(t - 5) = -8$$

$$3t - 15 = -8$$

$$3t = 7$$

$$t = \frac{7}{3}$$

7. $(2 + 4, -1 + 1) = (6, 0)$

$(6 + 4, 0 + 1) = (10, 1)$

$(2 - 4, -1 - 1) = (-2, -2)$

9. $(0, 0)$, $(0, 10)$

$$m = \frac{10 - 0}{0 - 0} = \frac{10}{0} \text{ undefined}$$

The line is vertical.

$$x = 0$$

11. $y - (-5) = \frac{3}{2}(x - 0)$

$\quad y + 5 = \frac{3}{2}x$

$\qquad y = \frac{3}{2}x - 5 \text{ or } 0 = 3x - 2y - 10$

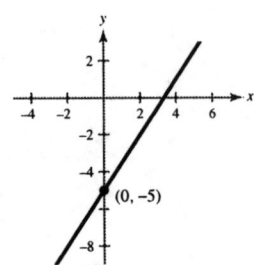

13. $5x - 4y = 8 \implies y = \frac{5}{4}x - 2 \text{ and } m = \frac{5}{4}$

(a) Parallel slope: $m = \frac{5}{4}$

$\quad y - (-2) = \frac{5}{4}(x - 3)$

$\qquad 4y + 8 = 5x - 15$

$\qquad\quad 0 = 5x - 4y - 23$

(b) Perpendicular slope: $m = -\frac{4}{5}$

$\quad y - (-2) = -\frac{4}{5}(x - 3)$

$\quad 5y + 10 = -4x + 12$

$4x + 5y - 2 = 0$

15. $(6, 12{,}500)$ $m = 850$

$y - 12{,}500 = 850(t - 6)$

$\qquad y = 850t - 5100 + 12{,}500$

$\qquad y = 850t + 7400$

17. The distance between $(-2, -5)$ and (x, y) equals the distance between $(6, 3)$ and (x, y).

$$\sqrt{(x + 2)^2 + (y + 5)^2} = \sqrt{(x - 6)^2 + (y - 3)^2}$$

$$(x + 2)^2 + (y + 5)^2 = (x - 6)^2 + (y - 3)^2$$

$$x^2 + 4x + 4 + y^2 + 10y + 25 = x^2 - 12x + 36 + y^2 - 6y + 9$$

$$4x + 10y + 29 = -12x - 6y + 45$$

$$16x + 16y - 16 = 0$$

$$x + y - 1 = 0$$

This line is the perpendicular bisector of the line segment joining the two points.

19. (2, 160,000), (3, 185,000)

$$m = \frac{185,000 - 160,000}{3 - 2} = 25,000$$

$$S - 160,000 = 25,000(t - 2)$$

$$S = 25,000t + 110,000$$

For the fourth quarter let $t = 4$. Then we have

$$S = 25,000(4) + 110,000 = \$210,000.$$

21. $A = \{10, 20, 30, 40\}$ and $B = \{0, 2, 4, 6\}$

(a) 20 is matched with two elements in the range so it is not a function.

(b) function

(c) function

(d) 30 is not matched with any element of B so it is not a function.

23. $16x - y^4 = 0$

$$y^4 = 16x$$

$$y = \pm 2\sqrt[4]{x}$$

y is **not** a function of x. Some x-values correspond to two y-values.

25. $y = \sqrt{1 - x}$

Each x-value, $x \le 1$, corresponds to only one y-value so y **is** a function of x.

27. $f(x) = x^2 + 1$

(a) $f(2) = (2)^2 + 1 = 5$

(b) $f(-4) = (-4)^2 + 1 = 17$

(c) $f(t^2) = (t^2)^2 + 1 = t^4 + 1$

(d) $-f(x) = -(x^2 + 1) = -x^2 - 1$

29. $f(x) = \sqrt{25 - x^2}$

Domain: $25 - x^2 \ge 0$

$$(5 + x)(5 - x) \ge 0$$

Critical Numbers: $x = \pm 5$

Test intervals: $(-\infty, -5)$, $(-5, 5)$, $(5, \infty)$

Solution set: $[-5, 5]$

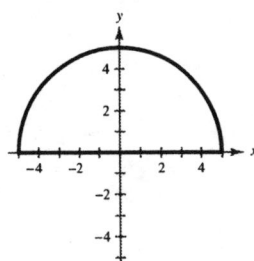

31. $g(s) = \dfrac{5}{3s - 9} = \dfrac{5}{3(s - 3)}$

Domain: All real numbers except $s = 3$.

33. $h(x) = \dfrac{x}{x^2 - x - 6} = \dfrac{x}{(x + 2)(x - 3)}$

Domain: All real numbers except $x = -2, 3$.

35. $f(x) = \dfrac{3x}{2(3-x)}$

The second setting shows the most complete graph.

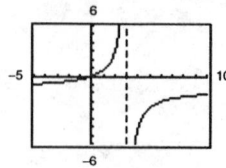

37. $g(x) = |x+2| - |x-2|$

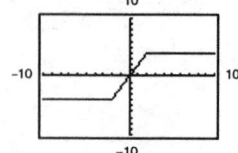

    (a) Increasing on $(-2, 2)$

        Constant on $(-\infty, -2] \cup [2, \infty)$

    (b) The graph has origin symmetry so the function is odd.

39. $h(x) = 4x^3 - x^4$

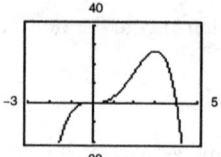

    (a) Increasing on $(-\infty, 3)$

        Decreasing on $(3, \infty)$

    (b) Neither odd nor even

41. $v(t) = -32t + 48$

    (a) $v(1) = 16$ ft/sec

    (b) $0 = -32t + 48$

        $t = \dfrac{48}{32} = 1.5$ sec

    (c) $v(2) = -16$ ft/sec

43.

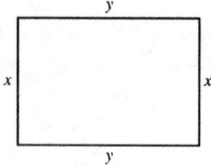

    (a) $2x + 2y = 24$

          $y = 12 - x$

       $A = xy = x(12 - x)$

    (b) Since x and y cannot be negative, we have $0 < x < 12$. The domain is $(0, 12)$.

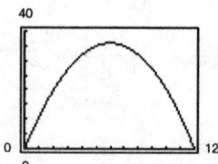

    (c) The maximum area of 36 occurs when $x = 6$ and the rectangle is a 6×6 square.

45. (a) Revenue = (number of people) (rate per person)

        $R(n) = n[8.00 - 0.05(n - 80)], \; n \geq 80$

    (b)

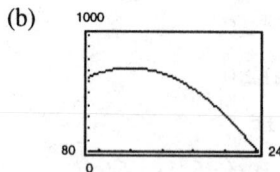

The revenue will be maximized ($720) when 120 passengers take the bus.

47. (a)
$$f(x) = \tfrac{1}{2}x - 3$$
$$y = \tfrac{1}{2}x - 3$$
$$x = \tfrac{1}{2}y - 3$$
$$x + 3 = \tfrac{1}{2}y$$
$$2(x + 3) = y$$
$$f^{-1}(x) = 2x + 6$$

(b)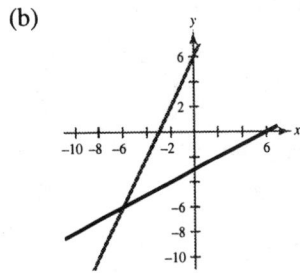

(c)
$$f^{-1}(f(x)) = f^{-1}\left(\tfrac{1}{2}x - 3\right)$$
$$= 2\left(\tfrac{1}{2}x - 3\right) + 6$$
$$= x - 6 + 6$$
$$= x$$

$$f(f^{-1}(x)) = f(2x + 6)$$
$$= \tfrac{1}{2}(2x + 6) - 3$$
$$= x + 3 - 3$$
$$= x$$

49. (a)
$$f(x) = \sqrt{x + 1}$$
$$y = \sqrt{x + 1}$$
$$x = \sqrt{y + 1}$$
$$x^2 = y + 1$$
$$x^2 - 1 = y$$
$$f^{-1}(x) = x^2 - 1, \ x \geq 0$$

(b)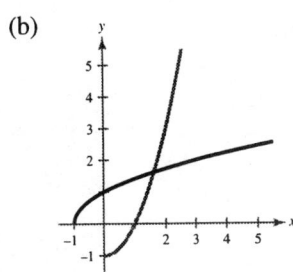

Note: The inverse must have a restricted domain.

(c)
$$f^{-1}(f(x)) = f^{-1}\left(\sqrt{x + 1}\right)$$
$$= \left(\sqrt{x + 1}\right)^2 - 1$$
$$= x + 1 - 1$$
$$= x$$

$$f(f^{-1}(x)) = f(x^2 - 1)$$
$$= \sqrt{(x^2 - 1) + 1}$$
$$= \sqrt{x^2} = x \text{ for } x \geq 0.$$

51. $f(x) = 2(x - 4)^2$ is increasing on $[4, \infty)$.

Let $\quad f(x) = 2(x - 4)^2, \ x \geq 4$ and $y \geq 0$

$\qquad y = 2(x - 4)^2$

$\qquad x = 2(y - 4)^2, \ x \geq 0, \ y \geq 4$

$\qquad \dfrac{x}{2} = (y - 4)^2$

$\qquad \sqrt{\dfrac{x}{2}} = y - 4$

$\qquad \sqrt{\dfrac{x}{2}} + 4 = y$

$\qquad f^{-1}(x) = \sqrt{\dfrac{x}{2}} + 4$

53. $(f - g)(4) = f(4) - g(4)$

$\qquad\qquad = [3 - 2(4)] - \sqrt{4}$

$\qquad\qquad = -5 - 2$

$\qquad\qquad = -7$

55. $(h \circ g)(7) = h(g(7))$

$\qquad\qquad = h(\sqrt{7})$

$\qquad\qquad = 3(\sqrt{7})^2 + 2$

$\qquad\qquad = 23$

57. $(f \circ h)(3) = f(h(3))$

$\qquad\qquad = f(3 \cdot 3^2 + 2)$

$\qquad\qquad = f(29)$

$\qquad\qquad = 3 - 2(29)$

$\qquad\qquad = -55$

❏ Practice Test for Chapter 2

1. Find the equation of the line through $(2, 4)$ and $(3, -1)$.

2. Find the equation of the line with slope $m = 4/3$ and y-intercept $b = -3$.

3. Find the equation of the line through $(4, 1)$ perpendicular to the line $2x + 3y = 0$.

4. If it costs a company \$32 to produce 5 units of a product and \$44 to produce 9 units, how much does it cost to produce 20 units? (Assume that the cost function is linear.)

5. Given $f(x) = x^2 - 2x + 1$, find $f(x - 3)$.

6. Given $f(x) = 4x - 11$, find $\dfrac{f(x) - f(3)}{x - 3}$

7. Find the domain and range of $f(x) = \sqrt{36 - x^2}$.

8. Which equations determine y as a function of x?

   (a) $6x - 5y + 4 = 0$

   (b) $x^2 + y^2 = 9$

   (c) $y^3 = x^2 + 6$

9. Sketch the graph of $f(x) = x^2 - 5$.

10. Sketch the graph of $f(x) = |x + 3|$.

11. Sketch the graph of $f(x) = \begin{cases} 2x + 1 & \text{if } x \geq 0, \\ x^2 - x & \text{if } x < 0. \end{cases}$

12. Use the graph of $f(x) = |x|$ to graph the following:

    (a) $f(x + 2)$

    (b) $-f(x) + 2$

13. Given $f(x) = 3x + 7$ and $g(x) = 2x^2 - 5$, find the following:

    (a) $(g - f)(x)$

    (b) $(fg)(x)$

14. Given $f(x) = x^2 - 2x + 16$ and $g(x) = 2x + 3$, find $f(g(x))$.

15. Given $f(x) = x^3 + 7$, find $f^{-1}(x)$.

16. Which of the following functions have inverses?

   (a) $f(x) = |x - 6|$

   (b) $f(x) = ax + b,\ a \neq 0$

   (c) $f(x) = x^3 - 19$

17. Given $f(x) = \sqrt{\dfrac{3 - x}{x}},\ 0 < x \leq 3$, find $f^{-1}(x)$.

Exercises 18–20, true or false?

18. $y = 3x + 7$ and $y = \frac{1}{3}x - 4$ are perpendicular.

19. $(f \circ g)^{-1} = g^{-1} \circ f^{-1}$

20. If a function has an inverse, then it must pass both the vertical line test and the horizontal line test.

CHAPTER 3
Zeros of Polynomial Functions

CHAPTER 3
Zeros of Polynomial Functions

Section 3.1   Quadratic Functions

You should know the following facts about parabolas.

- ■ $f(x) = ax^2 + bx + c,\ a \neq 0$, is a quadratic function, and its graph is a parabola.
- ■ If $a > 0$, the parabola opens upward and the vertex is the minimum point. If $a < 0$, the parabola opens downward and the vertex is the maximum point.
- ■ The vertex is $(-b/2a, f(-b/2a))$.
- ■ To find the x-intercepts (if any), solve
  $$ax^2 + bx + c = 0.$$
- ■ The standard form of the equation of a parabola is
  $$f(x) = a(x - h)^2 + k$$
  where $a \neq 0$.
  - (a) The vertex is (h, k).
  - (b) The axis is the vertical line $x = h$.

Solutions to Odd-Numbered Exercises

1. $f(x) = (x - 2)^2$ opens upward and has vertex $(2, 0)$. Matches graph (g).

3. $f(x) = x^2 - 2$ opens upward and has vertex $(0, -2)$. Matches graph (b).

5. $f(x) = 4 - (x - 2)^2 = -(x - 2)^2 + 4$ opens downward and has vertex $(2, 4)$. Matches graph (f).

7. $f(x) = x^2 + 3$ opens upward and has vertex $(0, 3)$. Matches graph (e).

9. (a) $y = \frac{1}{2}x^2$

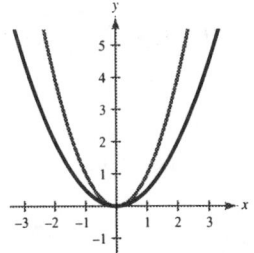

Vertical shrink

(b) $y = -\frac{1}{8}x^2$

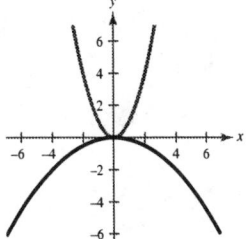

Vertical shrink and reflection in the x-axis

(c) $y = \frac{3}{2}x^2$

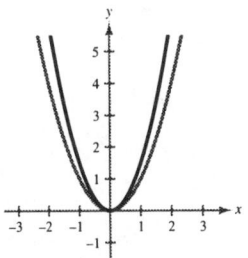

Vertical stretch

(d) $y = -3x^2$

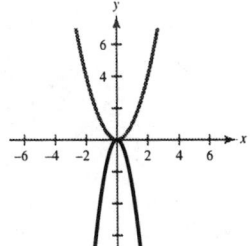

Vertical stretch and reflection in the x-axis

11. (a) $y = (x - 1)^2$

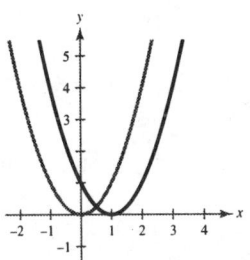

Horizontal translation one unit to the right

(b) $y = (x + 1)^2$

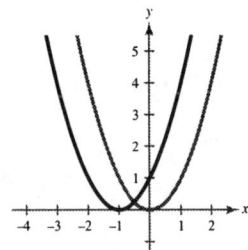

Horizontal translation one unit to the left

(c) $y = (x - 3)^2$

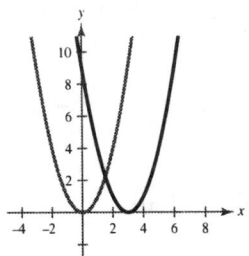

Horizontal translation three units to the right

(d) $y = (x + 3)^2$

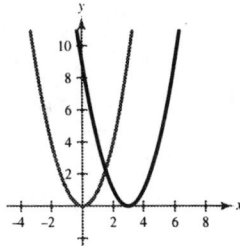

Horizontal translation three units to the left

13. $f(x) = x^2 - 5$

Vertex: $(0, -5)$

Intercepts: $\left(-\sqrt{5}, 0\right)$, $(0, -5)$, $\left(\sqrt{5}, 0\right)$

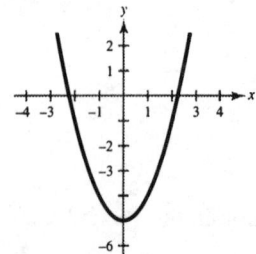

15. $f(x) = 16 - x^2$

Vertex: $(0, 16)$

Intercepts: $(-4, 0)$, $(0, 16)$, $(4, 0)$

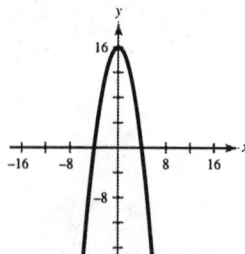

17. $f(x) = (x + 5)^2 - 6$

Vertex: $(-5, -6)$

Intercepts: $\left(-5 - \sqrt{6}, 0\right)$, $\left(-5 + \sqrt{6}, 0\right)$,
$(0, 19)$

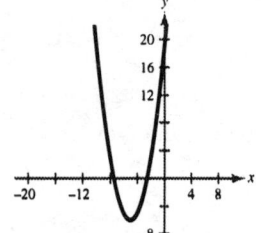

19. $h(x) = x^2 - 8x + 16 = (x - 4)^2$

Vertex: $(4, 0)$

Intercepts: $(0, 16)$, $(4, 0)$

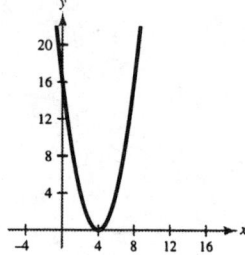

21. $f(x) = x^2 - x + \frac{5}{4} = \left(x - \frac{1}{2}\right)^2 + 1$

Vertex: $\left(\frac{1}{2}, 1\right)$

Intercept: $\left(0, \frac{5}{4}\right)$

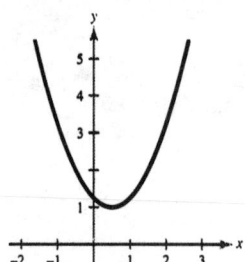

23. $f(x) = -x^2 + 2x + 5 = -(x - 1)^2 + 6$

Vertex: $(1, 6)$

Intercepts: $\left(1 - \sqrt{6}, 0\right)$, $(0, 5)$, $\left(1 + \sqrt{6}, 0\right)$

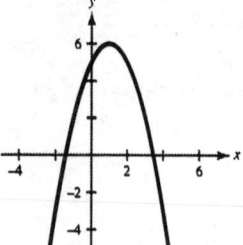

25. $h(x) = 4x^2 - 4x + 21 = 4\left(x - \frac{1}{2}\right)^2 + 20$

Vertex: $\left(\frac{1}{2}, 20\right)$

Intercept: $(0, 21)$

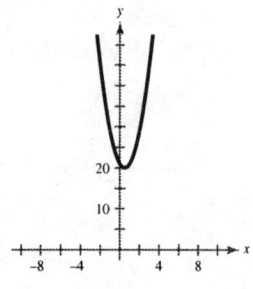

27. $f(x) = -(x^2 + 2x - 3) = -(x + 1)^2 + 4$

Vertex: $(-1, 4)$

Intercepts: $(-3, 0)$, $(0, 3)$, $(1, 0)$

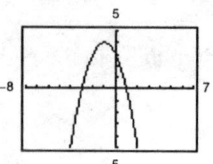

29. $f(x) = 2x^2 - 16x + 31$

$\quad = 2(x - 4)^2 - 1$

Vertex: $(4, -1)$

Intercepts: $\left(4 \pm \frac{1}{2}\sqrt{2}, 0\right),\ (0, 31)$

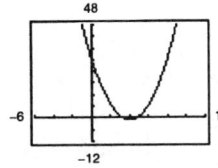

31. $(1, 0)$ is the vertex.

$\quad f(x) = a(x - 1)^2 + 0 = a(x - 1)^2$

Since the graph passes through the point $(0, 1)$ we have:

$\quad 1 = a(0 - 1)^2$

$\quad 1 = a$

$\quad f(x) = 1(x - 1)^2 = (x - 1)^2$

33. $(-1, 4)$ is the vertex.

$\quad f(x) = a(x + 1)^2 + 4$

Since the graph passes through the point $(1, 0)$, we have:

$\quad 0 = a(1 + 1)^2 + 4$

$\quad -4 = 4a$

$\quad -1 = a$

$\quad f(x) = -1(x + 1)^2 + 4 = -(x + 1)^2 + 4.$

35. $(-2, 2)$ is the vertex.

$\quad f(x) = a(x + 2)^2 + 2$

Since the graph passes through the point $(-1, 0)$, we have:

$\quad 0 = a(-1 + 2)^2 + 2$

$\quad -2 = a$

$\quad f(x) = -2(x + 2)^2 + 2$

37. $(-2, 5)$ is the vertex.

$\quad f(x) = a(x + 2)^2 + 5$

Since the graph passes through the point $(0, 9)$, we have:

$\quad 9 = a(0 + 2)^2 + 5$

$\quad 4 = 4a$

$\quad 1 = a$

$\quad f(x) = 1(x + 2)^2 + 5 = (x + 2)^2 + 5$

39. $(3, 4)$ is the vertex.

$\quad f(x) = a(x - 3)^2 + 4$

Since the graph passes through the point $(1, 2)$, we have:

$\quad 2 = a(1 - 3)^2 + 4$

$\quad -2 = 4a$

$\quad -\frac{1}{2} = a$

$\quad f(x) = -\frac{1}{2}(x - 3)^2 + 4$

41. $(5, 12)$ is the vertex.

$\quad f(x) = a(x - 5)^2 + 12$

Since the graph passes through the point $(7, 15)$, we have:

$\quad 15 = a(7 - 5)^2 + 12$

$\quad 3 = 4a \quad \Rightarrow \quad a = \frac{3}{4}$

$\quad f(x) = \frac{3}{4}(x - 5)^2 + 12$

43. $y = x^2 - 16$

$\quad$ x-intercepts: $(\pm 4, 0)$

$0 = x^2 - 16$

$x^2 = 16$

$x = \pm 4$

45. $y = x^2 - 4x - 5$

$\quad$ x-intercepts: $(5, 0),\ (-1, 0)$

$0 = x^2 - 4x - 5$

$0 = (x - 5)(x + 1)$

$x = 5 \text{ or } x = -1$

47. $y = x^2 - 4x$

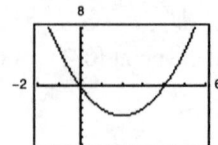

$0 = x^2 - 4x$

$0 = x(x - 4)$

$x = 0$ or $x = 4$

x-intercepts: $(0, 0)$, $(4, 0)$

49. $y = 2x^2 - 7x - 30$

$0 = 2x^2 - 7x - 30$

$0 = (2x + 5)(x - 6)$

$x = -\frac{5}{2}$ or $x = 6$

x-intercepts: $\left(-\frac{5}{2}, 0\right)$, $(6, 0)$

51. $f(x) = [x - (-1)](x - 3)$ opens upward

$= (x + 1)(x - 3)$

$= x^2 - 2x - 3$

$f(x) = -[x - (-1)](x - 3)$ opens downward

$= -(x + 1)(x - 3)$

$= -(x^2 - 2x - 3)$

$= -x^2 + 2x + 3$

Note: $f(x) = a(x + 1)(x - 3)$ has *x*-intercepts $(-1, 0)$ and $(3, 0)$ for all real numbers $a \neq 0$.

53. $f(x) = (x - 0)(x - 10)$ opens upward

$= x^2 - 10x$

$f(x) = -(x - 0)(x - 10)$ opens downward

$= -x^2 + 10x$

Note: $f(x) = a(x - 0)(x - 10) = ax(x - 10)$ has *x*-intercepts $(0, 0)$ and $(10, 0)$ for all real numbers $a \neq 0$.

55. $f(x) = [x - (-3)]\left[x - \left(-\frac{1}{2}\right)\right](2)$ opens upward

$= (x + 3)\left(x + \frac{1}{2}\right)(2)$

$= (x + 3)(2x + 1)$

$= 2x^2 + 7x + 3$

$f(x) = -(2x^2 + 7x + 3)$ opens downward

$= -2x^2 - 7x - 3$

Note: $f(x) = a(x + 3)(2x + 1)$ has *x*-intercepts $(-3, 0)$ and $\left(-\frac{1}{2}, 0\right)$ for all real numbers $a \neq 0$.

57. Let $x =$ the first number and $y =$ the second number. Then the sum is

$$x + y = 110 \implies y = 110 - x.$$

The product is $P(x) = xy = x(110 - x) = 110x - x^2$.

$P(x) = -x^2 + 110x$

$= -(x^2 - 110x + 3025 - 3025)$

$= -[(x - 55)^2 - 3025]$

$= -(x - 55)^2 + 3025$

The maximum value of the product occurs at the vertex of $P(x)$ and is 3025. This happens when $x = y = 55$.

59. Let $x =$ the first number and $y =$ the second number. Then the sum is

$$x + 2y = 24 \implies y = \frac{24 - x}{2}.$$

The product is $P(x) = xy = x\left(\frac{24 - x}{2}\right)$.

$P(x) = \frac{1}{2}(-x^2 + 24x)$

$= -\frac{1}{2}(x^2 - 24x + 144 - 144)$

$= -\frac{1}{2}[(x - 12)^2 - 144] = -\frac{1}{2}(x - 12)^2 + 72$

The maximum value of the product occurs at the vertex of $P(x)$ and is 72. This happens when $x = 12$ and $y = (24 - 12)/2 = 6$. Thus, the numbers are 12 and 6.

61.

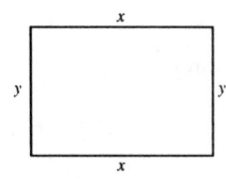

$2x + 2y = 100$

$y = 50 - x$

(a) $A(x) = xy = x(50 - x)$

Domain: $0 < x < 50$

(b)

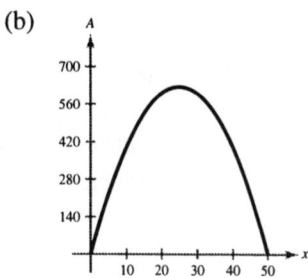

(c) The area is maximum (625 square feet) when $x = y = 25$. The rectangle has dimensions 25 ft × 25 ft.

63. (a) $4x + 3y = 200 \implies y = \frac{1}{3}(200 - 4x)$

x	y	Area
2	$\frac{1}{3}[200 - 4(2)]$	$2xy = 256$
4	$\frac{1}{3}[200 - 4(4)]$	$2xy \approx 491$
6	$\frac{1}{3}[200 - 4(6)]$	$2xy = 704$
8	$\frac{1}{3}[200 - 4(8)]$	$2xy = 896$
10	$\frac{1}{3}[200 - 4(10)]$	$2xy \approx 1067$
12	$\frac{1}{3}[200 - 4(12)]$	$2xy = 1216$

(b)

x	y	Area
20	$\frac{1}{3}[200 - 4(20)]$	$2xy = 1600$
22	$\frac{1}{3}[200 - 4(22)]$	$2xy \approx 1643$
24	$\frac{1}{3}[200 - 4(24)]$	$2xy = 1664$
26	$\frac{1}{3}[200 - 4(26)]$	$2xy = 1664$
28	$\frac{1}{3}[200 - 4(28)]$	$2xy \approx 1643$
30	$\frac{1}{3}[200 - 4(30)]$	$2xy = 1600$

(c) $A = 2xy = 2x\left(\dfrac{200 - 4x}{3}\right) = \dfrac{2x(4)(50 - x)}{3}$

$= \dfrac{8x(50 - x)}{3}$

(d)

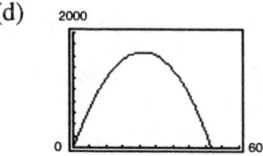

This area is maximum when $x = 25$ feet and $y = \frac{100}{3} = 33\frac{1}{3}$ feet.

(e) $A = \frac{8}{3}x(50 - x)$

$= -\frac{8}{3}(x^2 - 50x)$

$= -\frac{8}{3}(x^2 - 50x + 625 - 625)$

$= -\frac{8}{3}[(x - 25)^2 - 625]$

$= -\frac{8}{3}(x - 25)^2 + \frac{5000}{3}$

The maximum area occurs at the vertex and is 5000/3 square feet. This happens when $x = 25$ feet and $y = (200 - 4(25))/3 = 100/3$ feet. The dimensions are $2x = 50$ feet by $33\frac{1}{3}$ feet.

65. $R = 900x - 0.1x^2 = -0.1x^2 + 900x$

The vertex occurs at $x = -\dfrac{b}{2a} = -\dfrac{900}{2(-0.1)} = 4500$. The revenue is maximum when $x = 4500$ units.

67. $C = 800 - 10x + 0.25x^2 = 0.25x^2 - 10x + 800$

The vertex occurs at $x = -\dfrac{b}{2a} = -\dfrac{-10}{2(0.25)} = 20$. The cost is minimum when $x = 20$ fixtures.

69. $P = -0.0002x^2 + 140x - 250{,}000$

The vertex occurs at $x = -\dfrac{b}{2a} = -\dfrac{140}{2(-0.0002)} = 350{,}000$.

The profit is maximum when $x = 350{,}000$ units.

71. $y = -\dfrac{1}{12}x^2 + 2x + 4$

(a) When $x = 0$, $y = 4$ feet.

(b) The vertex occurs at $x = -\dfrac{b}{2a} = -\dfrac{2}{2\left(-\frac{1}{12}\right)} = 12$. The maximum height is

$y = -\dfrac{1}{12}(12)^2 + 2(12) + 4 = 16$ feet.

(c) When the ball strikes the ground, $y = 0$.

$$0 = -\frac{1}{12}x^2 + 2x + 4$$

$0 = x^2 - 24x - 48$ Multiply both sides by -12.

$$x = \frac{-(-24) \pm \sqrt{(-24)^2 - 4(1)(-48)}}{2(1)}$$

$$= \frac{24 \pm \sqrt{768}}{2} = \frac{24 \pm 16\sqrt{3}}{2} = 12 \pm 8\sqrt{3}$$

Using the positive value for x, we have $x = 12 + 8\sqrt{3} \approx 25.86$ feet.

73. $V = 0.77x^2 - 1.32x - 9.31$, $5 \le x \le 40$

(a)
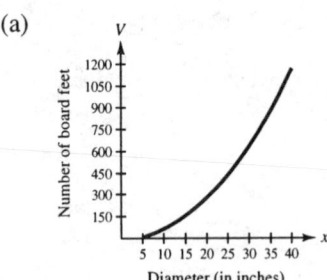

(b) $V(16) = 166.69$ board feet

(c) $500 = 0.77x^2 - 1.32x - 9.31$

$0 = 0.77x^2 - 1.32x - 509.31$

Using the Quadratic Formula and selecting the positive value for x, we have $x \approx 26.6$ inches in diameter.

75. $C = 4024.5 + 51.4t - 3.1t^2, \ -10 \le t \le 30$

(a)

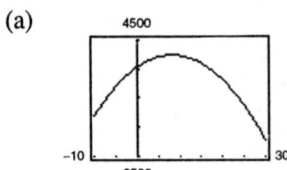

(b) $-\dfrac{b}{2a} = \dfrac{-51.4}{2(-3.1)} \approx 8.29$

The vertex occurs when $y \approx 4238$ which is the maximum average annual consumption. The warnings may not have had an immediate effect, but over time they and other findings about the health risks of cigarettes have had an effect.

(c) $C(0) = 4024.5$; annually $= \dfrac{116{,}530{,}000(4024.5)}{48{,}500{,}000} \approx 9670$; daily $= \dfrac{9670}{366} \approx 26$

77. If $f(x) = ax^2 + bx + c$ has two real zeros, then by the Quadratic Formula they are

$$x = \frac{-b \pm \sqrt{b^2 - 4ac}}{2a}.$$

The average of the zeros of f is

$$\frac{\dfrac{-b - \sqrt{b^2 - 4ac}}{2a} + \dfrac{-b + \sqrt{b^2 - 4ac}}{2a}}{2} = \frac{\dfrac{-2b}{2a}}{2} = -\frac{b}{2a}.$$

This is the x-coordinate of the vertex of the graph.

79. $(-4, 3)$ and $(2, 1)$

$$m = \frac{1 - 3}{2 - (-4)} = \frac{-2}{6} = -\frac{1}{3}$$

$$y - 1 = -\frac{1}{3}(x - 2)$$

$$3y - 3 = -x + 2$$

$$x + 3y - 5 = 0$$

81. $4x + 5y = 10 \ \Rightarrow \ y = -\frac{4}{5}x + 2$ and $m = -\frac{4}{5}$

The slope of the perpendicular line through $(0, 3)$ is $m = \frac{5}{4}$ and the y-intercept is $b = 3$.

$$y = \frac{5}{4}x + 3$$

$$0 = 5x - 4y + 12$$

Section 3.2   Polynomial Functions of Higher Degree

- ■ You should know the following basic principles about polynomials.
- ■ $f(x) = a_n x^n + a_{n-1} x^{n-1} + \cdots + a_2 x^2 + a_1 x + a_0$ is a polynomial function of degree n.
- ■ If f is of odd degree and
  - (a) $a_n > 0$, then
    - 1. $f(x) \to \infty$ as $x \to \infty$.
    - 2. $f(x) \to -\infty$ as $x \to -\infty$.
  - (b) $a_n < 0$, then
    - 1. $f(x) \to -\infty$ as $x \to \infty$.
    - 2. $f(x) \to \infty$ as $x \to -\infty$.
- ■ If f is of even degree and
  - (a) $a_n > 0$, then
    - 1. $f(x) \to \infty$ as $x \to \infty$.
    - 2. $f(x) \to \infty$ as $x \to -\infty$.
  - (b) $a_n < 0$, then
    - 1. $f(x) \to -\infty$ as $x \to \infty$.
    - 2. $f(x) \to -\infty$ as $x \to -\infty$.
- ■ The following are equivalent for a polynomial function.
  - (a) $x = a$ is a zero of a function.
  - (b) $x = a$ is a solution of the polynomial equation $f(x) = 0$.
  - (c) $(x - a)$ is a factor of the polynomial.
  - (d) $(a, 0)$ is an x-intercept of the graph of f.
- ■ A polynomial of degree n has at most n distinct zeros.
- ■ If f is a polynomial function such that $a < b$ and $f(a) \neq f(b)$, then f takes on every value between $f(a)$ and $f(b)$ in the interval $[a, b]$.
- ■ If you can find a value where a polynomial is positive and another value where it is negative, then there is at least one real zero between the values.

Solutions to Odd-Numbered Exercises

1. $f(x) = -2x + 3$ is a line with y-intercept $(0, 3)$. Matches graph (c).

3. $f(x) = -2x^2 - 5x$ is a parabola with x-intercepts $(0, 0)$ and $\left(-\frac{5}{2}, 0\right)$ and opens downward. Matches graph (h).

5. $f(x) = -\frac{1}{4}x^4 + 3x^2$ has intercepts $(0, 0)$ and $\left(\pm 2\sqrt{3}, 0\right)$. Matches graph (a).

7. $f(x) = x^4 + 2x^3$ has intercepts $(0, 0)$ and $(-2, 0)$. Matches graph (d).

9. $y = x^3$

(a) $f(x) = (x - 2)^3$

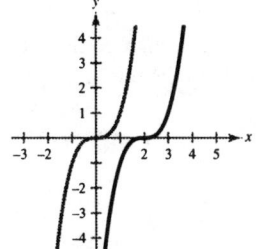

Horizontal shift two units to the right

(b) $f(x) = x^3 - 2$

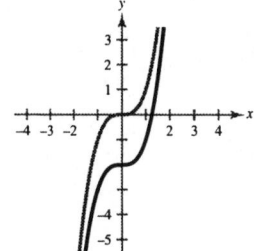

Vertical shift two units downward

(c) $f(x) = -\frac{1}{2}x^3$

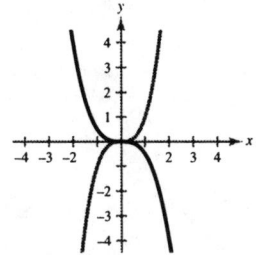

Reflection in the *x*-axis and a vertical shrink

(d) $f(x) = (x - 2)^3 - 2$

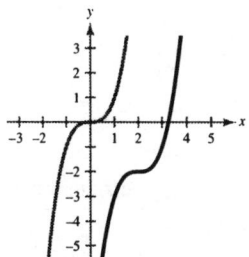

Horizontal shift two units to the right and a vertical shift two units downward

11. $y = x^4$

(a) $f(x) = (x + 3)^4$

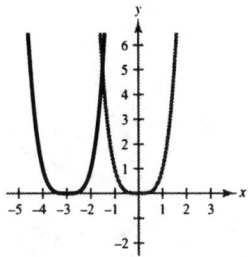

Horizontal shift three units to the left

(b) $f(x) = x^4 - 3$

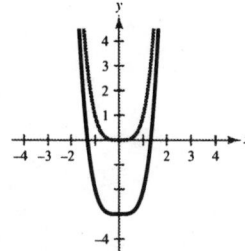

Vertical shift three units downward

(c) $f(x) = 4 - x^4$

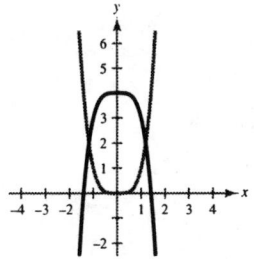

Reflection in the *x*-axis and then a vertical shift four units upward

(d) $f(x) = \frac{1}{2}(x - 1)^4$

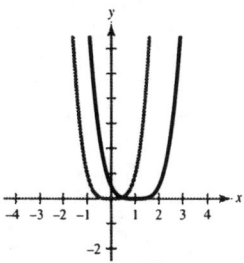

Horizontal shift one unit to the right and a vertical shrink

13. $f(x) = \frac{1}{3}x^3 + 5x$

Degree: 3

Leading coefficient: $\frac{1}{3}$

The degree is odd and the leading coefficient is positive. The graph falls to the left and rises to the right.

15. $g(x) = 5 - \frac{7}{2}x - 3x^2$

Degree: 2

Leading coefficient: -3

The degree is even and the leading coefficient is negative. The graph falls to the left and right.

17. $f(x) = -2.1x^5 + 4x^3 - 2$

Degree: 5

Leading coefficient: -2.1

The degree is odd and the leading coefficient is negative. The graph rises to the left and falls to the right.

19. $f(x) = 6 - 2x + 4x^2 - 5x^3$

Degree: 3

Leading coefficient: -5

The degree is odd and the leading coefficient is negative. The graph rises to the left and falls to the right.

21. $h(t) = -\frac{2}{3}(t^2 - 5t + 3)$

Degree: 2

Leading coefficient: $-\frac{2}{3}$

The degree is even and the leading coefficient is negative. The graph falls to the left and right.

23. $f(x) = 3x^3 - 9x + 1;\ g(x) = 3x^3$

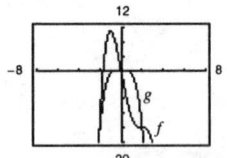

25. $f(x) = -(x^4 - 4x^3 + 16x);\ g(x) = -x^4$

27. $f(x) = x^2 - 25$

$0 = (x + 5)(x - 5)$

$x = \pm 5$

29. $h(t) = t^2 - 6t + 9$

$0 = (t - 3)^2$

$t = 3$

31. $f(x) = x^2 + x - 2$

$0 = (x + 2)(x - 1)$

$x = -2, 1$

33. $f(x) = 3x^2 - 12x + 3$

$0 = 3(x^2 - 4x + 1)$

$x = \dfrac{4 \pm \sqrt{16 - 4}}{2} = 2 \pm \sqrt{3}$

35. $f(t) = t^3 - 4t^2 + 4t$

$0 = t(t - 2)^2$

$t = 0, 2$

37. $g(t) = \frac{1}{2}t^4 - \frac{1}{2}$

$0 = \frac{1}{2}(t + 1)(t - 1)(t^2 + 1)$

$t = \pm 1$

39. $f(x) = 2x^4 - 2x^2 - 40$

$0 = 2(x^2 + 4)(x + \sqrt{5})(x - \sqrt{5})$

$x = \pm\sqrt{5}$

41. $f(x) = 5x^4 + 15x^2 + 10$

$0 = 5(x^4 + 3x^2 + 2)$

$0 = 5(x^2 + 2)(x^2 + 1)$

No real zeros

43. $y = 4x^3 - 20x^2 + 25x$ $0 = 4x^3 - 20x^2 + 25x$

$$0 = x(2x - 5)^2$$

$$x = 0 \text{ or } x = \tfrac{5}{2}$$

x-intercepts: $(0, 0), \left(\tfrac{5}{2}, 0\right)$

45. $y = x^5 - 5x^3 + 4x$ $0 = x^5 - 5x^3 + 4x$

$$0 = x(x^2 - 1)(x^2 - 4)$$

$$0 = x(x + 1)(x - 1)(x + 2)(x - 2)$$

$$x = 0, \ \pm 1, \ \pm 2$$

x-intercepts: $(0, 0), (\pm 1, 0), (\pm 2, 0)$

47. $f(x) = (x - 0)(x - 10)$

$f(x) = x^2 - 10x$

Note: $f(x) = a(x - 0)(x - 10) = ax(x - 10)$ has zeros 0 and 10 for all real numbers $a \neq 0$.

49. $f(x) = (x - 2)(x - (-6))$

$\qquad = (x - 2)(x + 6)$

$\qquad = x^2 + 4x - 12$

Note: $f(x) = a(x - 2)(x + 6)$ has zeros 2 and -6 for all real numbers $a \neq 0$.

51. $f(x) = (x - 0)(x - (-2))(x - (-3))$

$\qquad = x(x + 2)(x + 3)$

$\qquad = x^3 + 5x^2 + 6x$

Note: $f(x) = ax(x + 2)(x + 3)$ has zeros 0, -2, -3 for all real numbers $a \neq 0$.

53. $f(x) = (x - 4)(x + 3)(x - 3)(x - 0)$

$\qquad = (x - 4)(x^2 - 9)x$

$\qquad = x^4 - 4x^3 - 9x^2 + 36x$

Note: $f(x) = a(x^4 - 4x^3 - 9x^2 + 36x)$ has these zeros for all real numbers $a \neq 0$.

55. $f(x) = \left[x - \left(1 + \sqrt{3}\right)\right]\left[x - \left(1 - \sqrt{3}\right)\right]$

$\qquad = \left[(x - 1) - \sqrt{3}\right]\left[(x - 1) + \sqrt{3}\right]$

$\qquad = (x - 1)^2 - \left(\sqrt{3}\right)^2$

$\qquad = x^2 - 2x + 1 - 3$

$\qquad = x^2 - 2x - 2$

Note: $f(x) = a(x^2 - 2x - 2)$ has these zeros for all real numbers $a \neq 0$.

57. $f(x) = x^3 - 3x^2 + 3$

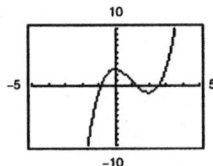

The functions has three zeros. They are in the intervals $(-1, 0)$, $(1, 2)$ and $(2, 3)$.

59. $g(x) = 3x^4 + 4x^3 - 3$

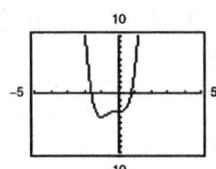

The function has two zeros. They are in the intervals $(-2, -1)$ and $(0, 1)$.

61. $f(x) = -\frac{3}{2}$

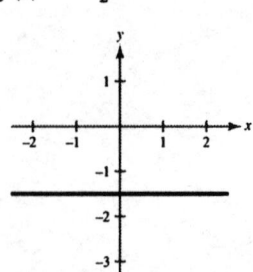

Horizontal line

63. $f(t) = \frac{1}{4}(t^2 - 2t + 15)$
$= \frac{1}{4}(t - 1)^2 + \frac{7}{2}$

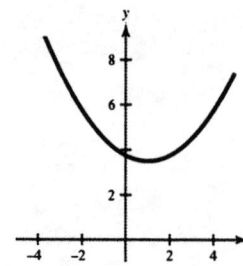

Parabola; opens upward
Vertex: $\left(1, \frac{7}{2}\right)$

65. $f(x) = x^3 - 3x^2 = x^2(x - 3)$

Zeros: 0 and 3

Right: Moves up

Left: Moves down

x	0	1	2	3	-1
$f(x)$	0	-2	-4	0	-4

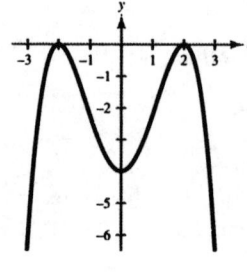

67. $g(t) = -\frac{1}{4}(t - 2)^2(t + 2)^2$

Zeros: 2 and -2

Right: Moves down

Left: Moves down

t	-3	-2	-1	0	1	2	3
$g(t)$	$-\frac{25}{4}$	0	$-\frac{9}{4}$	-4	$-\frac{9}{4}$	0	$-\frac{25}{4}$

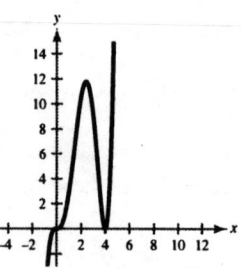

69. $h(x) = \frac{1}{3}x^3(x - 4)^2$

Zeros: 0 and 4

Right: Moves up

Left: Moves down

x	-1	0	1	2	3	4	5
$h(x)$	$-\frac{25}{3}$	0	3	$\frac{32}{3}$	9	0	$\frac{125}{3}$

71. $f(x) = 1 - x^6$

x-intercepts: $(\pm 1, 0)$

y-intercept: $(0, 1)$

x	0	± 1	± 2
$f(x)$	1	0	-63

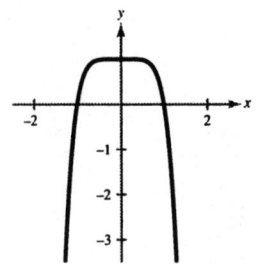

73. $f(x) = x^3 - 4x = x(x + 2)(x - 2)$

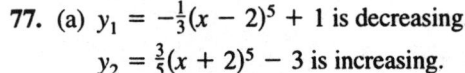

75. $g(x) = \frac{1}{5}(x + 1)^2(x - 3)(2x - 9)$

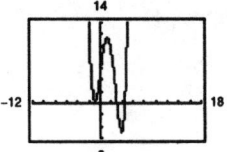

77. (a) $y_1 = -\frac{1}{3}(x - 2)^5 + 1$ is decreasing.

$y_2 = \frac{3}{5}(x + 2)^5 - 3$ is increasing.

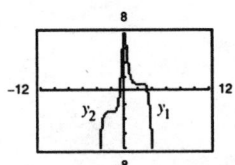

(b) If $a > 0$, $g(x)$ will always be increasing.

If $a < 0$, $g(x)$ will always be decreasing.

(c) $H(x) = x^5 - 3x^3 + 2x + 1$

Since $H(x)$ is not always increasing
or always decreasing, $H(x) \neq a(x - h)^5 + k$.

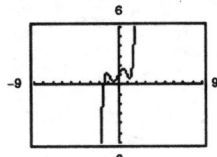

(d) $g(x) = a(x - h)^n + k$ is always increasing or
always decreasing if n is an odd natural number.

79. $f(x) = x^4$; $f(x)$ is even.

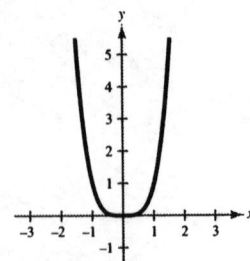

(a) $g(x) = f(x) + 2$

Vertical shift two units upward

$g(-x) = f(-x) + 2$

$= f(x) + 2$

$= g(x)$

Even

(b) $g(x) = f(x + 2)$

Horizontal shift two units to the left

Neither odd nor even

(c) $g(x) = f(-x) = (-x)^4 = x^4$

Reflection in the y-axis. The graph looks the same.

Even

(d) $g(x) = -f(x) = -x^4$

Reflection in the x-axis

Even

(e) $g(x) = f\left(\frac{1}{2}x\right) = \frac{1}{16}x^4$

Horizontal stretch

Even

(f) $g(x) = \frac{1}{2}f(x) = \frac{1}{2}x^4$

Vertical shrink

Even

(g) $g(x) = f(x^{3/4}) = (x^{3/4})^4 = x^3$

Odd

(h) $g(x) = (f \circ f)(x) = f(f(x)) = f(x^4) = (x^4)^4 = x^{16}$

Even

81. The point of diminishing returns (where the graph changes from curving upward to curving downward) occurs when $x = 200$. The point is $(200, 160)$ which corresponds to spending \$2,000,000 on advertising to obtain a revenue of \$160 million.

83. (a) $(f \circ g)(3) = f(g(3)) = f(1) = 0$

(b) The graph does not pass the horizontal line test.

(c) Since $g(3) = 1$, $g^{-1}(1) = 3$.

(d) $(g^{-1} \circ f)(0) = g^{-1}(f(0)) = g^{-1}(0) = 1$, since $g(1) = 0$.

85. $3x^2 - y = 4$

$y = 3x^2 - 4$

Yes, y is a function of x. Each x-value corresponds to only one y-value.

87. $g(x) = x^2 - 1$

$$\frac{g(x) - g(4)}{x - 4} = \frac{(x^2 - 1) - 15}{x - 4}$$

$$= \frac{x^2 - 16}{x - 4}$$

$$= \frac{(x + 4)(x - 4)}{x - 4}$$

$$= x + 4, \quad x \neq 4$$

Section 3.3   Polynomial and Synthetic Division

You should know the following basic techniques and principles of polynomial division.

■   The Division Algorithm (Long Division of Polynomials)

■   Synthetic Division

■ $f(k)$ is equal to the remainder of $f(x)$ divided by $(x - k)$.

■ $f(k) = 0$ if and only if $(x - k)$ is a factor of $f(x)$.

Solutions to Odd-Numbered Exercises

1. $y_2 = 4 + \dfrac{4}{x - 1}$

$= \dfrac{4(x - 1) + 4}{x - 1}$

$= \dfrac{4x - 4 + 4}{x - 1}$

$= \dfrac{4x}{x - 1}$

$= y_1$

3. $y_2 = x - 2 + \dfrac{4}{x + 2}$

$= \dfrac{(x - 2)(x + 2) + 4}{x + 2}$

$= \dfrac{x^2 - 4 + 4}{x + 2}$

$= \dfrac{x^2}{x + 2}$

$= y_1$

5. $y_2 = x^3 - 4x + \dfrac{4x}{x^2 + 1}$

$= \dfrac{(x^3 - 4x)(x^2 + 1) + 4x}{x^2 + 1}$

$= \dfrac{x^5 + x^3 - 4x^3 - 4x + 4x}{x^2 + 1}$

$= \dfrac{x^5 - 3x^3}{x^2 + 1}$

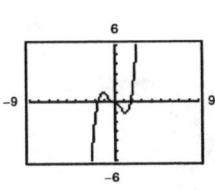

7.

$$
\begin{array}{r}
2x + 4 \\
x + 3 \overline{)\ 2x^2 + 10x + 12} \\
-(2x^2 + 6x) \\
\hline
4x + 12 \\
-(4x + 12) \\
\hline
0
\end{array}
$$

$\dfrac{2x^2 + 10x + 12}{x + 3} = 2x + 4$

9.

$$
\begin{array}{r}
x^2 - 3x + 1 \\
4x + 5 \overline{)\ 4x^3 - 7x^2 - 11x + 5} \\
-(4x^3 + 5x^2) \\
\hline
-12x^2 - 11x \\
-(-12x^2 - 15x) \\
\hline
4x + 5 \\
-(4x + 5) \\
\hline
0
\end{array}
$$

$\dfrac{4x^3 - 7x^2 - 11x + 5}{4x + 5} = x^2 - 3x + 1$

11.
$$
\begin{array}{r}
x^3 + 3x^2 \qquad\quad - 1 \\
x + 2 \overline{)\; x^4 + 5x^3 + 6x^2 \;-x - 2\;} \\
\underline{-\,(x^4 + 2x^3)} \\
3x^3 + 6x^2 \\
\underline{-\,(3x^3 + 6x^2)} \\
-x - 2 \\
\underline{-(-x - 2)} \\
0
\end{array}
$$

$$
\frac{x^4 + 5x^3 + 6x^2 - x - 2}{x + 2} = x^3 + 3x^2 - 1
$$

13.
$$
\begin{array}{r}
7 \\
x + 2 \overline{)\; 7x + 3\;} \\
\underline{-\,(7x + 14)} \\
-11
\end{array}
$$

$$
\frac{7x + 3}{x + 2} = 7 - \frac{11}{x + 2}
$$

15.
$$
\begin{array}{r}
3x + 5 \\
2x^2 + 0x + 1 \overline{)\; 6x^3 + 10x^2 + \; x + 8\;} \\
\underline{-\,(6x^3 + \; 0x^2 + 3x)} \\
10x^2 - 2x + 8 \\
\underline{-\,(10x^2 + 0x \; + 5)} \\
-2x + 3
\end{array}
$$

$$
\frac{6x^3 + 10x^2 + x + 8}{2x^2 + 1} = 3x + 5 - \frac{2x - 3}{2x^2 + 1}
$$

17.
$$
\begin{array}{r}
x^2 + 2x + \; 4 \\
x^2 - 2x + 3 \overline{)\; x^4 + 0x^3 + 3x^2 + 0x + 1\;} \\
\underline{-\,(x^4 - 2x^3 + 3x^2)} \\
2x^3 + 0x^2 + 0x \\
\underline{-\,(2x^3 - 4x^2 + 6x)} \\
4x^2 - 6x \; + \; 1 \\
\underline{-\,(4x^2 - 8x + 12)} \\
2x - 11
\end{array}
\quad\Longrightarrow\quad
$$

$$
\frac{x^4 + 3x^2 + 1}{x^2 - 2x + 3} = x^2 + 2x + 4 + \frac{2x - 11}{x^2 - 2x + 3}
$$

19.
$$
\begin{array}{r}
x + 3 \\
x^3 - 3x^2 + 3x - 1 \overline{)\; x^4 + 0x^3 + 0x^2 + 0x \; + 0\;} \\
\underline{-\,(x^4 - 3x^3 + 3x^2 - \; x)} \\
3x^3 - 3x^2 + \; x \; + 0 \\
\underline{-\,(3x^3 - 9x^2 + 9x \; - 3)} \\
6x^2 - 8x \; + 3
\end{array}
$$

$$
\frac{x^4}{(x - 1)^3} = x + 3 + \frac{6x^2 - 8x + 3}{(x - 1)^3}
$$

21.

$$
x^n + 3 \overline{\smash)\begin{array}{r} x^{2n} + 6x^n + 9 \\ x^{3n} + 9x^{2n} + 27x^n + 27 \end{array}}
$$

$$
\begin{array}{r}
- (x^{3n} + 3x^{2n}) \\
\hline
6x^{2n} + 27x^n \\
- (6x^{2n} + 18x^n) \\
\hline
9x^n + 27 \\
- (9x^n + 27) \\
\hline
0
\end{array}
$$

$$
\frac{x^{3n} + 9x^{2n} + 27x^n + 27}{x^n + 3} = x^{2n} + 6x^n + 9
$$

23.

$$
5 \;\big|\;\begin{array}{rrrr}
3 & -17 & 15 & -25 \\
& 15 & -10 & 25 \\
\hline
3 & -2 & 5 & 0
\end{array}
$$

$$
\frac{3x^3 - 17x^2 + 15x - 25}{x - 5} = 3x^2 - 2x + 5
$$

25.

$$
-2 \;\big|\;\begin{array}{rrrr}
4 & 8 & -9 & -18 \\
& -8 & 0 & 18 \\
\hline
4 & 0 & -9 & 0
\end{array}
$$

$$
\frac{4x^3 + 8x^2 - 9x - 18}{x + 2} = 4x^2 - 9
$$

27.

$$
-10 \;\big|\;\begin{array}{rrrr}
-1 & 0 & 75 & -250 \\
& 10 & -100 & 250 \\
\hline
-1 & 10 & -25 & 0
\end{array}
$$

$$
\frac{-x^3 + 75x - 250}{x + 10} = -x^2 + 10x - 25
$$

29.

$$
4 \;\big|\;\begin{array}{rrrr}
5 & -6 & 0 & 8 \\
& 20 & 56 & 224 \\
\hline
5 & 14 & 56 & 232
\end{array}
$$

$$
\frac{5x^3 - 6x^2 + 8}{x - 4} = 5x^2 + 14x + 56 + \frac{232}{x - 4}
$$

31.

$$
6 \;\big|\;\begin{array}{rrrrr}
10 & -50 & 0 & 0 & -800 \\
& 60 & 60 & 360 & 2160 \\
\hline
10 & 10 & 60 & 360 & 1360
\end{array}
$$

$$
\frac{10^4 - 50x^3 - 800}{x - 6} = 10x^3 + 10x^2 + 60x + 360 + \frac{1360}{x - 6}
$$

33.

$$
-8 \;\big|\;\begin{array}{rrrr}
1 & 0 & 0 & 512 \\
& -8 & 64 & -512 \\
\hline
1 & -8 & 64 & 0
\end{array}
$$

$$
\frac{x^3 + 512}{x + 8} = x^2 - 8x + 64
$$

35.

$$
2 \;\big|\;\begin{array}{rrrrr}
-3 & 0 & 0 & 0 & 0 \\
& -6 & -12 & -24 & -48 \\
\hline
-3 & -6 & -12 & -24 & -48
\end{array}
$$

$$
\frac{-3x^4}{x - 2} = -3x^3 - 6x^2 - 12x - 24 - \frac{48}{x - 2}
$$

37.

$$
6 \,\big|\!\begin{array}{rrrrr} -1 & 0 & 0 & 180 & 0 \\ & -6 & -36 & -216 & -216 \\ \hline -1 & -6 & -36 & -36 & -216 \end{array}
$$

$$\frac{180x - x^4}{x - 6} = -x^3 - 6x^2 - 36x - 36 - \frac{216}{x - 6}$$

39.

$$
-\tfrac{1}{2} \,\big|\!\begin{array}{rrrr} 4 & 16 & -23 & -15 \\ & -2 & -7 & 15 \\ \hline 4 & 14 & -30 & 0 \end{array}
$$

$$\frac{4x^3 + 16x^2 - 23x - 15}{x + \frac{1}{2}} = 4x^2 + 14x - 30$$

41. A divisor divided evenly into the dividend if the remainder is zero.

43.

$$
5 \,\big|\!\begin{array}{rrrr} 1 & 4 & -3 & c \\ & 5 & 45 & 210 \\ \hline 1 & 9 & 42 & c + 210 \end{array}
$$

For $c + 210$ to equal zero, c must equal -210.

45. $f(x) = x^3 - x^2 - 14x + 11, \ k = 4$

$$
4 \,\big|\!\begin{array}{rrrr} 1 & -1 & -14 & 11 \\ & 4 & 12 & -8 \\ \hline 1 & 3 & -2 & 3 \end{array}
$$

$f(x) = (x - 4)(x^2 + 3x - 2) + 3$

$f(4) = (0)(26) + 3 = 3$

47. $f(x) = x^3 + 3x^2 - 2x - 14, \ k = \sqrt{2}$

$$
\sqrt{2} \,\big|\!\begin{array}{rrrr} 1 & 3 & -2 & -14 \\ & \sqrt{2} & 2 + 3\sqrt{2} & 6 \\ \hline 1 & 3 + \sqrt{2} & 3\sqrt{2} & -8 \end{array}
$$

$f(x) = \left(x - \sqrt{2}\right)\left[x^2 + (3 + \sqrt{2})x + 3\sqrt{2}\right] - 8$

$f\!\left(\sqrt{2}\right) = (0)\left(4 + 6\sqrt{2}\right) - 8 = -8$

49. $f(x) = 4x^3 - 13x + 10$

(a)
$$
1 \,\big|\!\begin{array}{rrrr} 4 & 0 & -13 & 10 \\ & 4 & 4 & -9 \\ \hline 4 & 4 & -9 & \underline{1} = f(1) \end{array}
$$

(b)
$$
-2 \,\big|\!\begin{array}{rrrr} 4 & 0 & -13 & 10 \\ & -8 & 16 & -6 \\ \hline 4 & -8 & 3 & 4 \quad \underline{4} = f(-2) \end{array}
$$

(c)
$$
\tfrac{1}{2} \,\big|\!\begin{array}{rrrr} 4 & 0 & -13 & 10 \\ & 2 & 1 & -6 \\ \hline 4 & 2 & -12 & \underline{4} = f\!\left(\tfrac{1}{2}\right) \end{array}
$$

(d)
$$
8 \,\big|\!\begin{array}{rrrr} 4 & 0 & -13 & 10 \\ & 32 & 256 & 1944 \\ \hline 4 & 32 & 243 & \underline{1954} = f(8) \end{array}
$$

51. $h(x) = 3x^3 + 5x^2 - 10x + 1$

(a)
$$
3 \,\big|\!\begin{array}{rrrr} 3 & 5 & -10 & 1 \\ & 9 & 42 & 96 \\ \hline 3 & 14 & 32 & \underline{97} = h(3) \end{array}
$$

(b)
$$
\tfrac{1}{3} \,\big|\!\begin{array}{rrrr} 3 & 5 & -10 & 1 \\ & 1 & 2 & -\tfrac{8}{3} \\ \hline 3 & 6 & -8 & -\tfrac{5}{3} = h\!\left(\tfrac{1}{3}\right) \end{array}
$$

(c)
$$
-2 \,\big|\!\begin{array}{rrrr} 3 & 5 & -10 & 1 \\ & -6 & 2 & 16 \\ \hline 3 & -1 & -8 & \underline{17} = h(-2) \end{array}
$$

(d)
$$
-5 \,\big|\!\begin{array}{rrrr} 3 & 5 & -10 & 1 \\ & -15 & 50 & -200 \\ \hline 3 & -10 & 40 & \underline{-199} = h(-5) \end{array}
$$

53. $f(x) = (x + 3)^2(x - 3)(x + 1)^3$

The remainder when $k = -3$ is zero since $(x + 3)$ is a factor of $f(x)$.

55.

$$
\begin{array}{r|rrrr}
2 & 1 & 0 & -7 & 6 \\
  &   & 2 & 4 & -6 \\
\hline
  & 1 & 2 & -3 & 0
\end{array}
$$

$x^3 - 7x + 6 = (x - 2)(x^2 + 2x - 3)$

$\qquad\qquad\quad = (x - 2)(x + 3)(x - 1)$

Zeros: $2, -3, 1$

57.

$$
\begin{array}{r|rrrr}
\frac{1}{2} & 2 & -15 & 27 & -10 \\
  &   & 1 & -7 & 10 \\
\hline
  & 2 & -14 & 20 & 0
\end{array}
$$

$2x^3 - 15x^2 + 27x - 10$

$\qquad = \left(x - \frac{1}{2}\right)(2x^2 - 14x + 20)$

$\qquad = (2x - 1)(x - 2)(x - 5)$

Zeros: $\frac{1}{2}, 2, 5$

59.

$$
\begin{array}{r|rrrr}
\sqrt{3} & 1 & 2 & -3 & -6 \\
  &   & \sqrt{3} & 3 + 2\sqrt{3} & 6 \\
\hline
  & 1 & 2 + \sqrt{3} & 2\sqrt{3} & 0
\end{array}
$$

$$
\begin{array}{r|rrr}
-\sqrt{3} & 1 & 2 + \sqrt{3} & 2\sqrt{3} \\
  &   & -\sqrt{3} & -2\sqrt{3} \\
\hline
  & 1 & 2 & 0
\end{array}
$$

$x^3 + 2x^2 - 3x - 6 = \left(x - \sqrt{3}\right)\left(x + \sqrt{3}\right)(x + 2)$

Zeros: $\pm\sqrt{3}, -2$

61.

$$
\begin{array}{r|rrrr}
1 + \sqrt{3} & 1 & -3 & 0 & 2 \\
  &   & 1 + \sqrt{3} & 1 - \sqrt{3} & -2 \\
\hline
  & 1 & -2 + \sqrt{3} & 1 - \sqrt{3} & 0
\end{array}
$$

$$
\begin{array}{r|rrr}
1 - \sqrt{3} & 1 & -2 + \sqrt{3} & 1 - \sqrt{3} \\
  &   & 1 - \sqrt{3} & -1 + \sqrt{3} \\
\hline
  & 1 & -1 & 0
\end{array}
$$

$x^3 - 3x^2 + 2 = \left[x - \left(1 + \sqrt{3}\right)\right]\left[x - \left(1 - \sqrt{3}\right)\right](x - 1)$

$\qquad\qquad = (x - 1)\left(x - 1 - \sqrt{3}\right)\left(x - 1 + \sqrt{3}\right)$

Zeros: $1, 1 \pm \sqrt{3}$

63. $f(x) = x^3 - 2x^2 - 5x + 10$

(a) The zeros of f are 2 and $\approx \pm 2.236$.

(b)

$$
\begin{array}{r|rrrr}
2 & 1 & -2 & -5 & 10 \\
  &   & 2 & 0 & -10 \\
\hline
  & 1 & 0 & -5 & 0
\end{array}
$$

$f(x) = (x - 2)(x^2 - 5)$

$\quad = (x - 2)(x - \sqrt{5})(x + \sqrt{5})$

65. $h(t) = t^3 - 2t^2 - 7t + 2$

(a) The zeros of h are -2, ≈ 3.732, ≈ 0.268.

(b)

$$
\begin{array}{r|rrrr}
-2 & 1 & -2 & -7 & 2 \\
   &   & -2 & 8 & -2 \\
\hline
   & 1 & -4 & 1 & 0
\end{array}
$$

$h(t) = (t + 2)(t^2 - 4t + 1)$

By the Quadratic Formula, the zeros of

$$t^2 - 4t + 1 \text{ are } 2 \pm \sqrt{3}.$$

$$\text{Thus, } h(t) = (t + 2)\left[t - (2 + \sqrt{3})\right]\left[t - (2 - \sqrt{3})\right].$$

67.

$$
\begin{array}{r|rrrr}
\frac{3}{2} & 4 & -8 & 1 & 3 \\
  &   & 6 & -3 & -3 \\
\hline
  & 4 & -2 & -2 & 0
\end{array}
$$

$4x^3 - 8x^2 + x + 3 = \left(x - \frac{3}{2}\right)(4x^2 - 2x - 2)$

$\qquad\qquad\qquad\quad = \left(x - \frac{3}{2}\right)(2)(2x^2 - x - 1)$

$\qquad\qquad\qquad\quad = (2x - 3)(2x^2 - x - 1)$

Thus,

$$\frac{4x^3 - 8x^2 + x + 3}{2x - 3} = 2x^2 - x - 1 = (2x + 1)(x - 1).$$

69.

$$
\begin{array}{r|rrrr}
-1 & 1 & 3 & -1 & -3 \\
   &   & -1 & -2 & 3 \\
\hline
   & 1 & 2 & -3 & 0
\end{array}
$$

$\dfrac{x^3 + 3x^2 - x - 3}{x + 1} = x^2 + 2x - 3$

$\qquad\qquad\qquad\quad = (x + 3)(x - 1)$

71.

$$
\begin{array}{r|rrrrr}
-1 & 1 & 6 & 11 & 6 & 0 \\
   &   & -1 & -5 & -6 & 0 \\
\hline
   & 1 & 5 & 6 & 0 & 0
\end{array}
$$

$$
\begin{array}{r|rrrr}
-2 & 1 & 5 & 6 & 0 \\
   &   & -2 & -6 & 0 \\
\hline
   & 1 & 3 & 0 & 0
\end{array}
$$

$\dfrac{x^4 + 6x^3 + 11x^2 + 6x}{(x + 1)(x + 2)} = x^2 + 3x$

$\qquad\qquad\qquad\qquad = x(x + 3)$

73. Since $y = 110$ when $x = 5$, we have $x = 5$ as a zero to the equation $f(x) = y - 110$.

$$f(x) = y - 110, \quad 1 \le x \le 5$$

$$= -1.42x^3 + 5.04x^2 + 32.45x - 110.75$$

$$
\begin{array}{r|rrrr}
5 & -1.42 & 5.04 & 32.45 & -110.75 \\
  &   & -7.10 & -10.30 & 110.75 \\
\hline
  & -1.42 & -2.06 & 22.15 & 0
\end{array}
$$

Thus, $f(x) = (x - 5)(-1.42x^2 - 2.06x + 22.15)$. Using the Quadratic Formula to find the other two zeros yields

$$x = \frac{-(-2.06) \pm \sqrt{(-2.06)^2 - 4(-1.42)(22.15)}}{2(-1.42)}$$

$$= \frac{2.06 \pm \sqrt{130.0556}}{-2.84}$$

$$x \approx -4.74 \text{ or } x \approx 3.29.$$

Since $1 \le x \le 5$, we choose $x = 3.29$ which corresponds to 3290 rpm.

75. $f(x) = (x - k)q(x) + r$

(a) $k = 2$, $r = 5$, $q(x) = $ any quadratic $ax^2 + bx + c$ where $a > 0$.

One example: $f(x) = (x - 2)x^2 + 5 = x^3 - 2x^2 + 5$

(b) $k = -3$, $r = 1$, $q(x) = $ any quadratic $ax^2 + bx + c$ where $a < 0$.

One example: $f(x) = (x + 3)(-x^2) + 1 = -x^3 - 3x^2 + 1$

77. $f(x) = \sqrt[3]{x + 2}$

$y = \sqrt[3]{x + 2}$

$x = \sqrt[3]{y + 2}$

$x^3 = y + 2$

$x^3 - 2 = y$

$f^{-1}(x) = x^3 - 2$

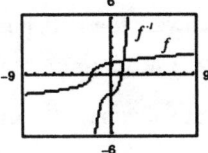

79. $f(x) = \sqrt{x}$, $g(x) = x^2 - 25$

$(f \circ g)(x) = f(g(x)) = f(x^2 - 25) = \sqrt{x^2 - 25}$

Domain: $x^2 - 25 \ge 0$

$(x + 5)(x - 5) \ge 0$

Critical numbers: $x = \pm 5$

Test intervals: $(-\infty, -5)(-5, 5)(5, \infty)$

Test: Is $x^2 - 25 \ge 0$?

Solution set: $[-\infty, -5] \cup [5, \infty)$ or $x \le -5, x \ge 5$

Section 3.4   Real Zeros of Polynomial Functions

■ You should know Descartes's Rule of Signs.

(a) The number of positive real zeros of f is either equal to the number of variations of sign of f or is less than that number by an even integer.

(b) The number of negative real zeros of f is either equal to the number of variations in sign of $f(-x)$ or is less than that number by an even integer.

(c) When there is only one variation in sign, there is exactly one positive (or negative) real zero.

■ You should know the Rational Zero Test.

■ You should know shortcuts for the Rational Zero Test.

(a) Use a graphing or programmable calculator.

(b) Sketch a graph.

(c) After finding a root, use synthetic division to reduce the degree of the polynomial.

■ You should be able to observe the last row obtained from synthetic division in order to determine upper or lower bounds.

(a) If the test value is positive and all of the entries in the last row are positive or zero, then the test value is an upper bound.

(b) If the test value is negative and the entries in the last row alternate from positive to negative, then the test value is a lower bound. (Zero entries count as positive or negative.)

Solutions to Odd-Numbered Exercises

1. $f(x) = x^3 + 3$

Sign variations: 0, positive zeros: 0

$f(-x) = -x^3 + 3$

Sign variations: 1, negative zeros: 1

3. $g(x) = 5x^5 + 10x = 5x(x^4 + 2)$

Let $f(x) = x^4 + 2$.

Sign variations: 0, positive zeros: 0

$f(-x) = x^4 + 2$

Sign variations: 0, negative zeros: 0

5. $h(x) = 3x^4 + 2x^2 + 1$

Sign variations: 0, positive zeros: 0

$h(-x) = 3x^4 + 2x^2 + 1$

Sign variations: 0, negative zeros: 0

7. $g(x) = 2x^3 - 3x^2 - 3$

Sign variations: 1, positive zeros: 1

$g(-x) = -2x^3 - 3x^2 - 3$

Sign variations: 0, negative zeros: 0

9. $f(x) = -5x^3 + x^2 - x + 5$

Sign variations: 3, positive zeros: 3 or 1

$f(-x) = 5x^3 + x^2 + x + 5$

Sign variations: 0, negative zeros: 0

11. $f(x) = x^3 + 3x^2 - x - 3$

Possible rational zeros: ± 1, ± 3

Zeros shown on graph: -3, -1, 1

13. $f(x) = 2x^4 - 17x^3 + 35x^2 + 9x - 45$

Possible rational zeros: ± 1, ± 3, ± 5, ± 9, ± 15, ± 45, $\pm \frac{1}{2}$, $\pm \frac{3}{2}$, $\pm \frac{5}{2}$, $\pm \frac{9}{2}$, $\pm \frac{15}{2}$, $\pm \frac{45}{2}$

Zeros shown on graph: -1, $\frac{3}{2}$, 3, 5

15. $f(x) = 20x^4 + 144x^3 - 253x^2 - 900x + 800$

Possible rational zeros: $\pm 1,\ \pm 2,\ \pm 4,\ \pm 5,\ \pm 8,\ \pm 10,\ \pm 16,\ \pm 20,\ \pm 25,$

$\pm 32,\ \pm 40,\ \pm 50,\ \pm 80,\ \pm 100,\ \pm 160,\ \pm 200,\ \pm 400,$

$\pm 800,\ \pm\frac{1}{2},\ \pm\frac{5}{2},\ \pm\frac{25}{2},\ \pm\frac{1}{4},\ \pm\frac{5}{4},\ \pm\frac{25}{4},\ \pm\frac{1}{5},\ \pm\frac{2}{5},\ \pm\frac{4}{5},$

$\pm\frac{8}{5},\ \pm\frac{16}{5},\ \pm\frac{32}{5},\ \pm\frac{1}{10},\ \pm\frac{1}{20}$

Zeros shown on graph: $-8,\ -\frac{5}{2},\ \frac{4}{5},\ \frac{5}{2}$

17. $f(x) = x^3 - 6x^2 + 11x - 6$

Possible rational zeros: $\pm 1,\ \pm 2,\ \pm 3$

$$
\begin{array}{r|rrrr}
1 & 1 & -6 & 11 & -6 \\
  &   & 1 & -5 & 6 \\
\hline
  & 1 & -5 & 6 & 0
\end{array}
$$

$x^3 - 6x^2 + 11x - 6 = (x - 1)(x^2 - 5x + 6) = (x - 1)(x - 2)(x - 3).$

Thus, the real zeros are 1, 2, and 3.

19. $g(x) = x^3 - 4x^2 - x + 4 = x^2(x - 4) - 1(x - 4) = (x - 4)(x^2 - 1)$

$\qquad = (x - 4)(x - 1)(x + 1)$

Thus, the zeros of $g(x)$ are 4 and ± 1.

21. $h(t) = t^3 + 12t^2 + 21t + 10$

Possible rational zeros: $\pm 1,\ \pm 2,\ \pm 5,\ \pm 10$

$$
\begin{array}{r|rrrr}
-1 & 1 & 12 & 21 & 10 \\
   &   & -1 & -11 & -10 \\
\hline
   & 1 & 11 & 10 & 0
\end{array}
$$

$t^3 + 12t^2 + 21t + 10 = (t + 1)(t^2 + 11t + 10)$

$\qquad\qquad\qquad\qquad = (t + 1)(t + 1)(t + 10)$

$\qquad\qquad\qquad\qquad = (t + 1)^2(t + 10)$

Thus, the zeros are -1 and -10.

23. $f(x) = x^3 - 4x^2 + 5x - 2$

Possible rational zeros: $\pm 1,\ \pm 2$

$$
\begin{array}{r|rrrr}
1 & 1 & -4 & 5 & -2 \\
  &   & 1 & -3 & 2 \\
\hline
  & 1 & -3 & 2 & 0
\end{array}
$$

$x^3 - 4x^2 + 5x - 2 = (x - 1)(x^2 - 3x + 2)$

$\qquad\qquad\qquad\quad = (x - 1)(x - 1)(x - 2)$

$\qquad\qquad\qquad\quad = (x - 1)^2(x - 2)$

Thus, the zeros are 1 and 2.

25. $C(x) = 2x^3 + 3x^2 - 1$

Possible rational zeros: $\pm 1,\ \pm\frac{1}{2}$

$$
\begin{array}{r|rrrr}
-1 & 2 & 3 & 0 & -1 \\
   &   & -2 & -1 & 1 \\
\hline
   & 2 & 1 & -1 & 0
\end{array}
$$

$2x^3 + 3x^2 - 1 = (x + 1)(2x^2 + x - 1)$

$\qquad\qquad\qquad = (x + 1)(x + 1)(2x - 1)$

$\qquad\qquad\qquad = (x + 1)^2(2x - 1)$

Thus, the zeros are -1 and $\frac{1}{2}$.

27. $f(x) = 9x^4 - 9x^3 - 58x^2 + 4x + 24$

Possible rational zeros: $\pm 1, \pm 2, \pm 3, \pm 4, \pm 6, \pm 8, \pm 12, \pm 24, \pm\frac{1}{3}, \pm\frac{2}{3},$
$\pm\frac{4}{3}, \pm\frac{8}{3}, \pm\frac{1}{9}, \pm\frac{2}{9}, \pm\frac{4}{9}$

$$
\begin{array}{r|rrrrr}
-2 & 9 & -9 & -58 & 4 & 24 \\
   &   & -18 & 54 & 8 & -24 \\
\hline
   & 9 & -27 & -4 & 12 & 0
\end{array}
$$

$$
\begin{array}{r|rrrr}
3 & 9 & -27 & -4 & 12 \\
  &   & 27 & 0 & -12 \\
\hline
  & 9 & 0 & -4 & 0
\end{array}
$$

$9x^4 - 9x^3 - 58x^2 + 4x - 24 = (x + 2)(x - 3)(9x^2 - 4)$

$\qquad\qquad\qquad\qquad\quad = (x + 2)(x - 3)(3x - 2)(3x + 2)$

Thus, the zeros are -2, 3, and $\pm\frac{2}{3}$.

29. $z^4 - z^3 - 2z - 4 = 0$

Possible rational zeros: $\pm 1, \pm 2, \pm 4$

$$
\begin{array}{r|rrrrr}
-1 & 1 & -1 & 0 & -2 & -4 \\
   &   & -1 & 2 & -2 & 4 \\
\hline
   & 1 & -2 & 2 & -4 & 0
\end{array}
$$

$$
\begin{array}{r|rrrr}
2 & 1 & -2 & 2 & -4 \\
  &   & 2 & 0 & 4 \\
\hline
  & 1 & 0 & 2 & 0
\end{array}
$$

$z^4 - z^3 - 2z - 4 = (x + 1)(x - 2)(x^2 + 2)$

The only real zeros are -1 and 2.

31. $2y^4 + 7y^3 - 26y^2 + 23y - 6 = 0$

Possible rational zeros: ± 1, ± 2, ± 3, ± 6, $\pm\frac{1}{2}$, $\pm\frac{3}{2}$

$$
\begin{array}{r|rrrrr}
1 & 2 & 7 & -26 & 23 & -6 \\
  &   & 2 & 9 & -17 & 6 \\
\hline
  & 2 & 9 & -17 & 6 & 0
\end{array}
$$

$$
\begin{array}{r|rrrr}
-6 & 2 & 9 & -17 & 6 \\
   &   & -12 & 18 & -6 \\
\hline
   & 2 & -3 & 1 & 0
\end{array}
$$

$$
\begin{aligned}
2y^4 + 7y^3 - 26y^2 + 23y - 6 &= (y - 1)(y + 6)(2y^2 - 3y + 1) \\
&= (y - 1)(y + 6)(2y - 1)(y - 1) \\
&= (y - 1)^2(y + 6)(2y - 1)
\end{aligned}
$$

The only real zeros are 1, -6, and $\frac{1}{2}$.

33. $f(x) = x^3 + x^2 - 4x - 4$

(a) Possible rational zeros: ± 1, ± 2, ± 4

(b)

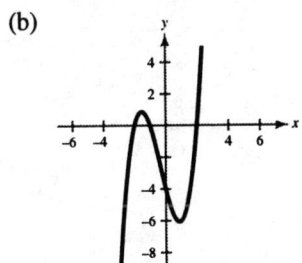

(c) $-2, -1, 2$ on graph

35. $f(x) = -4x^3 + 15x^2 - 8x - 3$

(a) Possible rational zeros: $\pm\frac{1}{4}$, $\pm\frac{1}{2}$, $\pm\frac{3}{4}$, ± 1, $\pm\frac{3}{2}$, ± 3

(b)

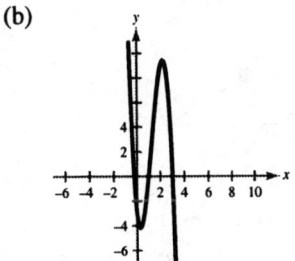

(c) $-\frac{1}{4}$, 1, 3 on graph

37. $f(x) = -2x^4 + 13x^3 - 21x^2 + 2x + 8$

(a) Possible rational zeros: $\pm\frac{1}{2}$, ± 1, ± 2, ± 4, ± 8

(b)

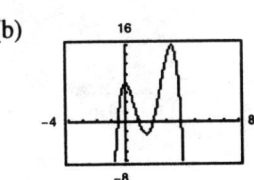

(c) $-\frac{1}{2}$, 1, 2, 4 on graph

39. $f(x) = 32x^3 - 52x^2 + 17x + 3$

(a) Possible rational zeros: ± 1, ± 3, $\pm\frac{1}{2}$, $\pm\frac{3}{2}$, $\pm\frac{1}{4}$, $\pm\frac{3}{4}$, $\pm\frac{1}{8}$, $\pm\frac{3}{8}$,

$$\pm\tfrac{1}{16}, \ \pm\tfrac{3}{16}, \ \pm\tfrac{1}{32}, \ \pm\tfrac{3}{32}$$

(b)

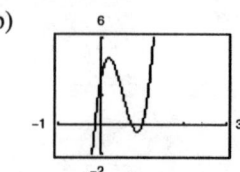

(c) $-\frac{1}{8}$, $\frac{3}{4}$, 1 on graph

41. $f(x) = x^4 - 3x^2 + 2$

(a) From the calculator we have

$$x = \pm 1 \text{ and } x \approx \pm 1.414.$$

(b)

$$
\begin{array}{r|rrrrr}
1 & 1 & 0 & -3 & 0 & 2 \\
  &   & 1 & 1 & -2 & -2 \\
\hline
  & 1 & 1 & -2 & -2 & 0 \\
\end{array}
$$

$$
\begin{array}{r|rrrr}
-1 & 1 & 1 & -2 & -2 \\
   &   & -1 & 0 & 2 \\
\hline
   & 1 & 0 & -2 & 0 \\
\end{array}
$$

$$f(x) = (x - 1)(x + 1)(x^2 - 2)$$

$$= (x - 1)(x + 1)\left(x - \sqrt{2}\right)\left(x + \sqrt{2}\right)$$

The exact roots are $x = \pm 1$, $\pm\sqrt{2}$.

43. $h(x) = x^5 - 7x^4 + 10x^3 + 14x^2 - 24x$

(a) $h(x) = x(x^4 - 7x^3 + 10x^2 + 14x - 24)$

From the calculator we have

$$x = 0, \ 3, \ 4 \text{ and } x \approx \pm 1.414.$$

(b)

$$
\begin{array}{r|rrrrr}
3 & 1 & -7 & 10 & 14 & -24 \\
  &   & 3 & -12 & -6 & 24 \\
\hline
  & 1 & -4 & -2 & 8 & 0 \\
\end{array}
$$

$$
\begin{array}{r|rrrr}
4 & 1 & -4 & -2 & 8 \\
  &   & 4 & 0 & -8 \\
\hline
  & 1 & 0 & -2 & 0 \\
\end{array}
$$

$$f(x) = x(x - 3)(x - 4)(x^2 - 2)$$

$$= x(x - 3)(x - 4)\left(x - \sqrt{2}\right)\left(x + \sqrt{2}\right)$$

The exact roots are $x = 0$, 3, 4, $\pm\sqrt{2}$.

45. $f(x) = x^4 - 4x^3 + 15$

(a)

$$
\begin{array}{r|rrrrr}
4 & 1 & -4 & 0 & 0 & 15 \\
  &   & 4 & 0 & 0 & 0 \\
\hline
  & 1 & 0 & 0 & 0 & 15 \\
\end{array}
$$

4 is an upper bound.

(b)

$$
\begin{array}{r|rrrrr}
-1 & 1 & -4 & 0 & 0 & 15 \\
   &   & -1 & 5 & -5 & 5 \\
\hline
   & 1 & -5 & 5 & -5 & 20 \\
\end{array}
$$

-1 is a lower bound.

47. $f(x) = x^4 - 4x^3 + 16x - 16$

(a)

$$
\begin{array}{r|rrrrr}
5 & 1 & -4 & 0 & 16 & -16 \\
  &   & 5 & 5 & 25 & 205 \\
\hline
  & 1 & 1 & 5 & 41 & 189 \\
\end{array}
$$

5 is an upper bound.

(b)

$$
\begin{array}{r|rrrrr}
-3 & 1 & -4 & 0 & 16 & -16 \\
   &   & -3 & 21 & -63 & 141 \\
\hline
   & 1 & -7 & 21 & -47 & 125 \\
\end{array}
$$

-3 is a lower bound.

49. $f(x) = 4x^3 - 3x - 1$

Possible rational zeros: $\pm 1,\ \pm\frac{1}{2},\ \pm\frac{1}{4}$

$$\begin{array}{r|rrrr}
1 & 4 & 0 & -3 & -1 \\
 & & 4 & 4 & 1 \\
\hline
 & 4 & 4 & 1 & 0
\end{array}$$

$4x^3 - 3x - 1 = (x - 1)(4x^2 + 4x + 1) = (x - 1)(2x + 1)^2$

Thus, the zeros are 1 and $-\frac{1}{2}$.

51. $f(y) = 4y^3 + 3y^2 + 8y + 6$

Possible rational zeros: $\pm 1,\ \pm 2,\ \pm 3,\ \pm 6,\ \pm\frac{1}{2},\ \pm\frac{3}{2},\ \pm\frac{1}{4},\ \pm\frac{3}{4}$

$$\begin{array}{r|rrrr}
-\frac{3}{4} & 4 & 3 & 8 & 6 \\
 & & -3 & 0 & -6 \\
\hline
 & 4 & 0 & 8 & 0
\end{array}$$

$4y^3 + 3y^2 + 8y + 6 = \left(y + \frac{3}{4}\right)(4y^2 + 8) = \left(y + \frac{3}{4}\right)4(y^2 + 2) = (4y + 3)(y^2 + 2)$

Thus, the only real zero is $-\frac{3}{4}$.

53. $P(x) = x^4 - \frac{25}{4}x^2 + 9$

$\quad = \frac{1}{4}(4x^4 - 25x^2 + 36)$

$\quad = \frac{1}{4}(4x^2 - 9)(x^2 - 4)$

$\quad = \frac{1}{4}(2x + 3)(2x - 3)(x + 2)(x - 2)$

The zeros are $\pm\frac{3}{2}$ and ± 2.

55. $f(x) = x^3 - \frac{1}{4}x^2 - x + \frac{1}{4}$

$\quad = \frac{1}{4}(4x^3 - x^2 - 4x + 1)$

$\quad = \frac{1}{4}[x^2(4x - 1) - 1(4x - 1)]$

$\quad = \frac{1}{4}(4x - 1)(x^2 - 1)$

$\quad = \frac{1}{4}(4x - 1)(x + 1)(x - 1)$

The zeros are $\frac{1}{4}$ and ± 1.

57. $f(x) = x^3 - 1 = (x - 1)(x^2 + x + 1)$

Rational zeros: $1\ (x = 1)$

Irrational zeros: 0

Matches (d).

59. $f(x) = x^3 - x = x(x + 1)(x - 1)$

Rational zeros: $3\ (x = 0, \pm 1)$

Irrational zeros: 0

Matches (b).

61. (a)

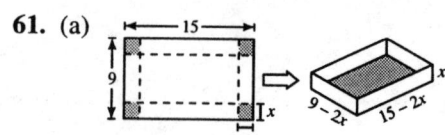

(b) $V = l \cdot w \cdot h = (15 - 2x)(9 - 2x)x$
$= x(9 - 2x)(15 - 2x)$

Since length, width, and height cannot be negative, we have $0 < x < \frac{9}{2}$ for the domain.

(c)

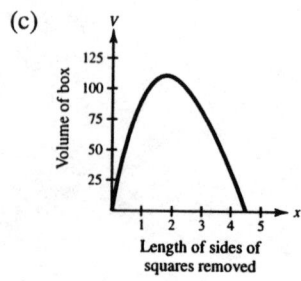

Length of sides of
squares removed

The volume is maximum when $x \approx 1.82$.

The dimensions are: length = $15 - 2(1.82) = 11.36$

width = $9 - 2(1.82) = \;\;\;5.36$

height = $x = 1.82$

1.82 cm × 5.36 cm × 11.36 cm

(d) $56 = x(9 - 2x)(15 - 2x)$

$56 = 135x - 48x^2 + 4x^3$

$0 = 4x^3 - 48x^2 + 135x - 56$

The zeros of this polynomial are $\frac{1}{2}, \frac{7}{2}$, and 8.
x cannot equal 8 since it is not in the domain of V.
[The length cannot equal -1 and the width cannot equal -7. The product of $(8)(-1)(-7) = 56$ so it showed up as an extraneous solution.]

63. $g(x) = -f(x)$. This function would have the same zeros as $f(x)$ so r_1, r_2, and r_3 are also zeros of $g(x)$.

65. $g(x) = f(x - 5)$. The graph of $g(x)$ is a horizontal shift of the graph of $f(x)$ five units to the right so the zeros of $g(x)$ are $5 + r_1$, $5 + r_2$, and $5 + r_3$.

67. $g(x) = 3 + f(x)$. Since $g(x)$ is a vertical shift of the graph of $f(x)$, the zeros of $g(x)$ cannot be determined.

69.
$$P = -76x^3 + 4830x^2 - 320,000, \; 0 \leq x \leq 60$$
$$2,500,000 = -76x^3 + 4830x^2 - 320,000$$
$$76x^3 - 4830x^2 + 2,820,000 = 0$$

The zeros of this equation are $x \approx 46.1$, $x \approx 38.4$, and $x \approx -21.0$. Since $0 \leq x \leq 60$, we disregard $x \approx -21.0$. The smaller remaining solution is $x \approx 38.4$.

71. $C = 100\left(\dfrac{200}{x^2} + \dfrac{x}{x + 30}\right), \; 1 \leq x$

C is minimum when $3x^3 - 40x^2 - 2400x - 36000 = 0$. The only real zero is $x \approx 40$.

73. $g(x) = f(x - 2)$

Horizontal shift two units
to the right

75. $g(x) = 2f(x)$

Vertical stretch

77. $g(x) = f(2x)$

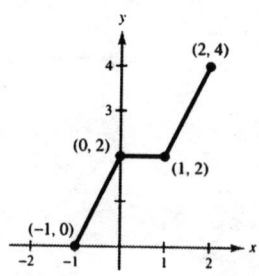

Horizontal shrink

Section 3.5   The Fundamental Theorem of Algebra

- You should know that if f is a polynomial of degree $n > 0$, then f has at least one zero in the complex number system.
- You should know that if $a + bi$ is a complex zero of a polynomial f, with real coefficients, then $a - bi$ is also a complex zero of f.
- You should know the difference between a factor that is irreducible over the rationals (such as $x^2 - 7$) and a factor that is irreducible over the reals (such as $x^2 + 9$).

Solutions to Odd-Numbered Exercises

1. $f(x) = x(x - 6)^2 = x(x - 6)(x - 6)$

The three zeros are: $x = 0$, $x = 6$, and $x = 6$.

3. $h(t) = (t - 3)(t - 2)(t - 3i)(t + 3i)$

The four zeros are:
$t = 3$, $t = 2$, $t = 3i$, and $t = -3i$.

5. $f(x) = x^3 - 4x^2 + x - 4 = x^2(x - 4) + 1(x - 4)$

$\quad = (x - 4)(x^2 + 1)$

The only real zero of $f(x)$ is $x = 4$. This corresponds to the x-intercept of $(4, 0)$ on the graph.

7. $f(x) = x^4 + 4x^2 + 4 = (x^2 + 2)^2$

$f(x)$ has no real zeros and the graph of $f(x)$ has no x-intercepts.

9. $f(x) = x^2 + 25$

$\quad = (x + 5i)(x - 5i)$

The zeros of $f(x)$ are $x = \pm 5i$.

11. $h(x) = x^2 - 4x + 1$

h has no rational zeros.

By the Quadratic Formula, the zeros are $x = \dfrac{4 \pm \sqrt{16 - 4}}{2} = 2 \pm \sqrt{3}$.

$h(x) = \left[x - (2 + \sqrt{3})\right]\left[x - (2 - \sqrt{3})\right] = (x - 2 - \sqrt{3})(x - 2 + \sqrt{3})$

13. $f(x) = x^4 - 81$

$\quad = (x^2 - 9)(x^2 + 9)$

$\quad = (x + 3)(x - 3)(x + 3i)(x - 3i)$

The zeros of $f(x)$ are $x = \pm 3$ and $x = \pm 3i$.

15. $f(z) = z^2 - 2z + 2$

f has no rational zeros.

By the Quadratic Formula, the zeros are $z = \dfrac{2 \pm \sqrt{4 - 8}}{2} = 1 \pm i$.

$f(z) = [z - (1 + i)][z - (1 - i)] = (z - 1 - i)(z - 1 + i)$

17. $g(x) = x^3 - 6x^2 + 13x - 10$

Possible rational zeros: $\pm 1, \pm 2, \pm 5, \pm 10$

$$
\begin{array}{r|rrrr}
2 & 1 & -6 & 13 & -10 \\
  &   & 2 & -8 & 10 \\
\hline
  & 1 & -4 & 5 & 0
\end{array}
$$

By the Quadratic Formula, the zeros of

$x^2 - 4x + 5$ are $x = \dfrac{4 \pm \sqrt{16 - 20}}{2} = 2 \pm i.$

The zeros of $g(x)$ are $x = 2$ and $x = 2 \pm i.$

$g(x) = (x - 2)[x - (2 + i)][x - (2 - i)]$

$\quad = (x - 2)(x - 2 - i)(x - 2 + i)$

19. $f(t) = t^3 - 3t^2 - 15t + 125$

Possible rational zeros: $\pm 1, \pm 5, \pm 25, \pm 125$

$$
\begin{array}{r|rrrr}
-5 & 1 & -3 & -15 & 125 \\
   &   & -5 & 40 & -125 \\
\hline
   & 1 & -8 & 25 & 0
\end{array}
$$

By the Quadratic Formula, the zeros of

$t^2 - 8t + 25$ are $t = \dfrac{8 \pm \sqrt{64 - 100}}{2} = 4 \pm 3i.$

The zeros of $f(t)$ are $t = -5$ and $t = 4 \pm 3i.$

$f(t) = [t - (-5)][t - (4 + 3i)][t - (4 - 3i)]$

$\quad = (t + 5)(t - 4 - 3i)(t - 4 + 3i)$

21. $h(x) = x^3 - x + 6$

Possible rational zeros: $\pm 1, \pm 2, \pm 3, \pm 6$

$$
\begin{array}{r|rrrr}
-2 & 1 & 0 & -1 & 6 \\
   &   & -2 & 4 & -6 \\
\hline
   & 1 & -2 & 3 & 0
\end{array}
$$

By the Quadratic Formula, the zeros of $x^2 - 2x + 3$ are

$x = \dfrac{2 \pm \sqrt{4 - 12}}{2} = 1 \pm \sqrt{2}\, i.$

The zeros of $h(x)$ are $x = -2$ and $x = 1 \pm \sqrt{2}\, i.$

$h(x) = [x - (-2)]\left[x - \left(1 + \sqrt{2}\, i\right)\right]\left[x - \left(1 - \sqrt{2}\, i\right)\right]$

$\quad = (x + 2)\left(x - 1 - \sqrt{2}\, i\right)\left(x - 1 + \sqrt{2}\, i\right)$

23. $f(x) = 5x^3 - 9x^2 + 28x + 6$

Possible rational zeros: $\pm 1, \pm 2, \pm 3, \pm 6, \pm\frac{1}{5}, \pm\frac{2}{5}, \pm\frac{3}{5}, \pm\frac{6}{5}$

$$
\begin{array}{r|rrrr}
-\frac{1}{5} & 5 & -9 & 28 & 6 \\
             &   & -1 & 2 & -6 \\
\hline
             & 5 & -10 & 30 & 0
\end{array}
$$

By the Quadratic Formula, the zeros of $5x^2 - 10x + 30 = 5(x^2 - 2x + 6)$ are

$x = \dfrac{2 \pm \sqrt{4 - 24}}{2} = 1 \pm \sqrt{5}\, i.$

The zeros of $f(x)$ are $x = -\frac{1}{5}$ and $x = 1 \pm \sqrt{5}\, i.$

$f(x) = \left[x - \left(-\frac{1}{5}\right)\right](5)\left[x - \left(1 + \sqrt{5}\, i\right)\right]\left[x - \left(1 - \sqrt{5}\, i\right)\right]$

$\quad = (5x + 1)\left(x - 1 - \sqrt{5}\, i\right)\left(x - 1 + \sqrt{5}\, i\right)$

25. $g(x) = x^4 - 4x^3 + 8x^2 - 16x + 16$

Possible rational zeros: $\pm 1, \pm 2, \pm 4, \pm 8, \pm 16$

$$
\begin{array}{r|rrrrr}
2 & 1 & -4 & 8 & -16 & 16 \\
  &   & 2 & -4 & 8 & -16 \\
\hline
2 & 1 & -2 & 4 & -8 \\
  &   & 2 & 0 & 8 \\
\hline
  & 1 & 0 & 4 & 0
\end{array}
$$

$g(x) = (x - 2)(x - 2)(x^2 + 4) = (x - 2)^2(x + 2i)(x - 2i)$

The zeros of $g(x)$ are 2 and $\pm 2i$.

27. $f(x) = x^4 + 10x^2 + 9$

$\quad = (x^2 + 1)(x^2 + 9)$

$\quad = (x + i)(x - i)(x + 3i)(x - 3i)$

The zeros of $f(x)$ are $x = \pm i$ and $x = \pm 3i$.

29. $f(x) = x^3 + 24x^2 + 214x + 740$

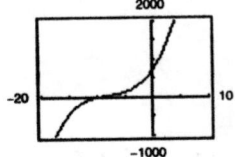

The graph reveals one zero at $x = -10$.

$$
\begin{array}{r|rrrr}
-10 & 1 & 24 & 214 & 740 \\
    &   & -10 & -140 & -740 \\
\hline
    & 1 & 14 & 74 & 0
\end{array}
$$

By the Quadratic Formula, the zeros of $x^2 + 14x + 74$ are

$$x = \frac{-14 \pm \sqrt{196 - 296}}{2} = -7 \pm 5i.$$

The zeros of $f(x)$ are $x = -10$ and $x = -7 \pm 5i$.

31. $f(x) = 16x^3 - 20x^2 - 4x + 15$

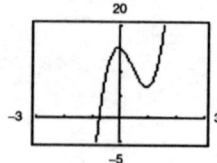

The graph reveals one zero at $x = -\frac{3}{4}$.

$$
\begin{array}{r|rrrr}
-\frac{3}{4} & 16 & -20 & -4 & 15 \\
 & & -12 & 24 & -15 \\
\hline
 & 16 & -32 & 20 & 0
\end{array}
$$

By the Quadratic Formula, the zeros of $16x^2 - 32x + 20 = 4(4x^2 - 8x + 5)$ are

$$x = \frac{8 \pm \sqrt{64 - 80}}{8} = 1 \pm \frac{1}{2}i.$$

The zeros of $f(x)$ are $x = -\frac{3}{4}$ and $x = 1 \pm \frac{1}{2}i$.

33. $f(x) = 2x^4 + 5x^3 + 4x^2 + 5x + 2$

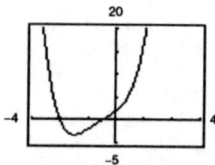

The graph reveals one zero at $x = -2$ and $x = -\frac{1}{2}$.

$$
\begin{array}{r|rrrrr}
-2 & 2 & 5 & 4 & 5 & 2 \\
 & & -4 & -2 & -4 & -2 \\
\hline
 & 2 & 1 & 2 & 1 & 0
\end{array}
$$

$$
\begin{array}{r|rrrr}
-\frac{1}{2} & 2 & 1 & 2 & 1 \\
 & & -1 & 0 & -1 \\
\hline
 & 2 & 0 & 2 & 0
\end{array}
$$

The zeros of $2x^2 + 2 = 2(x^2 + 1)$ are $x = \pm i$.

The zeros of $f(x)$ are $x = -2$, $x = -\frac{1}{2}$, and $x = \pm i$.

35. $f(x) = (x - 1)(x - 5i)(x + 5i)$

$\qquad = (x - 1)(x^2 + 25)$

$\qquad = x^3 - x^2 + 25x - 25$

Note: $f(x) = a(x^3 - x^2 + 25x - 25)$, where a is any nonzero real number, has the zeros 1 and $\pm 5i$.

37. $f(x) = (x - 6)[x - (-5 + 2i)][x - (-5 - 2i)]$

$\qquad = (x - 6)[(x + 5) - 2i][(x + 5) + 2i]$

$\qquad = (x - 6)[(x + 5)^2 - (2i)^2]$

$\qquad = (x - 6)(x^2 + 10x + 25 + 4)$

$\qquad = (x - 6)(x^2 + 10x + 29)$

$\qquad = x^3 + 4x^2 - 31x - 174$

Note: $f(x) = a(x^3 + 4x^2 - 31x - 174)$, where a is any nonzero real number, has the zeros 6, and $-5 \pm 2i$.

39. $f(x) = (x - i)(x + i)(x - 6i)(x + 6i)$

$\quad = (x^2 + 1)(x^2 + 36)$

$\quad = x^4 + 37x^2 + 36$

Note: $f(x) = a(x^4 + 37x^2 + 36)$, where a is any nonzero real number, has the zeros $\pm i$ and $\pm 6i$.

41. If $3 + \sqrt{2}\,i$ is a zero, so is its conjugate, $3 - \sqrt{2}i$.

$f(x) = (3x - 2)(x + 1)\left[x - (3 + \sqrt{2}\,i)\right]\left[x - (3 - \sqrt{2}\,i)\right]$

$\quad = (3x - 2)(x + 1)\left[(x - 3) - \sqrt{2}\,i\right]\left[(x - 3) + \sqrt{2}\,i\right]$

$\quad = (3x^2 + x - 2)\left[(x - 3)^2 - (\sqrt{2}\,i)^2\right]$

$\quad = (3x^2 + x - 2)(x^2 - 6x + 9 + 2)$

$\quad = (3x^2 + x - 2)(x^2 - 6x + 11)$

$\quad = 3x^4 - 17x^3 + 25x^2 + 23x - 22$

Note: $f(x) = a(3x^4 - 17x^3 + 25x^2 + 23x - 22)$, where a is any nonzero real number, has the zeros $\frac{2}{3}$, -1, and $3 \pm \sqrt{2}\,i$.

43. If $-\frac{1}{2} + i$ is a zero, so is its conjugate, $-\frac{1}{2} - i$.

$f(x) = 4\left(x - \frac{3}{4}\right)(x + 2)\,2\left[x - \left(-\frac{1}{2} + i\right)\right]2\left[x - \left(-\frac{1}{2} - i\right)\right]$

$\quad = (4x - 3)(x + 2)[(2x + 1) - 2i][(2x + 1) + 2i]$

$\quad = (4x^2 + 5x - 6)[(2x + 1)^2 - (2i)^2]$

$\quad = (4x^2 + 5x - 6)(4x^2 + 4x + 1 + 4)$

$\quad = (4x^2 + 5x - 6)(4x^2 + 4x + 5)$

$\quad = 16x^4 + 36x^3 + 16x^2 + x - 30$

Note: $f(x) = a(16x^4 + 36x^3 + 16x^2 + x - 30)$, where a is any nonzero real number, has the zeros $x = \frac{3}{4}$, $x = -2$, and $x = -\frac{1}{2} \pm i$. In fact, we used constant multiples of the linear factors in this equation to simplify the fractions. Without these constants, you would have

$f(x) = \left(x - \frac{3}{4}\right)(x + 2)\left(x + \frac{1}{2} - i\right)\left(x + \frac{1}{2} + i\right)$

$\quad = x^4 + \frac{9}{4}x^3 + x^2 + \frac{1}{16}x - \frac{15}{8}.$

45. $f(x) = x^4 + 6x^2 - 27$

(a) $f(x) = (x^2 + 9)(x^2 - 3)$

(b) $f(x) = (x^2 + 9)(x + \sqrt{3})(x - \sqrt{3})$

(c) $f(x) = (x + 3i)(x - 3i)(x + \sqrt{3})(x - \sqrt{3})$

47.

$$
\begin{array}{r}
x^2 - 2x + 3 \\
x^2 - 2x - 2 \overline{)\,x^4 - 4x^3 + 5x^2 - 2x - 6} \\
\underline{x^4 - 2x^3 - 2x^2} \\
-2x^3 + 7x^2 - 2x \\
\underline{-2x^3 + 4x^2 + 4x} \\
3x^2 - 6x - 6 \\
\underline{3x^2 - 6x - 6} \\
0
\end{array}
$$

$$f(x) = (x^2 - 2x - 2)(x^2 - 2x + 3)$$

(a) $f(x) = (x^2 - 2x - 2)(x^2 - 2x + 3)$

(b) $f(x) = \left(x - 1 + \sqrt{3}\right)\left(x - 1 - \sqrt{3}\right)(x^2 - 2x + 3)$

(c) $f(x) = \left(x - 1 + \sqrt{3}\right)\left(x - 1 - \sqrt{3}\right)\left(x - 1 + \sqrt{2}\,i\right)\left(x - 1 - \sqrt{2}\,i\right)$

Note: Use the Quadratic Formula for (b) and (c).

49. $f(x) = 2x^3 + 3x^2 + 50x + 75$

Since $5i$ is a zero, so is $-5i$.

$$
\begin{array}{r|rrrr}
5i & 2 & 3 & 50 & 75 \\
 & & 10i & -50 + 15i & -75 \\
\hline
 & 2 & 3 + 10i & 15i & 0
\end{array}
$$

$$
\begin{array}{r|rrr}
-5i & 2 & 3 + 10i & 15i \\
 & & -10i & -15i \\
\hline
 & 2 & 3 & 0
\end{array}
$$

The zero of $2x + 3$ is $x = -\frac{3}{2}$.

The zeros of $f(x)$ are $x = -\frac{3}{2}$ and $x = \pm 5i$.

<u>Alternate Solution</u>

Since $x = \pm 5i$ are zeros of $f(x)$, $(x + 5i)(x - 5i) = x^2 + 25$ is a factor of $f(x)$.
By long division we have:

$$
\begin{array}{r}
2x + 3 \\
x^2 + 0x + 25 \overline{)\,2x^3 + 3x^2 + 50x + 75} \\
\underline{2x^3 + 0x^2 + 50x} \\
3x^2 + 0x + 75 \\
\underline{3x^2 + 0x + 75} \\
0
\end{array}
$$

Thus, $f(x) = (x^2 + 25)(2x + 3)$ and the zeros of f are $x = \pm 5i$ and $x = -\frac{3}{2}$.

51. $f(x) = 2x^4 - x^3 + 7x^2 - 4x - 4$

Since $2i$ is a zero, so is $-2i$.

$$
\begin{array}{r|rrrrr}
2i & 2 & -1 & 7 & -4 & -4 \\
   &   & 4i & -8-2i & 4-2i & 4 \\
\hline
   & 2 & -1+4i & -1-2i & -2 & 0
\end{array}
$$

$$
\begin{array}{r|rrrr}
-2i & 2 & -1+4i & -1-2i & -2i \\
    &   & -4i & 2i & 2i \\
\hline
    & 2 & -1 & -1 & 0
\end{array}
$$

The zeros of $2x^2 - x - 1 = (2x + 1)(x - 1)$ are $x = -\frac{1}{2}$ and $x = 1$.

The zeros of $f(x)$ are $x = \pm 2i,\ x = -\frac{1}{2},$ and $x = 1$.

Alternate Solution

Since $x = \pm 2i$ are zeros of $f(x)$, $(x + 2i)(x - 2i) = x^2 + 4$ is a factor of $f(x)$.
By long division we have:

$$
\require{enclose}
\begin{array}{r}
2x^2 - x - 1 \\[-2pt]
x^2 + 0x + 4 \enclose{longdiv}{2x^4 - x^3 + 7x^2 - 4x - 4} \\
\underline{2x^4 + 0x^3 + 8x^2} \\
-x^3 - x^2 - 4x \\
\underline{-x^3 + 0x^2 - 4x} \\
-x^2 + 0x - 4 \\
\underline{-x^2 + 0x - 4} \\
0
\end{array}
$$

Thus, $f(x) = (x^2 + 4)(2x^2 - x - 1)$

$\qquad\qquad = (x + 2i)(x - 2i)(2x + 1)(x - 1)$

and the zeros of $f(x)$ are $x = \pm 2i,\ x = -\frac{1}{2},$ and $x = 1$.

53. $g(x) = 4x^3 + 23x^2 + 34x - 10$

Since $-3 + i$ is a zero, so is $-3 - i$.

$-3 + i$	4	23	34	-10
		$-12 + 4i$	$-37 - i$	10
	4	$11 + 4i$	$-3 - i$	0

$-3 - i$	4	$11 + 4i$	$-3 - i$
		$-12 - 4i$	$3 + i$
	4	-1	0

The zero of $4x - 1$ is $x = \frac{1}{4}$. The zeros of $g(x)$ are $x = -3 \pm i$ and $x = \frac{1}{4}$.

Alternate Solution

Since $-3 \pm i$ are zeros of $g(x)$,
$$[x - (-3 + i)][x - (-3 - i)] = [(x + 3) - i][(x + 3) + i]$$
$$= (x + 3)^2 - i^2$$
$$= x^2 + 6x + 10$$

is a factor of $g(x)$. By long division we have:

$$
\begin{array}{r}
4x - 1 \\
x^2 + 6x + 10 \overline{)\,4x^3 + 23x^2 + 34x - 10} \\
\underline{4x^3 + 24x^2 + 40x} \\
-x^2 - 6x - 10 \\
\underline{-x^2 - 6x - 10} \\
0
\end{array}
$$

Thus, $g(x) = (x^2 + 6x + 10)(4x - 1)$ and the zeros of $g(x)$ are $x = -3 \pm i$ and $x = \frac{1}{4}$.

55. Since $-3 + \sqrt{2}\,i$ is a zero, so is $-3 - \sqrt{2}\,i$, and
$$\left[x - \left(-3 + \sqrt{2}\,i\right)\right]\left[x - \left(-3 - \sqrt{2}\,i\right)\right]$$
$$= \left[(x + 3) - \sqrt{2}\,i\right]\left[(x + 3) + \sqrt{2}\,i\right]$$
$$= (x + 3)^2 - \left(\sqrt{2}\,i\right)^2$$
$$= x^2 + 6x + 11$$

is a factor of $f(x)$. By long division, we have:

$$
\begin{array}{r}
x^2 - 3x + 2 \\
x^2 + 6x + 11 \overline{)\,x^4 + 3x^3 - 5x^2 - 21x + 22} \\
\underline{x^4 + 6x^3 + 11x^2} \\
-3x^3 - 16x^2 - 21x \\
\underline{-3x^3 - 18x^2 - 33x} \\
2x^2 + 12x + 22 \\
\underline{2x^2 + 12x + 22} \\
0
\end{array}
$$

Thus, $f(x) = (x^2 + 6x + 11)(x^2 - 3x + 2)$
$$= (x^2 + 6x + 11)(x - 1)(x - 2)$$
and the zeros of f are $x = -3 \pm \sqrt{2}\,i$, $x = 1$, and $x = 2$.

57. Since $\frac{1}{2}\left(1 - \sqrt{5}\,i\right)$ is a zero, so is $\frac{1}{2}\left(1 + \sqrt{5}\,i\right)$, and
$$\left[x - \tfrac{1}{2}\left(1 - \sqrt{5}\,i\right)\right]\left[x - \tfrac{1}{2}\left(1 + \sqrt{5}\,i\right)\right]$$
$$= \left[\left(x - \tfrac{1}{2}\right) + \tfrac{1}{2}\sqrt{5}\,i\right]\left[\left(x - \tfrac{1}{2}\right) - \tfrac{1}{2}\sqrt{5}\,i\right]$$
$$= \left(x - \tfrac{1}{2}\right)^2 - \left(\tfrac{1}{2}\sqrt{5}\,i\right)^2$$
$$= x^2 - x + \tfrac{1}{4} + \tfrac{5}{4}$$
$$= x^2 - x + \tfrac{3}{2}$$

is a factor of $h(x)$. By long division, we have:

$$
\begin{array}{r}
8x - 6 \\
x^2 - x + \tfrac{3}{2} \overline{)\,8x^3 - 14x^2 + 18x - 9} \\
\underline{8x^3 - 8x^2 + 12x} \\
-6x^2 + 6x - 9 \\
\underline{-6x^2 + 6x - 9} \\
0
\end{array}
$$

Thus, $h(x) = \left(x^2 - x + \tfrac{3}{2}\right)(8x - 6)$ and the zeros of $h(x)$ are

$$x = \frac{3}{4} \text{ and } x = \frac{1}{2} \pm \frac{\sqrt{5}}{2}i.$$

59. $f(x) = x^3 + ix^2 + ix - 1$

(a)

$$
\begin{array}{c|cccc}
i & 1 & i & i & -1 \\
  &   & i & -2 & -1-2i \\
\hline
  & 1 & 2i & -2+i & -2-2i
\end{array}
$$

Since the remainder is not zero, $x = i$ is not a zero of f.

(b) The theorem that states that complex zeros occur in conjugate pairs has the condition that the coefficients of $f(x)$ must be real numbers. This polynomial has complex coefficients for x^2 and x.

61. $f(x) = x^4 - 4x^2 + k$

$$x^2 = \frac{-(-4) \pm \sqrt{(-4)^2 - 4(1)(k)}}{2(1)} = \frac{4 \pm 2\sqrt{4-k}}{2} = 2 \pm \sqrt{4-k}$$

$$x = \pm\sqrt{2 \pm \sqrt{4-k}}$$

(a) For there to be four distinct real roots, both $4 - k$ and $2 \pm \sqrt{4-k}$ must be positive. This occurs when $0 < k < 4$. Thus, some possible k-values are $k = 1$, $k = 2$, $k = 3$, $k = \frac{1}{2}$, $k = \sqrt{2}$, etc.

(b) For there to be two real roots, each of multiplicity 2, $4 - k$ must equal zero. Thus, $k = 4$.

(c) For there to be two real zeros and two complex zeros, $2 + \sqrt{4-k}$ must be positive and $2 - \sqrt{4-k}$ must be negative. This occurs when $k < 0$. Thus, some possible k-values are $k = -1$, $k = -2$, $k = -\frac{1}{2}$, etc.

(d) For there to be four complex zeros, $2 \pm \sqrt{4-k}$ must be complex. This occurs when $k > 4$. Some possible k-values are $k = 5$, $k = 6$, $k = 7.4$, etc.

63. $h = -16t^2 + 48t, \ 0 \le t \le 3$

$$= -16(t^2 - 3t)$$

$$= -16\left(t^2 - 3t + \frac{9}{4} - \frac{9}{4}\right)$$

$$= -16\left[\left(t - \frac{3}{2}\right)^2 - \frac{9}{4}\right]$$

$$= -16\left(t - \frac{3}{2}\right)^2 + 36$$

The maximum height that the baseball reaches is 36 feet when $t = 1.5$ seconds. No, it is not possible for the ball to reach a height of 64 feet.

Alternate Solution

Let $h = 64$ and solve for t.

$$64 = -16t^2 + 48t$$

$$16t^2 - 48t + 64 = 0$$

$$16(t^2 - 3t + 4) = 0$$

$$t^2 - 3t + 4 = 0$$

$$t = \frac{3 \pm \sqrt{9 - 16}}{2} = \frac{3 \pm \sqrt{7}\,i}{2}$$

No, it is not possible since solving this equation yields imaginary roots.

65. (a) $f(x) = (x - \sqrt{b}\,i)(x + \sqrt{b}\,i) = x^2 + b$

    (b) $f(x) = [x - (a + bi)][x - (a - bi)]$

            $= [(x - a) - bi][(x - a) + bi]$

            $= (x - a)^2 - (bi)^2$

            $= x^2 - 2ax + a^2 + b^2$

67.

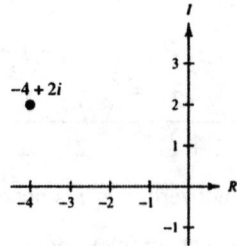

69. $(1.50)(420) = 630$

71. $\dfrac{x}{10} = \dfrac{3}{1.25}$

       $1.25x = 30$

          $x = 24$ feet

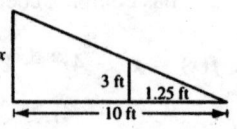

Section 3.6   Mathematical Modeling

You should know the following the following terms and formulas.

- Direct Variation (varies directly, directly proportional)
  - (a) $y = mx$
  - (b) $y = kx^n$ (as nth power)
- Inverse Variation (varies inversely, inversely proportional)
  - (a) $y = k/x$
  - (b) $y = k/(x^n)$ (as nth power)
- Joint Variation (varies jointly, jointly proportional)
  - (a) $z = kxy$
  - (b) $z = kx^n y^m$ (as nth power of x and mth power of y)
- k is called the constant of proportionality.
- Least Squares Regression Line $y = ax + b$ where

$$a = \frac{n\sum_{i=1}^{n} x_i y_i - \sum_{i=1}^{n} x_i \sum_{i=1}^{n} y_i}{n\sum_{i=1}^{n} x_i^2 - \left(\sum_{i=1}^{n} x_i\right)^2}$$

$$b = \frac{1}{n}\left(\sum_{i=1}^{n} y_i - a\sum_{i=1}^{n} x_i\right).$$

Solutions to Odd-Numbered Exercises

1. $y = 113{,}336.2 + 1265t, \ 7 \le t \le 13$

$t = 0$ represents 1980.

Year	1987	1988	1989	1990	1991	1992	1993
Actual Number	121,602	123,378	125,557	126,424	126,867	128,548	129,525
Model	122,191	123,456	124,721	125,986	127,251	128,516	129,781

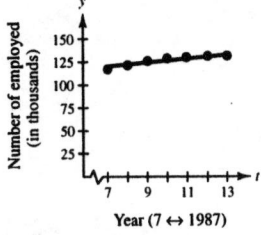

Year (7 ↔ 1987)

The model is a "good fit" for the actual data.

3. The graph appears to represent $y = 4/x$ which is an inverse variation.

5.

x	2	4	6	8	10
$y = x^2$	4	16	36	64	100

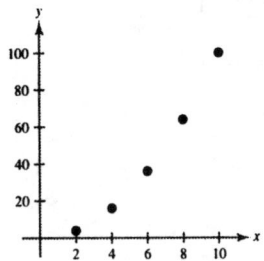

7.

x	2	4	6	8	10
$y = \frac{1}{2}x^2$	2	8	18	32	50

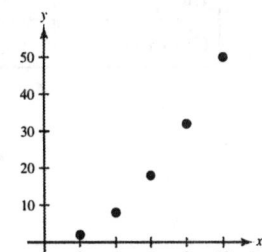

9.

x	2	4	6	8	10
$y = \dfrac{2}{x^2}$	$\dfrac{1}{2}$	$\dfrac{1}{8}$	$\dfrac{1}{18}$	$\dfrac{1}{32}$	$\dfrac{1}{50}$

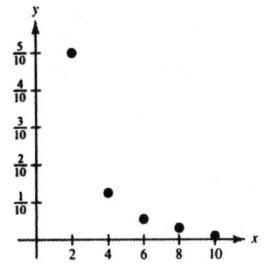

11.

x	2	4	6	8	10
$y = \dfrac{10}{x^2}$	$\dfrac{5}{2}$	$\dfrac{5}{8}$	$\dfrac{5}{18}$	$\dfrac{5}{32}$	$\dfrac{1}{10}$

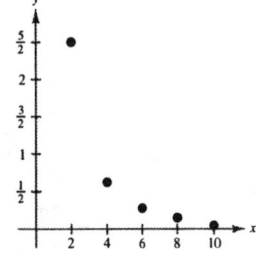

13. The chart represents the equation $y = \dfrac{5}{x}$.

15. $y = kx$

     $-7 = k(10)$

     $-\frac{7}{10} = k$

     $y = -\frac{7}{10}x$

This equation checks with the other points given in the chart.

17. $y = mx$

     $12 = m(5)$

     $\frac{12}{5} = m$

     $y = \frac{12}{5}x$

19. $y = mx$

     $2050 = m(10)$

     $205 = m$

     $y = 205x$

21. $I = kP$

     $187.59 = k(2500)$

     $0.075 = k$

     $I = 0.075P$

23. $y = kx$

     $1840 = k(50{,}000)$

     $0.0368 = k$

     $y = 0.0368x$

When $x = 85{,}000$ we have

     $y = 0.0368(85{,}000) = \$3128.$

25. $y = kx$

     $33 = k(13)$

     $\frac{33}{13} = k$

     $y = \frac{33}{13}x$

Inches	5	10	20	25	30
Centimeters	12.7	25.4	50.8	63.5	76.2

27. $d = kF$

     $0.15 = k(265)$

     $\frac{3}{5300} = k$

     $d = \frac{3}{5300}F$

(a) $d = \frac{3}{5300}(90) \approx 0.05$ meter

(b) $0.1 = \frac{3}{5300}F$

     $\frac{530}{3} = F$

     $F = 176\frac{2}{3}$ newtons

29. $d = kF$

     $1.9 = k(25) \implies k = 0.076$

     $d = 0.076F$

When the distance compressed is 3 inches, we have

     $3 = 0.076F$

     $F \approx 39.47.$

No child over 39.47 pounds should use the toy.

31. $A = kr^2$

33. $y = \dfrac{k}{x^2}$

35. $z = k\sqrt[3]{u}$

37. $z = kuv$

39. $F = \dfrac{kg}{r^2}$

41. $P = \dfrac{k}{V}$

43. $F = \dfrac{km_1 m_2}{r^2}$

45. $A = \frac{1}{2}bh$

The area of a triangle is jointly proportional to the magnitude of the base and the height.

47. $V = \frac{4}{3}\pi r^3$

The volume of a sphere varies directly as the cube of its radius.

49. $r = \frac{d}{t}$

Average speed is directly proportional to the distance and inversely proportional to the time.

51. $A = kr^2$

$9\pi = k(3)^2$

$\pi = k$

$A = \pi r^2$

53. $y = \frac{k}{x}$

$3 = \frac{k}{25}$

$75 = k$

$y = \frac{75}{x}$

55. $h = \frac{k}{t^3}$

$\frac{3}{16} = \frac{k}{(4)^3}$

$\frac{3}{16} = \frac{k}{64}$

$k = 12$

$h = \frac{12}{t^3}$

57. $z = kxy$

$64 = k(4)(8)$

$2 = k$

$z = 2xy$

59. $F = krs^3$

$4158 = k(11)(3)^3$

$k = 14$

$F = 14rs^3$

61. $z = \frac{kx^2}{y}$

$6 = \frac{k(6)^2}{4}$

$\frac{24}{36} = k$

$\frac{2}{3} = k$

$z = \frac{(2/3)x^2}{y} = \frac{2x^2}{3y}$

63. $S = \frac{kL}{L - S}$

$4 = \frac{k(6)}{6 - 4}$

$4 = 3k$

$k = \frac{4}{3}$

$S = \frac{4/3 L}{L - S} = \frac{4L}{3(L - S)}$

65. $d = kv^2$

$0.02 = k\left(\frac{1}{4}\right)^2$

$k = 0.32$

$d = 0.32v^2$

$0.12 = 0.32v^2$

$v^2 = \frac{0.12}{0.32} = \frac{3}{8}$

$v = \frac{\sqrt{3}}{2\sqrt{2}} = \frac{\sqrt{6}}{4} \approx 0.61 \text{ mi/hr}$

67. $r = \frac{kl}{A}, \quad A = \pi r^2 = \frac{\pi d^2}{4}$

$r = \frac{4kl}{\pi d^2}$

$66.17 = \frac{4(1000)k}{\pi(0.0126/12)^2}$

$k \approx 5.73 \times 10^{-8}$

$r = \frac{4(5.73 \times 10^{-8})l}{\pi(0.0126/12)^2}$

$33.5 = \frac{4(5.73 \times 10^{-8})l}{\pi(0.0126/12)^2}$

$\frac{33.5\pi(0.0126/12)^2}{4(5.73 \times 10^{-8})} = l$

$l \approx 506 \text{ feet}$

69. $S = kt^2$

$144 = k(3)^2$

$16 = k$

$S = 16t^2$

$S = 16(5)^2 = 400$ feet

71. $P = kA = k(\pi r^2) = k\pi\left(\dfrac{d}{2}\right)^2$

$8.78 = k\pi\left(\dfrac{9}{2}\right)^2$

$\dfrac{4(8.78)}{81\pi} = k$

$k \approx 0.138$

However, we do not obtain \$11.78 when $d = 12$ inches.

$P = 0.138\pi\left(\dfrac{12}{2}\right)^2 \approx \$15.61$

Instead, $k = \dfrac{11.78}{36\pi} \approx 0.104$.

For the 15 inch pizza, we have $k = \dfrac{4(14.18)}{225\pi} \approx 0.080$.

The price is not directly proportional to the surface area. The best buy is the 15-inch pizza.

73. $v = \dfrac{k}{A}$

(a) $v = \dfrac{k}{0.75A} = \dfrac{4}{3}\left(\dfrac{k}{A}\right)$

The velocity is four-thirds the original.

(b) $v = \dfrac{k}{(1 + \frac{1}{3})A} = \dfrac{k}{\frac{4}{3}A} = \dfrac{3}{4}\left(\dfrac{k}{A}\right)$

The velocity is decreased by one-fourth.

75. (a)

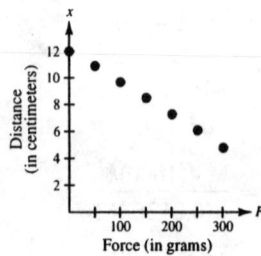

(b) It appears to fit Hooke's Law.

$k \approx \left|\dfrac{12.00 - 4.84}{0 - 300}\right| \approx |-0.0238| = 0.0238$

(c) $x = kF$

$9 = 0.0238F$

$F \approx 378$ grams

77. $y = \dfrac{262.76}{x^{2.12}}$

(a)

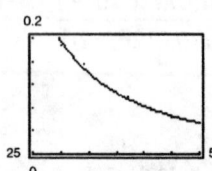

(b) $y = \dfrac{262.76}{(25)^{2.12}} \approx 0.2857$ microwatts per square centimeter

79. The data shown could be represented by a linear model which would be a good fit.

81. The points do not follow a linear pattern. A linear model would not be a good approximation.

83.

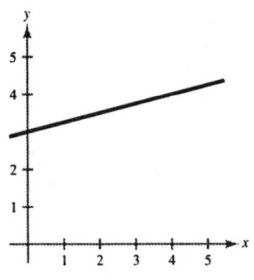

Using the points (0, 3) and (4, 4), we have
$y = \frac{1}{4}x + 3$.

85.

Using the points (2, 2) and (4, 1), we have
$y = -\frac{1}{2}x + 3$.

87. (a) (7, 203), (8, 239), (9, 295), (10, 352), (11, 415), and (12, 488)

$$a = \frac{6(19929) - (57)(1992)}{6(559) - (57)^2} \approx 57.4$$

$$b = \frac{1}{6}[1992 - a(57)] \approx -213.6$$

$$y \approx 57.4t - 213.6$$

(b)

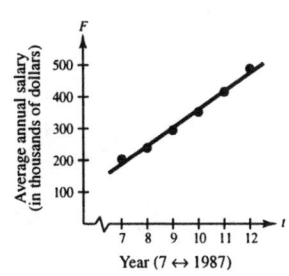

(c)

Year	t	y
1994	14	590.0
1995	15	647.4
1996	16	704.8

89. (a) Using the linear regression capabilities of a calculator yields $y \approx 1.145t + 124.425$.

(b)

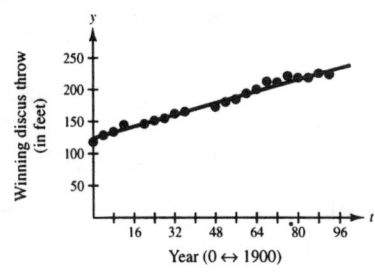

(c) Using the model with $t = 96$ yields 234.345 feet.

91. (a) Using the linear regression capabilities of a calculator yields $y \approx -0.74x + 106$.

(b)

(c) Using the model with $x = 62$ yields 60.12 pounds.

(d) For each 1 pound increase in per capita consumption of poultry, the per capita consumption of beef decreases by an average of 0.74 pounds.

❏ **Review Exercises for Chapter 3**

Solutions to Odd-Numbered Exercises

1. $f(x) = \left(x + \frac{3}{2}\right)^2 + 1$

Vertex: $\left(-\frac{3}{2}, 1\right)$

y-intercept: $\left(0, \frac{13}{4}\right)$

No x-intercepts

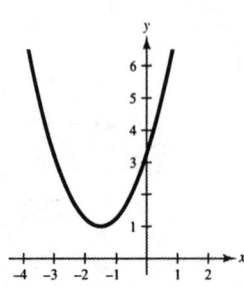

3. $f(x) = \frac{1}{3}(x^2 + 5x - 4)$

$= \frac{1}{3}\left(x^2 + 5x + \frac{25}{4} - \frac{25}{4} - 4\right)$

$= \frac{1}{3}\left[\left(x - \frac{5}{2}\right)^2 - \frac{41}{4}\right]$

$= \frac{1}{3}\left(x - \frac{5}{2}\right)^2 - \frac{41}{12}$

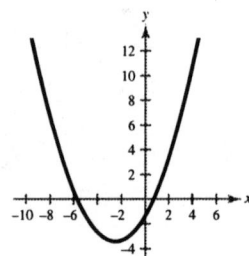

Vertex: $\left(\frac{5}{2}, -\frac{41}{12}\right)$

y-intercept: $\left(0, -\frac{3}{4}\right)$

y-intercept: $0 = \frac{1}{3}(x^2 + 5x - 4)$

$0 = x^2 + 5x - 4$

$x = \dfrac{-5 \pm \sqrt{41}}{2}$ Use the Quadratic Formula.

$\left(\dfrac{-5 \pm \sqrt{41}}{2}, 0\right)$

5. Vertex: $(4, 1) \implies f(x) = a(x - 4)^2 + 1$

Point: $(2, -1) \implies -1 = a(2 - 4)^2 + 1$

$-2 = 4a$

$-\frac{1}{2} = a$

Thus, $f(x) = -\frac{1}{2}(x - 4)^2 + 1$.

7. Vertex: $(1, -4) \implies f(x) = a(x - 1)^2 - 4$

Point: $(2, -3) \implies -3 = a(2 - 1)^2 - 4$

$1 = a$

Thus, $f(x) = (x - 1)^2 - 4$.

9. (a) $y = 2x^2$
Vertical stretch

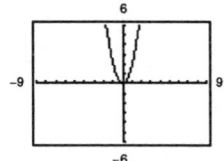

(b) $y = -2x^2$
Vertical stretch and a reflection in the x-axis

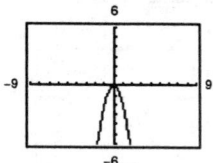

(c) $y = x^2 + 2$
Vertical shift two units upward

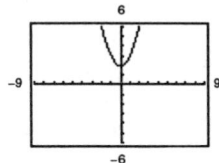

(d) $y = (x + 2)^2$
Horizontal shift two units to the left

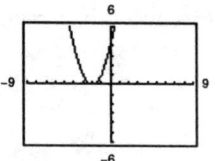

11. $g(x) = x^2 - 2x$
$$= x^2 - 2x + 1 - 1$$
$$= (x - 1)^2 - 1$$

The minimum occurs at the vertex $(1, -1)$.

13. $f(x) = 6x - x^2$
$$= -(x^2 - 6x + 9 - 9)$$
$$= -(x - 3)^2 + 9$$

The maximum occurs at the vertex $(3, 9)$.

15. $f(t) = -2t^2 + 4t + 1$
$$= -2(t^2 - 2t + 1 - 1) + 1$$
$$= -2[(t - 1)^2 - 1] + 1$$
$$= -2(t - 1)^2 + 3$$

The maximum occurs at the vertex $(1, 3)$.

17. $h(x) = x^2 + 5x - 4$
$$= x^2 + 5x + \frac{25}{4} - \frac{25}{4} - 4$$
$$= \left(x + \frac{5}{2}\right)^2 - \frac{25}{4} - \frac{16}{4}$$
$$= \left(x + \frac{5}{2}\right)^2 - \frac{41}{4}$$

The minimum occurs at the vertex $\left(-\frac{5}{2}, -\frac{41}{4}\right)$.

19. (a)

x	y	Area
1	$4 - \frac{1}{2}(1)$	$(1)[4 - \frac{1}{2}(1)] = \frac{7}{2}$
2	$4 - \frac{1}{2}(2)$	$(2)[4 - \frac{1}{2}(2)] = 6$
3	$4 - \frac{1}{2}(3)$	$(3)[4 - \frac{1}{2}(3)] = \frac{15}{2}$
4	$4 - \frac{1}{2}(4)$	$(4)[4 - \frac{1}{2}(4)] = 8$
5	$4 - \frac{1}{2}(5)$	$(5)[4 - \frac{1}{2}(5)] = \frac{15}{2}$
6	$4 - \frac{1}{2}(6)$	$(6)[4 - \frac{1}{2}(6)] = 6$

(b) The dimensions that will produce a maximum area are $x = 4$ and $y = 2$.

(c) $A = xy = x\left(\dfrac{8 - x}{2}\right)$ since $x + 2y - 8 = 0 \implies y = \dfrac{8 - x}{2}$.

Since the figure is in the first quadrant and x and y must be positive, the

domain of $A = x\left(\dfrac{8 - x}{2}\right)$ is $0 < x < 8$.

(d)

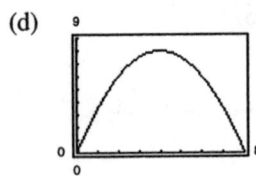

The maximum area of 8 occurs at the vertex when
$x = 4$ and $y = \dfrac{8 - 4}{2} = 2$.

(e) $A = x\left(\dfrac{8 - x}{2}\right)$

$= \dfrac{1}{2}(8x - x^2)$

$= -\dfrac{1}{2}(x^2 - 8x)$

$= -\dfrac{1}{2}(x^2 - 8x + 16 - 16)$

$= -\dfrac{1}{2}[(x - 4)^2 - 16]$

$= -\dfrac{1}{2}(x - 4)^2 + 8$

The maximum area of 8 occurs when $x = 4$ and $y = \dfrac{8 - 4}{2} = 2$.

21. Let x = the number of \$30 increases in rent. Then,
Rent $= 540 + 30x$
Number of occupied units $= 50 - x$
Revenue $=$ (Number of occupied units)(rent)
$\qquad R = (50 - x)(540 + 30x)$
Cost $=$ (Number of occupied units)(\$18)
$\qquad C = (50 - x)(18)$
Profit $=$ Revenue $-$ Cost
$\qquad P = (50 - x)(540 + 30x) - (50 - x)(18)$
$\qquad\quad = 27{,}000 + 960x - 30x^2 - 900 + 18x$
$\qquad\quad = -30x^2 + 978x + 26{,}100$

The maximum profit occurs at the vertex.
$$-\frac{b}{2a} = \frac{-978}{2(-30)} = 16.3 \approx 16 \text{ increases}$$
The corresponding rent is $540 + 30(16) = \$1020$.

23. $f(x) = -x^2 + 6x + 9$

The degree is even and the leading coefficient is negative. The graph falls to the left and right.

25. $f(x) = \frac{3}{4}(x^4 + 3x^2 + 2)$

The degree is even and the leading coefficient is positive. The graph rises to the left and right.

27. $f(x) = \frac{1}{2}x^3 - 2x + 1$; $g(x) = \frac{1}{2}x^3$

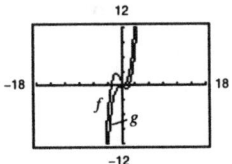

29. $f(x) = -x^4 + 2x^3$; $g(x) = -x^4$

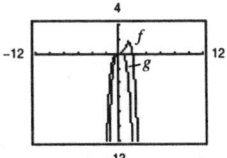

31. $f(x) = -(x - 2)^3$

This is the graph of $y = x^3$ reflected in the x-axis and shifted two units to the right.

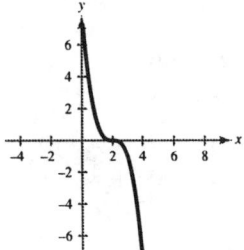

33. $f(t) = t^3 - 3t = t(t^2 - 3)$

Intercepts: $(0, 0)(\pm\sqrt{3}, 0)$

The graph rises to the right and falls to the left.

x	-2	-1	0	1	2
y	-2	2	0	-2	2

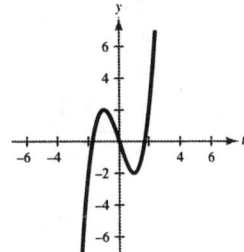

35. (a) The combined length and girth is

$$y + 4x = 216$$
$$y = 216 - 4x.$$

The volume is

$$V = x^2y = x^2(216 - 4x).$$

(b)
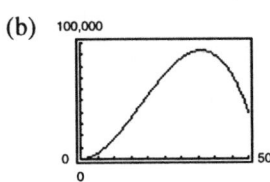

The volume is maximum when $x = 36$ centimeters and $y = 216 - 4(36) = 72$ centimeters.

37. $y_1 = \dfrac{x^2}{x - 2}$

$$y_2 = x + 2 + \frac{4}{x - 2}$$
$$= \frac{(x + 2)(x - 2)}{x - 2} + \frac{4}{x - 2}$$
$$= \frac{x^2 - 4}{x - 2} + \frac{4}{x - 2}$$
$$= \frac{x^2}{x - 2}$$
$$= y_1$$

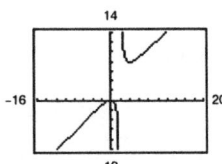

39.

$$
\begin{array}{r}
8x + 5 \\
3x - 2 \overline{)24x^2 - x - 8} \\
\underline{24x^2 - 16x} \\
15x - 8 \\
\underline{15x - 10} \\
2
\end{array}
$$

Thus, $\dfrac{24x^2 - x - 8}{3x - 2} = 8x + 5 + \dfrac{2}{3x - 2}.$

41.

$$
\begin{array}{r}
5x + 2 \\
x^2 - 3x + 1 \overline{)5x^3 - 13x^2 - x + 2} \\
\underline{5x^3 - 15x^2 + 5x} \\
2x^2 - 6x + 2 \\
\underline{2x^2 - 6x + 2} \\
0
\end{array}
$$

Thus, $\dfrac{5x^3 - 13x^2 - x + 2}{x^2 - 3x + 1} = 5x + 2.$

43.

$$
\begin{array}{r}
x^2 - 3x + 2 \\
x^2 + 0x + 2 \overline{)x^4 - 3x^3 + 4x^2 - 6x + 3} \\
\underline{x^4 + 0x^3 + 2x^2} \\
-3x^3 + 2x^2 - 6x \\
\underline{-3x^3 + 0x^2 - 6x} \\
2x^2 + 0x + 3 \\
\underline{2x^2 + 0x + 4} \\
-1
\end{array}
$$

Thus, $\dfrac{x^4 - 3x^3 + 4x^2 - 6x + 3}{x^2 + 2} = x^2 - 3x + 2 - \dfrac{1}{x^2 + 2}.$

45.

$$
\begin{array}{r|rrrrr}
\tfrac{2}{3} & 6 & -4 & -27 & 18 & 0 \\
 & & 4 & 0 & -18 & 0 \\
\hline
 & 6 & 0 & -27 & 0 & 0
\end{array}
$$

Thus, $\dfrac{6x^4 - 4x^3 - 27x^2 + 18x}{x - (2/3)} = 6x^3 - 27x.$

47.

$$
\begin{array}{r|rrrr}
1 + 2i & 2 & -5 & 12 & -5 \\
 & & 2 + 4i & -11 - 2i & 5 \\
\hline
 & 2 & -3 + 4i & 1 - 2i & 0
\end{array}
$$

Thus, $\dfrac{2x^3 - 5x^2 + 12x - 5}{x - (1 + 2i)} = 2x^2 + (-3 + 4i)x + (1 - 2i)$

$\phantom{Thus, \dfrac{2x^3 - 5x^2 + 12x - 5}{x - (1 + 2i)}} = 2x^2 - (3 - 4i)x + (1 - 2i).$

49. $f(x) = 20x^4 + 9x^3 - 14x^2 - 3x$

(a)
$$
\begin{array}{r|rrrrr}
-1 & 20 & 9 & -14 & -3 & 0 \\
 & & -20 & 11 & 3 & 0 \\
\hline
 & 20 & -11 & -3 & 0 & 0
\end{array}
$$

Yes, $x = -1$ is a zero of f.

(b)
$$
\begin{array}{r|rrrrr}
\tfrac{3}{4} & 20 & 9 & -14 & -3 & 0 \\
 & & 15 & 18 & 3 & 0 \\
\hline
 & 20 & 24 & 4 & 0 & 0
\end{array}
$$

Yes, $x = \tfrac{3}{4}$ is a zero of f.

(c)
$$
\begin{array}{r|rrrrr}
0 & 20 & 9 & -14 & -3 & 0 \\
 & & 0 & 0 & 0 & 0 \\
\hline
 & 20 & 9 & -14 & -3 & 0
\end{array}
$$

Yes, $x = 0$ is a zero of f.

(d)
$$
\begin{array}{r|rrrrr}
1 & 20 & 9 & -14 & -3 & 0 \\
 & & 20 & 29 & 15 & 12 \\
\hline
 & 20 & 29 & 15 & 12 & 12
\end{array}
$$

No, $x = 1$ is not a zero of f.

51. $f(x) = x^4 + 10x^3 - 24x^2 + 20x + 44$

(a)

$$
\begin{array}{r|rrrrr}
-3 & 1 & 10 & -24 & 20 & 44 \\
   &   & -3 & -21 & 135 & -465 \\
\hline
   & 1 & 7 & -45 & 155 & -421
\end{array}
$$

Thus, $f(-3) = -421$.

(b)

$$
\begin{array}{r|rrrrr}
\sqrt{2}i & 1 & 10 & -24 & 20 & 44 \\
   &   & \sqrt{2}i & -2 + 10\sqrt{2}i & -20 - 26\sqrt{2}i & 52 \\
\hline
   & 1 & 10 + \sqrt{2}i & -26 + 10\sqrt{2}i & -26\sqrt{2}i & 96
\end{array}
$$

Thus, $f(\sqrt{2}i) = 96$.

53. $f(x) = 6(x + 1)^2\left(x - \tfrac{1}{3}\right)\left(x + \tfrac{1}{2}\right)$ Multiply by 6 to clear the fractions.

$$= (x + 1)^2 \, 3\left(x - \tfrac{1}{3}\right) 2\left(x + \tfrac{1}{2}\right)$$

$$= (x^2 + 2x + 1)(3x - 1)(2x + 1)$$

$$= (x^2 + 2x + 1)(6x^2 + x - 1)$$

$$= 6x^4 + 13x^3 + 7x^2 - x - 1$$

Note: $f(x) = a(6x^4 + 13x^3 + 7x^2 - x - 1)$, where a is any real nonzero number, has zeros $-1, -1, \tfrac{1}{3},$ and $-\tfrac{1}{2}$.

55. $f(x) = 3\left(x - \tfrac{2}{3}\right)(x - 4)(x - \sqrt{3}i)(x + \sqrt{3}i)$ \qquad Multiply by 3 to clear the fraction.

$$= (3x - 2)(x - 4)(x^2 + 3)$$

$$= (3x^2 - 14x + 8)(x^2 + 3)$$

$$= 3x^4 - 14x^3 + 17x^2 - 42x + 24$$

Note: $f(x) = a(3x^4 - 14x^3 + 17x^2 - 42x + 24)$, where a is any real nonzero number, has zeros $\tfrac{2}{3}, 4,$ and $\pm\sqrt{3}i$.

57. $f(x) = 5x^3 + 3x^2 - 6x + 9$

$f(x)$ has two variations in sign so f has either two or zero positive real zeros.

$f(-x) = -5x^3 + 3x^2 + 6x + 9$

$f(-x)$ has one variation in sign so f has one negative real zero.

59. $f(x) = -4x^3 + 8x^2 - 3x + 15$

Possible rational zeros: $\pm 1, \pm 3, \pm 5, \pm 15, \pm\tfrac{1}{2}, \pm\tfrac{3}{2}, \pm\tfrac{5}{2}, \pm\tfrac{15}{2}, \pm\tfrac{1}{4}, \pm\tfrac{3}{4}, \pm\tfrac{5}{4}, \pm\tfrac{15}{4}$

61. $f(x) = 4x^3 - 11x^2 + 10x - 3$

Possible rational zeros: $\pm 1, \pm 3, \pm\tfrac{1}{2}, \pm\tfrac{3}{2}, \pm\tfrac{1}{4}, \pm\tfrac{3}{4}$

$$
\begin{array}{r|rrrr}
1 & 4 & -11 & 10 & -3 \\
  &   & 4 & -7 & 3 \\
\hline
  & 4 & -7 & 3 & 0
\end{array}
$$

$4x^3 - 11x^2 + 10x - 3 = (x - 1)(4x^2 - 7x + 3) = (x - 1)^2(4x - 3)$

Thus, the zeros of $f(x)$ are $x = 1$ and $x = \tfrac{3}{4}$.

63. $f(x) = 6x^4 - 25x^3 + 14x^2 + 27x - 18$

Possible rational zeros: $\pm 1, \pm 2, \pm 3, \pm 6, \pm 9, \pm 18, \pm\frac{1}{2}, \pm\frac{3}{2}, \pm\frac{9}{2}, \pm\frac{1}{3}, \pm\frac{2}{3}, \pm\frac{1}{6}$

$$
\begin{array}{r|rrrrr}
-1 & 6 & -25 & 14 & 27 & -18 \\
   &   & -6 & 31 & -45 & 18 \\
\hline
   & 6 & -31 & 45 & -18 & 0
\end{array}
$$

$$
\begin{array}{r|rrrr}
3 & 6 & -31 & 45 & -18 \\
  &   & 18 & -39 & 18 \\
\hline
  & 6 & -13 & 6 & 0
\end{array}
$$

$6x^4 - 25x^3 + 14x^2 + 27x - 18 = (x + 1)(x - 3)(6x^2 - 13x + 6)$

$\qquad\qquad\qquad\qquad\qquad\qquad = (x + 1)(x - 3)(3x - 2)(2x - 3)$

Thus, the zeros of $f(x)$ are $x = -1$, $x = 3$, $x = \frac{2}{3}$, and $x = \frac{3}{2}$.

65. $f(x) = x^4 + 2x + 1$

(a)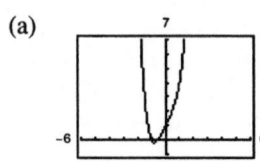

(b) The graph has two x-intercepts, so there are two real zeros.

(c) The zeros are $x = -1$ and $x \approx -0.54$.

67. $h(x) = x^3 - 6x^2 + 12x - 10$

(a)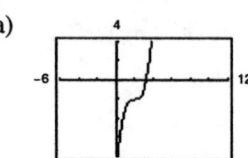

(b) The graph has one x-intercept, so there is one real zero.

(c) $x \approx 3.26$

69. (a) $S = 1.2087 + 0.2896t + 0.1762t^2 - 0.0309t^3 + 0.0013t^4$

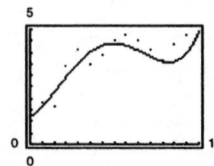

The model is a fairly "good fit."

(b) One explanation may be a recession. The model also shows a downturn in sales.

(c) $S(9) - S(11) \approx 0.47$

The actual decrease of 0.90 was more than this.

(d) $S(15) \approx \$6.72$ billion

71. $F = kx\sqrt{y}$

$6 = k(9)\sqrt{4}$

$6 = 18k$

$\frac{1}{3} = k$

$F = \frac{1}{3}x\sqrt{y}$

73. $\qquad P = kS^3$

$750 = k(27)^3$

$k = \frac{250}{6561}$

$P = \frac{250}{6561}S^3$

$P = \frac{250}{6561}(40)^3 \approx 2438.7$ kilowatts

75. $y = kx$

$4 = k(2.5)$

$k = \frac{8}{5}$

$y = \frac{8}{5}x$

Miles, x	2	5	10	12
Kilometers, y	3.2	8	16	19.2

❑ **Practice Test for Chapter 3**

1. Sketch the graph of $f(x) = x^2 - 6x + 5$ and identify the vertex and the intercepts.

2. Find the number of units x that produce a minimum cost C if
   $C = 0.01x^2 - 90x + 15{,}000$.

3. Find the quadratic function that has a maximum at $(1, 7)$ and passes through the point $(2, 5)$.

4. Find two quadratic functions that have x-intercepts $(2, 0)$ and $\left(\frac{4}{3}, 0\right)$.

5. Use the leading coefficient test to determine the right and left end behavior of the graph of the polynomial function $f(x) = -3x^5 + 2x^3 - 17$.

6. Find all the real zeros of $f(x) = x^5 - 5x^3 + 4x$.

7. Find a polynomial function with 0, 3, and -2 as zeros.

8. Sketch $f(x) = x^3 - 12x$.

9. Divide $3x^4 - 7x^2 + 2x - 10$ by $x - 3$ using long division.

10. Divide $x^3 - 11$ by $x^2 + 2x - 1$.

11. Use synthetic division to divide $3x^5 + 13x^4 + 12x - 1$ by $x + 5$.

12. Use synthetic division to find $f(-6)$ given $f(x) = 7x^3 + 40x^2 - 12x + 15$.

13. Find the real zeros of $f(x) = x^3 - 19x - 30$.

14. Find the real zeros of $f(x) = x^4 + x^3 - 8x^2 - 9x - 9$.

15. List all possible rational zeros of the function $f(x) = 6x^3 - 5x^2 + 4x - 15$.

16. Find the rational zeros of the polynomial $f(x) = x^3 - \frac{20}{3}x^2 + 9x - \frac{10}{3}$.

17. Write $f(x) = x^4 + x^3 + 5x - 10$ as a product of linear factors.

18. Find a polynomial with real coefficients that has 2, $3 + i$, and $3 - 2i$ as zeros.

19. Use synthetic division to show that $3i$ is a zero of $f(x) = x^3 + 4x^2 + 9x + 36$.

20. Find a mathematical model for the statement, "z varies directly as the square of x and inversely as the square root of y."

CHAPTER 4
Rational Functions and Conics

CHAPTER 4
Rational Functions and Conics

Section 4.1   Rational Functions and Asymptotes

■ You should know the following basic facts about rational functions.

(a) A function of the form $f(x) = P(x)/Q(x)$, $Q(x) \neq 0$, where $P(x)$ and $Q(x)$ are polynomials, is called a rational function.

(b) The domain of a rational function is the set of all real numbers except those which make the denominator zero.

(c) If $f(x) = P(x)/Q(x)$ is in reduced form, and a is a value such that $Q(a) = 0$, then the line $x = a$ is a vertical asymptote of the graph of f. $f(x) \to \infty$ or $f(x) \to -\infty$ as $x \to a$.

(d) The line $y = b$ is a horizontal asymptote of the graph of f if $f(x) \to b$ as $x \to \infty$ or $x \to -\infty$.

(e) Let $f(x) = \dfrac{P(x)}{Q(x)} = \dfrac{a_n x^n + a_{n-1} x^{n-1} + \cdots + a_1 x + a_0}{b_m x^m + b_{m-1} x^{m-1} + \cdots + b_1 x + b_0}$ where $P(x)$ and $Q(x)$ have no common factors.

1. If $n < m$, then the x-axis ($y = 0$) is a horizontal asymptote.

2. If $n = m$, then $y = \dfrac{a_n}{b_m}$ is a horizontal asymptote.

3. If $n > m$, then there are no horizontal asymptotes.

Solutions to Odd-Numbered Exercises

1. $f(x) = \dfrac{1}{x-1}$

(a)

x	$f(x)$
0.5	-2
0.9	-10
0.99	-100
0.999	-1000

x	$f(x)$
1.5	2
1.1	10
1.01	100
1.001	1000

x	$f(x)$
5	0.25
10	$0.\overline{1}$
100	$0.\overline{01}$
1000	$0.\overline{001}$

(b) The zero of the denominator is $x = 1$, so $x = 1$ is a vertical asymptote. The degree of the numerator is less than the degree of the denominator so the x-axis, or $y = 0$ is a horizontal asymptote.

(c) The domain is all real numbers except $x = 1$.

3. $f(x) = \dfrac{3x}{|x-1|}$

(a)

x	$f(x)$
0.5	3
0.9	27
0.99	297
0.999	2997

x	$f(x)$
1.5	9
1.1	33
1.01	303
1.001	3003

x	$f(x)$
5	3.75
10	$3.\overline{33}$
100	$3.\overline{03}$
1000	$3.\overline{003}$

(b) The zero of the denominator is $x = 1$, so $x = 1$ is a vertical asymptote. Since $f(x) \to 3$ as $x \to \infty$ and $f(x) \to -3$ as $x \to -\infty$, both $y = 3$ and $y = -3$ are horizontal asymptotes.

(c) The domain is all real numbers except $x = 1$.

5. $f(x) = \dfrac{3x^2}{x^2 - 1}$

(a)

x	$f(x)$
0.5	-1
0.9	-12.79
0.99	-148.8
0.999	-1498

x	$f(x)$
1.5	5.4
1.1	17.29
1.01	152.3
1.001	1502

x	$f(x)$
5	3.125
10	$3.\overline{03}$
100	$3.\overline{0003}$
1000	3

(b) The zeros of the denominator are $x = \pm 1$ so both $x = 1$ and $x = -1$ are vertical asymptotes. Since the degree of the numerator equals the degree of the denominator, $y = \frac{3}{1} = 3$ is a horizontal asymptote.

(c) The domain is all real numbers except $x = \pm 1$.

7. $f(x) = \dfrac{1}{x^2}$ x

Domain: all real numbers except $x = 0$
Vertical asymptote: $x = 0$
Horizontal asymptote: $y = 0$
[Degree of $p(x) <$ degree of $q(x)$]

9. $f(x) = \dfrac{2+x}{2-x} = \dfrac{x+2}{-x+2}$

Domain: all real numbers except $x = 2$
Vertical asymptote: $x = 2$
Horizontal asymptote: $y = -1$
[Degree of $p(x) =$ degree of $q(x)$]

11. $f(x) = \dfrac{x^3}{x^2 - 1}$

Domain: all real numbers except $x = \pm 1$
Vertical asymptotes: $x = \pm 1$
Horizontal asymptote: None
[Degree of $p(x) >$ degree of $q(x)$]

13. $f(x) = \dfrac{3x^2 + 1}{x^2 + x + 9}$

Domain: All real numbers. The denominator has no real zeros. [Try the Quadratic Formula on the denominator.]
Vertical asymptote: None
Horizontal asymptote: $y = 3$
[Degree of $p(x) =$ degree of $q(x)$]

15. $f(x) = \dfrac{2}{x + 2}$

Vertical asymptote: $y = -2$
Matches graph (d).

17. $f(x) = \dfrac{4x + 1}{x}$

Horizontal asymptote: $y = 4$
Matches graph (f).

19. $f(x) = \dfrac{x - 2}{x - 4}$

Vertical asymptote: $x = 4$
Horizontal asymptote: $y = 1$
Matches graph (e).

21. $f(x) = \dfrac{x^2 - 4}{x + 2}$, $g(x) = x - 2$

(a) Domain of f: all real numbers except -2
Domain of g: all real numbers

(b) Since $x + 2$ is a common factor of both the numerator and the denominator of $f(x)$, $x = -2$ is not a vertical asymptote of f. f has no vertical asymptotes.

(c)

x	-4	-3	-2.5	-2	-1.5	-1	0
$f(x)$	-6	-5	-4.5	Undef.	-3.5	-3	-2
$g(x)$	-6	-5	-4.5	-4	-3.5	-3	-2

(d) f and g differ only where f is undefined.

23. $f(x) = \dfrac{x - 3}{x^2 - 3x}$, $g(x) = \dfrac{1}{x}$

(a) Domain of f: all real number except 0 and 3
Domain of g: all real numbers except 0

(b) Since $x - 3$ is a common factor of both the numerator and the denominator of f, $x = 3$ is not a vertical asymptote of f. The only vertical asymptote is $x = 0$.

(c)

x	-1	-0.5	0	0.5	2	3	4
$f(x)$	-1	-2	Undef.	2	$\frac{1}{2}$	Undef.	$\frac{1}{4}$
$g(x)$	-1	-2	Undef	2	$\frac{1}{2}$	$\frac{1}{3}$	$\frac{1}{4}$

(d) They differ only at $x = 3$, where f is undefined and g is defined.

25. $f(x) = \dfrac{1}{(x + 2)(x - 1)} = \dfrac{1}{x^2 + x - 2}$

27. $f(x) = \dfrac{2x^2}{x^2 + 1}$

29. $f(x) = 4 - \dfrac{1}{x}$

(a) As $x \to \pm\infty$, $f(x) \to 4$

(b) As $x \to \infty$, $f(x) \to 4$ but is less than 4.

(c) As $x \to -\infty$, $f(x) \to 4$ but is greater than 4.

31. $f(x) = \dfrac{2x - 1}{x - 3}$

(a) As $x \to \pm\infty$, $f(x) \to 2$

(b) As $x \to \infty$, $f(x) \to 2$ but is greater than 2.

(c) As $x \to -\infty$, $f(x) \to 2$ but is less than 2.

33. $f(x) = \dfrac{x^2 - 4}{x + 1} = \dfrac{(x + 2)(x - 2)}{x + 1}$

The zeros of f correspond to the zeros of the numerator and are $x = \pm 2$.

35. $f(x) = 1 - \dfrac{2}{x - 3} = \dfrac{x - 5}{x - 3}$

The zero of f corresponds to the zero of the numerator and is $x = 5$.

37. $t = \dfrac{38M + 16{,}695}{10(M + 5000)}$

M	200	400	600	800	1000
t	0.472	0.596	0.710	0.817	0.916

M	1200	1400	1600	1800	2000
t	1.009	1.096	1.178	1.255	1.328

The greater the mass, the more time required per oscillation. Also, the model is a "good fit" to the actual data.

39. $C = \dfrac{255p}{100 - p}, \ 0 \le p < 100$

(a) $C(10) = \dfrac{255(10)}{100 - 10} \approx 28.33$ million dollars

(b) $C(40) = \dfrac{255(40)}{100 - 40} = 170$ million dollars

(c) $C(75) = \dfrac{255(75)}{100 - 75} = 765$ million dollars

(d) $C \to \infty$ as $x \to 100$. No, it would not be possible to remove 100% of the pollutants.

41. $N = \dfrac{20(5 + 3t)}{1 + 0.04t}, \ 0 \le t$

(a) $N(5) \approx 333$ deer

$N(10) = 500$ deer

$N(25) = 800$ deer

(b) The herd is limited by the horizontal asymptote: $N = \dfrac{60}{0.04} = 1500$ deer

43. $P = \dfrac{0.5 + 0.9(n - 1)}{1 + 0.9(n - 1)}, \ 0 < n$

(a)

n	1	2	3	4	5	6	7	8	9	10
P	0.50	0.74	0.82	0.86	0.89	0.91	0.92	0.93	0.94	0.95

P approaches 1 as n increases.

(b) $P = \dfrac{0.9n - 0.4}{0.9n + 0.1}$

The percentage of correct responses is limited by a horizontal asymptote:

$P = \dfrac{0.9}{0.9} = 1 = 100\%$

45. $y = \dfrac{8216 - 585t}{1000 + 3t + 29t^2}$

(a)

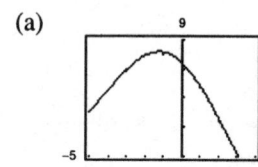

(b) For 1994, use $t = 4$. $\quad y(4) \approx 3.98\%$

(c) Since the model approaches zero as t increases, it would not be good for estimating yields for future years.

Section 4.2   Graphs of Rational Functions

■ You should be able to graph $f(x) = \dfrac{p(x)}{q(x)}$.

(a) Find the x-and y-intercepts.

(b) Find any vertical or horizontal asymptotes.

(c) Plot additional points.

(d) If the degree of the numerator is one more than the degree of the denominator, use long division to find the slant asymptote.

Solutions to Odd-Numbered Exercises

1. $g(x) = \dfrac{2}{x} + 1$

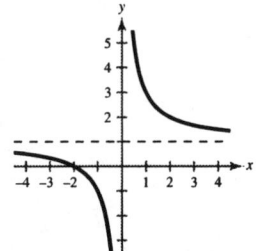

Vertical shift one unit upward

3. $g(x) = -\dfrac{2}{x}$

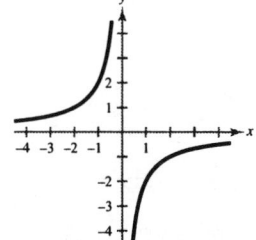

Reflection in the x-axis

5. $g(x) = \dfrac{2}{x^2} - 2$

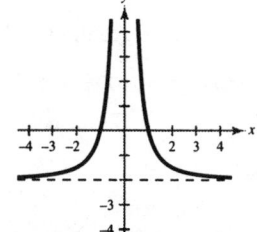

Vertical shift two units downward

7. $g(x) = \dfrac{2}{(x-2)^2}$

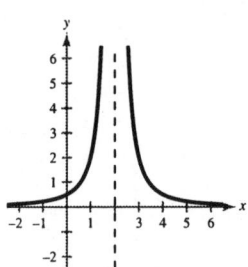

Horizontal shift two units to the right

9. $g(x) = \dfrac{4}{(x+2)^3}$

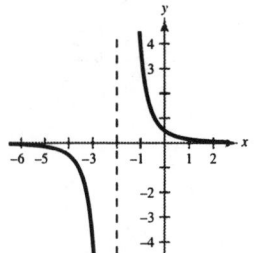

Horizontal shift two units to the left

11. $g(x) = -\dfrac{4}{x^3}$

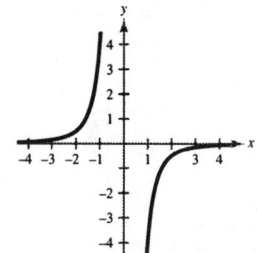

Reflection in the x-axis

13. $f(x) = \dfrac{1}{x+2}$

y-intercept: $\left(0, \frac{1}{2}\right)$

Vertical asymptote: $x = -2$

Horizontal asymptote: $y = 0$

x	-4	-3	-1	0	1
y	$-\frac{1}{2}$	-1	1	$\frac{1}{2}$	$\frac{1}{3}$

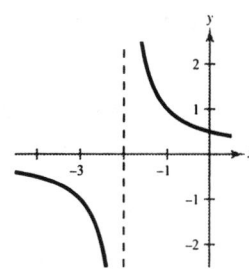

15. $h(x) = -\dfrac{1}{x+2}$

y-intercept: $\left(0, -\frac{1}{2}\right)$
Vertical asymptote: $x = -2$
Horizontal asymptote: $y = 0$

x	−4	−3	−1	0
y	$\frac{1}{2}$	1	−1	$-\frac{1}{2}$

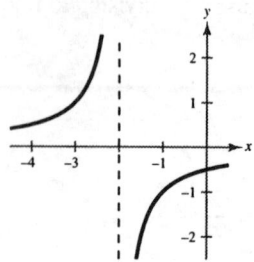

Note: This is the graph of $f(x) = \dfrac{1}{x+2}$
(Exercise 13) reflected about the x-axis.

17. $C(x) = \dfrac{5+2x}{1+x} = \dfrac{2x+5}{x+1}$

x-intercept: $\left(-\frac{5}{2}, 0\right)$
y-intercept: $(0, 5)$
Vertical asymptote: $x = -1$
Horizontal asymptote: $y = 2$

x	−4	−3	−2	0	1	2
C(x)	1	$\frac{1}{2}$	−1	5	$\frac{7}{2}$	3

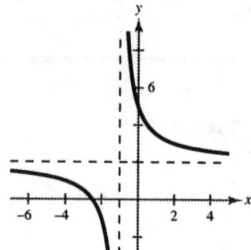

19. $g(x) = \dfrac{1}{x+2} + 2 = \dfrac{2x+5}{x+2}$

Intercepts: $\left(-\frac{5}{2}, 0\right)$, $\left(0, \frac{5}{2}\right)$
Vertical asymptote: $x = -2$
Horizontal asymptote: $y = 2$

x	−4	−3	−1	0	1
y	$\frac{3}{2}$	1	3	$\frac{5}{2}$	$\frac{7}{3}$

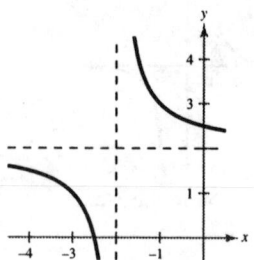

Note: This is the graph of $f(x) = \dfrac{1}{x+2}$
(Exercise 13) shifted upward two units.

21. $f(x) = \dfrac{x^2}{x^2+9}$

Intercept: $(0, 0)$
Horizontal asymptote: $y = 1$
y-axis symmetry

x	±1	±2	±3
y	$\frac{1}{10}$	$\frac{4}{13}$	$\frac{1}{2}$

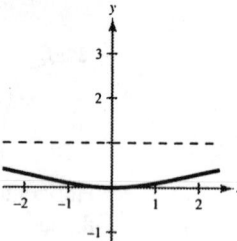

23. $h(x) = \dfrac{x^2}{x^2 - 9}$

Intercept: $(0, 0)$

Vertical asymptotes: $x = \pm 3$

Horizontal asymptote: $y = 1$

y-axis symmetry

x	± 5	± 4	± 2	± 1	0
y	$\frac{25}{16}$	$\frac{16}{7}$	$-\frac{4}{5}$	$-\frac{1}{8}$	0

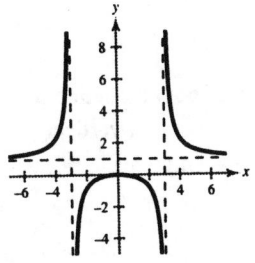

25. $g(s) = \dfrac{s}{s^2 + 1}$

Intercept: $(0, 0)$

Horizontal asymptote: $y = 0$

Origin symmetry

s	-2	-1	0	1	2
$g(s)$	$-\frac{2}{5}$	$-\frac{1}{2}$	0	$\frac{1}{2}$	$\frac{2}{5}$

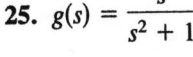

27. $g(x) = \dfrac{4(x + 1)}{x(x - 4)}$

Intercept: $(-1, 0)$

Vertical asymptotes: $x = 0$ and $x = 4$

Horizontal asymptote: $y = 0$

x	-2	-1	1	2	3	5	6
y	$-\frac{1}{3}$	0	$-\frac{8}{3}$	-3	$-\frac{16}{3}$	$\frac{24}{5}$	$\frac{7}{3}$

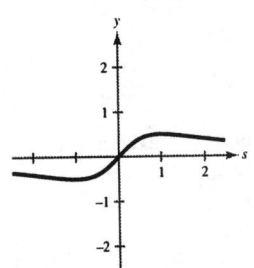

29. $f(x) = \dfrac{3x}{x^2 - x - 2} = \dfrac{3x}{(x + 1)(x - 2)}$

Intercept: $(0, 0)$

Vertical asymptotes: $x = -1$ and $x = 2$

Horizontal asymptote: $y = 0$

x	-3	0	1	3	4
y	$-\frac{9}{10}$	0	$-\frac{3}{2}$	$\frac{9}{4}$	$\frac{6}{5}$

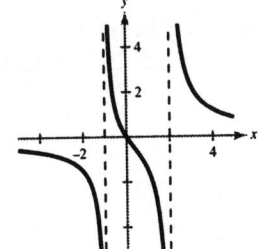

31. $f(x) = \dfrac{2 + x}{1 - x} = -\dfrac{x + 2}{x - 1}$

x-intercept: $(-2, 0)$

y-intercept: $(0, 2)$

Vertical asymptote: $x = 1$

Horizontal asymptote: $y = -1$

x	-2	-1	0	2	3
y	0	$\frac{1}{2}$	2	-4	$-\frac{5}{2}$

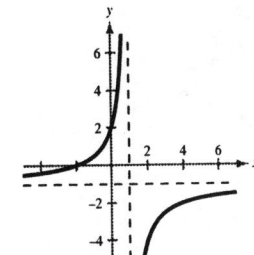

33. $f(t) = \dfrac{3t + 1}{t}$

t-intercept: $\left(-\frac{1}{3}, 0\right)$
Vertical asymptote: $t = 0$
Horizontal asymptote: $y = 3$

t	-2	-1	1	2
$f(t)$	$\frac{5}{2}$	2	4	$\frac{7}{2}$

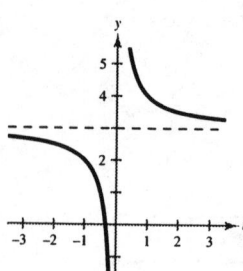

35. $f(x) = \dfrac{x^2 - 1}{x + 1}$, $g(x) = x - 1$

(a) Domain of f: all real numbers except -1
Domain of g: all real numbers

(b) Because $(x + 1)$ is a factor of both the numerator and the denominator of f, $x = -1$ is not a vertical asymptote. f has no vertical asymptotes.

(c)

x	-3	-2	-1.5	-1	-0.5	0	1
$f(x)$	-4	-3	-2.5	Undef.	-1.5	-1	0
$g(x)$	-4	-3	-2.5	-2	-1.5	-1	0

(d)

```
           1
    -4 |--------| 2
       |       /|
       |      / |
       |     /  |
       |    /   |
       |   /    |
       |  /     |
          -5
```

(e) Because there are only a finite number of pixels, the utility may not attempt to evaluate the function where it does not exist.

37. $f(x) = \dfrac{x - 2}{x^2 - 2x}$, $g(x) = \dfrac{1}{x}$

(a) Domain of f: all real numbers except 0 and 2
Domain of g: all real numbers except 0

(b) Because $(x - 2)$ is a factor of both the numerator and the denominator of f, $x = 2$ is not a vertical asymptote. The only vertical asymptote of f is $x = 0$.

(c)

x	-0.5	0	0.5	1	1.5	2	3
$f(x)$	-2	Undef.	2	1	$\frac{2}{3}$	Undef.	$\frac{1}{3}$
$g(x)$	-2	Undef.	2	1	$\frac{2}{3}$	$\frac{1}{2}$	$\frac{1}{3}$

(d)

```
           3
    -3 |--------| 3
       |    |___
       |    |
       |___ |
       |    |
          -3
```

(e) Because there are only a finite number of pixels, the utility may not attempt to evaluate the function where it does not exist.

39. $h(t) = \dfrac{4}{t^2 + 1}$

Domain: all real numbers
Horizontal asymptote: $y = 0$

t	± 2	± 1	0
$h(t)$	$\frac{4}{5}$	2	4

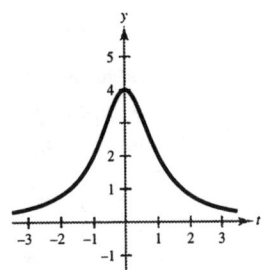

41. $f(t) = \dfrac{2t^2}{t^2 - 4}$

Domain: all real numbers except ± 2,
Vertical asymptotes: $t = \pm 2$
Horizontal asymptote: $y = 2$

t	± 4	± 3	± 1	0
$f(t)$	$\frac{8}{3}$	$\frac{18}{5}$	$-\frac{2}{3}$	0

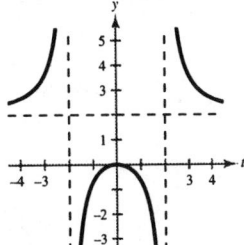

43. $f(x) = \dfrac{20x}{x^2 + 1} - \dfrac{1}{x} = \dfrac{19x^2 - 1}{x(x^2 + 1)}$

Domain: all real numbers except 0,
Vertical asymptote: $x = 0$
Horizontal asymptote: $y = 0$
Origin symmetry

x	-2	-1	1	2
y	$-\frac{15}{2}$	-9	9	$\frac{15}{2}$

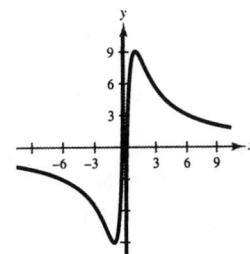

45. $h(x) = \dfrac{6x}{\sqrt{x^2 + 1}}$

Horizontal asymptotes: $y = \pm 6$

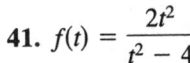

47. $f(x) = \dfrac{4(x - 1)^2}{x^2 - 4x + 5}$

Horizontal asymptote: $y = 4$

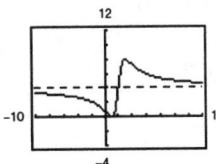

49. $h(x) = \dfrac{6 - 2x}{3 - x} = \dfrac{2(3 - x)}{3 - x}$

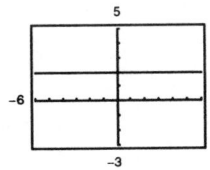

Since $h(x)$ is not reduced and $(3 - x)$ is a factor of both the numerator and the denominator, $x = 3$ is not a horizontal asymptote.

51. False. The graph would have two distinct branches that are separated by the vertical asymptote.

53. $f(x) = \dfrac{2x^2 + 1}{x} = 2x + \dfrac{1}{x}$

Vertical asymptote: $x = 0$
Slant asymptote: $y = 2x$
Origin symmetry

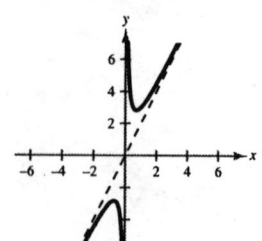

55. $g(x) = \dfrac{x^2 + 1}{x} = x + \dfrac{1}{x}$

Vertical asymptote: $x = 0$
Slant asymptote: $y = x$
Origin symmetry

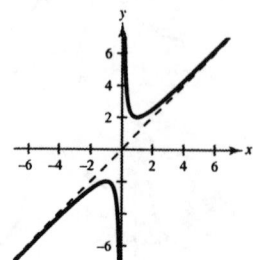

57. $f(x) = \dfrac{x^3}{x^2 - 1} = x + \dfrac{x}{x^2 - 1}$

Intercept: $(0, 0)$
Vertical asymptotes: $x = \pm 1$
Slant asymptote: $y = x$
Origin symmetry

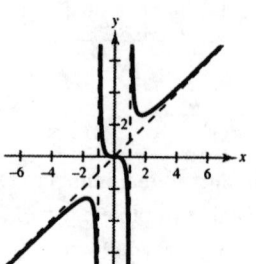

59. $f(x) = \dfrac{x^2 - x + 1}{x - 1} = x + \dfrac{1}{x - 1}$

y-intercept: $(0, -1)$
Vertical asymptote: $x = 1$
Slant asymptote: $y = x$

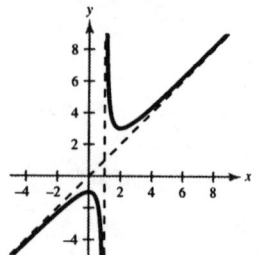

61. $f(x) = \dfrac{x^2 + 5x + 8}{x + 3} = x + 2 + \dfrac{2}{x + 3}$

Domain: all real numbers except -3
y-intercept: $\left(0, \frac{8}{3}\right)$
Vertical asymptote: $x = -3$
Slant asymptote: $y = x + 2$

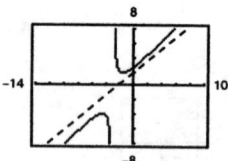

63. $g(x) = \dfrac{1 + 3x^2 - x^3}{x^2} = \dfrac{1}{x^2} + 3 - x = -x + 3 + \dfrac{1}{x^2}$

Domain: all real numbers except 0
Vertical asymptote: $x = 0$
Slant asymptote: $y = -x + 3$

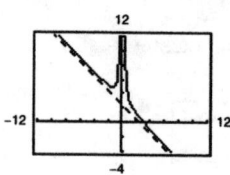

65. (a) x-intercept: $(-1, 0)$

    (b) $\quad 0 = \dfrac{x + 1}{x - 3}$

$\qquad\quad 0 = x + 1$

$\qquad\quad -1 = x$

67. (a) x-intercepts: $(\pm 1, 0)$

    (b) $\quad 0 = \dfrac{1}{x} - x$

$\qquad\quad x = \dfrac{1}{x}$

$\qquad\quad x^2 = 1$

$\qquad\quad x = \pm 1$

69. $y = \dfrac{1}{x+5} + \dfrac{4}{x}$

(a)

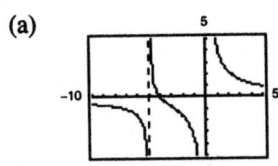

x-intercept: $(-4, 0)$

(b)
$$0 = \frac{1}{x+5} + \frac{4}{x}$$
$$-\frac{4}{x} = \frac{1}{x+5}$$
$$-4(x+5) = x$$
$$-4x - 20 = x$$
$$-5x = 20$$
$$x = -4$$

71. $y = x - \dfrac{6}{x-1}$

(a)

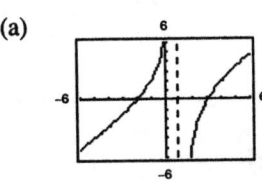

x-intercepts: $(-2, 0),\ (3, 0)$

(b)
$$0 = x - \frac{6}{x-1}$$
$$\frac{6}{x-1} = x$$
$$6 = x(x-1)$$
$$0 = x^2 - x - 6$$
$$0 = (x+2)(x-3)$$
$$x = -2,\quad x = 3$$

73. (a) $0.25(50) + 0.75(x) = C(50 + x)$

$$C = \frac{12.50 + 0.75x}{50 + x} \cdot \frac{4}{4}$$

$$C = \frac{50 + 3x}{4(50 + x)} = \frac{3x + 50}{4(x + 50)}$$

(b) Domain: $x > 0$ and $x \le 1000 - 50$
Thus, $0 \le x \le 950$ OR $[0, 950]$.

(c)

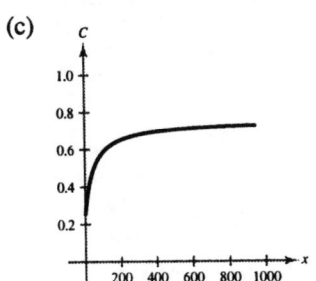

As the tank is filled, the rate at which the concentration is increasing slows down.
It approaches the horizontal asymptote of $C = \frac{3}{4} = 0.75$.

75. (a) $A = xy$ and

$$(x - 4)(y - 2) = 30$$

$$y - 2 = \frac{30}{x - 4}$$

$$y = 2 + \frac{30}{x - 4} = \frac{2x + 22}{x - 4}$$

Thus, $A = xy = x\left(\frac{2x + 22}{x - 4}\right) = \frac{2x(x + 11)}{x - 4}$.

(b) Domain: Since the margins on the left and right are each 2 inches, $x > 4$, OR $(4, \infty)$.

(c)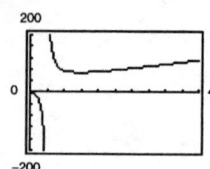
The area is minimum when $x \approx 11.75$ inches and $y \approx 5.89$ inches.

77. $f(x) = \frac{3(x + 1)}{x^2 + x + 1}$

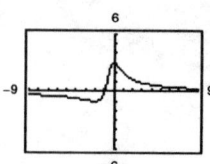

Minimum: $(-2, -1)$
Maximum: $(0, 3)$

79. $C = 100\left(\frac{200}{x^2} + \frac{x}{x + 30}\right)$, $1 \le x$

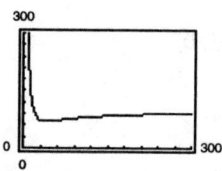

The minimum occurs when $x \approx 40.4 \approx 40$.

81. $C = \frac{3t^2 + t}{t^3 + 50}$, $0 \le t$

(a) The horizontal asymptote is the t-axis, or $C = 0$. This indicates that the chemical eventually dissipates.

(b)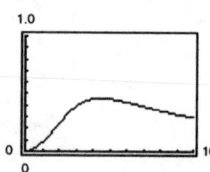
The maximum occurs when $t \approx 4.5$.

83. $y = x + 1 + \frac{a}{x - 2}$ This has a slant asymptote of $x + 1$ and a vertical asymptote of $x = 2$.

$0 = -2 + 1 + \frac{a}{-2 - 2}$ Since $x = -2$ is a zero, $(-2, 0)$ is on the graph. Use this point to solve for a.

$1 = \frac{a}{-4}$

$-4 = a$

Thus, $y = x + 1 - \frac{4}{x - 2} = \frac{(x + 1)(x - 2) - 4}{x - 2} = \frac{x^2 - x - 6}{x - 2}$.

85. $10 - 3x \le 0$

$\quad -3x \le -10$

$\quad x \ge \frac{10}{3}$

$\quad x \ge 3\frac{1}{3}$

87. $|4(x - 2)| \le 20$

$\quad -20 < 4(x - 2) < 20$

$\quad -5 < x - 2 < 5$

$\quad -3 < x < 7$

Section 4.3   Partial Fractions

■ You should know how to decompose a rational function $\dfrac{N(x)}{D(x)}$ into partial fractions.

(a) If the fraction is improper, divide to obtain

$$\frac{N(x)}{D(x)} = p(x) + \frac{N_1(x)}{D(x)}$$

where $p(x)$ is a polynomial.

(b) Factor the denominator completely into linear and irreducible (over the reals) quadratic factors.

(c) For each factor of the form $(px + q)^m$, the partial fraction decomposition includes the terms

$$\frac{A_1}{(px + q)} + \frac{A_2}{(px + q)^2} + \cdots + \frac{A_m}{(px + q)^m}.$$

(d) For each factor of the form $(ax^2 + bx + c)^n$, the partial fraction decomposition includes the terms

$$\frac{B_1x + C_1}{ax^2 + bx + c} + \frac{B_2x + C_2}{(ax^2 + bx + c)^2} + \cdots + \frac{B_nx + C_n}{(ax^2 + bx + c)^n}.$$

■ You should know how to determine the values of the constants in the numerators.

(a) Set $\dfrac{N_1(x)}{D(x)}$ = partial fraction decomposition.

(b) Multiply both sides by $D(x)$. This is called the basic equation.

(c) For distinct linear factors, substitute the roots of the distinct linear factors into the basic equation.

(d) For repeated linear factors, use the coefficients found in part (c) to rewrite the basic equation. Then use other values of x to solve for the remaining coefficients.

(e) For quadratic factors, expand the basic equation, collect like terms, and then equate the coefficients of like terms.

Solutions to Odd-Numbered Exercises

1. $\dfrac{7}{x^2 - 14x} = \dfrac{7}{x(x-14)} = \dfrac{A}{x} + \dfrac{B}{x-14}$

3. $\dfrac{12}{x^3 - 10x^2} = \dfrac{12}{x^2(x-10)} = \dfrac{A}{x} + \dfrac{B}{x^2} + \dfrac{C}{x-10}$

5. $\dfrac{2x-3}{x^3 + 10x} = \dfrac{2x-3}{x(x^2+10)} = \dfrac{A}{x} + \dfrac{Bx+C}{x^2+10}$

7. $\dfrac{1}{x^2 - 1} = \dfrac{A}{x+1} + \dfrac{B}{x-1}$

$1 = A(x-1) + B(x+1)$

Let $x = -1$: $1 = -2A \implies A = -\dfrac{1}{2}$

Let $x = 1$: $1 = 2B \implies B = \dfrac{1}{2}$

$\dfrac{1}{x^2-1} = \dfrac{1/2}{x-1} - \dfrac{1/2}{x+1} = \dfrac{1}{2}\left(\dfrac{1}{x-1} - \dfrac{1}{x+1}\right)$

9. $\dfrac{1}{x^2 + x} = \dfrac{A}{x} + \dfrac{B}{x+1}$

$1 = A(x+1) + Bx$

Let $x = 0$: $1 = A$

Let $x = -1$: $1 = -B \implies B = -1$

$\dfrac{1}{x^2+x} = \dfrac{1}{x} - \dfrac{1}{x+1}$

11. $\dfrac{1}{2x^2 + x} = \dfrac{A}{2x+1} + \dfrac{B}{x}$

$1 = Ax + B(2x+1)$

Let $x = -\dfrac{1}{2}$: $1 = -\dfrac{1}{2}A \implies A = -2$

Let $x = 0$: $1 = B$

$\dfrac{1}{2x^2+x} = \dfrac{1}{x} - \dfrac{2}{2x+1}$

13. $\dfrac{3}{x^2 + x - 2} = \dfrac{A}{x-1} + \dfrac{B}{x+2}$

$3 = A(x+2) + B(x-1)$

Let $x = 1$: $3 = 3A \implies A = 1$

Let $x = -2$: $3 = -3B \implies B = -1$

$\dfrac{3}{x^2+x-2} = \dfrac{1}{x-1} - \dfrac{1}{x+2}$

15. $\dfrac{x^2 + 12x + 12}{x^3 - 4x} = \dfrac{A}{x} + \dfrac{B}{x+2} + \dfrac{C}{x-2}$

$x^2 + 12x + 12 = A(x+2)(x-2) + Bx(x-2) + Cx(x+2)$

Let $x = 0$: $12 = -4A \implies A = -3$

Let $x = -2$: $-8 = 8B \implies B = -1$

Let $x = 2$: $40 = 8C \implies C = 5$

$\dfrac{x^2 + 12x + 12}{x^3 - 4x} = -\dfrac{3}{x} - \dfrac{1}{x+2} + \dfrac{5}{x-2}$

17. $\dfrac{4x^2 + 2x - 1}{x^2(x+1)} = \dfrac{A}{x} + \dfrac{B}{x^2} + \dfrac{C}{x+1}$

$4x^2 + 2x - 1 = Ax(x+1) + B(x+1) + Cx^2$

Let $x = 0$: $-1 = B$

Let $x = -1$: $1 = C$

Let $x = 1$: $5 = 2A + 2B + C$

$\qquad\qquad\quad 5 = 2A - 2 + 1$

$\qquad\qquad\quad 6 = 2A$

$\qquad\qquad\quad 3 = A$

$\dfrac{4x^2 + 2x - 1}{x^2(x+1)} = \dfrac{3}{x} - \dfrac{1}{x^2} + \dfrac{1}{x+1}$

19. $\dfrac{3x}{(x-3)^2} = \dfrac{A}{x-3} + \dfrac{B}{(x-3)^2}$

$$3x = A(x-3) + B$$

Let $x = 3$: $9 = B$

Let $x = 0$: $0 = -3A + B$

$$0 = -3A + 9$$

$$3 = A$$

$$\frac{3x}{(x-3)^2} = \frac{3}{x-3} + \frac{9}{(x-3)^2}$$

21. $\dfrac{x^2-1}{x(x^2+1)} = \dfrac{A}{x} + \dfrac{Bx+C}{x^2+1}$

$$x^2 - 1 = A(x^2+1) + (Bx+C)x$$

Let $x = 0$: $-1 = A$

$$x^2 - 1 = Ax^2 + A + Bx^2 + Cx$$
$$= -x^2 - 1 + Bx^2 + Cx$$
$$= x^2(B-1) + Cx - 1$$

Equating coefficients of like powers:

$$1 = B - 1$$
$$2 = B \ \text{ and } \ 0 = C$$

$$\frac{x^2-1}{x(x^2+1)} = -\frac{1}{x} + \frac{2x}{x^2+1}$$

23. $\dfrac{x^2}{x^4 - 2x^2 - 8} = \dfrac{x^2}{(x^2-4)(x^2+2)} = \dfrac{A}{x+2} + \dfrac{B}{x-2} + \dfrac{Cx+D}{x^2+2}$

$$x^2 = A(x-2)(x^2+2) + B(x+2)(x^2+2) + (Cx+D)(x^2-4)$$

Let $x = -2$: $4 = -24A \implies A = -\dfrac{1}{6}$

Let $x = 2$: $4 = 24B \implies B = \dfrac{1}{6}$

$$x^2 = -\frac{1}{6}(x-2)(x^2+2) + \frac{1}{6}(x+2)(x^2+2) + (Cx+D)(x^2-4)$$

$$x^2 = -\frac{1}{6}x^3 + \frac{1}{3}x^2 - \frac{1}{3}x + \frac{2}{3} + \frac{1}{6}x^3 + \frac{1}{3}x^2 + \frac{1}{3}x + \frac{2}{3} + Cx^3 + Dx^2 - 4Cx - 4D$$

$$x^2 = Cx^3 + \left(\frac{2}{3} + D\right)x^2 - 4Cx + \left(\frac{4}{3} - 4D\right)$$

Equating coefficients of like powers:

$$C = 0$$

$$1 = \frac{2}{3} + D \implies D = \frac{1}{3}$$

$$\frac{x^2}{x^4 - 2x^2 - 8} = -\frac{1}{6(x+2)} + \frac{1}{6(x-2)} + \frac{1}{3(x^2+2)}$$

25. $\dfrac{x}{16x^4 - 1} = \dfrac{A}{2x+1} + \dfrac{B}{2x-1} + \dfrac{Cx+D}{4x^2+1}$

$$x = A(2x-1)(4x^2+1) + B(2x+1)(4x^2+1) + (Cx+D)(2x+1)(2x-1)$$

Let $x = -\dfrac{1}{2}$: $-\dfrac{1}{2} = -4A \implies A = \dfrac{1}{8}$

Let $x = \dfrac{1}{2}$: $\dfrac{1}{2} = 4B \implies B = \dfrac{1}{8}$

Let $x = 0$: $0 = -A + B - D$

$$0 = -\frac{1}{8} + \frac{1}{8} - D$$

$$0 = D$$

— **CONTINUED** —

25. — CONTINUED —

Let $x = 1$: $1 = 5A + 15B + 3C + 3D$

$$1 = \frac{5}{8} + \frac{15}{8} + 3C + 0$$

$$-\frac{1}{2} = C$$

$$\frac{x}{16x^4 - 1} = \frac{1/8}{2x + 1} + \frac{1/8}{2x - 1} - \frac{x/2}{4x^2 + 1} = \frac{1}{8(2x + 1)} + \frac{1}{8(2x - 1)} - \frac{x}{2(4x^2 + 1)}$$

27. $\dfrac{x^2 + 5}{(x + 1)(x^2 - 2x + 3)} = \dfrac{A}{x + 1} + \dfrac{Bx + C}{x^2 - 2x + 3}$

$$x^2 + 5 = A(x^2 - 2x + 3) + (Bx + C)(x + 1)$$

Let $x = -1$: $6 = 6A \implies A = 1$

$$x^2 + 5 = x^2 - 2x + 3 + Bx^2 + Bx + Cx + C$$

$$= x^2(1 + B) + x(-2 + B + C) + (3 + C)$$

Equating coefficients of like powers:

$1 = 1 + B$, $\quad 0 = -2 + B + C$, $\quad$ and $\quad 5 = 3 + C$

$0 = B \qquad\qquad 0 = -2 + 0 + C \qquad\qquad 2 = C$

$\qquad\qquad\qquad\quad 2 = C$

$$\frac{x^2 + 5}{(x + 1)(x^2 - 2x + 3)} = \frac{1}{x + 1} + \frac{2}{x^2 - 2x + 3}$$

29. $\qquad \dfrac{x^4}{(x - 1)^3} = \dfrac{x^4}{x^3 - 3x^2 + 3x - 1} = x + 3 + \dfrac{6x^2 - 8x + 3}{(x - 1)^3}$

$$\frac{6x^2 - 8x + 3}{(x - 1)^3} = \frac{A}{x - 1} + \frac{B}{(x - 1)^2} + \frac{C}{(x - 1)^3}$$

$$6x^2 - 8x + 3 = A(x - 1)^2 + B(x - 1) + C$$

Let $x = 1$: $1 = C$

$$6x^2 - 8x + 3 = Ax^2 - 2Ax + A + Bx - B + 1$$

$$6x^2 - 8x + 3 = Ax^2 + (-2A + B)x + (A - B + 1)$$

Equating coefficients of like powers: .

$6 = A$, $\quad -8 = -2A + B$ $\quad$ and $\quad 3 = A - B + 1$

$\qquad\qquad -8 = -12 + B \qquad\qquad 3 = 6 - B + 1$

$\qquad\qquad\quad 4 = B \qquad\qquad\qquad\quad 4 = B$

$$\frac{x^4}{(x - 1)^3} = x + 3 + \frac{6}{x - 1} + \frac{4}{(x - 1)^2} + \frac{1}{(x - 1)^3}$$

31. $\dfrac{5 - x}{2x^2 + x - 1} = \dfrac{A}{2x - 1} + \dfrac{B}{x + 1}$

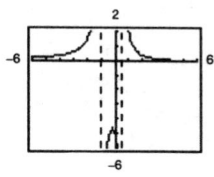

$$-x + 5 = A(x + 1) + B(2x - 1)$$

Let $x = \dfrac{1}{2}$: $\dfrac{9}{2} = \dfrac{3}{2}A \implies A = 3$

Let $x = -1$: $6 = -3B \implies B = -2$

$$\dfrac{5 - x}{2x^2 + x - 1} = \dfrac{3}{2x - 1} - \dfrac{2}{x + 1}$$

33. $\dfrac{x - 1}{x^3 + x^2} = \dfrac{A}{x} + \dfrac{B}{x^2} + \dfrac{C}{x + 1}$

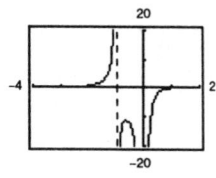

$$x - 1 = Ax(x + 1) + B(x + 1) + Cx^2$$

Let $x = -1$: $-2 = C$

Let $x = 0$: $-1 = B$

Let $x = 1$: $0 = 2A + 2B + C$

$$0 = 2A - 2 - 2$$

$$2 = A$$

$$\dfrac{x - 1}{x^3 + x^2} = \dfrac{2}{x} - \dfrac{1}{x^2} - \dfrac{2}{x + 1}$$

35. $\dfrac{x^2 + x + 2}{(x^2 + 2)^2} = \dfrac{Ax + B}{x^2 + 2} + \dfrac{Cx + D}{(x^2 + 2)^2}$

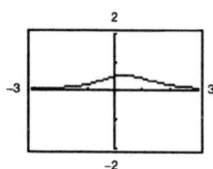

$$x^2 + x + 2 = (Ax + B)(x^2 + 2) + Cx + D$$

$$x^2 + x + 2 = Ax^3 + Bx^2 + (2A + C)x + (2B + D)$$

Equating coefficients of like powers:

$$0 = A$$

$$1 = B$$

$$1 = 2A + C \implies C = 1$$

$$2 = 2B + D \implies D = 0$$

$$\dfrac{x^2 + x + 2}{(x^2 + 2)^2} = \dfrac{1}{x^2 + 2} + \dfrac{x}{(x^2 + 2)^2}$$

37. $\dfrac{2x^3 - 4x^2 - 15x + 5}{x^2 - 2x - 8} = 2x + \dfrac{x + 5}{(x + 2)(x - 4)}$

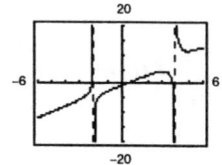

$$\dfrac{x + 5}{(x + 2)(x - 4)} = \dfrac{A}{x + 2} + \dfrac{B}{x - 4}$$

$$x + 5 = A(x - 4) + B(x + 2)$$

Let $x = -2$: $3 = -6A \implies A = -\dfrac{1}{2}$

Let $x = 4$: $9 = 6B \implies B = \dfrac{3}{2}$

$$\dfrac{2x^3 - 4x^2 - 15x + 5}{x^2 - 2x - 8} = 2x + \dfrac{1}{2}\left(\dfrac{3}{x - 4} - \dfrac{1}{x + 2}\right)$$

39. $\dfrac{1}{a^2 - x^2} = \dfrac{A}{a + x} + \dfrac{B}{a - x}$, a is a constant.

$$1 = A(a - x) + B(a + x)$$

Let $x = -a$: $1 = 2aA \implies A = \dfrac{1}{2a}$

Let $x = a$: $1 = 2aB \implies B = \dfrac{1}{2a}$

$$\dfrac{1}{a^2 - x^2} = \dfrac{1}{2a}\left(\dfrac{1}{a + x} + \dfrac{1}{a - x}\right)$$

41. $\dfrac{1}{y(a - y)} = \dfrac{A}{y} + \dfrac{B}{a - y}$

$$1 = A(a - y) + By$$

Let $y = 0$: $1 = aA \implies A = \dfrac{1}{a}$

Let $y = a$: $1 = aB \implies B = \dfrac{1}{a}$

$$\dfrac{1}{y(a - y)} = \dfrac{1}{a}\left(\dfrac{1}{y} + \dfrac{1}{a - y}\right)$$

43. $\dfrac{x - 12}{x(x - 4)} = \dfrac{A}{x} + \dfrac{B}{x - 4}$

$$x - 12 = A(x - 4) + Bx$$

Let $x = 0$: $-12 = -4A \implies A = 3$

Let $x = 4$: $-8 = 4B \implies B = -2$

$$\dfrac{x - 12}{x(x - 4)} = \dfrac{3}{x} - \dfrac{2}{x - 4}$$

$y = \dfrac{x - 12}{x(x - 4)}$

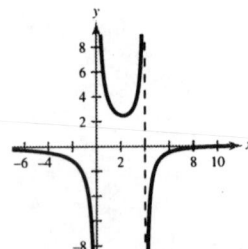

Vertical asymptotes: $x = 0$ and $x = 4$

$y = \dfrac{3}{x}$

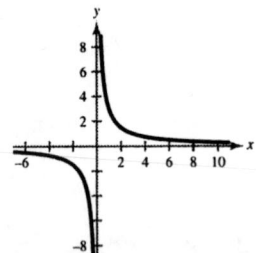

Vertical asymptote: $x = 0$

$y = -\dfrac{2}{x - 4}$

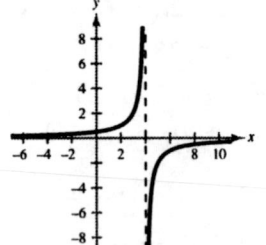

Vertical asymptote: $x = 4$

The combination of the vertical asymptotes of the terms of the decompositions are the same as the vertical asymptotes of the rational function.

45. $\dfrac{2(4x - 3)}{x^2 - 9} = \dfrac{A}{x - 3} + \dfrac{B}{x + 3}$

$2(4x - 3) = A(x + 3) + B(x - 3)$

Let $x = 3$: $18 = 6A \implies A = 3$

Let $x = -3$: $-30 = -6B \implies B = 5$

$\dfrac{2(4x - 3)}{x^2 - 9} = \dfrac{3}{x - 3} + \dfrac{5}{x + 3}$

$y = \dfrac{2(4x - 3)}{x^2 - 9}$

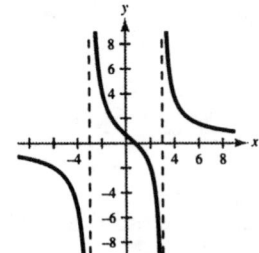

Vertical asymptotes: $x = \pm 3$

$y = \dfrac{3}{x - 3}$

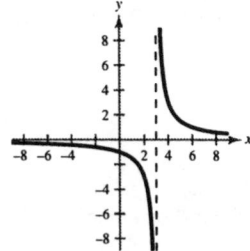

Vertical asymptote: $x = 3$

$y = \dfrac{5}{x + 3}$

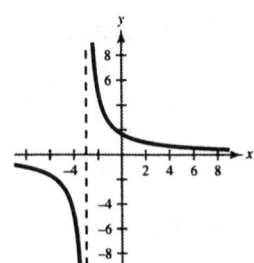

Vertical asymptote: $x = -3$

The combination of the vertical asymptotes of the terms of the decompositions are the same as the vertical asymptotes of the rational function.

47. (a) $\dfrac{2000(4 - 3x)}{(11 - 7x)(7 - 4x)} = \dfrac{A}{11 - 7x} + \dfrac{B}{7 - 4x}, \ 0 \le x \le 1$

$2000(4 - 3x) = A(7 - 4x) + B(11 - 7x)$

Let $x = \dfrac{11}{7}$: $-\dfrac{10{,}000}{7} = \dfrac{5}{7}A \implies A = -2000$

Let $x = \dfrac{7}{4}$: $-2500 = -\dfrac{5}{4}B \implies B = 2000$

$\dfrac{2000(4 - 3x)}{(11 - 7x)(7 - 4x)} = \dfrac{-2000}{11 - 7x} + \dfrac{2000}{7 - 4x} = \dfrac{2000}{7 - 4x} - \dfrac{2000}{11 - 7x}, \ 0 \le x \le 1$

(b) $y_1 = \dfrac{2000}{7 - 4x}$

$y_2 = \dfrac{2000}{11 - 7x}$

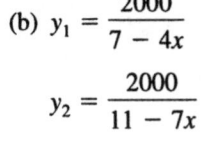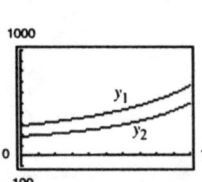

Section 4.4   Conics

You should know the following basic definitions of conic sections.

■ A parabola is the set of all points (x, y) that are equidistant from a fixed line (directrix) and a fixed point (focus) not on the line.

    (a) Standard Equation with Vertex $(0, 0)$ and Directrix $y = -p$ (vertical axis): $x^2 = 4py$

    (b) Standard Equation with Vertex $(0, 0)$ and Directrix $x = -p$ (horizontal axis): $y^2 = 4px$

    (c) The focus lies on the axis p units (directed distance) from the vertex.

■ An ellipse is the set of all points (x, y) the sum of whose distances from two distinct fixed points (foci) is constant.

    (a) Standard Equation of an Ellipse with Center $(0, 0)$, Major Axis Length $2a$, and Minor Axis Length $2b$:

         1. Horizontal Major Axis: $\dfrac{x^2}{a^2} + \dfrac{y^2}{b^2} = 1$

         2. Vertical Major Axis: $\dfrac{x^2}{b^2} + \dfrac{y^2}{a^2} = 1$

    (b) The foci lie on the major axis, c units from the center, where a, b, and c are related by the equation $c^2 = a^2 - b^2$.

    (c) The vertices and endpoints of the minor axis are:

         1. Horizontal Axis: $(\pm a, 0)$ and $(0, \pm b)$

         2. Vertical Axis: $(0, \pm a)$ and $(\pm b, 0)$

■ A hyperbola is the set of all points (x, y) the difference of whose distances from two distinct fixed points (foci) is constant.

    (a) Standard Equation of Hyperbola with Center $(0, 0)$

         1. Horizontal Transverse Axis: $\dfrac{x^2}{a^2} - \dfrac{y^2}{b^2} = 1$

         2. Vertical Transverse Axis: $\dfrac{y^2}{a^2} - \dfrac{x^2}{b^2} = 1$

    (b) The vertices and foci are a and c units from the center and $b^2 = c^2 - a^2$.

    (c) The asymptotes of the hyperbola are:

         1. Horizontal Transverse Axis: $y = \pm \dfrac{b}{a} x$

         2. Vertical Transverse Axis: $y = \pm \dfrac{a}{b} x$

Solutions to Odd-Numbered Exercises

1. $x^2 = 2y$

Parabola opening upward
Not shown

3. $y^2 = 2x$

Parabola opening to the right
Matches (e).

5. $9x^2 + y^2 = 9$

$\dfrac{x^2}{1} + \dfrac{y^2}{9} = 1$

Ellipse with vertical major axis
Not shown

7. $9x^2 - y^2 = 9$

$\dfrac{x^2}{1} - \dfrac{y^2}{9} = 1$

Hyperbola with horizontal transverse axis
Matches (f).

9. $y = \frac{1}{2}x^2$

$x^2 = 2y = 4\left(\frac{1}{2}\right)y; \ p = \frac{1}{2}$

Vertex: $(0, 0)$

Focus: $\left(0, \frac{1}{2}\right)$

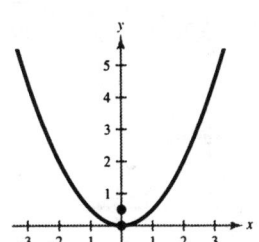

11. $y^2 = -6x$

$y^2 = 4\left(-\frac{3}{2}\right)x; \ p = -\frac{3}{2}$

Vertex: $(0, 0)$

Focus: $\left(-\frac{3}{2}, 0\right)$

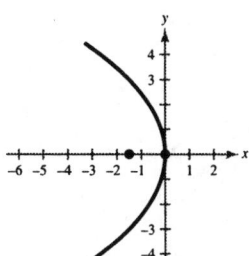

13. $x^2 + 8y = 0$

$x^2 = 4(-2)y; \ p = -2$

Vertex: $(0, 0)$

Focus: $(0, -2)$

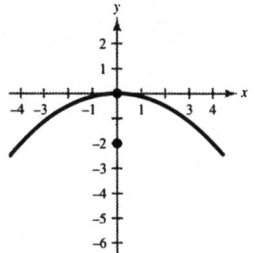

15. $y^2 - 8 = 0$ and $x - y + 2 = 0$

$\qquad y^2 = 8x \qquad\qquad y_3 = x + 2$

$\qquad y_1 = \sqrt{8x}$

$\qquad y_2 = -\sqrt{8x}$

The point of tangency is $(2, 4)$.

17. Focus: $\left(0, -\frac{3}{2}\right)$

$x^2 = 4\left(-\frac{3}{2}\right)y$

$x^2 = -6y$

19. Focus: $(-2, 0)$

$y^2 = 4(-2)x$

$y^2 = -8x$

21. Directrix: $y = -1$

$x^2 = 4(1)y$

$x^2 = 4y$

23. Directrix: $y = 2$

$x^2 = 4(-2)y$

$x^2 = -8y$

25. $y^2 = 4px$

$6^2 = 4p(4)$

$36 = 16p$

$p = \frac{9}{4}$

$y^2 = 4\left(\frac{9}{4}\right)x$

$y^2 = 9x$

27. $x^2 = 4py$

$3^2 = 4p(6)$

$9 = 24p$

$\frac{3}{8} = p$

$x^2 = 4\left(\frac{3}{8}\right)y$

$x^2 = \frac{3}{2}y \quad \text{OR} \quad y = \frac{2}{3}x^2$

Focus: $\left(0, \frac{3}{8}\right)$

29. $x^2 = 4py, \ p = 3.5$

$x^2 = 4(3.5)y$

$x^2 = 14y$

$y = \frac{1}{14}x^2$

31. (a) $x^2 = 4py$

$32^2 = 4p\left(\frac{1}{12}\right)$

$1024 = \frac{1}{3}p$

$3072 = p$

$x^2 = 4(3072)y$

$y = \frac{x^2}{12,288}$

(b) $\dfrac{1}{24} = \dfrac{x^2}{12,288}$

$\dfrac{12,288}{24} = x^2$

$512 = x^2$

$x \approx 22.6$ feet

33. False. If the graph crossed the directrix there would exist points closer to the directrix than to the focus.

35. $\dfrac{x^2}{25} + \dfrac{y^2}{16} = 1$

Horizontal major axis

$a = 5, \ b = 4$

Center: $(0, 0)$

Vertices: $(\pm 5, 0)$

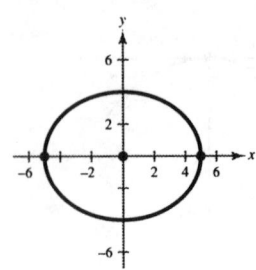

37. $\dfrac{x^2}{16} + \dfrac{y^2}{25} = 1$

Vertical major axis

$a = 5, \ b = 4$

Center: $(0, 0)$

Vertices: $(0, \pm 5)$

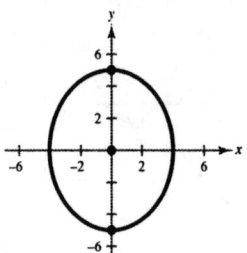

39. $\dfrac{x^2}{9} + \dfrac{y^2}{5} = 1$

Horizontal major axis

$a = 3, \ b = \sqrt{5}$

Center: $(0, 0)$

Vertices: $(\pm 3, 0)$

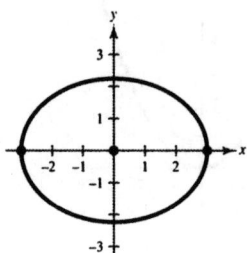

41. $5x^2 + 3y^2 = 15$

$$3y^2 = 15 - 5x^2$$

$$y^2 = 5 - \tfrac{5}{3}x^2$$

$$y = \pm\sqrt{5 - \tfrac{5}{3}x^2}$$

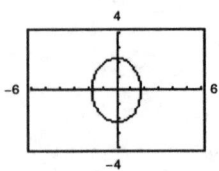

43. Since x is negative, $x = -\frac{3}{2}\sqrt{4 - y^2}$ represents the left half of the ellipse.

45. Vertices: $(0, \pm 2) \ \Rightarrow \ a = 2$

Minor axis of length 2 $\ \Rightarrow \ b = 1$

Vertical major axis

$$\dfrac{x^2}{b^2} + \dfrac{y^2}{a^2} = 1$$

$$\dfrac{x^2}{1} + \dfrac{y^2}{4} = 1$$

47. Vertices: $(\pm 5, 0) \ \Rightarrow \ a = 5$

Foci: $(\pm 2, 0) \ \Rightarrow \ b = \sqrt{5^2 - 2^2} = \sqrt{21}$

Horizontal major axis

$$\dfrac{x^2}{a^2} + \dfrac{y^2}{b^2} = 1$$

$$\dfrac{x^2}{25} + \dfrac{y^2}{21} = 1$$

49. Foci: $(\pm 5, 0) \ \Rightarrow \ c = 5$

Major axis of length 12 $\ \Rightarrow \ a = 6$

$b = \sqrt{6^2 - 5^2} = \sqrt{11}$

Horizontal major axis

$$\dfrac{x^2}{a^2} + \dfrac{y^2}{b^2} = 1$$

$$\dfrac{x^2}{36} + \dfrac{y^2}{11} = 1$$

51. Vertices: $(0, \pm 5) \ \Rightarrow \ a = 5$

Vertical major axis

$$\dfrac{x^2}{b^2} + \dfrac{y^2}{25} = 1$$

$$\dfrac{(4)^2}{b^2} + \dfrac{(2)^2}{25} = 1$$

$$\dfrac{400}{21} = b^2$$

$$\dfrac{21x^2}{400} + \dfrac{y^2}{25} = 1$$

53. (a) The length of the string is $2a$.

(b) The path is an ellipse since the sum of the distances from the two fixed points (thumbtacks) is constant ($2a$).

55.

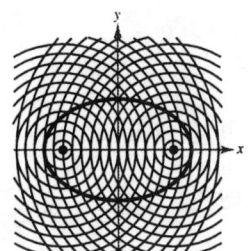

57. (a) $a + b = 20 \implies b = 20 - a$

$A = \pi ab = \pi a(20 - a)$

(b) $264 = \pi a(20 - a)$

$\pi a^2 - 20\pi a + 264 = 0$

$a \approx 14$ OR $a \approx 6$ by the Quadratic Formula

$b = 6$ $b = 14$

Since $a > b$ we choose $a = 14$ and $b = 6$.

$\dfrac{x^2}{14^2} + \dfrac{y^2}{6^2} = 1$

$\dfrac{x^2}{196} + \dfrac{y^2}{36} = 1$

(c)

a	8	9	10	11	12	13
A	301.6	311.0	314.2	311.0	301.6	285.9

(d)

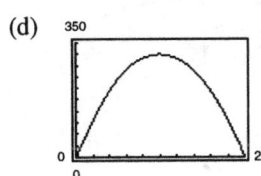

The area is maximum when $a = b = 10$ and it is a circle.

59. No it is not an ellipse. The exponent on y is not 2.

61. $\dfrac{x^2}{a^2} + \dfrac{y^2}{4^2} = 1$, $1 \le a \le 8$

The shape changes from an ellipse with a vertical major axis of length 8 and a minor axis of length 2 to a circle with a diameter of 8 and then to an ellipse with a horizontal major axis of length 16 and a minor axis of length 8.

63. $\dfrac{x^2}{4} + \dfrac{y^2}{1} = 1$

$a = 2$, $b = 1$, $c = \sqrt{3}$

Points on the ellipse: $(\pm 2, 0)$, $(0, \pm 1)$

Length of latus recta: $\dfrac{2b^2}{a} = 1$

Additional points: $\left(\sqrt{3}, \pm\dfrac{1}{2}\right)$, $\left(-\sqrt{3}, \pm\dfrac{1}{2}\right)$

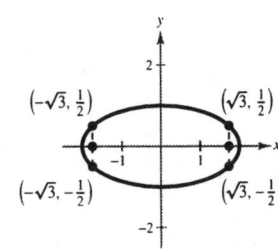

65. $9x^2 + 4y^2 = 36$

$\dfrac{x^2}{4} + \dfrac{y^2}{9} = 1$

Points on the ellipse: $(\pm 2, 0), (0, \pm 3)$

Length of latus recta: $\dfrac{2b^2}{a} = \dfrac{2 \cdot 2^2}{3} = \dfrac{8}{3}$

Additional points: $\left(\pm\dfrac{4}{3}, -\sqrt{5}\right), \left(\pm\dfrac{4}{3}, \sqrt{5}\right)$

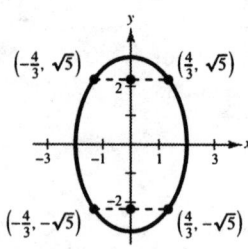

67. $x^2 - y^2 = 1$

$a = 1, \ b = 1$

Center: $(0, 0)$

Vertices: $(\pm 1, 0)$

Asymptotes: $y = \pm x$

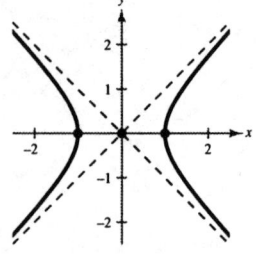

69. $\dfrac{y^2}{1} - \dfrac{x^2}{4} = 1$

$a = 1, \ b = 1$

Center: $(0, 0)$

Vertices: $(0, \pm 1)$

Asymptotes: $y = \pm\dfrac{1}{2}x$

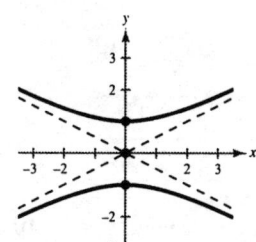

71. $\dfrac{y^2}{25} - \dfrac{x^2}{144} = 1$

$a = 5, \ b = 12$

Center: $(0, 0)$

Vertices: $(0, \pm 5)$

Asymptotes: $y = \pm\dfrac{5}{12}x$

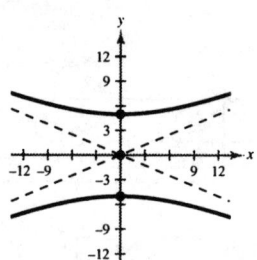

73. $2x^2 - 3y^2 = 6$

$y^2 = \dfrac{1}{3}(2x^2 - 6)$

$y = \pm\sqrt{\dfrac{2}{3}(x^2 - 3)}$

Asymptotes: $y = \pm\sqrt{\dfrac{2}{3}}x$

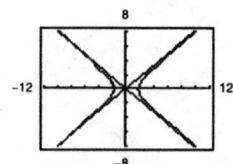

75. Since y is negative, $y = -\dfrac{2}{3}\sqrt{x^2 - 9}$ represents the bottom half of the hyperbola.

77. Vertices: $(0, \pm 2) \implies a = 2$

Foci: $(0, \pm 4) \implies c = 4$

$b^2 = c^2 - a^2 = 12$

Vertical transverse axis

$\dfrac{y^2}{a^2} - \dfrac{x^2}{b^2} = 1$

$\dfrac{y^2}{4} - \dfrac{x^2}{12} = 1$

79. Vertices: $(\pm 1, 0) \implies a = 1$

Asymptotes: $y = \pm 3x$

Horizontal transverse axis

$3 = \dfrac{b}{a} = \dfrac{b}{1} \implies b = 3$

$\dfrac{x^2}{a^2} - \dfrac{y^2}{b^2} = 1$

$\dfrac{x^2}{1} - \dfrac{y^2}{9} = 1$

81. Foci: $(0, \pm 8) \implies c = 8$
Asymptotes: $y = \pm 4x$
Vertical transverse axis

$$4 = \frac{a}{b} \implies a = 4b$$

$$16b^2 + b^2 = (8)^2$$

$$b^2 = \frac{64}{17} \implies a^2 = \frac{1024}{17}$$

$$\frac{y^2}{a^2} - \frac{x^2}{b^2} = 1$$

$$\frac{y^2}{\frac{1024}{17}} - \frac{x^2}{\frac{64}{17}} = 1$$

$$\frac{17y^2}{1024} - \frac{17x^2}{64} = 1$$

83. Vertices: $(0, \pm 3) \implies a = 3$
Vertical transverse axis

$$\frac{y^2}{9} - \frac{x^2}{b^2} = 1$$

$$\frac{5^2}{9} - \frac{(-2)^2}{b^2} = 1$$

$$b^2 = \frac{9}{4}$$

$$\frac{y^2}{9} - \frac{x^2}{\frac{9}{4}} = 1$$

85. Center: $(0, 0)$
Focus: $(24, 0)$
$b^2 = c^2 - a^2 = 24^2 - a^2 = 576 - a^2$

$$\frac{x^2}{a^2} - \frac{y^2}{576 - a^2} = 1$$

$$\frac{24^2}{a^2} - \frac{24^2}{576 - a^2} = 1$$

$$\frac{576}{a^2} - \frac{576}{576 - a^2} = 1$$

$$576(576 - a^2) - 576a^2 = a^2(576 - a^2)$$

$$a^4 - 1728a^2 + 331{,}776 = 0$$

$$a \approx \pm 38.83 \text{ OR } a \approx \pm 14.83$$

Since $a < c$ and $c = 24$, we choose $a = 14.83$. The vertex is approximately at $(14.83, 0)$. [Note: By the Quadratic Formula, the exact value of a is $a = 12(\sqrt{5} - 1)$.]

87. Let (x, y) be such that the sum of the distance from $(c, 0)$ and $(-c, 0)$ is $2a$. (Note that this is only deriving the standard form for the ellipse with horizontal major axis.)

$$2a = \sqrt{(x - c)^2 + y^2} + \sqrt{(x + c)^2 + y^2}$$

$$2a - \sqrt{(x + c)^2 + y^2} = \sqrt{(x - c) + y^2}$$

$$4a^2 - 4a\sqrt{(x + c)^2 + y^2} + (x + c)^2 + y^2 \overset{\cdot}{=} (x - c)^2 + y^2$$

$$4a^2 + 4cx = 4a\sqrt{(x + c)^2 + y^2}$$

$$a^2 + cx = a\sqrt{(x + c)^2 + y^2}$$

$$a^4 + 2a^2cx + c^2x^2 = a^2(x^2 + 2cx + c^2 + y^2)$$

$$a^4 + c^2x^2 = a^2x^2 + a^2c^2 + a^2y^2$$

$$a^2(a^2 - c^2) = (a^2 - c^2)x^2 + a^2y^2$$

Let $b^2 = a^2 - c^2$. Then we have

$$a^2b^2 = b^2x^2 + a^2y^2 \implies 1 = \frac{x^2}{a^2} + \frac{y^2}{b^2}.$$

89. $f(x) = (x - 3)[x - (2 + i)][x - (2 - i)]$

$= (x - 3)[(x - 2) - i][(x - 2) + i]$

$= (x - 3)[(x - 2)^2 - i^2]$

$= (x - 3)(x^2 - 4x + 5)$

$= x^3 - 7x^2 + 17x - 15$

91. $g(x) = 6x^4 + 7x^3 - 29x^2 - 28x + 20$

Possible rational zeros:

$\pm 1, \pm 2, \pm 4, \pm 5, \pm 10, \pm 20, \pm \frac{1}{2}, \pm \frac{5}{2}$

$\pm \frac{1}{3}, \pm \frac{2}{3}, \pm \frac{4}{3}, \pm \frac{5}{3}, \pm \frac{10}{3}, \pm \frac{20}{3}, \pm \frac{1}{6}, \pm \frac{5}{6}$

[Note: The actual zeros are $\pm 2, \frac{1}{2}$, and $-\frac{5}{3}$.]

Section 4.5   Translations of Conics

You should know the following basic facts about conic sections.

■ Parabola with Vertex (h, k)

(a) Vertex Axis

1. Standard Equation: $(x - h)^2 = 4p(y - k)$

2. Focus: $(h, k + p)$

3. Directrix: $y = k - p$

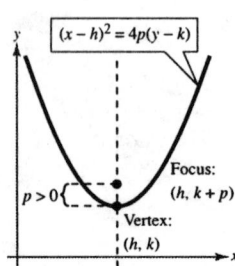

(b) Horizontal Axis

1. Standard Equation: $(y - k)^2 = 4p(x - h)$

2. Focus: $(h + p, k)$

3. Directrix: $x = h - p$

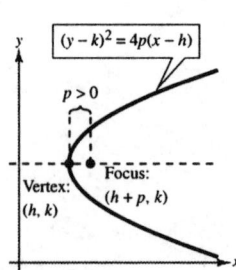

■ Circle with Center (h, k) and Radius r

Standard Equation: $(x - h)^2 + (y - k)^2 = r^2$

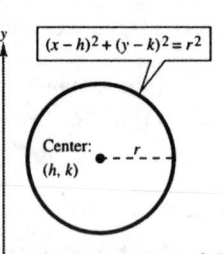

■ Ellipse with Center (h, k)

(a) Horizontal Major Axis:

1. Standard Equation:

$$\frac{(x - h)^2}{a^2} + \frac{(y - k)^2}{b^2} = 1$$

2. Vertices: $(h \pm a, k)$

3. Foci: $(h \pm c, k)$

4. $c^2 = a^2 - b^2$

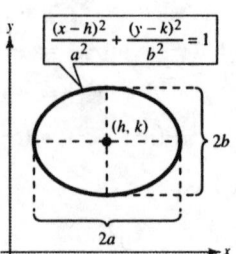

continued

continued

    (b) Vertical Major Axis:

        1. Standard Equation:

$$\frac{(x-h)^2}{b^2} + \frac{(y-k)^2}{a^2} = 1$$

        2. Vertices: $(h, k \pm a)$

        3. Foci: $(h, k \pm c)$

        4. $c^2 = a^2 - b^2$

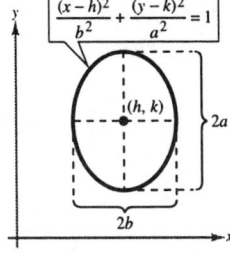

■ Hyperbola with Center (h, k)

    (a) Horizontal Transverse Axis:

        1. Standard Equation:

$$\frac{(x-h)^2}{a^2} - \frac{(y-k)^2}{b^2} = 1$$

        2. Vertices: $(h \pm a, k)$

        3. Foci: $(h \pm c, k)$

        4. Asymptotes: $y - k = \pm\frac{b}{a}(x - h)$

        5. $c^2 = a^2 + b^2$

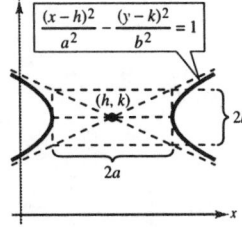

    (b) Vertical Transverse Axis:

        1. Standard Equation:

$$\frac{(y-k)^2}{a^2} - \frac{(x-h)^2}{b^2} = 1$$

        2. Vertices: $(h, k \pm a)$

        3. Foci: $(h, k \pm c)$

        4. Asymptotes: $y - k = \pm\frac{b}{a}(x - h)$

        5. $c^2 = a^2 + b^2$

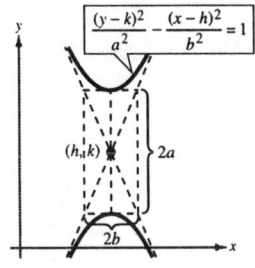

Solutions to Odd-Numbered Exercises

1. $(x - 1)^2 + 8(y + 2) = 0$

$(x - 1)^2 = 4(-2)(y + 2); p = -2$

Vertex: $(1, -2)$

Focus: $(1, -4)$

Directrix: $y = 0$

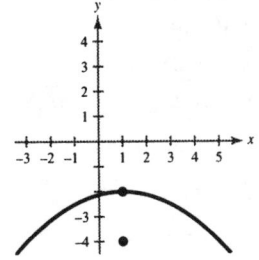

3. $\left(y + \frac{1}{2}\right)^2 = 2(x - 5)$

$\left(y + \frac{1}{2}\right)^2 = 4\left(\frac{1}{2}\right)(x - 5); p = \frac{1}{2}$

Vertex: $\left(5, -\frac{1}{2}\right)$

Focus: $\left(\frac{11}{2}, -\frac{1}{2}\right)$

Directrix: $x = \frac{9}{2}$

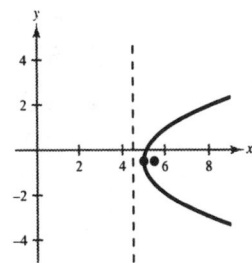

5. $y = \frac{1}{4}(x^2 - 2x + 5)$

$y = \frac{1}{4}(x - 1)^2 + 1$

$4(y - 1) = (x - 1)^2;\ p = 1$

Vertex: $(1, 1)$

Focus: $(1, 2)$

Directrix: $y = 0$

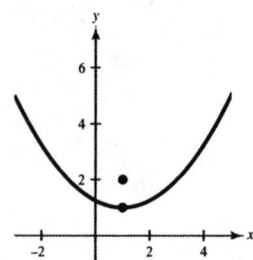

7. $y^2 + 6y + 8x + 25 = 0$

$(y + 3)^2 = 4(-2)(x + 2);\ p = -2$

Vertex: $(-2, -3)$

Focus: $(-4, -3)$

Directrix: $x = 0$

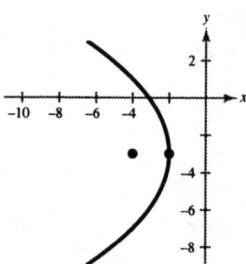

9. $y = -\frac{1}{6}(x^2 + 4x - 2)$

$y = -\frac{1}{6}(x + 2)^2 + 1$

$4\left(-\frac{3}{2}\right)(y - 1) = (x + 2)^2;\ p = -\frac{3}{2}$

Vertex: $(-2, 1)$

Focus: $\left(-2, -\frac{1}{2}\right)$

Directrix: $y = \frac{5}{2}$

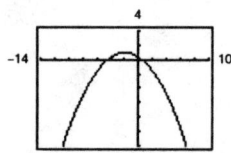

11. $y^2 + x + y = 0$

$\left(y + \frac{1}{2}\right)^2 = 4\left(-\frac{1}{4}\right)\left(x - \frac{1}{4}\right);\ p = -\frac{1}{4}$

Vertex: $\left(\frac{1}{4}, -\frac{1}{2}\right)$

Focus: $\left(0, -\frac{1}{2}\right)$

Directrix: $x = \frac{1}{2}$

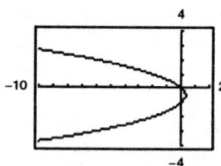

Note: Use $y_1 = -\frac{1}{2} + \sqrt{\frac{1}{4} - x}$ and
$y_2 = -\frac{1}{2} - \sqrt{\frac{1}{4} - x}$ in your graphing utility.

13. Vertex: $(3, 1)$ and opens downward

Passes through $(2, 0)$ and $(4, 0)$

$y = -(x - 2)(x - 4)$

$\quad = -x^2 + 6x - 8$

$\quad = -(x - 3)^2 + 1$

$(x - 3)^2 = -(y - 1)$

15. Vertex: $(-2, 0)$ and opens to the right

Passes through $(0, 2)$

$(y - 0)^2 = 4p(x + 2)$

$\quad 2^2 = 4p(0 + 2)$

$\quad \frac{1}{2} = p$

$y^2 = 4\left(\frac{1}{2}\right)(x + 2)$

$y^2 = 2(x + 2)$

17. Vertex: $(3, 2)$

Focus: $(1, 2)$

Horizontal axis

$p = 1 - 3 = -2$

$(y - 2)^2 = 4(-2)(x - 3)$

$(y - 2)^2 = -8(x - 3)$

19. Vertex: $(0, 4)$

Directrix: $y = 2$

Vertical axis

$p = 4 - 2 = 2$

$(x - 0)^2 = 4(2)(y - 4)$

$\quad x^2 = 8(y - 4)$

21. Focus: $(2, 2)$

Directrix: $x = -2$

Horizontal axis

Vertex: $(0, 2)$

$p = 2 - 0 = 2$

$(y - 2)^2 = 4(2)(x - 0)$

$(y - 2)^2 = 8x$

23. $(y - 3)^2 = 6(x + 1)$

For the upper half of the parabola:

$$y - 3 = +\sqrt{6(x + 1)}$$
$$y = \sqrt{6(x + 1)} + 3$$

25. (a) $V = 17,500\sqrt{2}$ mi/hr

$\approx 24,750$ mi/hr

(b) $p = -4100$, $(h, k) = (0, 4100)$

$$(x - 0)^2 = 4(-4100)(y - 4100)$$
$$x^2 = -16,400(y - 4100)$$

27. $y = -0.08x^2 + x + 4$

(a)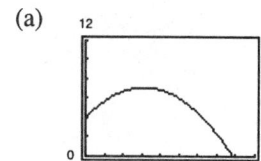

(b) The maximum is at the point (6.25, 7.125).
The range is the x-intercept of ≈ 15.69 feet.

29. $\dfrac{(x - 1)^2}{9} + \dfrac{(y - 5)^2}{25} = 1$

$a = 5$, $b = 3$, $c = \sqrt{a^2 - b^2} = 4$

Center: $(1, 5)$
Foci: $(1, 1)$, $(1, 9)$
Vertices: $(1, 0)$, $(1, 10)$

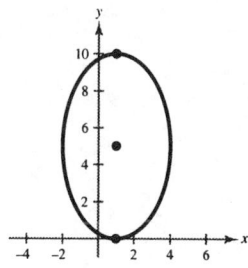

31. $9x^2 + 4y^2 + 36x - 24y + 36 = 0$

$$9(x^2 + 4x + 4) + 4(y^2 - 6y + 9) = 36$$

$$\frac{(x + 2)^2}{4} + \frac{(y - 3)^2}{9} = 1$$

$a = 3$, $b = 2$, $c = \sqrt{a^2 - b^2} = \sqrt{5}$

Vertical major axis
Center: $(-2, 3)$
Foci: $\left(-2, 3 - \sqrt{5}\right)$, $\left(-2, 3 + \sqrt{5}\right)$
Vertices: $(-2, 0)$, $(-2, 6)$

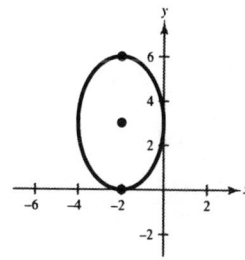

33. $16x^2 + 25y^2 - 32x + 50y + 16 = 0$

$$16(x^2 - 2x + 1) + 25(y^2 + 2y + 1) = 25$$

$$\frac{(x - 1)^2}{25/16} + (y + 1)^2 = 1$$

$a = \dfrac{5}{4}$, $b = 1$, $c = \sqrt{a^2 - b^2} = \dfrac{3}{4}$

Horizontal major axis
Center: $(1, -1)$

Foci: $\left(\dfrac{1}{4}, -1\right)$, $\left(\dfrac{7}{4}, -1\right)$

Vertices: $\left(-\dfrac{1}{4}, -1\right)$, $\left(\dfrac{9}{4}, -1\right)$

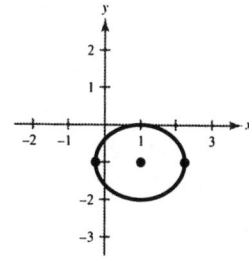

35. $12x^2 + 20y^2 - 12x + 40y - 37 = 0$

$$12\left(x^2 - x + \frac{1}{4}\right) + 20(y^2 + 2y + 1) = 60$$

$$\frac{\left(x - \frac{1}{2}\right)^2}{5} + \frac{(y + 1)^2}{3} = 1$$

$a = \sqrt{5},\ b = \sqrt{3},\ c = \sqrt{a^2 - b^2} = \sqrt{2}$

Horizontal major axis

Center: $\left(\dfrac{1}{2}, -1\right)$

Foci: $\left(\dfrac{1}{2} - \sqrt{2}, -1\right),\ \left(\dfrac{1}{2} + \sqrt{2}, -1\right)$

Vertices: $\left(\dfrac{1}{2} - \sqrt{5}, -1\right),\ \left(\dfrac{1}{2} + \sqrt{5}, -1\right)$

Note: Use $y_1 = -1 + \sqrt{3 - \frac{3}{5}\left(x - \frac{1}{2}\right)^2}$ and

$y_2 = -1 - \sqrt{3 - \frac{3}{5}\left(x - \frac{1}{2}\right)^2}$ in your graphing utility.

37. Center: $(2, 3)$
Vertical major axis
$a = 3,\ b = 1$

$$\frac{(x - 2)^2}{1} + \frac{(y - 3)^2}{9} = 1$$

39. Center: $(2, 2)$
Horizontal major axis
$a = 3,\ b = 2$

$$\frac{(x - 2)^2}{9} + \frac{(y - 2)^2}{4} = 1$$

41. Vertices: $(0, 2),\ (4, 2)$
Minor axis of length 2

$a = \dfrac{4 - 0}{2} = 2,\ b = \dfrac{2}{2} = 1$

Center: $(2, 2)$
Horizontal major axis

$$\frac{(x - 2)^2}{4} + \frac{(y - 2)^2}{1} = 1$$

43. Foci: $(0, 0),\ (0, 8)$
Major axis of length 16
$a = 8,\ c = 4,\ b^2 = 48$
Center: $(0, 4)$
Vertical major axis

$$\frac{x^2}{48} + \frac{(y - 4)^2}{64} = 1$$

45. Vertices: $(3, 1),\ (3, 9)$
Minor axis of length 6
$a = 4,\ b = 3$
Center: $(3, 5)$
Vertical major axis

$$\frac{(x - 3)^2}{9} + \frac{(y - 5)^2}{16} = 1$$

47. Center: $(0, 4)$
$a = 2c$
Vertices: $(-4, 4),\ (4, 4)$
$a = 4,\ c = 2,\ b^2 = 12$
Horizontal major axis

$$\frac{x^2}{16} + \frac{(y - 4)^2}{12} = 1$$

49. $\dfrac{(x-3)^2}{9} + \dfrac{y^2}{4} = 1$

$$\dfrac{(x-3)^2}{9} = 1 - \dfrac{y^2}{4}$$

$$(x-3)^2 = 9\left(\dfrac{4-y^2}{4}\right)$$

$$x - 3 = \pm\dfrac{3}{2}\sqrt{4 - y^2}$$

$$x = 3 \pm \dfrac{3}{2}\sqrt{4 - y^2}$$

$$x = \dfrac{3}{2}\left(2 \pm \sqrt{4 - y^2}\right)$$

The right half of the ellipse is $x = \frac{3}{2}\left(2 + \sqrt{4 - y^2}\right)$.

51. Vertices: $(\pm 5, 0)$

$$e = \dfrac{c}{a} = \dfrac{3}{5}$$

$a = 5,\ c = 3,\ b = 4$

$$\dfrac{x^2}{25} + \dfrac{y^2}{16} = 1$$

53. $a = 3.666 \times 10^9$

$$e = \dfrac{c}{a} = 0.248$$

$c = 909{,}168{,}000$

Smallest distance: $a - c = 2{,}756{,}832{,}000$

Greatest distance: $a + c = 4{,}575{,}168{,}000$

55. Least distance: $a - c = 6378 + 212 = 6590$

Greatest distance: $a + c = 6378 + 938 = 7316$

$a = 6953$ and $c = 363$

$$e = \dfrac{c}{a} = \dfrac{363}{6953} \approx 0.052$$

57. $\dfrac{(x-1)^2}{4} - \dfrac{(y+2)^2}{1} = 1$

$a = 2,\ b = 1,\ c = \sqrt{a^2 + b^2} = \sqrt{5}$

Center: $(1, -2)$

Horizontal transverse axis

Vertices: $(-1, -2),\ (3, -2)$

Foci: $\left(1 \pm \sqrt{5}, -2\right)$

Asymptotes: $y = \pm\dfrac{1}{2}(x - 1) - 2$

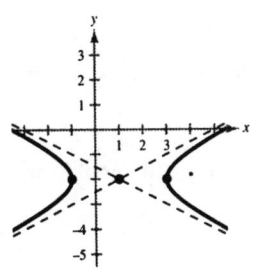

59. $(x+6)^2 - (x-2)^2 = 1$

$a = 1,\ b = 1,\ c = \sqrt{a^2 + b^2} = \sqrt{2}$

Center: $(2, -6)$

Vertical transverse axis

Vertices: $(2, -5),\ (2, -7)$

Foci: $\left(2, -6 \pm \sqrt{2}\right)$

Asymptotes: $y = \pm(x - 2) - 6$

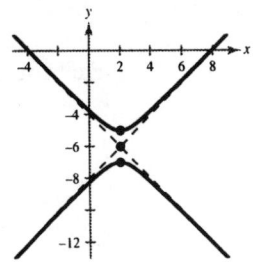

61.
$$9x^2 - y^2 - 36x - 6y + 18 = 0$$
$$9(x^2 - 4x + 4) - (y^2 + 6y + 9) = 9$$
$$(x - 2)^2 - \frac{(y + 3)^2}{9} = 1$$
$$a = 1, \ b = 3, \ c = \sqrt{a^2 + b^2} = \sqrt{10}$$
Center: $(2, -3)$
Horizontal transverse axis
Vertices: $(1, -3), \ (3, -3)$
Foci: $\left(2 \pm \sqrt{10}, -3\right)$

Asymptotes: $y = \pm 3(x - 2) - 3$

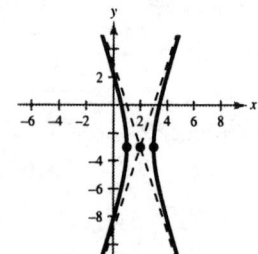

63.
$$x^2 - 9y^2 + 2x - 54y - 80 = 0$$
$$(x^2 + 2x + 1) - 9(y^2 + 6y + 9) = 0$$
$$9(y + 3)^2 = (x + 1)^2$$
$$y = \pm \frac{1}{3}(x + 1) - 3$$

The graph of this equation is two lines intersecting at $(-1, -3)$.

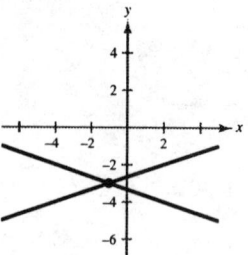

65.
$$9y^2 - x^2 + 2x + 54y + 62 = 0$$
$$9(y^2 + 6y + 9) - (x^2 - 2x + 1) = 18$$
$$\frac{(y + 3)^2}{2} - \frac{(x - 1)^2}{18} = 1$$
$$a = \sqrt{2}, \ b = 3\sqrt{2}, \ c = \sqrt{a^2 + b^2} = 2\sqrt{5}$$
Center: $(1, -3)$
Vertical transverse axis
Vertices: $\left(1, -3 \pm \sqrt{2}\right)$
Foci: $\left(1, -3 \pm 2\sqrt{5}\right)$

Asymptotes: $y = \pm \frac{1}{3}(x - 1) - 3$

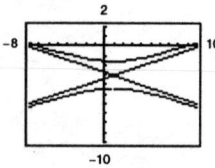

Note: Use $\quad y_1 = \frac{1}{3}(x - 1) - 3$
$$y_2 = -\frac{1}{3}(x - 1) - 3$$
$$y_3 = -3 + \frac{1}{3}\sqrt{(x - 1)^2 + 18}$$
and $\quad \cdot y_4 = -3 - \frac{1}{3}\sqrt{(x - 1)^2 + 18}$ in your graphing utility.

67. Vertices: $(0, 0)$, $(0, 2)$
Passes through $\left(\sqrt{3}, 3\right)$
Center: $(0, 1)$
Vertical transverse axis
$a = 1$
$$\frac{(y-1)^2}{1} - \frac{x^2}{b^2} = 1$$
$$\frac{(3-1)^2}{1} - \frac{\left(\sqrt{3}\right)^2}{b^2} = 1$$
$$b^2 = 1$$
$$(y-1)^2 - x^2 = 1$$

69. Vertices: $(1, 2)$, $(5, 2)$
Center: $(3, 2)$
Passes through $(0, 0)$
Horizontal transverse axis
$a = 2$
$$\frac{(x-3)^2}{4} - \frac{(y-2)^2}{b^2} = 1$$
$$\frac{(0-3)^2}{4} - \frac{(0-2)^2}{b^2} = 1$$
$$b^2 = \frac{16}{5}$$
$$\frac{(x-3)^2}{4} - \frac{(y-2)^2}{16/5} = 1$$

71. Vertices: $(2, 0)$, $(6, 0)$
Foci: $(0, 0)$, $(8, 0)$
Center: $(4, 0)$
Horizontal transverse axis
$a = 2$, $c = 4$, $b^2 = c^2 - a^2 = 12$
$$\frac{(x-4)^2}{4} - \frac{y^2}{12} = 1$$

73. Vertices: $(4, 1)$, $(4, 9)$
Foci: $(4, 0)$, $(4, 10)$
Center: $(4, 5)$
Vertical transverse axis
$a = 4$, $c = 5$, $b^2 = c^2 - a^2 = 9$
$$\frac{(y-5)^2}{16} - \frac{(x-4)^2}{9} = 1$$

75. Vertices: $(2, 3)$, $(2, -3)$
Passes through the point $(0, 5)$
Center: $(2, 0)$
Vertical transverse axis
$a = 3$
$$\frac{y^2}{9} - \frac{(x-2)^2}{b^2} = 1$$
$$\frac{5^2}{9} - \frac{(0-2)^2}{b^2} = 1$$
$$b^2 = \frac{9}{4}$$
$$\frac{y^2}{9} - \frac{4(x-2)^2}{9} = 1$$

77. Vertices: $(0, 2)$, $(6, 2)$
Asymptotes: $y = \frac{2}{3}x$, $y = 4 - \frac{2}{3}x$
Center: $(3, 2)$
Horizontal transverse axis
$a = 3$, $b = 2$
$$\frac{(x-3)^2}{9} - \frac{(y-2)^2}{4} = 1$$

79. $\dfrac{(x-3)^2}{4} - \dfrac{(y-1)^2}{9} = 1$
$$(x-3)^2 = \frac{4}{9}[9 + (y-1)^2]$$
$$x - 3 = \pm\frac{2}{3}\sqrt{9 + (y-1)^2}$$
$$x = 3 \pm \frac{2}{3}\sqrt{9 + (y-1)^2}$$
The left half of the hyperbola is
$x = 3 - \frac{2}{3}\sqrt{9 + (y-1)^2}$.

81. $x^2 + y^2 - 6x + 4y + 9 = 0$
$$(x-3)^2 + (y+2)^2 = 4$$
Circle

83. $4x^2 - y^2 - 4x - 3 = 0$

$$4\left(x - \frac{1}{2}\right)^2 - y^2 = 4$$

$$\left(x - \frac{1}{2}\right)^2 - \frac{y^2}{4} = 1$$

Hyperbola

85. $4x^2 + 3y^2 + 8x - 24y + 51 = 0$

$$4(x + 1)^2 + 3(y - 4)^2 = 1$$

$$\frac{(x + 1)^2}{1/4} + \frac{(y - 4)^2}{1/3} = 1$$

Ellipse

87. $25x^2 - 10x - 200y - 119 = 0$

$$25\left(x - \tfrac{1}{5}\right)^2 = 200\left(y + \tfrac{3}{5}\right)$$

$$\left(x - \tfrac{1}{5}\right)^2 = 8\left(y + \tfrac{3}{5}\right)$$

Parabola

89. $x^2 + 9y^2 = 9 \;\Rightarrow\; y = \pm\frac{1}{3}\sqrt{9 - x^2}$

$$y = x^2 - 4$$

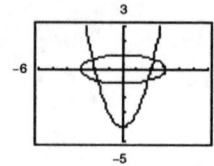

The approximate points of intersection are $(\pm 2.166, 0.692)$ and $(\pm 1.788, -0.803)$.

91. Center: $(0, 0)$

$2a = 36.23$

$a = 18.115 \;\Rightarrow\; a^2 \approx 328.15$

$e = \dfrac{c}{a} \;\Rightarrow\; 0.97 = \dfrac{c}{18.115}$

$c = 17.57155$

$c^2 = a^2 - b^2 \;\Rightarrow\; b^2 = a^2 - c^2 \approx 19.39$

Thus, the equation of the ellipse is approximately $\dfrac{x^2}{328.15} + \dfrac{y^2}{19.39} = 1$.

93. $(a + 4) - (a + 4) = 0$

Additive Inverse Property

95. $(x + 3)(a + b) = x(a + b) + 3(a + b)$

Distributive Property

❑ Review Exercises for Chapter 4

Solutions to Odd Numbered Exercises

1. $f(x) = \dfrac{4}{x + 3}$

Domain: All real numbers except -3
Vertical asymptote: $x = -3$
Horizontal asymptote: $y = 0$

3. $g(x) = \dfrac{x^2}{x^2 - 4}$

Domain: All real numbers except ± 2
Vertical asymptotes: $x = -2$, $x = 2$
Horizontal asymptote: $y = 1$

5. $f(x) = \dfrac{-5}{x^2}$

y-axis symmetry
Vertical asymptote: $x = 0$
Horizontal asymptote: $y = 0$

x	± 3	± 2	± 1
y	$-\frac{5}{9}$	$-\frac{5}{4}$	-5

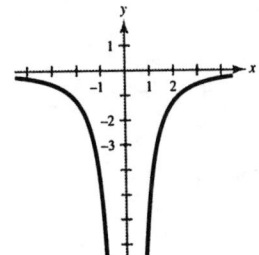

7. $g(x) = \dfrac{2 + x}{1 - x} = -\dfrac{x + 2}{x - 1}$

x-intercept: $(-2, 0)$
y-intercept: $(0, 2)$
Vertical asymptote: $x = 1$
Horizontal asymptote: $y = -1$

x	-1	0	2	3
y	$\frac{1}{2}$	2	-4	$-\frac{5}{2}$

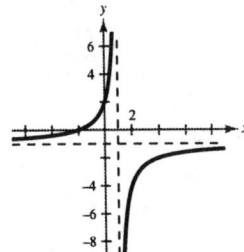

9. $p(x) = \dfrac{x^2}{x^2 + 1}$

Intercept: $(0, 0)$
y-axis symmetry
Horizontal asymptote: $y = 1$

x	± 3	± 2	± 1	0
y	$\frac{9}{10}$	$\frac{4}{5}$	$\frac{1}{2}$	0

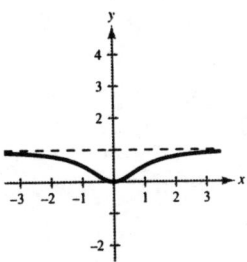

11. $f(x) = \dfrac{x}{x^2 + 1}$

Intercept: $(0, 0)$
Origin symmetry
Horizontal asymptote: $y = 0$

x	-2	-1	0	1	2
y	$-\frac{2}{5}$	$-\frac{1}{2}$	0	$\frac{1}{2}$	$\frac{2}{5}$

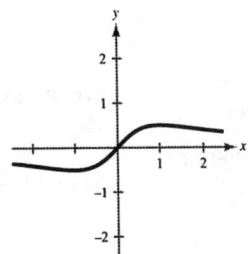

13. $f(x) = \dfrac{2x^3}{x^2 + 1} = 2x - \dfrac{2x}{x^2 + 1}$

Intercept: $(0, 0)$
Origin symmetry
Slant asymptote: $y = 2x$

x	-2	-1	0	1	2
y	$-\frac{16}{5}$	-1	0	1	$\frac{16}{5}$

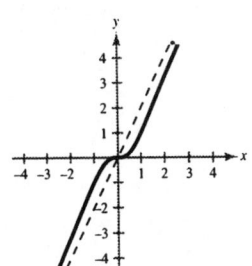

15. $y = \dfrac{x}{x^2 - 1}$

Intercept: $(0, 0)$

Origin symmetry

Vertical asymptotes: $x = -1,\ x = 1$

Horizontal asymptote: $y = 0$

x	-3	-2	0	2	3
y	$-\frac{3}{8}$	$-\frac{2}{3}$	0	$\frac{2}{3}$	$\frac{3}{8}$

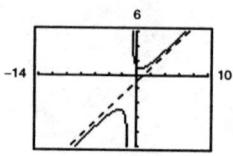

17. $s(x) = \dfrac{8x^2}{x^2 + 4}$

Intercept: $(0, 0)$

Horizontal asymptote: $y = 8$

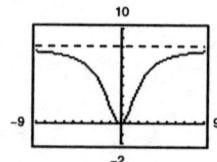

19. $g(x) = \dfrac{x^2 + 1}{x + 1} = x - 1 + \dfrac{2}{x + 1}$

Intercept: $(0, 1)$

Slant asymptote: $y = x - 1$

21. $f(x) = \dfrac{2x^2}{(x + 3)(x - 4)} = \dfrac{2x^2}{x^2 - x - 12}$

This answer is not unique.

23. $\overline{C} = \dfrac{C}{x} = \dfrac{0.5x + 500}{x},\ 0 < x$

$\quad\ = 0.5 + \dfrac{500}{x}$

As x increases, the cost approaches the horizontal asymptote, $\overline{C} = 0.5$.

25. $C = \dfrac{528p}{100 - p},\ 0 \le p < 100$

(a) When $p = 25,\ C = \dfrac{528(25)}{100 - 25} = \$176$ million.

(b) When $p = 50,\ C = \dfrac{528(50)}{100 - 50} = \$528$ million.

(c) When $p = 75,\ C = \dfrac{528(75)}{100 - 75} = \$1584$ million.

(d) As $p \to 100,\ C \to \infty$. No, it is not possible.

27. (a)

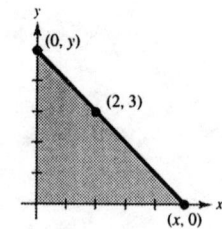

(b) $A = \frac{1}{2}bh$ where $b = x$ and $h = y$.

Slope: $\dfrac{y - 0}{0 - x} = \dfrac{0 - 3}{x - 2}$

$y = -x\left(\dfrac{-3}{x - 2}\right) = \dfrac{3x}{x - 2}$

Area: $A = \dfrac{1}{2}xy = \dfrac{1}{2}x\left(\dfrac{3x}{x - 2}\right) = \dfrac{3x^2}{2(x - 2)},\ x > 2$

— **CONTINUED** —

27. — CONTINUED —

(c)

x	2.5	3	3.5	4	4.5
y	18.75	13.50	12.25	12	12.15

The area is minimum when

$$x = 4 \text{ and } y = \frac{3(4)}{4 - 2} = 6.$$

(d)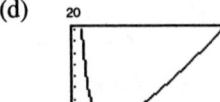

The area is minimum (12) when $x = 4$ and $y = 6$.

(e) $A = \dfrac{3x^2}{2x - 4} = \dfrac{3}{2}x + 3 + \dfrac{12}{2x - 4} = \dfrac{3}{2}(x + 2) + \dfrac{6}{x - 2}$

The slant asymptote is $y = \frac{3}{2}(x + 2)$. The area increases without bound as x increases.

29. $y = \dfrac{18.47x - 2.96}{0.23x + 1},\ 0 < x$

The limiting amount of CO_2 uptake is determined by the horizontal asymptote.

$$y = \frac{18.47}{0.23} \approx 80.3 \text{ mg/dc}^2/\text{hr}$$

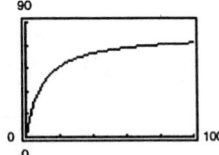

31. $\dfrac{4 - x}{x^2 + 6x + 8} = \dfrac{A}{x + 2} + \dfrac{B}{x + 4}$

$4 - x = A(x + 4) + B(x + 2)$

Let $x = -2$: $6 = 2A \implies A = 3$

Let $x = -4$: $8 = -2B \implies B = -4$

$$\frac{4 - x}{x^2 + 6x + 8} = \frac{3}{x + 2} - \frac{4}{x + 4}$$

33. $\dfrac{x^2}{x^2 + 2x - 15} = 1 - \dfrac{2x - 15}{x^2 + 2x - 15}$

$\dfrac{-2x + 15}{(x + 5)(x - 3)} = \dfrac{A}{x + 5} + \dfrac{B}{x - 3}$

$-2x + 15 = A(x - 3) + B(x + 5)$

Let $x = -5$: $25 = -8A \implies A = -\dfrac{25}{8}$

Let $x = 3$: $9 = 8B \implies B = \dfrac{9}{8}$

$$\frac{x^2}{x^2 + 2x - 15} = 1 + \frac{9}{8(x - 3)} - \frac{25}{8(x + 5)}$$

35. $\dfrac{x^2 + 2x}{x^3 - x^2 + x - 1} = \dfrac{A}{x - 1} + \dfrac{Bx + C}{x^2 + 1}$

$$x^2 + 2x = A(x^2 + 1) + (Bx + C)(x - 1)$$

Let $x = 1$: $3 = 2A \implies A = \dfrac{3}{2}$

Let $x = 0$: $0 = A - C \implies C = \dfrac{3}{2}$

Let $x = 2$: $8 = 5A + 2B + C$

$$8 = \left(\frac{15}{2}\right) + 2B + \left(\frac{3}{2}\right) \implies B = -\frac{1}{2}$$

$$\frac{x^2 + 2x}{x^3 - x^2 + x - 1} = \frac{3/2}{x - 1} + \frac{-(1/2)x + 3/2}{x^2 + 1}$$

$$= \frac{1}{2}\left(\frac{3}{x - 1} - \frac{x - 3}{x^2 + 1}\right)$$

37. $\dfrac{3x^3 + 4x}{(x^2 + 1)^2} = \dfrac{Ax + B}{x^2 + 1} + \dfrac{Cx + D}{(x^2 + 1)^2}$

$3x^3 + 4x = (Ax + B)(x^2 + 1) + Cx + D$
$\qquad\qquad = Ax^3 + Bx^2 + (A + C)x + (B + D)$

Equating coefficients of like powers:

$3 = A$
$0 = B$
$4 = 3 + C \implies C = 1$
$0 = B + D \implies D = 0$

$\dfrac{3x^3 + 4x}{(x^2 + 1)^2} = \dfrac{3x}{x^2 + 1} + \dfrac{x}{(x^2 + 1)^2}$

39. $4x = y^2$
Parabola
Vertex: $(0, 0)$
Horizontal axis

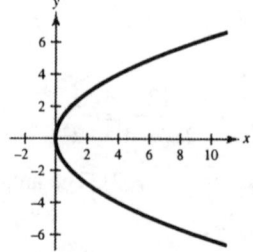

41. $x^2 - 6x + 2y + 9 = 0$
$\qquad\qquad (x - 3)^2 = -2y$

Parabola
Vertex: $(3, 0)$
Focus: $\left(3, -\frac{1}{2}\right)$

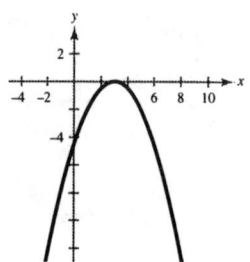

43. $\qquad x^2 + y^2 - 2x - 4y + 5 = 0$
$(x^2 - 2x + 1) + (y^2 - 4y + 4) = -5 + 1 + 4$
$\qquad\qquad (x - 1)^2 + (y - 2)^2 = 0$

Point: $(1, 2)$

Note: This a degenerate conic, a circle of radius zero.

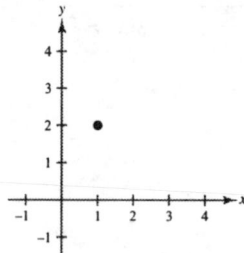

45. $4x^2 + y^2 = 16$

$\dfrac{x^2}{4} + \dfrac{y^2}{16} = 1$

Ellipse
Center: $(0, 0)$
Vertices: $(0, -4),\ (0, 4)$

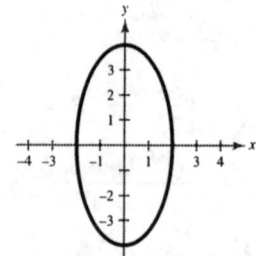

47. $x^2 + 9y^2 + 10x - 18y + 25 = 0$

$$(x + 5)^2 + 9(y - 1)^2 = 9$$

$$\frac{(x + 5)^2}{9} + (y - 1)^2 = 1$$

Ellipse

Center: $(-5, 1)$

Vertices: $(-8, 1), (-2, 1)$

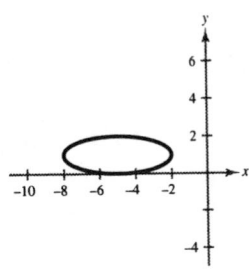

51. $x^2 - 10xy + y^2 + 1 = 0$

$$y^2 - 10xy + (x^2 + 1) = 0$$

$$y = \frac{-(-10x) \pm \sqrt{(-10x)^2 - 4(1)(x^2 + 1)}}{2(1)}$$

$$= \frac{10x \pm \sqrt{96x^2 - 4}}{2}$$

$$= \frac{10x \pm 2\sqrt{24x^2 - 1}}{2}$$

$$= 5x \pm \sqrt{24x^2 - 1}$$

The graph reveals a hyperbola.

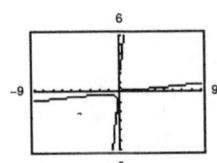

49. $5y^2 - 4x^2 = 20$

$$\frac{y^2}{4} - \frac{x^2}{5} = 1$$

Hyperbola

Vertical transverse axis

Center: $(0, 0)$

Vertices: $(0, \pm 2)$

Asymptotes: $y = \pm\frac{2\sqrt{5}}{5}x$

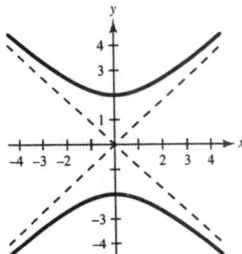

53. Vertex: $(-6, 4)$

Passes through $(0, 0)$

Vertical axis

$$(x + 6)^2 = 4p(y - 4)$$

$$(0 + 6)^2 = 4p(0 - 4)$$

$$36 = -16p$$

$$-\frac{9}{4} = p$$

$$(x + 6)^2 = 4\left(-\frac{9}{4}\right)(y - 4)$$

$$(x + 6)^2 = -9(y - 4)$$

55. Vertex: $(4, 2)$

Focus: $(4, 0)$

Vertical axis, $p = -2$

$$(x - 4)^2 = 4(-2)(y - 2)$$

$$(x - 4)^2 = -8(y - 2)$$

57. Vertex: $(0, 2)$

Horizontal axis

Passes through $(-1, 0)$

$$(y - 2)^2 = 4p(x - 0)$$

$$(0 - 2)^2 = 4p(-1 - 0)$$

$$4 = -4p$$

$$-1 = p$$

$$(y - 2)^2 = 4(-1)(x - 0)$$

$$(y - 2)^2 = -4x$$

59. Vertices: $(0, 3), (10, 3)$

Passes through $(5, 0)$

Center: $(5, 3)$

Horizontal major axis

$a = 5$, $b = 3$

$$\frac{(x - 5)^2}{25} + \frac{(y - 3)^2}{9} = 1$$

61. Vertices: $(-3, 0)$, $(7, 0)$

Foci: $(0, 0)$, $(4, 0)$

Horizontal major axis

Center: $(2, 0)$

$a = 5, c = 2,$

$b = \sqrt{25 - 4} = \sqrt{21}$

$\dfrac{(x - h)^2}{a^2} + \dfrac{(y - k)^2}{b^2} = 1$

$\dfrac{(x - 2)^2}{25} + \dfrac{y^2}{21} = 1$

63. Vertices: $(0, \pm 6)$

Passes through $(2, 2)$

Vertical major axis

Center: $(0, 0)$

$a = 6$

$\dfrac{(x - h)^2}{b^2} + \dfrac{(y - k)^2}{a^2} = 1$

$\dfrac{(2 - 0)^2}{b^2} + \dfrac{(2 - 0)^2}{6^2} = 1$

$b^2 = \dfrac{9}{2}$

$\dfrac{2x^2}{9} + \dfrac{y^2}{36} = 1$

65. Vertices: $(\pm 1, 0)$

Horizontal transverse axis

Center: $(0, 0)$

$a = 1$

$\pm\dfrac{b}{a} = \pm 2 \implies b = 2$

$\dfrac{x^2}{1} - \dfrac{y^2}{4} = 1$

67. Vertices: $(0, \pm 1)$

Foci: $(0, \pm 3)$

Vertical transverse axis

Center: $(0, 0)$

$a = 1, c = 3,$

$b = \sqrt{9 - 1} = \sqrt{8}$

$\dfrac{y^2}{1} - \dfrac{x^2}{8} = 1$

69. Foci: $(0, 0)$, $(8, 0)$

Asymptotes: $y = \pm 2(x - 4)$

Horizontal transverse axis

Center: $(4, 0) \implies c = 4$

$\dfrac{b}{a} = 2 \implies b = 2a$

$a^2 + b^2 = c^2$

$a^2 + (2a)^2 = 4^2$

$a^2 = \dfrac{16}{5}$

$b^2 = \dfrac{64}{5}$

$\dfrac{(x - h)^2}{a^2} - \dfrac{(y - k)^2}{b^2} = 1$

$\dfrac{5(x - 4)^2}{16} - \dfrac{5y^2}{64} = 1$

71. $y = \dfrac{x^2}{200}, \quad -100 \le x \le 100$

Vertex: $(0, 0)$

$x^2 = 200y$

$4p = 200$

$p = 50$

Focus: $(0, 50)$

73. (a) Parabola: Vertex: $(0, 4)$

  Passes through $(\pm 4, 0)$

$$x^2 = 4p(y - 4)$$
$$16 = 4p(0 - 4)$$
$$16 = -16p$$
$$-1 = p$$
$$x^2 = -4(y - 4)$$

  Circle: Passes through $(\pm 4, 0)$

  Radius: $r = 8$

  Center: $(0, k)$

$$x^2 + (y - k)^2 = 8^2$$
$$(\pm 4)^2 + (0 - k)^2 = 8^2$$
$$16 + k^2 = 64$$
$$k^2 = 48$$
$$k = -\sqrt{48} = -4\sqrt{3}$$
$$x^2 + \left(y + 4\sqrt{3}\right)^2 = 64$$

(b) Parabola: $x^2 = -4(y - 4) \implies y = -\frac{1}{4}x^2 + 4$

  Circle:

$$x^2 + \left(y + 4\sqrt{3}\right)^2 = 64 \implies y = \sqrt{64 - x^2} - 4\sqrt{3}$$
$$d = \left(-\frac{1}{4}x^2 + 4\right) - \left(\sqrt{64 - x^2} - 4\sqrt{3}\right)$$
$$= -\frac{1}{4}x^2 - \sqrt{64 - x^2} + 4 + 4\sqrt{3}$$

x	0	1	2	3	4
d	2.928	2.741	2.182	1.262	0

75. $a = 5, b = 4, c = \sqrt{a^2 - b^2} = 3$

The foci should be placed 3 feet on either side of center and have the same height as the pillars.

❏ Practice Test for Chapter 4

1. Sketch the graph of $f(x) = \dfrac{x-1}{2x}$ and label all intercepts and asymptotes.

2. Sketch the graph of $f(x) = \dfrac{3x^2 - 4}{x}$ and label all intercepts and asymptotes.

3. Find all the asymptotes of $f(x) = \dfrac{8x^2 - 9}{x^2 + 1}$.

4. Find all the asymptotes of $f(x) = \dfrac{4x^2 - 2x + 7}{x - 1}$.

5. Sketch the graph of $f(x) = \dfrac{x-5}{(x-5)^2}$.

For Exercises 6–9, write the partial fraction decomposition for the rational expression.

6. $\dfrac{1 - 2x}{x^2 + x}$

7. $\dfrac{6x}{x^2 - x - 2}$

8. $\dfrac{6x - 17}{(x-3)^2}$

9. $\dfrac{3x^2 - x + 8}{x^3 + 2x}$

10. Find the vertex, focus, and directrix of the parabola $x^2 = 20y$.

11. Find the equation of the parabola with vertex $(0, 0)$ and focus $(7, 0)$.

12. Find the center, foci, and vertices of the ellipse $\dfrac{x^2}{144} + \dfrac{y^2}{25} = 1$.

13. Find the equation of the ellipse with foci $(\pm 4, 0)$ and minor axis of length 6.

14. Find the center, vertices, foci, and asymptotes of the hyperbola $\dfrac{y^2}{144} - \dfrac{x^2}{169} = 1$.

15. Find the equation of the hyperbola with vertices $(\pm 4, 0)$ and asymptotes $y = \pm\dfrac{1}{2}x$.

16. Find the equation of the parabola with vertex $(6, -1)$ and focus $(6, 3)$.

17. Find the center, foci, and vertices of the ellipse $16x^2 + 9y^2 - 96x + 36y + 36 = 0$.

18. Find the equation of the ellipse with vertices $(-1, 1)$ and $(7, 1)$ and minor axis of length 2.

19. Find the center, vertices, foci, and asymptotes of the hyperbola $4(x + 3)^2 - 9(y - 1)^2 = 1$.

20. Find the equation of the hyperbola with vertices $(3, 4)$ and $(3, -4)$ and foci $(3, 7)$ and $(3, -7)$.

CHAPTER 5
Exponential and Logarithmic Functions

CHAPTER 5
Exponential and Logarithmic Functions

Section 5.1 Exponential Functions and Their Graphs

- You should know that a function of the form $y = a^x$, where $a > 0$, $a \neq 1$, is called an exponential function with base a.
- You should be able to graph exponential functions.
- You should know formulas for compound interest.

  (a) For n compoundings per year: $A = P\left(1 + \dfrac{r}{n}\right)^{nt}$.

  (b) For continuous compoundings: $A = Pe^{rt}$.

Solutions to Odd-Numbered Exercises

1. $(3.4)^{5.6} \approx 946.852$

3. $(1.005)^{400} \approx 7.352$

5. $5^{-\pi} \approx 0.006$

7. $100^{\sqrt{2}} \approx 673.639$

9. $e^{-3/4} \approx 0.472$

11. $f(x) = 3^{x-2}$

$= 3^x 3^{-2}$

$= 3^x\left(\dfrac{1}{3^2}\right)$

$= \dfrac{1}{9}(3^x)$

$= h(x)$

Thus, $f(x) \neq g(x)$, but $f(x) = h(x)$.

13. $f(x) = 16(4^{-x})$ and $f(x) = 16(4^{-x})$

$= 4^2(4^{-x})$ $= 16(2^2)^{-x}$

$= 4^{2-x}$ $= 16(2^{-2x})$

$= \left(\dfrac{1}{4}\right)^{-(2-x)}$ $= h(x)$

$= \left(\dfrac{1}{4}\right)^{x-2}$

$= g(x)$

Thus, $f(x) = g(x) = h(x)$.

15. $f(x) = 2^x$ rises to the right.

Asymptote: $y = 0$

Intercept: $(0, 1)$

Matches graph (d).

17. $f(x) = 2^{-x}$ falls to the right.

Asymptote: $y = 0$

Intercept: $(0, 1)$

Matches graph (a).

19. $g(x) = 5^x$

x	-2	-1	0	1	2
$g(x)$	$\frac{1}{25}$	$\frac{1}{5}$	1	5	25

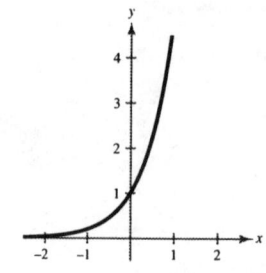

240

21. $f(x) = \left(\dfrac{1}{5}\right)^x = 5^{-x}$

x	-2	-1	0	1	2
y	25	5	1	$\frac{1}{5}$	$\frac{1}{25}$

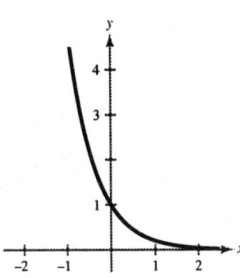

23. $h(x) = 5^{x-2}$

x	-1	0	1	2	3
y	$\frac{1}{125}$	$\frac{1}{25}$	$\frac{1}{5}$	1	5

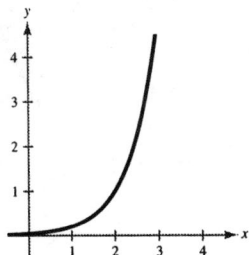

25. $g(x) = 5^{-x} - 3$

x	-1	0	1	2
y	2	-2	$-2\frac{4}{5}$	$-2\frac{24}{25}$

Asymptote: $y = -3$

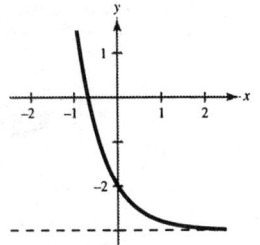

27. $y = 2^{-x^2}$

x	± 2	± 1	0
y	$\frac{1}{16}$	$\frac{1}{2}$	1

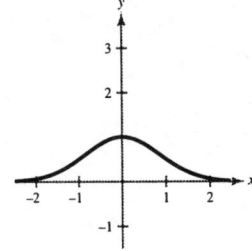

29. $f(x) = 3^{x-2} + 1$

x	-1	0	1	2	3	4
y	$1\frac{1}{27}$	$1\frac{1}{9}$	$1\frac{1}{3}$	2	4	10

Asymptote: $y = 1$

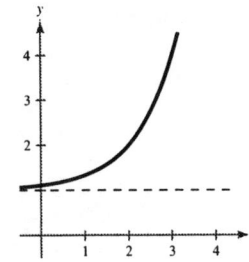

31. $y = 1.08^{-5x}$

x	-1	0	1	2
y	1.47	1	0.68	0.46

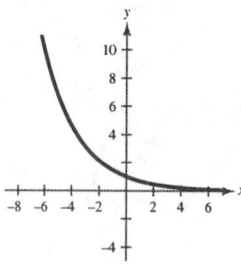

33. $s(t) = 2e^{0.12t}$

t	-4	0	4	8
$s(t)$	1.24	2	3.23	5.22

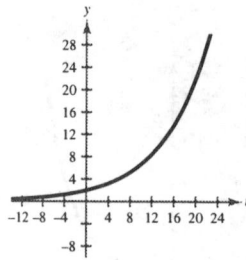

35. $g(x) = 1 + e^{-x}$

x	-2	-1	0	1	2
y	8.39	3.72	2	1.37	1.14

Asymptote: $y = 1$

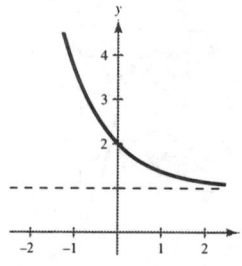

37. $y = 3^x$ and $y = 4^x$

x	-2	-1	0	1	2
3^x	$\frac{1}{9}$	$\frac{1}{3}$	1	3	9
4^x	$\frac{1}{16}$	$\frac{1}{4}$	1	4	16

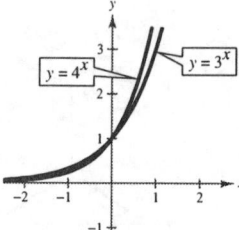

(a) $4^x < 3^x$ when $x < 0$.

(b) $4^x > 3^x$ when $x > 0$.

39. $f(x) = 3^x$

(a) $g(x) = f(x - 2) = 3^{x-2}$ (b) $h(x) = -\frac{1}{2}f(x) = -\frac{1}{2}(3^x)$ (c) $q(x) = f(-x) + 3 = 3^{-x} + 3$

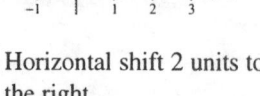

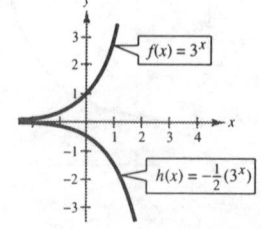

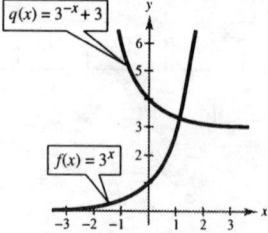

Horizontal shift 2 units to the right   Vertical shrink and a reflection about the x-axis   Reflection about the y-axis and a vertical translation 3 units upward

41. (a) $f(x) = x^2 e^{-x}$

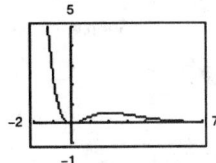

Decreasing: $(-\infty, 0)$, $(2, \infty)$
Increasing: $(0, 2)$
Relative maximum: $(2, 4e^{-2})$
Relative minimum: $(0, 0)$

(b) $g(x) = x2^{3-x}$

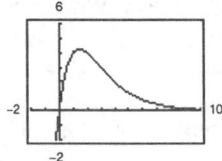

Decreasing: $(1.44, \infty)$
Increasing: $(-\infty, 1.44)$
Relative maximum: $(1.44, 4.25)$

43. The exponential function, $y = e^x$, increases at a faster rate than the polynomial function, $y = x^n$.

45. $f(x) = \left(1 + \dfrac{0.5}{x}\right)^x$ and $g(x) = e^{0.5}$ (Horizontal line)

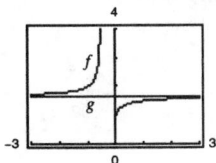

As $x \to \infty$, $f(x) \to g(x)$.

47. $P = \$2500$, $r = 12\%$, $t = 10$ years

Compounded n times per year: $A = 2500\left(1 + \dfrac{0.12}{n}\right)^{10n}$

Compounded continuously: $A = 2500e^{0.12(10)}$

n	1	2	4	12	365	Continuous Compounding
A	\$7,764.62	\$8,017.84	\$8,155.09	\$8250.97	\$8298.66	\$8,300.29

49. $P = \$2500$, $r = 12\%$, $t = 20$ years

Compounded n times per year: $A = 2500\left(1 + \dfrac{0.12}{n}\right)^{20n}$

Compounded continuously: $A = 2500e^{0.12(20)}$

n	1	2	4	12	365	Continuous Compounding
A	\$24,115.73	\$25,714.29	\$26,602.23	\$27,231.38	\$27,547.07	\$27,557.94

51. $A = Pe^{rt}$

$100,000 = Pe^{0.09t}$

$\dfrac{100,000}{e^{0.09t}} = P$

$P = 100,000e^{-0.09t}$

t	1	10	20	30	40	50
P	\$91,393.12	\$40,656.97	\$16,529.89	\$6,720.55	\$2,732.37	\$1,110.90

53. $A = 25,000e^{(0.0875)(25)} \approx \$222,822.57$

55. (a) The graph that is increasing faster represents 7% compounded annually. When interest is compounded, you earn interest on that interest. With simple interest there is no compounding so the growth is linear.

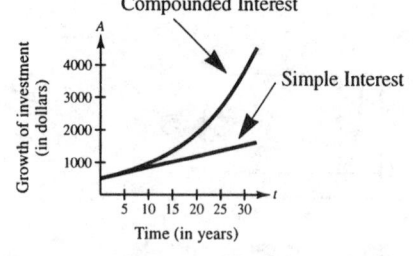

(b) Compound interest formula: $A = 500\left(1 + \dfrac{0.07}{1}\right)^{(1)t}$

$$= 500(1.07)^t$$

Simple interest formula: $A = Prt + P$

$$= 500(0.07)t + 500$$

57. $C(10) = 23.95(1.04)^{10} \approx \$35.45$

59. $P(t) = 100e^{0.2197t}$

(a) $P(0) = 100$

(b) $P(5) \approx 300$

(c) $P(10) \approx 900$

61. $Q = 25\left(\frac{1}{2}\right)^{t/1620}$

(a) When $t = 0$, $Q = 25\left(\frac{1}{2}\right)^{0/1620} = 25(1) = 25$ units.

(b) When $t = 1000$, $Q = 25\left(\frac{1}{2}\right)^{1000/1620} \approx 16.30$ units.

(c)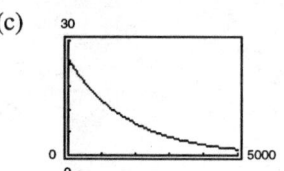

63. $P = 10{,}958e^{-0.15h}$

(a)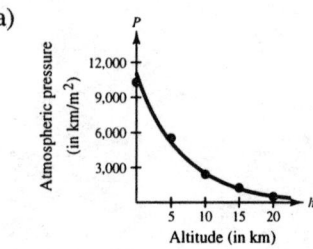

(b)

h	0	5	10	15	20
P	10,958	5176	2445	1155	546

The model is a "good fit."

(c) $P(8) \approx 3300$ kg/m^2

(d) $2000 = 10{,}958e^{-0.15h}$ when $x \approx 11.3$ km.

65. False, $e \neq \dfrac{271{,}801}{99{,}990}$.

Since e is an irrational number it cannot equal a rational number.

67. Since $\sqrt{2} \approx 1.414$ we know that $1 < \sqrt{2} < 2$.

Thus, $2^1 < 2^{\sqrt{2}} < 2^2$

$\quad 2 < 2^{\sqrt{2}} < 4$.

69. $y_4 = 1 + \dfrac{x}{1!} + \dfrac{x^2}{2!} + \dfrac{x^3}{3!} + \dfrac{x^4}{4!}$

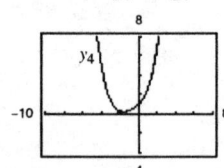

$e^x = 1 + \dfrac{x}{1!} + \dfrac{x^2}{2!} + \dfrac{x^3}{3!} + \dfrac{x^4}{4!} + \dfrac{x^5}{5!} + \cdots$

71. $2x - 7y + 14 = 0$

$\quad 2x + 14 = 7y$

$\quad \frac{1}{7}(2x + 14) = y$

73. $x^2 + y^2 = 25$

$\quad y^2 = 25 - x^2$

$\quad y = \pm\sqrt{25 - x^2}$

Section 5.2   Logarithmic Functions and Their Graphs

- ■ You should know that a function of the form $y = \log_a x$, where $a > 0$, $a \neq 1$, and $x > 0$, is called a logarithm of x to base a.
- ■ You should be able to convert from logarithmic form to exponential form and vice versa.

  $$y = \log_a x \iff a^y = x$$

- ■ You should know the following properties of logarithms.
  - (a) $\log_a 1 = 0$ since $a^0 = 1$.
  - (b) $\log_a a = 1$ since $a^1 = a$.
  - (c) $\log_a a^x = x$ since $a^x = a^x$.
  - (d) If $\log_a x = \log_a y$, then $x = y$.
- ■ You should know the definition of the natural logarithmic function.

  $$\log_e x = \ln x, x > 0$$

- ■ You should know the properties of the natural logarithmic function.
  - (a) $\ln 1 = 0$ since $e^0 = 1$.
  - (b) $\ln e = 1$ since $e^1 = e$.
  - (c) $\ln e^x = x$ since $e^x = e^x$.
  - (d) If $\ln x = \ln y$, then $x = y$.
- ■ You should be able to graph logarithmic functions.

Solutions to Odd-Numbered Exercises

1. $\log_4 64 = 3 \implies 4^3 = 64$

3. $\log_7 \frac{1}{49} = -2 \implies 7^{-2} = \frac{1}{49}$

5. $\log_{32} 4 = \frac{2}{5} \implies 32^{2/5} = 4$

7. $\ln 1 = 0 \implies e^0 = 1$

9. $5^3 = 125 \implies \log_5 125 = 3$

11. $81^{1/4} = 3 \implies \log_{81} 3 = \frac{1}{4}$

13. $6^{-2} = \frac{1}{36} \implies \log_6 \frac{1}{36} = -2$

15. $e^3 = 20.0855\ldots \implies \ln 20.0855\ldots = 3$

17. $e^x = 4 \implies \ln 4 = x$

19. $\log_2 16 = \log_2 2^4 = 4$

21. $\log_{16} 4 = \log_{16} 16^{1/2} = \frac{1}{2}$

23. $\log_7 1 = \log_7 7^0 = 0$

25. $\log_{10} 0.01 = \log_{10} 10^{-2} = -2$

27. $\log_8 32 = \log_8 8^{5/3} = \frac{5}{3}$

29. $\ln e^3 = 3$

31. $\log_{10} 345 \approx 2.538$

33. $\log_{10} 145 \approx 2.161$

35. $\ln 18.42 \approx 2.913$

37. $\ln(1 + \sqrt{3}) \approx 1.005$

39. $\ln 0.32 \approx -1.139$

41. $f(x) = 3^x$, $g(x) = \log_3 x$

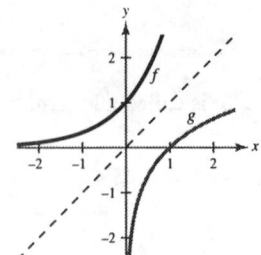

f and g are inverses. Their graphs are reflected about the line $y = x$.

43. $f(x) = e^x$, $g(x) = \ln x$

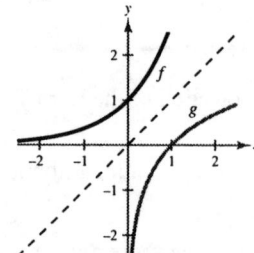

f and g are inverses. Their graphs are reflected about the line $y = x$.

45. $f(x) = \log_3 x + 2$
Asymptote: $x = 0$
Point on graph: $(1, 2)$
Matches graph (c).

47. $f(x) = -\log_3(x + 2)$
Asymptote: $x = -2$
Point on graph: $(-1, 0)$
Matches graph (d).

49. $f(x) = \log_3(1 - x)$
Asymptote: $x = 1$
Point on graph: $(0, 0)$
Matches graph (b).

51. $f(x) = \log_4 x$

Domain: $x > 0 \implies$ The domain is $(0, \infty)$.

Vertical asymptote: $x = 0$

x-intercept: $(1, 0)$

$y = \log_4 x \implies 4^y = x$

x	$\frac{1}{4}$	1	4	2
y	-1	0	1	$\frac{1}{2}$

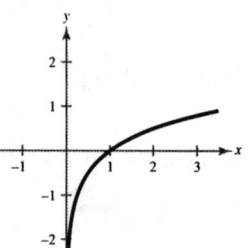

53. $y = -\log_3 x + 2$

Domain: $(0, \infty)$

Vertical asymptote: $x = 0$

x-intercept: $-\log_3 x + 2 = 0$

$$2 = \log_3 x$$
$$3^2 = x$$
$$9 = x$$

The x-intercept is $(9, 0)$.

$y = -\log_2 x + 2$

$\log_3 x = 2 - y \implies 3^{2-y} = x$

x	27	9	3	1	$\frac{1}{3}$
y	-1	0	1	2	3

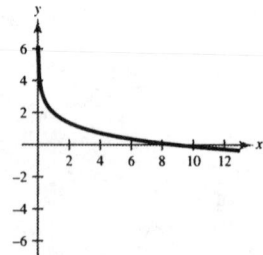

55. $f(x) = -\log_6(x + 2)$

Domain: $x + 2 > 0 \implies x > -2$
The domain is $(-2, \infty)$.
Vertical asymptote: $x + 2 = 0 \implies x = -2$
x-intercept: $0 = -\log_6(x + 2)$
$$0 = \log_6(x + 2)$$
$$6^0 = x + 2$$
$$1 = x + 2$$
$$-1 = x$$

The x-intercept is $(-1, 0)$.
$$y = -\log_6(x + 2)$$
$$-y = \log_6(x + 2)$$
$$6^{-y} - 2 = x$$

x	4	-1	$-1\frac{5}{6}$	$-1\frac{35}{36}$
y	-1	0	1	2

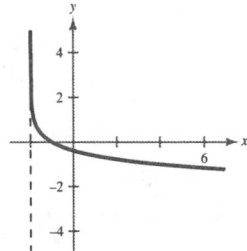

57. $y = \log_{10}\left(\dfrac{x}{5}\right)$

Domain: $\dfrac{x}{5} > 0 \implies x > 0$

The domain is $(0, \infty)$.

Vertical asymptote: $\dfrac{x}{5} = 0 \implies x = 0$

The vertical asymptote is the y-axis.

x-intercept: $\log_{10}\left(\dfrac{x}{5}\right) = 0$

$$\dfrac{x}{5} = 10^0$$

$$\dfrac{x}{5} = 1 \implies x = 5$$

The x-intercept is $(5, 0)$.

x	1	2	3	4	5	6	7
y	-0.70	-0.40	-0.22	-0.10	0	0.08	0.15

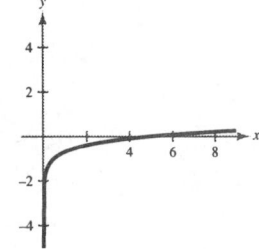

59. $f(x) = \ln(x - 2)$

Domain: $x - 2 > 0 \implies x > 2$
The domain is $(2, \infty)$.
Vertical asymptote: $x - 2 = 0 \implies x = 2$
x-intercept: $0 = \ln(x - 2)$
$$e^0 = x - 2$$
$$3 = x$$
The x-intercept is $(3, 0)$.

x	2.5	3	4	5
y	-0.69	0	0.69	1.10

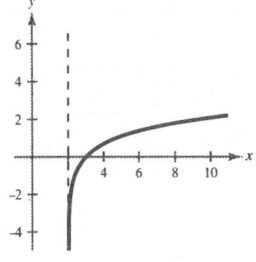

61. $g(x) = \ln(-x)$

Domain: $-x > 0 \implies x < 0$

The domain is $(-\infty, 0)$.

Vertical asymptote: $-x = 0 \implies x = 0$

x-intercept: $\quad 0 = \ln(-x)$

$$e^0 = -x$$

$$-1 = x$$

The x-intercept is $(-1, 0)$.

x	-0.5	-1	-2	-3
y	-0.69	0	0.69	1.10

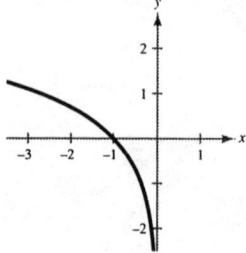

63. $f(x) = |\ln x|$

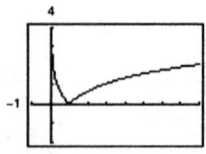

Increasing on $(1, \infty)$

Decreasing on $(0, 1)$

Relative minimum: $(1, 0)$

65. $f(x) = \dfrac{x}{2} - \ln \dfrac{x}{4}$

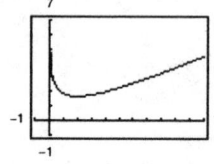

Increasing on $(2, \infty)$

Decreasing on $(0, 2)$

Relative minimum:

$\left(2, 1 - \ln \frac{1}{2}\right)$

67. (a) $f(x) = \ln x$

$g(x) = \sqrt{x}$

The natural log function grows at a slower rate than the square root function.

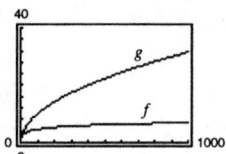

(b) $f(x) = \ln x$

$g(x) = \sqrt[4]{x}$

The natural log function grows at a slower rate than the fourth root function.

69. $y_1 = \ln x$

$y_2 = x - 1$

$y_3 = (x - 1) - \frac{1}{2}(x - 1)^2$

$y_4 = (x - 1) - \frac{1}{2}(x - 1)^2 + \frac{1}{3}(x - 1)^3$

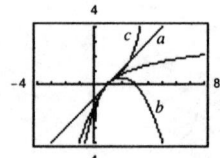

71. $f(t) = 80 - 17 \log_{10}(t + 1), \, 0 \leq t \leq 12$

(a) $f(0) = 80 - 17 \log_{10} 1 = 80.0$

(b) $f(4) = 80 - 17 \log_{10} 5 \approx 68.1$

(c) $f(10) = 80 - 17 \log_{10} 11 \approx 62.3$

73. $t = \dfrac{\ln 2}{r}$

r	0.005	0.01	0.015	0.02	0.025	0.03
t	138.6 yr	69.3 yr	46.2 yr	34.7 yr	27.7 yr	23.1 yr

75. $y = 80.4 - 11 \ln x$

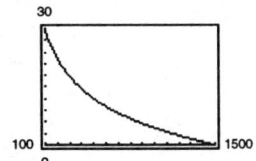

$y(300) = 80.4 - 11 \ln 300 \approx 17.66 \text{ ft}^3/\text{min}$

77. $W = 19{,}440(\ln 9 - \ln 3) \approx 21{,}357 \text{ ft-lb}$

79. $t = 10.042 \ln\left(\dfrac{1316.35}{1316.35 - 1250}\right) \approx 30 \text{ years}$

81. Total amount $= (1316.35)(30)(12) = \$473{,}886$

Interest $= 473{,}886 - 150{,}000 = \$323{,}886$

83. $f(x) = \dfrac{\ln x}{x}$

(a)
x	1	5	10	10^2	10^4	10^6
$f(x)$	0	0.322	0.230	0.046	0.00092	0.0000138

(b) As $x \to \infty$, $f(x) \to 0$.

(c)

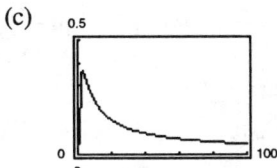

85. $8n - 3$

87. $83.95 + 37.50t$ Parts and labor

Section 5.3   Properties of Logarithms

■ You should know the following properties of logarithms.

(a) $\log_a x = \dfrac{\log_b x}{\log_b a}$

(b) $\log_a(uv) = \log_a u + \log_a v$ $\ln(uv) = \ln u + \ln v$

(c) $\log_a(u/v) = \log_a u - \log_a v$ $\ln(u/v) = \ln u - \ln v$

(d) $\log_a u^n = n \log_a u$ $\ln u^n = n \ln u$

■ You should be able to rewrite logarithmic expressions using these properties.

Solutions to Odd-Numbered Exercises

1. $f(x) = \log_{10} x$

$g(x) = \dfrac{\ln x}{\ln 10}$

$f(x) = g(x)$

3. $\log_3 5 = \dfrac{\log_{10} 5}{\log_{10} 3}$

5. $\log_2 x = \dfrac{\log_{10} x}{\log_{10} 2}$

7. $\log_3 5 = \dfrac{\ln 5}{\ln 3}$

9. $\log_2 x = \dfrac{\ln x}{\ln 2}$

11. $\log_3 7 = \dfrac{\log_{10} 7}{\log_{10} 3} = \dfrac{\ln 7}{\ln 3} \approx 1.771$

13. $\log_{1/2} 4 = \dfrac{\log_{10} 4}{\log_{10} (1/2)} = \dfrac{\ln 4}{\ln (1/2)} = -2.000$

15. $\log_9(0.4) = \dfrac{\log_{10} 0.4}{\log_{10} 9} = \dfrac{\ln 0.4}{\ln 9} \approx -0.417$

17. $\log_{15} 1250 = \dfrac{\log_{10} 1250}{\log_{10} 15} = \dfrac{\ln 1250}{\ln 15} \approx 2.633$

19. $\log_{10} 5x = \log_{10} 5 + \log_{10} x$

21. $\log_{10} \dfrac{5}{x} = \log_{10} 5 - \log_{10} x$

23. $\log_8 x^4 = 4 \log_8 x$

25. $\ln \sqrt{z} = \ln z^{1/2} = \tfrac{1}{2} \ln z$

27. $\ln xyz = \ln x + \ln y + \ln z$

29. $\ln \sqrt{a-1} = \tfrac{1}{2} \ln(a-1)$

31. $\ln z(z-1)^2 = \ln z + \ln(z-1)^2$

$\qquad\qquad\quad = \ln z + 2 \ln(z-1)$

33. $\ln \sqrt[3]{\dfrac{x}{y}} = \dfrac{1}{3} \ln \dfrac{x}{y}$

$\qquad\quad = \dfrac{1}{3}[\ln x - \ln y]$

$\qquad\quad = \dfrac{1}{3} \ln x - \dfrac{1}{3} \ln y$

35. $\ln \left(\dfrac{x^4 \sqrt{y}}{z^5} \right) = \ln x^4 \sqrt{y} - \ln z^5$

$\qquad\qquad\quad = \ln x^4 + \ln \sqrt{y} - \ln z^5$

$\qquad\qquad\quad = 4 \ln x + \dfrac{1}{2} \ln y - 5 \ln z$

37. $\log_b \left(\dfrac{x^2}{y^2 z^3} \right) = \log_b x^2 - \log_b y^2 z^3$

$\qquad\qquad\quad = \log_b x^2 - [\log_b y^2 + \log_b z^3]$

$\qquad\qquad\quad = 2 \log_b x - 2 \log_b y - 3 \log_b z$

39. $y_1 = \ln[x^3(x+4)]$

$y_2 = 3 \ln x + \ln(x+4)$

$y_1 = y_2$

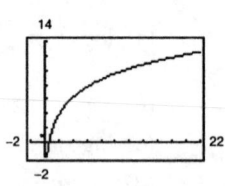

41. $\ln x + \ln 2 = \ln 2x$

43. $\log_4 z - \log_4 y = \log_4 \dfrac{z}{y}$

45. $2 \log_2(x+4) = \log_2(x+4)^2$

47. $\tfrac{1}{3} \log_3 5x = \log_3 (5x)^{1/3} = \log_3 \sqrt[3]{5x}$

49. $\ln x - 3 \ln(x+1) = \ln x - \ln(x+1)^3$

$\qquad\qquad\qquad\quad = \ln \dfrac{x}{(x+1)^3}$

51. $\ln(x-2) - \ln(x+2) = \ln \left(\dfrac{x-2}{x+2} \right)$

53. $\ln x - 2[\ln(x + 2) + \ln(x - 2)] = \ln x - 2\ln(x + 2)(x - 2)$

$$= \ln x - 2\ln(x^2 - 4)$$

$$= \ln x - \ln(x^2 - 4)^2$$

$$= \ln \frac{x}{(x^2 - 4)^2}$$

55. $\frac{1}{3}[2\ln(x + 3) + \ln x - \ln(x^2 - 1)] = \frac{1}{3}[\ln(x + 3)^2 + \ln x - \ln(x^2 - 1)]$

$$= \frac{1}{3}[\ln x(x + 3)^2 - \ln(x^2 - 1)]$$

$$= \frac{1}{3}\ln\frac{x(x + 3)^2}{x^2 - 1}$$

$$= \ln \sqrt[3]{\frac{x(x + 3)^2}{x^2 - 1}}$$

57. $\frac{1}{3}[\ln y + 2\ln(y + 4)] - \ln(y - 1) = \frac{1}{3}[\ln y + \ln(y + 4)^2] - \ln(y - 1)$

$$= \frac{1}{3}\ln y(y + 4)^2 - \ln(y - 1)$$

$$= \ln\sqrt[3]{y(y + 4)^2} - \ln(y - 1)$$

$$= \ln \frac{\sqrt[3]{y(y + 4)^2}}{y - 1}$$

59. $2\ln 3 - \frac{1}{2}\ln(x^2 + 1) = \ln 3^2 - \ln\sqrt{x^2 + 1}$

$$= \ln\frac{9}{\sqrt{x^2 + 1}}$$

61. $y_1 = 2[\ln 8 - \ln(x^2 + 1)]$

$y_2 = \ln\left[\dfrac{64}{(x^2 + 1)^2}\right]$

$y_1 = y_2$

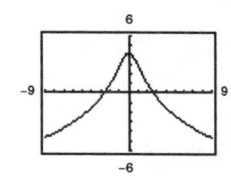

63. $y_1 = \ln x^2$

$y_2 = 2\ln x$

$y_1 = y_2$ for $x > 0$

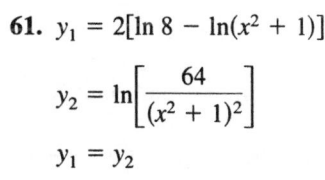

They are not equivalent. The domain of $f(x)$ is all real numbers except 0. The domain of $g(x)$ is $x > 0$.

65. $\log_2\left(\frac{32}{4}\right) = \log_2 32 - \log_2 4$ by Property 2.

67. $f(x) = \ln\dfrac{x}{2}$, $g(x) = \dfrac{\ln x}{\ln 2}$, $h(x) = \ln x - \ln 2$

$f(x) = h(x)$ by Property 2.

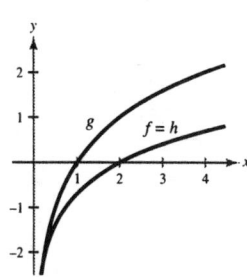

69. $\log_3 9 = 2\log_3 3 = 2$

71. $\log_4 16^{1.2} = 1.2(\log_4 16) = 1.2(2) = 2.4$

73. $\log_3(-9)$ is undefined. -9 is not in the domain of $\log_3 x$.

75. $\log_5 75 - \log_5 3 = \log_5 \frac{75}{3} = \log_5 25 = \log_5 5^2 = 2\log_5 5 = 2$

77. $\ln e^2 - \ln e^5 = 2 - 5 = -3$

79. $\log_{10} 0$ is undefined. 0 is not in the domain of $\log_{10} x$.

81. $\ln e^{4.5} = 4.5$

83. $\log_4 8 = \log_4 2^3 = 3\log_4 2 = 3\log_4 \sqrt{4} = 3\log_4 4^{1/2} = 3\left(\frac{1}{2}\right)\log_4 4 = \frac{3}{2}$

85. $\log_5 \frac{1}{250} = \log_5 1 - \log_5 250 = 0 - \log_5(125 \cdot 2)$

$\qquad = -\log_5(5^3 \cdot 2) = -[\log_5 5^3 + \log_5 2]$

$\qquad = -[3\log_5 5 + \log_5 2] = -3 - \log_5 2$

87. $\ln(5e^6) = \ln 5 + \ln e^6 = \ln 5 + 6 = 6 + \ln 5$

89. $f(t) = 90 - 15\log_{10}(t + 1), \ 0 \leq t \leq 12$

(a) $f(0) = 90$

(b) $f(6) \approx 77$

(c) $f(12) \approx 73$

(d) $\quad 75 = 90 - 15\log_{10}(t + 1)$

$\qquad -15 = -15\log_{10}(t + 1)$

$\qquad\quad 1 = \log_{10}(t + 1)$

$\qquad 10^1 = t + 1$

$\qquad\quad t = 9$ months

(e) $f(t) = 90 - \log_{10}(t + 1)^{15}$

(f)
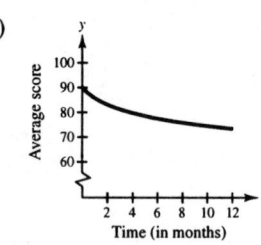

91. $f(x) = \ln x$

False, $f(0) \neq 0$ since 0 is not in the domain of $f(x)$. $f(1) = \ln 1 = 0$

93. False. $f(x) - f(2) = \ln x - \ln 2 = \ln \frac{x}{2} \neq \ln(x - 2)$

95. False. $f(u) = 2f(v) \implies \ln u = 2\ln v \implies \ln u = \ln v^2 \implies u = v^2$

97. Let $x = \log_b u$ and $y = \log_b v$, then $b^x = u$ and $b^y = v$.

$\qquad \dfrac{u}{v} = \dfrac{b^x}{b^y} = b^{x-y}$

$\qquad \log_b\left(\dfrac{u}{v}\right) = \log_b(b^{x-y}) = x - y = \log_b u - \log_b v$

99. $\dfrac{24xy^{-2}}{16x^{-3}y} = \dfrac{24xx^3}{16yy^2} = \dfrac{3x^4}{2y^3}, \ x \neq 0$

101. $(18x^3y^4)^{-3}(18x^3y^4)^3 = \dfrac{(18x^3y^4)^3}{(18x^3y^4)^3} = 1$ if $x \neq 0, y \neq 0$.

Section 5.4   Exponential and Logarithmic Equations

■   To solve an exponential equation, isolate the exponential expression, then take the logarithm of both sides. Then solve for the variable.

    1. $\log_a a^x = x$

    2. $\ln e^x = x$

■   To solve a logarithmic equation, rewrite it in exponential form. Then solve for the variable.

    1. $a^{\log_a x} = x$

    2. $e^{\ln x} = x$

■   If $a > 0$ and $a \neq 1$ we have the following:

    1. $\log_a x = \log_a y \implies x = y$

    2. $a^x = a^y \implies x = y$

Solutions to Odd-Numbered Exercises

1. $4^{2x-7} = 64$

   (a) $x = 5$

     $4^{2(5)-7} = 4^3 = 64$

     Yes, $x = 5$ is a solution.

   (b) $x = 2$

     $4^{2(2)-7} = 4^{-3} = \frac{1}{64} \neq 64$

     No, $x = 2$ is not a solution.

3. $3e^{x+2} = 75$

   (a) $x = -2 + e^{25}$

     $3e^{(-2+e^{25})+2} = 3e^{e^{25}} \neq 75$

     No, $x = -2 + e^{25}$ is not a solution.

   (b) $x = -2 + \ln 25$

     $3e^{(-2+\ln 25)+2} = 3e^{\ln 25} = 3(25) = 75$

     Yes, $x = -2 + \ln 25$ is a solution.

   (c) $x \approx 1.2189$

     $3e^{1.2189+2} = 3e^{3.2189} \approx 75$

     Yes, $x \approx 1.2189$ is a solution.

5. $\log_4(3x) = 3 \implies 3x = 4^3 \implies 3x = 64$

   (a) $x \approx 20.3560$

     $3(20.3560) = 61.0680 \neq 64$

     No, $x \approx 20.3560$ is not a solution.

   (b) $x = -4$

     $3(-4) = -12 \neq 64$

     No, $x = -4$ is not a solution.

   (c) $x = \frac{64}{3}$

     $3\left(\frac{64}{3}\right) = 64$

     Yes, $x = \frac{64}{3}$ is a solution.

7. $f(x) = g(x)$

    $2^x = 8$

    $2^x = 2^3$

    $x = 3$

   Point of intersection: $(3, 8)$

9. $f(x) = g(x)$

    $\log_3 x = 2$

    $x = 3^2$

    $x = 9$

   Point of intersection: $(9, 2)$

11. $4^x = 16$

    $4^x = 4^2$

    $x = 2$

13. $7^x = \frac{1}{49}$

    $7^x = 7^{-2}$

    $x = -2$

15. $\left(\frac{3}{4}\right)^x = \frac{27}{64}$

    $\left(\frac{3}{4}\right)^x = \left(\frac{3}{4}\right)^3$

    $x = 3$

17. $\log_4 x = 3$

    $x = 4^3$

    $x = 64$

19. $\log_{10} x = -1$

$\quad x = 10^{-1}$

$\quad x = \frac{1}{10}$

21. $\log_{10} 10^{x^2} = x^2$

23. $e^{\ln(5x+2)} = 5x + 2$

25. $e^{\ln x^2} = x^2$

27. $e^x = 10$

$\quad x = \ln 10 \approx 2.303$

29. $7 - 2e^x = 5$

$\quad -2e^x = -2$

$\quad e^x = 1$

$\quad x = \ln 1 = 0$

31. $e^{3x} = 12$

$\quad 3x = \ln 12$

$\quad x = \dfrac{\ln 12}{3} \approx 0.828$

33. $500e^{-x} = 300$

$\quad e^{-x} = \dfrac{3}{5}$

$\quad -x = \ln \dfrac{3}{5}$

$\quad x = -\ln \dfrac{3}{5} = \ln \dfrac{5}{3} \approx 0.511$

35. $\quad e^{2x} - 4e^x - 5 = 0$

$\quad (e^x - 5)(e^x + 1) = 0$

$\quad e^x = 5$ or $e^x = -1$ (No solution)

$\quad x = \ln 5 \approx 1.609$

37. $20(100 - e^{x/2}) = 500$

$\quad 100 - e^{x/2} = 25$

$\quad -e^{x/2} = -75$

$\quad e^{x/2} = 75$

$\quad \dfrac{x}{2} = \ln 75$

$\quad x = 2 \ln 75 \approx 8.635$

39. $10^x = 42$

$\quad x = \log_{10} 42 \approx 1.623$

41. $\quad 3^{2x} = 80$

$\quad \ln 3^{2x} = \ln 80$

$\quad 2x \ln 3 = \ln 80$

$\quad x = \dfrac{\ln 80}{2 \ln 3} \approx 1.994$

43. $5^{-t/2} = 0.20$

$\quad 5^{-t/2} = \dfrac{1}{5}$

$\quad 5^{-t/2} = 5^{-1}$

$\quad -\dfrac{t}{2} = -1$

$\quad t = 2$

45. $\quad 2^{3-x} = 565$

$\quad \ln 2^{3-x} = \ln 565$

$\quad (3 - x) \ln 2 = \ln 565$

$\quad 3 \ln 2 - x \ln 2 = \ln 565$

$\quad -x \ln 2 = \ln 565 - \ln 2^3$

$\quad x \ln 2 = \ln 8 - \ln 565$

$\quad x = \dfrac{\ln 8 - \ln 565}{\ln 2} \approx -6.142$

47. $g(x) = 6e^{1-x} - 25$

The zero is $x \approx -0.427$.

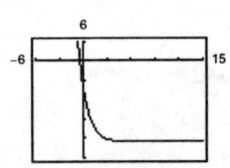

49. $g(t) = e^{0.09t} - 3$

The zero is $x \approx 12.207$.

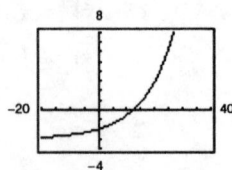

51. $8(10^{3x}) = 12$

$$10^{3x} = \frac{12}{8}$$

$$3x = \log_{10}\left(\frac{3}{2}\right)$$

$$x = \tfrac{1}{3}\log_{10}\left(\frac{3}{2}\right) \approx 0.059$$

53. $\left(1 + \frac{0.065}{365}\right)^{365t} = 4$

$$\ln\left(1 + \frac{0.065}{365}\right)^{365t} = \ln 4$$

$$365t \ln\left(1 + \frac{0.065}{365}\right) = \ln 4$$

$$t = \frac{\ln 4}{365 \ln\left(1 + \frac{0.065}{365}\right)} \approx 21.330$$

55. $\ln x = -3$

$$x = e^{-3} \approx 0.050$$

57. $\ln 2x = 2.4$

$$2x = e^{2.4}$$

$$x = \frac{e^{2.4}}{2} \approx 5.512$$

59. $\ln \sqrt{x + 2} = 1$

$$\sqrt{x + 2} = e^1$$

$$x + 2 = e^2$$

$$x = e^2 - 2 \approx 5.389$$

61. $\log_{10}(z - 3) = 2$

$$z - 3 = 10^2$$

$$z = 10^2 + 3 = 103$$

63. $\ln x + \ln(x - 2) = 1$

$$\ln[x(x - 2)] = 1$$

$$x(x - 2) = e^1$$

$$x^2 - 2x - e = 0$$

$$x = \frac{2 \pm \sqrt{4 + 4e}}{2}$$

$$= \frac{2 \pm 2\sqrt{1 + e}}{2}$$

$$= 1 \pm \sqrt{1 + e}$$

Using the positive value for x, we have
$x = 1 + \sqrt{1 + e} \approx 2.928$.

65. $\log_{10}(x + 4) - \log_{10} x = \log_{10}(x + 2)$

$$\log_{10}\left(\frac{x + 4}{x}\right) = \log_{10}(x + 2)$$

$$\frac{x + 4}{x} = x + 2$$

$$x + 4 = x^2 + 2x \qquad \text{Quadratic}$$

$$0 = x^2 + x - 4 \qquad \text{Formula}$$

$$x = \frac{-1 \pm \sqrt{17}}{2}$$

Choosing the positive value of x (the negative value is extraneous), we have

$$x = \frac{-1 + \sqrt{17}}{2} \approx 1.562.$$

67. $\log_3 x + \log_3(x^2 - 8) = \log_3 8x$

$$\log_3 x(x^2 - 8) = \log_3 8x$$

$$x(x^2 - 8) = 8x$$

$$x^3 - 8x = 8x$$

$$x^3 - 16x = 0$$

$$x(x + 4)(x - 4) = 0$$

$$x = 0, \ x = -4, \text{ or } x = 4$$

The only solution that is in the domain is $x = 4$.
Both $x = 0$ and $x = -4$ are extraneous.

69. $\ln (x + 5) = \ln(x - 1) - \ln(x + 1)$

$$\ln(x + 5) = \ln\left(\frac{x - 1}{x + 1}\right)$$

$$x + 5 = \frac{x - 1}{x + 1}$$

$$(x + 5)(x + 1) = x - 1$$

$$x^2 + 6x + 5 = x - 1$$

$$x^2 + 5x + 6 = 0$$

$$(x + 2)(x + 3) = 0$$

$$x = -2 \text{ or } x = -3$$

Both of these solutions are extraneous, so the equation has no solution.

71. $6 \log_3(0.5x) = 11$

$\quad\quad \log_3(0.5x) = \frac{11}{6}$

$\quad\quad\quad\quad 0.5x = 3^{11/6}$

$\quad\quad\quad\quad\quad\quad x = 2(3^{11/6}) \approx 14.988$

73. $2 \ln x = 7$

$\quad\quad \ln x = \frac{7}{2}$

$\quad\quad\quad x = e^{7/2} \approx 33.115$

75. $\ln x + \ln(x^2 + 1) = 8$

$\quad\quad \ln x(x^2 + 1) = 8$

$\quad\quad\quad x(x^2 + 1) = e^8$

$\quad\quad x^3 + x - e^8 = 0$

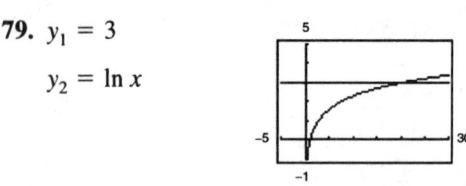

From the graph we have $x \approx 14.369$.

77. $y_1 = 7$

$\quad y_2 = 2^x$

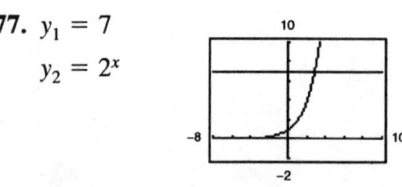

From the graph we have $x \approx 2.807$ when $y = 7$.

79. $y_1 = 3$

$\quad y_2 = \ln x$

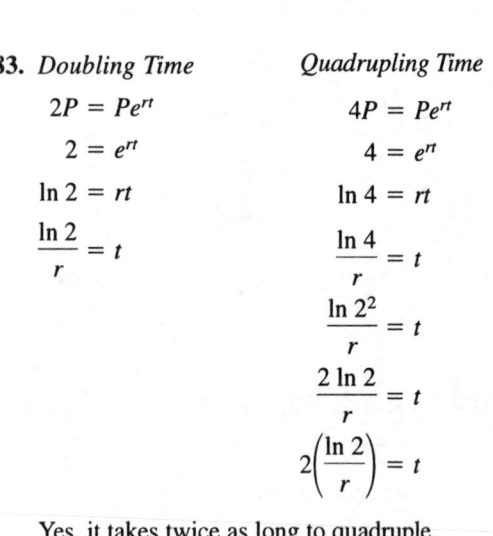

From the graph we have $x \approx 20.806$ when $y = 3$.

81. $\quad A = Pe^{rt}$

$\quad 2000 = 1000e^{0.085t}$

$\quad\quad\quad 2 = e^{0.085t}$

$\quad\quad \ln 2 = 0.085t$

$\quad \dfrac{\ln 2}{0.085} = t$

$\quad\quad\quad t \approx 8.2$ years

83. *Doubling Time*

$\quad 2P = Pe^{rt}$

$\quad\quad 2 = e^{rt}$

$\quad \ln 2 = rt$

$\quad \dfrac{\ln 2}{r} = t$

Quadrupling Time

$\quad 4P = Pe^{rt}$

$\quad\quad 4 = e^{rt}$

$\quad \ln 4 = rt$

$\quad \dfrac{\ln 4}{r} = t$

$\quad \dfrac{\ln 2^2}{r} = t$

$\quad \dfrac{2 \ln 2}{r} = t$

$\quad 2\left(\dfrac{\ln 2}{r}\right) = t$

Yes, it takes twice as long to quadruple.

85. $\quad A = Pe^{rt}$

$\quad 3000 = 1000e^{0.085t}$

$\quad\quad\quad 3 = e^{0.085t}$

$\quad\quad \ln 3 = 0.085t$

$\quad \dfrac{\ln 3}{0.085} = t$

$\quad\quad\quad t \approx 12.9$ years

87. $p = 500 - 0.5(e^{0.004x})$

(a) $\quad\quad p = 350$

$\quad\quad 350 = 500 - 0.5(e^{0.004x})$

$\quad\quad 300 = e^{0.004x}$

$\quad 0.004x = \ln 300$

$\quad\quad\quad x \approx 1426$ units

(b) $\quad\quad p = 300$

$\quad\quad 300 = 500 - 0.5(e^{0.004x})$

$\quad\quad 400 = e^{0.004x}$

$\quad 0.004x = \ln 400$

$\quad\quad\quad x \approx 1498$ units

89. $V = 6.7e^{-48.1/t}$, $t \geq 0$

(a)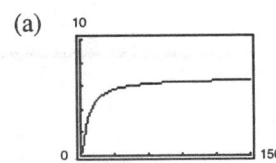

(b) As $x \to \infty$, $V \to 6.7$.

Horizontal asymptote: $y = 6.7$

The yield will approach 6.7 million cubic feet per acre.

(c) $1.3 = 6.7e^{-48.1/t}$

$\dfrac{1.3}{6.7} = e^{-48.1/t}$

$\ln\left(\dfrac{13}{67}\right) = \dfrac{-48.1}{t}$

$t = \dfrac{-48.1}{\ln(13/67)} \approx 29.3$ years

91. (a) From the graph shown in the textbook, we see horizontal asymptotes at $y = 0$ and $y = 100$. These represent the lower and upper percent bounds.

(b) Males

$$50 = \frac{100}{1 + e^{-0.6114(x - 69.71)}}$$

$1 + e^{-0.6114(x - 69.71)} = 2$

$e^{-0.6114(x - 69.71)} = 1$

$-0.6114(x - 69.71) = \ln 1$

$-0.6114(x - 69.71) = 0$

$x = 69.71$ inches

Females

$$50 = \frac{100}{1 + e^{-0.66607(x - 64.51)}}$$

$1 + e^{-0.66607(x - 64.51)} = 2$

$e^{-0.66607(x - 64.51)} = 1$

$-0.66607(x - 64.51) = \ln 1$

$-0.66607(x - 64.51) = 0$

$x = 64.51$ inches

93. $T = 20[1 + 7(2^{-h})]$

(a) From the graph in the textbook we see a horizontal asymptote at $y = 20$. This represents the room temperature.

(b) $100 = 20[1 + 7(2^{-h})]$

$5 = 1 + 7(2^{-h})$

$4 = 7(2^{-h})$

$\dfrac{4}{7} = 2^{-h}$

$\ln\left(\dfrac{4}{7}\right) = \ln 2^{-h}$

$\ln\left(\dfrac{4}{7}\right) = -h \ln 2$

$\dfrac{\ln\left(\frac{4}{7}\right)}{-\ln 2} = h$

$h \approx 0.81$ hour

95. $\sqrt{48x^2 y^5} = \sqrt{16x^2 y^4 3y} = 4|x|y^2\sqrt{3y}$

97. $\sqrt[3]{25}\sqrt[3]{15} = \sqrt[3]{375} = \sqrt[3]{125 \cdot 3} = 5\sqrt[3]{3}$

Section 5.5   Exponential and Logarithmic Models

■ You should be able to solve compound interest problems.

(a) Compound interest formulas:

1. $A = P\left(1 + \dfrac{r}{n}\right)^{nt}$

2. $A = Pe^{rt}$

(b) Doubling time:

1. $t = \dfrac{\ln 2}{n \ln[1 + (r/n)]}$, n compoundings per year

2. $t = \dfrac{\ln 2}{r}$, continuous compounding

■ You should be able to solve growth and decay problems.

(a) Exponential growth if $b > 0$ and $y = ae^{bx}$.

(b) Exponential decay if $b > 0$ and $y = ae^{-bx}$.

■ You should be able to use the Gaussian model

$y = ae^{-(x-b)^2/c}$.

■ You should be able to use the logistics growth model

$y = \dfrac{a}{1 + be^{-(x-c)/d}}$.

■ You should be able to use the logarithmic models

$y = \ln(ax + b)$ and $y = \log_{10}(ax + b)$.

Solutions to Odd-Numbered Exercises

1. $y = 2e^{x/4}$

This is an exponential growth model. Matches graph (c)

3. $y = \frac{1}{16}(x^2 + 8x + 32)$

This is a quadratic function. Its graph is a parabola. Matches graph (a)

5. $y = \ln(x + 1)$

This is a logarithmic model. Matches graph (d)

7. Since $A = 1000e^{0.12t}$, the time to double is given by $2000 = 1000e^{0.12t}$ and we have

$t = \dfrac{\ln 2}{0.12} \approx 5.78$ years.

Amount after 10 years: $A = 1000e^{1.2} \approx \$3320.12$

9. Since $A = 750e^{rt}$ and $A = 1500$ when $t = 7.75$, we have the following.

$1500 = 750e^{7.75r}$

$r = \dfrac{\ln 2}{7.75} \approx 0.0894 = 8.94\%$

Amount after 10 years: $A = 750e^{0.0894(10)} \approx \$1833.67$

11. Since $A = 500e^{rt}$ and $A = 1292.85$ when $t = 10$, we have the following.

$$1292.85 = 500e^{10r}$$

$$r = \frac{\ln(1292.85/500)}{10} \approx 0.9095 = 9.5\%$$

The time to double is given by

$$1000 = 500e^{0.095t}$$

$$t = \frac{\ln 2}{0.095} \approx 7.30 \text{ years.}$$

13. Since $A = Pe^{0.045t}$ and $A = 10,000.00$ when $t = 10$, we have the following.

$$10,000.00 = Pe^{0.045(10)}$$

$$\frac{10,000.00}{e^{0.045(10)}} = P \approx \$6376.28$$

The time to double is given by

$$t = \frac{\ln 2}{0.045} \approx 15.40 \text{ years.}$$

15. $500,000 = P\left(1 + \dfrac{0.075}{12}\right)12(20)$

$$P = \frac{500,000}{\left(1 + \dfrac{0.075}{12}\right)}12(20) = \$112,087.09$$

17. $P = 1000, r = 11\%$

(a) $n = 1$

$$t = \frac{\ln 2}{\ln(1 + 0.11)} \approx 6.642 \text{ years}$$

(b) $n = 12$

$$t = \frac{\ln 2}{12 \ln\left(1 + \frac{0.11}{12}\right)} \approx 6.330 \text{ years}$$

(c) $n = 365$

$$t = \frac{\ln 2}{365 \ln\left(1 + \frac{0.11}{365}\right)} \approx 6.302 \text{ years}$$

(d) Continuously

$$t = \frac{\ln 2}{0.11} \approx 6.301 \text{ years}$$

19. $3P = Pe^{rt}$

$3 = e^{rt}$

$\ln 3 = rt$

$\dfrac{\ln 3}{r} = t$

r	2%	4%	6%	8%	10%	12%
$t = \dfrac{\ln 3}{r}$	54.93	27.47	18.31	13.73	10.99	9.16

21. $3P = P(1 + r)^t$

$3 = (1 + r)^t$

$\ln 3 = \ln(1 + r)^t$

$\ln 3 = t \ln(1 + r)$

$\dfrac{\ln 3}{\ln(1 + r)} = t$

r	2%	4%	6%	8%	10%	12%
$t = \dfrac{\ln 3}{\ln(1 + r)}$	55.48	28.01	18.85	14.27	11.53	9.69

23. Continuous compounding results in faster growth.

$$A = 1 + 0.075[\![t]\!] \text{ and } A = e^{0.07t}$$

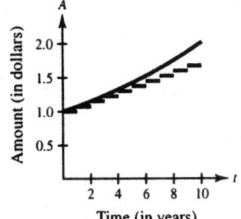

25. $\frac{1}{2}C = Ce^{k(1620)}$

$k = \dfrac{\ln 0.5}{1620}$

Given $C = 10$ grams, after 1000 years we have

$y = 10e^{[(\ln 0.5)/1620](1000)}$

≈ 6.52 grams.

27. $\frac{1}{2}C = Ce^{k(5730)}$

$k = \dfrac{\ln 0.5}{5730}$

Given $y = 2$ grams after 1000 years, we have

$2 = Ce^{[(\ln 0.5)/5730](1000)}$

$C \approx 2.26$ grams.

29. $\frac{1}{2}C = Ce^{k(24,360)}$

$k = \dfrac{\ln 0.5}{24,360}$

Given $y = 2.1$ grams after 1000 years, we have

$2.1 = Ce^{[(\ln 0.5)/24,360](1000)}$

$C \approx 2.16$ grams.

31. $y = ae^{bx}$

$1 = ae^{b(0)} \implies 1 = a$

$10 = e^{b(3)}$

$\ln 10 = 3b$

$\dfrac{\ln 10}{3} = b \qquad \implies b \approx 0.7675$

Thus, $y = e^{0.7675x}$.

33. $y = ae^{bx}$

$1 = ae^{b(0)} \implies 1 = a$

$\dfrac{1}{4} = e^{b(3)}$

$\ln\left(\dfrac{1}{4}\right) = 3b$

$\dfrac{\ln\left(\frac{1}{4}\right)}{3} = b \qquad \implies b \approx -0.4621$

Thus, $y = e^{-0.4621x}$.

35. $P = 105,300e^{0.015t}$

$150,000 = 105,300e^{0.015t}$

$\ln \frac{1500}{1053} = 0.015t$

$t \approx 23.59$

The population will reach 150,000 during 2013.
[Note: 1990 + 23.59]

37. For 1945, use $t = -45$.

$1350 = 2500d^{k(-45)}$

$\ln\left(\dfrac{1350}{2500}\right) = -45k \implies k \approx 0.0137$.

For 2010, use $t = 20$.

$P = 2500e^{0.0137(20)} \approx 3288$ people

39. $y = ae^{bt}$

$4.22 = ae^{b(0)} \implies a = 4.22$

$6.49 = 4.22e^{b(10)}$

$\dfrac{6.49}{4.22} = e^{10b}$

$\ln\left(\dfrac{6.49}{4.22}\right) = 10b \implies b \approx 0.0430$

$y = 4.22e^{0.0430t}$

When $t = 20$,
$y = 4.22e^{0.0430(20)} \approx 9.97$ million.

41. $y = ae^{bt}$

$3.00 = ae^{b(0)} \implies a = 3$

$2.74 = 3e^{b(10)}$

$\dfrac{2.74}{3} = e^{10b}$

$\ln\left(\dfrac{2.74}{3}\right) = 10b \implies b \approx -0.0091$

$y = 3e^{-0.0091t}$

When $t = 20$,
$y = 3e^{-0.0091(20)} \approx 2.50$ million.

43. b is determined by the growth rate. The greater the rate of growth, the greater the value of b.

45. $N = 100e^{kt}$

$300 = 100e^{5k}$

$k = \dfrac{\ln 3}{5} \approx 0.2197$

$N = 100e^{0.2197t}$

$200 = 100e^{0.2197t}$

$t = \dfrac{\ln 2}{0.2197} \approx 3.15$ hours

47. $y = Ce^{kt}$

$$\frac{1}{2}C = Ce^{(1620)k}$$

$$\ln\frac{1}{2} = 1620k$$

$$k = \frac{\ln(1/2)}{1620}$$

When $t = 100$, we have

$$y = Ce^{[\ln(1/2)/1620](100)} \approx 0.958C = 95.8\%C.$$

After 100 years, approximately 95.8% of the radioactive radium will remain.

49. (0, 22,000), (2, 13,000)

(a) $m = \dfrac{13,000 - 22,000}{2 - 0} = -4500$

$b = 22,000$

Thus, $V = -4500t + 22,000.$

(b) $a = 22,000$

$13,000 = 22,000e^{k(2)}$

$\dfrac{13}{22} = e^{2k}$

$\ln\left(\dfrac{13}{22}\right) = 2k \Rightarrow k \approx -0.263$

Thus, $V = 22,000e^{-0.263t}.$

(c) The exponential model depreciates faster in the first two years.

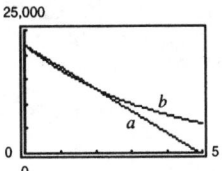

(d)

t	1	3
$V = -4500t + 22,000$	\$17,500	\$8500
$V = 22,000e^{-0.263t}$	\$16,912	\$9995

(e) The slope of the linear model means that the car depreciates \$4500 per year.

51. $S(t) = 100(1 - e^{kt})$

(a) $15 = 100(1 - e^{k(1)})$

$-85 = -100e^k$

$k = \ln 0.85$

$k \approx -0.1625$

$S(t) = 100(1 - e^{-0.1625t})$

(c) $S(5) = 100(1 - e^{-0.1625(5)})$

$\approx 55.625 = 55,625$ units

(b)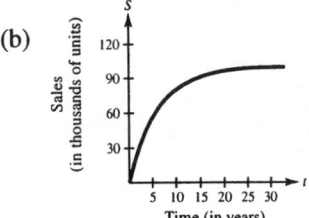

53. $S = 10(1 - e^{kx})$

$x = 5$ (in hundreds), $S = 2.5$ (in thousands)

(a) $2.5 = 10(1 - e^{k(5)})$

$0.25 = 1 - e^{5k}$

$e^{5k} = 0.75$

$5k = \ln 0.75$

$k \approx -0.0575$

$S = 10(1 - e^{-0.0575x})$

(b) When $x = 7$,

$S = 10(1 - e^{-0.0575(7)}) \approx 3.314$

which corresponds to 3314 units.

55. $N = 30(1 - e^{kt})$

(a) $N = 19, \; t = 20$

$19 = 30(1 - e^{20k})$

$20k = \ln \frac{11}{30}$

$k \approx -0.050$

$N = 30(1 - e^{-0.050t})$

(b) $N = 25$

$25 = 30(1 - e^{-0.05t})$

$\frac{5}{30} = e^{-0.05t}$

$t = -\frac{1}{0.05} \ln \frac{5}{30} \approx 36$ days

(c) No, this is not a linear function.

57. $R = \log_{10} \dfrac{I}{I_0} = \log_{10} I$ since $I_0 = 1$.

(a) $R = \log_{10} 80{,}500{,}000 \approx 7.91$

(b) $R = \log_{10} 48{,}275{,}000 \approx 7.68$

59. $\beta(I) = 10 \log_{10} \dfrac{I}{I_0}$ where $I_0 = 10^{-16}$ watt/cm^2.

(a) $\beta(10^{-14}) = 10 \log_{10} \dfrac{10^{-14}}{10^{-16}} = 10 \log_{10} 10^2 = 20$ decibels

(b) $\beta(10^{-9}) = 10 \log_{10} \dfrac{10^{-9}}{10^{-16}} = 10 \log_{10} 10^7 = 70$ decibels

(c) $\beta(10^{-6.5}) = 10 \log_{10} \dfrac{10^{-6.5}}{10^{-16}} = 10 \log_{10} 10^{9.5} = 95$ decibels

(d) $\beta(10^{-4}) = 10 \log_{10} \dfrac{10^{-4}}{10^{-16}} = 10 \log_{10} 10^{12} = 120$ decibels

61. $\beta = 10 \log_{10} \dfrac{I}{I_0}$

$10^{\beta/10} = \dfrac{I}{I_0}$

$I = I_0 10^{\beta/10}$

% decrease $= \dfrac{I_0 10^{9.3} - I_0 10^{8.0}}{I_0 10^{9.3}} \times 100 \approx 95\%$

63. pH $= -\log_{10}[\text{H}^+] = -\log_{10}[2.3 \times 10^{-5}] \approx 4.64$

65. $5.8 = -\log_{10}[\text{H}^+]$

$10^{-5.8} = \text{H}^+$

$\text{H}^+ \approx 1.58 \times 10^{-6}$ moles per liter

67. $2.5 = -\log_{10}[\text{H}^+]$

$10^{-2.5} = \text{H}^+$ for the fruit.

$9.5 = -\log_{10}[\text{H}^+]$

$10^{-9.5} = \text{H}^+$ for the antacid tablet.

$\dfrac{10^{-2.5}}{10^{-9.5}} = 10^7$

69. Interest: $u = M - \left(M - \dfrac{Pr}{12}\right)\left(1 + \dfrac{r}{12}\right)^{12t}$

Principle: $v = \left(M - \dfrac{Pr}{12}\right)\left(1 + \dfrac{r}{12}\right)^{12t}$

(a) $P = 120{,}000$, $t = 35$, $r = 0.095$, $M = 985.93$

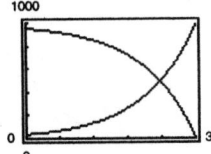

(b) In the early years of the mortgage, the majority of the monthly payment goes toward interest. The principle and interest are nearly equal when $t \approx 27.676 \approx 28$ years.

(c) $P = 120{,}000$, $t = 20$, $r = 0.095$, $M = 1118.56$

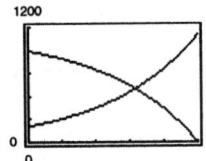

The interest is still the majority of the monthly payment in the early years. Now the principle and interest are nearly equal when $t \approx 12.675 \approx 12.7$ years.

71. $t_1 = 40.757 + 0.556s - 15.817 \ln s$

$t_2 = 1.2259 + 0.0023s^2$

(a) Linear model: $t_3 \approx 0.2729s - 6.0143$

Exponential model: $t_4 \approx 1.5385e^{1.0291s}$

(b)

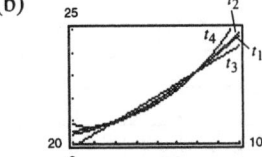

(c)

s	30	40	50	60	70	80	90
t_1	3.6	4.6	6.7	9.4	12.5	15.9	19.6
t_2	3.3	4.9	7.0	9.5	12.5	15.9	19.9
t_3	2.2	4.9	7.6	10.4	13.1	15.8	18.5
t_4	3.7	4.9	6.6	8.8	11.8	15.8	21.2

(d) Model t_1: $S_1 = |3.4 - 3.6| + |5 - 4.6| + |7 - 6.7| + |9.3 - 9.4| + |12 - 12.5| + |15.8 - 15.9| + |20 - 19.6| \doteq 2.0$

Model t_2: $S_2 = |3.4 - 3.3| + |5 - 4.9| + |7 - 7| + |9.3 - 9.5| + |12 - 12.5| + |15.8 - 15.9| + |20 - 19.9| = 1.1$

Model t_3: $S_3 = |3.4 - 2.2| + |5 - 4.9| + |7 - 7.6| + |9.3 - 10.4| + |12 - 13.1| + |15.8 - 15.8| + |20 - 18.5| = 5.6$

Model t_4: $S_4 = |3.4 - 3.7| + |5 - 4.9| + |7 - 6.6| + |9.3 - 8.8| + |12 - 11.8| + |15.8 - 15.8| + |20 - 21.2| = 2.7$

t_2, the Quadratic model, is the best fit with the data.

73. Answers will vary.

75.

$$-4 \mid \begin{array}{cccc} 4 & 4 & -39 & 36 \\ & -16 & 48 & -36 \\ \hline 4 & -12 & 9 & 0 \end{array}$$

Thus, $\dfrac{4x^3 + 4x^2 - 39x + 36}{x + 4} = 4x^2 - 12x + 9.$

77.

$$4 \mid \begin{array}{cccc} 2 & -8 & 3 & -9 \\ & 8 & 0 & 12 \\ \hline 2 & 0 & 3 & 3 \end{array}$$

Thus, $\dfrac{2x^3 - 8x^2 + 3x - 9}{x - 4} = 2x^2 + 3 + \dfrac{3}{x - 4}.$

❑ Review Exercises for Chapter 5

Solutions to Odd-Numbered Exercises

1. $f(x) = 4^x$

Intercept: $(0, 1)$

Horizontal asymptote: x-axis

Increasing on: $(-\infty, \infty)$

Matches graph (e)

3. $f(x) = -4^x$

Intercept: $(0, -1)$

Horizontal asymptote: x-axis

Decreasing on: $(-\infty, \infty)$

Matches graph (a)

5. $f(x) = \log_4 x$

Intercept: $(1, 0)$

Vertical asymptote: y-axis

Increasing on: $(0, \infty)$

Matches graph (d)

7. $f(x) = 0.3^x$

x	-2	-1	0	1	2
y	11.11	3.33	1	0.3	0.09

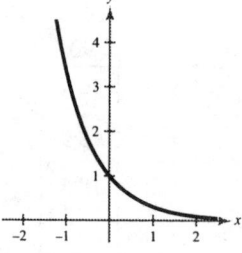

9. $h(x) = e^{-x/2}$

x	-2	-1	0	1	2
y	2.72	1.65	1	0.61	0.37

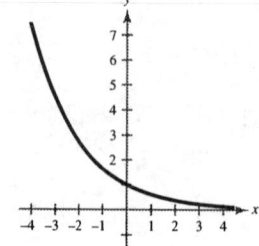

11. $f(x) = e^{x+2}$

x	-3	-2	-1	0	1
y	0.37	1	2.72	7.39	20.09

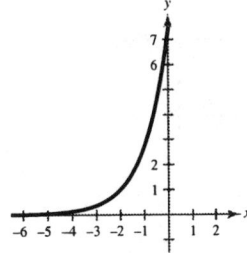

13. $g(x) = 200e^{4/x}$

As $x \to \infty$, $g(x) \to 200$ so we have a horizontal asymptote at $y = 200$.

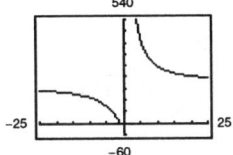

15. $A = 3500\left(1 + \dfrac{0.105}{n}\right)^{10n}$ or $A = 3500e^{(0.105)(10)}$

n	1	2	4	12	365	Continuous Compounding
A	\$9,499.28	\$9,738.91	\$9,867.22	\$9,956.20	\$10,000.27	\$10,001.78

17. $200{,}000 = Pe^{0.08t}$

$P = \dfrac{200{,}000}{e^{0.08t}}$

t	1	10	20	30	40	50
P	\$184,623.27	\$89,865.79	\$40,379.30	\$18,143.59	\$8,152.44	\$3,663.13

19. $F(t) = 1 - e^{-t/3}$
(a) $F\left(\frac{1}{2}\right) \approx 0.154$
(b) $F(2) \approx 0.487$
(c) $F(5) \approx 0.811$

21. (a) $A = 50{,}000e^{(0.0875)(35)} \approx \$1{,}069{,}047.14$
(b) The doubling time is
$\dfrac{\ln 2}{0.0875} \approx 7.9$ years.

23. $g(x) = \log_2 x \implies 2^y = x$
Domain: $(0, \infty)$
Vertical asymptote: $x = 0$

x	$\frac{1}{4}$	$\frac{1}{2}$	1	2	4
y	-2	-1	0	1	2

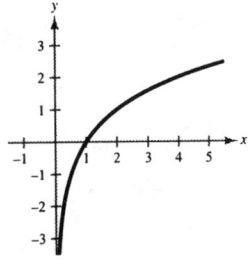

25. $f(x) = \ln x + 3$
Domain: $(0, \infty)$.
Vertical asymptote: $x = 0$

x	1	2	3	$\frac{1}{2}$	$\frac{1}{4}$
$f(x)$	3	3.69	4.10	2.31	1.61

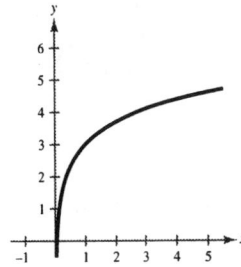

27. $h(x) = \ln(e^{x-1})$

$= (x - 1) \ln e$

$= x - 1$

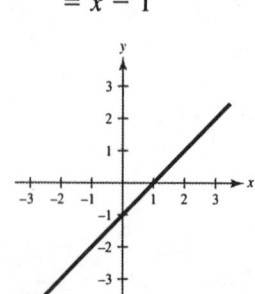

29. $y = \log_{10}(x^2 + 1)$

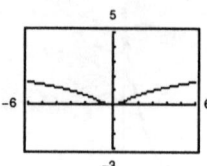

31. $\quad 4^3 = 64$

$\log_4 64 = 3$

33. $\log_{10} 1000 = \log_{10} 10^3 = 3$

35. $\ln e^7 = 7$

37. $\log_4 9 = \dfrac{\log_{10} 9}{\log_{10} 4} \approx 1.585$

$\log_4 9 = \dfrac{\ln 9}{\ln 4} \approx 1.585$

39. $\log_{12} 200 = \dfrac{\log_{10} 200}{\log_{10} 12} \approx 2.132$

$\log_{12} 200 = \dfrac{\ln 200}{\ln 12} \approx 2.132$

41. $\log_5 5x^2 = \log_5 5 + \log_5 x^2$

$= 1 + 2 \log_5 |x|$

43. $\log_{10} \dfrac{5\sqrt{y}}{x^2} = \log_{10} 5\sqrt{y} - \log_{10} x^2$

$= \log_{10} 5 + \log_{10} \sqrt{y} - \log_{10} x^2$

$= \log_{10} 5 + \dfrac{1}{2} \log_{10} y - 2 \log_{10} |x|$

45. $\log_2 5 + \log_2 x = \log_2 5x$

47. $\dfrac{1}{2} \ln|2x - 1| - 2 \ln|x + 1| = \ln \sqrt{|2x - 1|} - \ln|x + 1|^2$

$= \ln \dfrac{\sqrt{|2x - 1|}}{(x + 1)^2}$

49. True; by the inverse properties, $\log_b b^{2x} = 2x$.

51. False; $\ln x + \ln y = \ln(xy) \neq \ln(x + y)$

53. True, $\log\left(\dfrac{10}{x}\right) = \log 10 - \log x = 1 - \log x$.

55. $S = 25 - \dfrac{13 \ln(10/12)}{\ln 3} \approx 27.16$ miles

57. $e^x = 12$

$x = \ln 12 \approx 2.485$

59. $3e^{-5x} = 132$

$e^{-5x} = 44$

$-5x = \ln 44$

$x = \dfrac{\ln 44}{-5} \approx -0.757$

61. $e^{2x} - 7e^x + 10 = 0$

$(e^x - 2)(e^x - 5) = 0$

$e^x = 2 \quad$ or $\quad e^x = 5$

$x = \ln 2 \qquad x = \ln 5$

$x \approx 0.693 \qquad x \approx 1.609$

63. $\ln 3x = 8.2$

$3x = e^{8.2}$

$x = \dfrac{e^{8.2}}{3} \approx 1213.650$

65. $\ln x - \ln 3 = 2$

$\ln \dfrac{x}{3} = 2$

$\dfrac{x}{3} = e^2$

$x = 3e^2 \approx 22.167$

67. $\log(x - 1) = \log(x - 2) - \log(x + 2)$

$$\log(x - 1) = \log\left(\frac{x - 2}{x + 2}\right)$$

$$x - 1 = \frac{x - 2}{x + 2}$$

$$(x - 1)(x + 2) = x - 2$$

$$x^2 + x - 2 = x - 2$$

$$x^2 = 0$$

$$x = 0$$

Since $x = 0$ is not in the domain of $\ln(x - 1)$ or of $\ln(x - 2)$, it is an extraneous solution. The equation has no solution.

71. $2 \ln(x + 3) + 3x = 8$

Graph $y_1 = 2 \ln(x + 3) + 3x - 8$

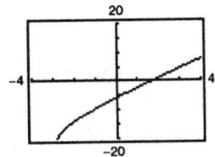

The x-intercept is at $x \approx 1.64$.

75. $p = 500 - 0.5e^{0.004x}$

(a) $p = 450$

$$450 = 500 - 0.5e^{0.004x}$$

$$0.5e^{0.004x} = 50$$

$$e^{0.004x} = 100$$

$$0.004x = \ln 100$$

$$x \approx 1151 \text{ units}$$

77. (a) $\dfrac{\ln 2}{r} = 5$

$$\ln 2 = 5r$$

$$r = \frac{\ln 2}{5} \approx 0.1386 = 13.86\%$$

(b) $A = 10,000e^{0.1386(1)} \approx \$11,486.65$

69. $2^{0.6x} - 3x = 0$

Graph $y_1 = 2^{0.6x} - 3x$

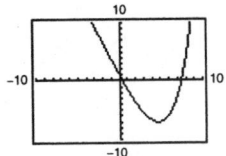

The x-intercepts are at $x \approx 0.39$ and at $x \approx 7.48$.

73. $y = ae^{bx}$

$$2 = ae^{b(0)} \implies a = 2$$

$$3 = 2e^{b(4)}$$

$$1.5 = e^{4b}$$

$$\ln 1.5 = 4b \implies b \approx 0.1014$$

Thus, $y \approx 2e^{0.1014x}$

(b) $p = 400$

$$400 = 500 - 0.5e^{0.004x}$$

$$0.5e^{0.004x} = 100$$

$$e^{0.004x} = 200$$

$$0.004x = \ln 200$$

$$x \approx 1325 \text{ units}$$

79. $R = \log_{10} I$ since $I_0 = 1$.

(a) $\log_{10} I = 8.4$

$$I = 10^{8.4}$$

(b) $\log_{10} I = 6.85$

$$I = 10^{6.85}$$

(c) $\log_{10} I = 9.1$

$$I = 10^{9.1}$$

❑ **Practice Test for Chapter 5**

1. Solve for x: $x^{3/5} = 8$.

2. Solve for x: $3^{x-1} = \frac{1}{81}$.

3. Graph $f(x) = 2^{-x}$.

4. Graph $g(x) = e^x + 1$.

5. If $5000 is invested at 9% interest, find the amount after three years if the interest is compounded

   (a) monthly   (b) quarterly   (c) continuously.

6. Write the equation in logarithmic form: $7^{-2} = \frac{1}{49}$.

7. Solve for x: $x - 4 = \log_2 \frac{1}{64}$.

8. Given $\log_b 2 = 0.3562$ and $\log_b 5 = 0.8271$, evaluate $\log_b \sqrt[4]{8/25}$.

9. Write $5 \ln x - \frac{1}{2} \ln y + 6 \ln z$ as a single logarithm.

10. Using your calculator and the change of base formula, evaluate $\log_9 28$.

11. Use your calculator to solve for N: $\log_{10} N = 0.6646$

12. Graph $y = \log_4 x$.

13. Determine the domain of $f(x) = \log_3(x^2 - 9)$.

14. Graph $y = \ln(x - 2)$.

15. True or false: $\dfrac{\ln x}{\ln y} = \ln(x - y)$

16. Solve for x: $5^x = 41$

17. Solve for x: $x - x^2 = \log_5 \frac{1}{25}$

18. Solve for x: $\log_2 x + \log_2(x - 3) = 2$

19. Solve for x: $\dfrac{e^x + e^{-x}}{3} = 4$

20. Six thousand dollars is deposited into a fund at an annual interest rate of 13%. Find the time required for the investment to double if the interest is compounded continuously.

C H A P T E R   6
Systems of Equations and Inequalities

CHAPTER 6
Systems of Equations and Inequalities

Section 6.1 Solving Systems of Equations

- You should be able to solve systems of equations by the method of substitution.
  1. Solve one of the equations for one of the variables.
  2. Substitute this expression into the other equation and solve.
  3. Back-substitute into the first equation to find the value of the other variable.
  4. Check your answer in each of the original equations.
- You should be able to find solutions graphically. (See Example 5 in textbook.)

Solutions to Odd-Numbered Exercises

1. $2x + y = 6$ Equation 1

$-x + y = 0$ Equation 2

Solve for y in Equation 1: $y = 6 - 2x$

Substitute for y in Equation 2: $-x + (6 - 2x) = 0$

Solve for x: $-3x + 6 = 0 \implies x = 2$

Back-substitute $x = 2$: $y = 6 - 2(2) = 2$

Answer: $(2, 2)$

3. $x - y = -4$ Equation 1

$x^2 - y = -2$ Equation 2

Solve for y in Equation 1: $y = x + 4$

Substitute for y in Equation 2: $x^2 - (x + 4) = -2$

Solve for x: $x^2 - x - 2 = 0 \implies (x + 1)(x - 2) = 0 \implies x = -1, 2$

Back-substitute $x = -1$: $y = -1 + 4 = 3$

Back-substitute $x = 2$: $y = 2 + 4 = 6$

Answers: $(-1, 3), (2, 6)$

5. $x - 3y = 15$ Equation 1

    $x^2 + y^2 = 25$ Equation 2

Solve for x in Equation 1: $x = 3y + 15$

Substitute for x in Equation 2: $(3y + 15)^2 + y^2 = 25$

Solve for y: $10y^2 + 90y + 200 = 0 \implies y^2 + 9y + 20 = 0 \implies (y + 5)(y + 4) = 0 \implies y = -5, -4$

Back-substitute $y = -5$: $x = 3(-5) + 15 = 0$

Back-substitute $y = -4$: $x = 3(-4) + 15 = 3$

Answers: $(0, -5), (3, -4)$

7. $x^2 + y = 0$ Equation 1

   $x^2 - 4x - y = 0$ Equation 2

Solve for y in Equation 1: $y = -x^2$

Substitute for y in Equation 2: $x^2 - 4x - (-x^2) = 0$

Solve for x: $2x^2 - 4x = 0 \implies 2x(x - 2) = 0 \implies x = 0, 2$

Back-substitute $x = 0$: $y = -0^2 = 0$

Back-substitute $x = 2$: $y = -2^2 = -4$

Answers: $(0, 0), (2, -4)$

9. $x - 6y = -8$ Equation 1

   $x^2 - 4y^3 = 0$ Equation 2

Solve for x in Equation 1: $x = 6y - 8$

Substitute for x in Equation 2: $(6y - 8)^2 - 4y^3 = 0$

Solve for y: $-4y^3 + 36y^2 - 96y + 64 = 0$

$$y^3 - 9y^2 + 24y - 16 = 0$$

$$(y - 1)(y - 4)^2 = 0 \implies y = 1, 4$$

Back-substitute $y = 1$: $x = 6(1) - 8 = -2$

Back-substitute $y = 4$: $x = 6(4) - 8 = 16$

Answers: $(-2, 1), (16, 4)$

11. $x - y = 0$ Equation 1

   $5x - 3y = 10$ Equation 2

Solve for y in Equation 1: $y = x$

Substitute for y in Equation 2: $5x - 3x = 10$

Solve for x: $2x = 10 \implies x = 5$

Back-substitute in Equation 1: $y = x = 5$

Answer: $(5, 5)$

13. $2x - y + 2 = 0$ Equation 1

$4x + y - 5 = 0$ Equation 2

Solve for y in Equation 1: $y = 2x + 2$

Substitute for y in Equation 2: $4x + (2x + 2) - 5 = 0$

Solve for x: $4x + (2x + 2) - 5 = 0 \implies 6x - 3 = 0 \implies x = \frac{1}{2}$

Back-substitute $x = \frac{1}{2}$: $y = 2x + 2 = 2\left(\frac{1}{2}\right) + 2 = 3$

Answer: $\left(\frac{1}{2}, 3\right)$

15. $30x - 40y - 33 = 0$ Equation 1

$10x + 20y - 21 = 0$ Equation 2

Solve for x in Equation 2: $x = -2y + \frac{21}{10}$

Substitute for x in Equation 1: $30\left(-2y + \frac{21}{10}\right) - 40y - 33 = 0$

Solve for y: $-100y + 30 = 0 \implies y = \frac{3}{10}$

Back-substitute $y = \frac{3}{10}$: $x = -2y + \frac{21}{10} = -2\left(\frac{3}{10}\right) + \frac{21}{10} = \frac{3}{2}$

Answer: $\left(\frac{3}{2}, \frac{3}{10}\right)$

17. $\frac{1}{5}x + \frac{1}{2}y = 8$ Equation 1

$x + y = 20$ Equation 2

Solve for x in Equation 2: $x = 20 - y$

Substitute for x in Equation 1: $\frac{1}{5}(20 - y) + \frac{1}{2}y = 8$

Solve for y: $4 + \frac{3}{10}y = 8 \implies y = \frac{40}{3}$

Back-substitute $y = \frac{40}{3}$: $x = 20 - y = 20 - \frac{40}{3} = \frac{20}{3}$

Answer: $\left(\frac{20}{3}, \frac{40}{3}\right)$

19. $2x + y = 4$ Equation 1

$-4x + 2y = -12$ Equation 2

Solve for y in Equation 1: $y = 2x - 4$

Substitute for y in Equation 2: $-4x + 2(2x - 4) = -12$

Solve for x: $-8 \neq -12$ Inconsistent

No Solution.

21. $x - y = 0$ Equation 1

$2x + y = 0$ Equation 2

Solve for y in Equation 1: $y = x$

Substitute for y in Equation 2: $2x + x = 0$

Solve for x: $3x = 0 \implies x = 0$

Back-substitute $x = 0$: $y = x = 0$

Answer: $(0, 0)$

23. $-x + 2y = 2$

$3x + y = 15$

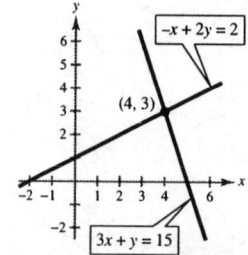

Point of intersection: $(4, 3)$

25. $x - 3y = -2$

$5x + 3y = 17$

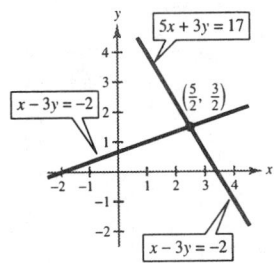

Point of intersection: $\left(\frac{5}{2}, \frac{3}{2}\right)$

27. $x + y = 4$

$x^2 + y^2 - 4x = 0$

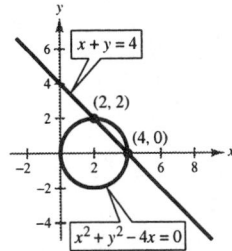

Points of intersection: $(2, 2)$, $(4, 0)$

29. $7x + 8y = 24 \implies y_1 = -\frac{7}{8}x + 3$

$x - 8y = \ 8 \implies y_2 = \frac{1}{8}x - 1$

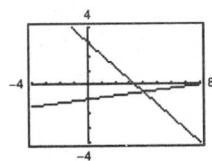

Point of intersection: $\left(4, -\frac{1}{2}\right)$

31. $3x - 2y = 0 \implies y_1 = \frac{3}{2}x$

$x^2 - y^2 = 4 \implies y_2 = \sqrt{x^2 - 4}$

$y_3 = -\sqrt{x^2 - 4}$

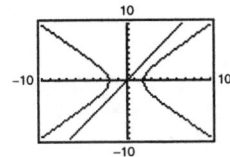

No points of intersection
Inconsistent
No solution

33. $x^2 + y^2 = 8 \implies y_1 = \sqrt{8 - x^2}$ and $y_2 = -\sqrt{8 - x^2}$

$y = x^2 \implies y_3 = x^2$

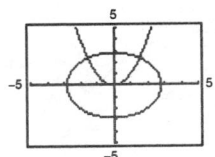

Points of intersection: $\left(\pm\sqrt{\dfrac{-1 + \sqrt{33}}{2}}, \dfrac{-1 + \sqrt{33}}{2}\right)$

$\approx (\pm 1.54, 2.37)$

Algebraically we have:

$x^2 + (x^2)^2 = 8$

$x^4 + x^2 - 8 = 0$

$x^2 = \dfrac{-1 \pm \sqrt{1^2 - 4(1)(-8)}}{2(1)}$

Use the positive value $\pm\sqrt{\dfrac{1 + \sqrt{33}}{2}}$.

$y = x^2 = \dfrac{-1 + \sqrt{33}}{2}$

35. $y = e^x$

$x - y + 1 = 0 \implies y = x + 1$

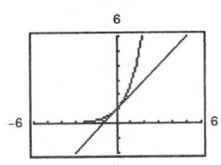

Point of intersection: $(0, 1)$

37. $y = \sqrt{x}$

$y = x$

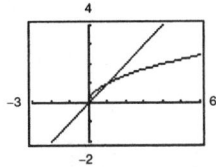

Points of intersection: $(0, 0)$, $(1, 1)$

39. $x^2 + y^2 = 169 \implies y_1 = \sqrt{169 - x^2}$ and $y_2 = -\sqrt{169 - x^2}$

$x^2 - 8y = 104 \implies y_3 = \frac{1}{8}x^2 - 13$

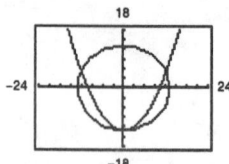

Points of intersection: $(0, -13)$, $(\pm 12, 5)$

41. $y = 2x$ Equation 1

     $y = x^2 + 1$ Equation 2

Substitute for y in Equation 2: $2x = x^2 + 1$

Solve for x: $x^2 - 2x + 1 = (x - 1)^2 = 0 \implies x = 1$

Back-substitute $x = 1$ in Equation 1: $y = 2x = 2$

Answer: $(1, 2)$

43. $3x - 7y + 6 = 0$ Equation 1

       $x^2 - y^2 = 4$ Equation 2

Solve for y in Equation 1: $y = \dfrac{3x + 6}{7}$

Solve for y in Equation 2: $x^2 - \left(\dfrac{3x + 6}{7}\right)^2 = 4$

Solve for x: $x^2 - \left(\dfrac{9x^2 + 36x + 36}{49}\right) = 4$

          $49x^2 - (9x^2 + 36x + 36) = 196$

         $40x^2 - 36x - 232 = 0$

$4(10x - 29)(x + 2) = 0 \implies x = \dfrac{29}{10}, -2$

Back-substitute $x = \dfrac{29}{10}$: $y = \dfrac{3x + 6}{7} = \dfrac{3(29/10) + 6}{7} = \dfrac{21}{10}$

Back-substitute $x = -2$: $y = \dfrac{3x + 6}{7} = 0$

Answers: $\left(\dfrac{29}{10}, \dfrac{21}{10}\right)$, $(-2, 0)$

45. $x - 2y = 4$ Equation 1

     $x^2 - y = 0$ Equation 2

Solve for y in Equation 2: $y = x^2$

Substitute for y in Equation 1: $x - 2x^2 = 4$

Solve for x: $0 = 2x^2 - x + 4$

No real solutions, the discriminant in the Quadratic Formula is negative.

Inconsistent

No real solution

47. $y - e^{-x} = 1 \implies y = e^{-x} + 1$

    $y - \ln x = 3 \implies y = \ln x + 3$

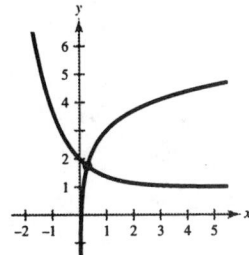

Point of intersection: Approximately $(0.287, 1.75)$

51. $xy - 1 = 0$ Equation 1

$2x - 4y + 7 = 0$ Equation 2

Solve for y in Equation 1: $y = \dfrac{1}{x}$

Substitute for y in Equation 2: $2x - 4\left(\dfrac{1}{x}\right) + 7 = 0$

Solve for x: $2x^2 - 4 + 7x = 0 \implies (2x - 1)(x + 4) = 0 \implies x = \dfrac{1}{2}, -4$

Back-substitute $x = \dfrac{1}{2}$: $y = \dfrac{1}{1/2} = 2$

Back-substitute $x = -4$: $y = \dfrac{1}{-4} = -\dfrac{1}{4}$

Answers: $\left(\dfrac{1}{2}, 2\right), \left(-4, -\dfrac{1}{4}\right)$

53. The system has no solution if you arrive at a false statement, ie. $4 = 8$, or you have a quadratic equation with a negative discriminant, which would yield imaginary solutions.

55. $C = 8650x + 250{,}000, \ R = 9950x$

     $R = C$

     $9950x = 8650x + 250{,}000$

     $1300x = 250{,}000$

        $x \approx 192$ units

57. $C = 2.65x + 350{,}000, \ R = 4.15x$

     $R = C$

     $4.15x = 2.65x + 350{,}000$

     $1.50x = 350{,}000$

        $x \approx 233{,}333$ units

49. $y = x^4 - 2x^2 + 1$ Equation 1

     $y = 1 - x^2$ Equation 2

Substitute for y in Equation 1: $1 - x^2 = x^4 - 2x^2 + 1$

Solve for x: $x^4 - x^2 = 0 \implies x^2(x^2 - 1) = 0$

                       $\implies x = 0, \pm 1$

Back-substitute $x = 0$: $1 - x^2 = 1$

Back-substitute $x = 1$: $1 - x^2 = 1 - 1^2 = 0$

Back-substitute $x = -1$: $1 - x^2 = 1 - (-1)^2 = 0$

Answers: $(0, 1), (\pm 1, 0)$

59. $C = 3.45x + 16{,}000, \ R = 5.95x$

(a) $R = C$

     $5.95x = 3.45x + 16{,}000$

     $2.50x = 16{,}000$

        $x \approx 6400$ units

(b) $P = R - C$

     $6000 = 5.95x - (3.45x + 16{,}000)$

     $6000 = 2.5x - 16{,}000$

   $22000 = 2.5x$

        $x \approx 8800$ units

61. (a)
$$x + y = 25{,}000$$
$$0.06x + 0.085y = 2000$$

(b) $y_1 = 25{,}000 - x$

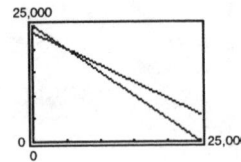

$$y_2 = \frac{2000 - 0.06x}{0.085}$$

As the amount at 6% increases, the amount at 8.5% decreases. The amount of interest is fixed at $2000.

(c) The point of intersection occurs when $x = 5000$, so the most that can be invested at 6% and still earn $2000 per year in interest is $5000.

63. $0.06x = 0.03x + 250$

$0.03x = 250$

$x \approx \$8333.33$

To make the straight commission offer the better offer, you would have to sell more than $8333.33 per week.

65. $V = (D - 4)^2, 5 \le D \le 40$ Doyle Log Rule

$V = 0.79D^2 - 2D - 4, 5 \le D \le 40$ Scribner Log Rule

(a)

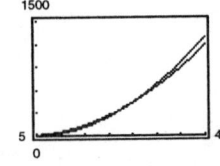

(b) The graphs intersect when $D \approx 24.7$ inches.

(c) For large logs, Doyle's Log Rule gives a greater volume for a given diameter.

67. $2l + 2w = 30 \implies l + w = 15$

$l = w + 3 \implies (w + 3) + 2 = 15$

$\qquad\qquad\qquad 2w = 12$

$\qquad\qquad\qquad w = 6$

$l = w + 3 = 9$

Dimensions: 6 meters × 9 meters

69. $2l + 2w = 42 \implies l + w = 21$

$w = \frac{3}{4}l \implies l + \frac{3}{4}l = 21$

$\qquad\qquad\qquad \frac{7}{4}l = 21$

$\qquad\qquad\qquad l = 12$

$w = \frac{3}{4}l = 9$

Dimensions: 9 inches × 12 inches

71. $2l + 2w = 40 \implies l + w = 20 \implies w = 20 - l$

$lw = 96 \implies l(20 - l) = 96$

$\qquad\qquad 20l - l^2 = 96$

$\qquad\qquad 0 = l^2 - 20l + 96$

$\qquad\qquad 0 = (l - 8)(l - 12)$

$\qquad\qquad l = 8 \text{ or } l = 12$

$w = 12, w = 8$

Since the length is supposed to be greater than the width, we have $l = 12$ kilometers and $w = 8$ kilometers.

73. The point lies on the line through $(0, 10)$ and $\left(\sqrt{61}, 0\right)$. This line's equation is $y = -\dfrac{10}{\sqrt{61}}x + 10$.

Substitute for y in $\dfrac{x^2}{25} - \dfrac{y^2}{36} = 1$.

$$\frac{x^2}{25} - \frac{\left(-10/\sqrt{61}\,x + 10\right)^2}{36} = 1$$

$$36x^2 - 25\left(-\frac{10}{\sqrt{61}}x + 10\right)^2 = 900$$

$$36x^2 - 25\left(\frac{100}{61}x^2 - \frac{200}{\sqrt{61}}x + 100\right) = 900$$

$$-\frac{304}{61}x^2 + \frac{5000}{\sqrt{61}}x - 3400 = 0$$

By the Quadratic Formula we have $x \approx 122.91$ or $x \approx 5.55$. Only $x \approx 5.55$ agrees with the graph. The point of intersection is approximately $(5.55, 2.89)$.

75. $(0, 6.6), (1, 6.8), (2, 7.1), (3, 7.1)$

(a) Linear model: $f(t) = 0.18t + 6.63$

Quadratic model: $g(t) = -0.05t^2 + 0.33t + 6.58$

(b)

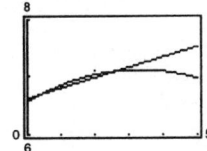

(c) Points of intersection: $(0.382, 6.70), (2.618, 7.10)$

(d) Linear model when
$t = 4$: $f(4) = 7.35$ million short tons.

Quadratic model when
$t = 4$: $g(4) = 7.1$ million short tons.

Since the sale of newsprint has been slowly decreasing (more papers are on-line now and many people would rather listen to the news than read it) the Quadratic model is probably more accurate.

77. $(-2, 7), (5, 5)$

$$m = \frac{5 - 7}{5 - (-2)} = -\frac{2}{7}$$

$$y - 7 = -\frac{2}{7}(x - (-2))$$

$$7y - 49 = -2x - 4$$

$$2x + 7y - 45 = 0$$

79. $(6, 3), (10, 3)$

$$m = \frac{3 - 3}{10 - 6} = 0 \quad \Rightarrow \quad \text{The line is horizontal.}$$

$$y = 3$$

81. $\left(\frac{3}{5}, 0\right), (4, 6)$

$$m = \frac{6 - 0}{4 - \frac{3}{5}} = \frac{6}{\frac{17}{5}} = \frac{30}{17}$$

$$y - 6 = \frac{30}{17}(x - 4)$$

$$17y - 102 = 30x - 120$$

$$0 = 30x - 17y - 18$$

Section 6.2 Two-Variable Linear Systems

- ■ You should be able to solve a linear system by the method of elimination.
  1. Obtain coefficients for either x or y that differ only in sign. This is done by multiplying all the terms of one or both equations by appropriate constants.
  2. Add the equations to eliminate one of the variables and then solve for the remaining variable.
  3. Use back-substitution into either original equation and solve for the other variable.
  4. Check your answer.
- ■ You should know that for a system of two linear equations, one of the following is true.
  1. There are infinitely many solutions; the lines are identical. The system is consistent.
  2. There is no solution; the lines are parallel. The system is inconsistent.
  3. There is one solution; the lines intersect at one point. The system is consistent.

Solutions to Odd-Numbered Exercises

1. $2x + y = 5$ Equation 1

   $x - y = 5$ Equation 2

   Add to eliminate y: $3x = 6 \implies x = 2$

   Substitute $x = 2$ in Equation 2: $2 - y = 1 \implies y = 1$

   Answer: $(2, 1)$

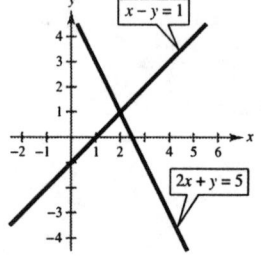

3. $x + y = 0$ Equation 1

   $3x + 2y = 1$ Equation 2

   Multiply Equation 1 by -2: $-2x - 2y = 0$

   Add this to Equation 2 to eliminate y: $x = 1$

   Substitute $x = 1$ in Equation 1: $1 + y = 0 \implies y = -1$

   Answer: $(1, -1)$

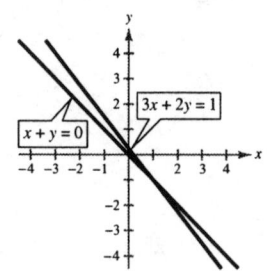

5. $x - y = 2$ Equation 1

   $-2x + 2y = 5$ Equation 2

   Multiply Equation 1 by 2: $2x - 2y = 4$

   Add this to Equation 2: $0 = 9$

   There are no solutions.

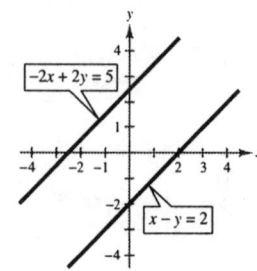

7. $3x - 2y = 5$ Equation 1

$-6x + 4y = -10$ Equation 2

Multiply Equation 1 by 2 and add to Equation 2: $0 = 0$

The equations are dependent. There are infinitely many solutions.

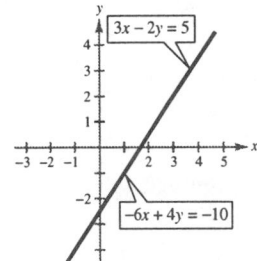

9. $9x + 3y = 1$ Equation 1

$3x - 6y = 5$ Equation 2

Multiply Equation 2 by (-3): $9x + 3y = 1$

$ -9x + 18y = -15$

Add to eliminate x: $21y = -14 \implies y = -\frac{2}{3}$

Substitute $y = -\frac{2}{3}$ in Equation 1: $9x + 3\left(-\frac{2}{3}\right) = 1$

$x = \frac{1}{3}$

Answer: $\left(\frac{1}{3}, -\frac{2}{3}\right)$

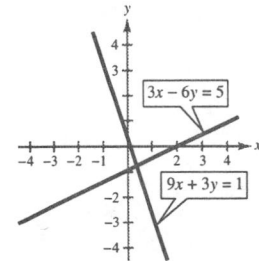

11. $x + 2y = 4$ Equation 1

$x - 2y = 1$ Equation 2

Add to eliminate y:

$2x = 5$

$x = \frac{5}{2}$

Substitute $x = \frac{5}{2}$ in Equation 1:

$\frac{5}{2} + 2y = 4 \implies y = \frac{3}{4}$

Answer: $\left(\frac{5}{2}, \frac{3}{4}\right)$

13. $2x + 3y = 18$ Equation 1

$5x - y = 11$ Equation 2

Multiply Equation 2 by 3: $15x - 3y = 33$

Add this to Equation 1 to eliminate y:

$17x = 51 \implies x = 3$

Substitute $x = 3$ in Equation 1:

$6 + 3y = 18 \implies y = 4$

Answer: $(3, 4)$

15. $3x + 2y = 10$ Equation 1

$2x + 5y = 3$ Equation 2

Multiply Equation 1 by 2 and Equation 2 by (-3):

$6x + 4y = 20$

$-6x - 15y = -9$

Add to eliminate x: $-11y = 11 \implies y = -1$

Substitute $y = -1$ in Equation 1:

$3x - 2 = 10 \implies x = 4$

Answer: $(4, -1)$

17. $2u + v = 120$ Equation 1

$u + 2v = 120$ Equation 2

Multiply Equation 2 by (-2):

$-2u - 4v = -240$

Add this to Equation 1 to eliminate u:

$-3v = -120$

$v = 40$

Substitute $v = 40$ in Equation 2:

$u + 80 = 120 \implies u = 40$

Answer: $(40, 40)$

19. $6r - 5s = 3$ Equation 1

$-12r + 10s = 5$ Equation 2

Multiply Equation 1 by 2: $12r - 10s = 6$

Add this to Equation 2 to eliminate r: $0 = 11$

Inconsistent

No solution

21. $\dfrac{x}{4} + \dfrac{y}{6} = 1$ Equation 1

$x - y = 3$ Equation 2

Multiply Equation 1 by 6: $\dfrac{3}{2}x + y = 6$

Add this to Equation 2 to eliminate *y*:

$\dfrac{5}{2}x = 9 \implies x = \dfrac{18}{5}$

Substitute $x = \dfrac{18}{5}$ in Equation 2:

$\dfrac{18}{5} - y = 3$

$y = \dfrac{3}{5}$

Answer: $\left(\dfrac{18}{5}, \dfrac{3}{5} \right)$

23. $\dfrac{x+3}{4} + \dfrac{y-1}{3} = 1$ Equation 1

$2x - y = 12$ Equation 2

Multiply Equation 1 by 12 and Equation 2 by 4:

$3x + 4y = 7$

$8x - 4y = 48$

Add to eliminate *y*: $11x = 55 \implies x = 5$

Substitute $x = 5$ into Equation 2:

$2(5) - y = 12 \implies y = -2$

Answer: $(5, -2)$

25. $2.5x - 3y = 1.5$ Equation 1

$10x - 12y = 6$ Equation 2

Multiply Equation 1 by (-4):

$-10x + 12y = -6$

Add this to Equation 2 to eliminate *x*:

$0 = 0$ (Dependent)

The solution set consists of all points lying

on the line $10x - 12y = 6$.

Let $x = a$, then $y = \dfrac{5}{6}a - \dfrac{1}{2}$.

Answer: $\left(a, \dfrac{5}{6}a - \dfrac{1}{2} \right)$, where *a* is any
real number.

27. $0.05x - 0.03y = 0.21$ Equation 1

$0.07x + 0.02y = 0.16$ Equation 2

Multiply Equation 1 by 200 and
Equation 2 by 300: $10x - 6y = 42$

$21x + 6y = 48$

Add to eliminate *y*: $31x = 90$

$x = \dfrac{90}{31}$

Substitute $x = \dfrac{90}{31}$ in Equation 2:

$0.07 \left(\dfrac{90}{31} \right) + 0.02y = 0.16$

$y = -\dfrac{67}{31}$

Answer: $\left(\dfrac{90}{31}, -\dfrac{67}{31} \right)$

29. $4b + 3m = 3$ Equation 1

$3b + 11m = 13$ Equation 2

Multiply Equation 1 by 3 and Equation 2 by (-4):

$12b + 9m = 9$

$-12b - 44m = -52$

Add to eliminate *b*: $-35m = -43$

$m = \dfrac{43}{35}$

Substitute $m = \dfrac{43}{35}$ in Equation 1: $5b + 3\left(\dfrac{43}{35} \right) = 3 \implies b = -\dfrac{6}{35}$

Answer: $\left(-\dfrac{6}{35}, \dfrac{43}{35} \right)$

31. $\dfrac{1}{5}x - \dfrac{1}{3}y = 1$

$-3x + 5y = 9$

The lines are parallel. The system is inconsistent.

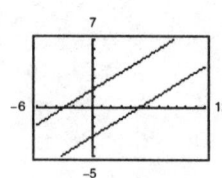

33. $2x - 5y = 0$

$\quad x - y = 3$

The system is consistent. There is one solution.

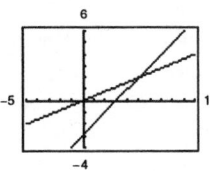

35. $8x + 9y = 42$

$\quad 6x - y = 16$

Answer: $(3, 2)$

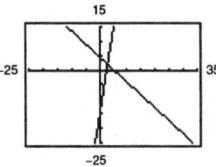

37. $\quad\quad 4y = -8$

$\quad 7x - 2y = 25$

Answer: $(3, -2)$

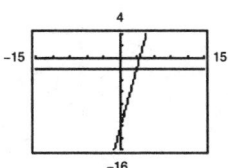

39. $3x - 5y = 7$ Equation 1

$\quad 2x + y = 9$ Equation 2

Multiply Equation 2 by 5:

$\quad 10x + 5y = 45$

Add this to Equation 1:

$\quad 13x = 52 \implies x = 4$

Back-substitute $x = 4$ into Equation 2:

$\quad 2(4) + y = 9 \implies y = 1$

Answer: $(4, 1)$

41. $y = 2x - 5$ Equation 1

$\quad y = 5x - 11$ Equation 2

Since both equations are solved for y, set them equal to one another and solve for x.

$\quad 2x - 5 = 5x - 11$

$\quad\quad 6 = 3x$

$\quad\quad 2 = x$

Back-substitute $x = 2$ into Equation 1:

$\quad y = 2(2) - 5 = -1$

Answer: $(2, -1)$

43. There are infinitely many systems that have the solution $\left(3, \frac{5}{2}\right)$. One possible system is:

$\quad 2(3) + 2\left(\frac{5}{2}\right) = 11 \implies 2x + 2y = 11$

$\quad 3 - 4\left(\frac{5}{2}\right) = -7 \implies x - 4y = -7$

45. $100y - x = 200$ Equation 1

$\quad 99y - x = -198$ Equation 2

Subtract Equation 2 from Equation 1 to eliminate x: $y = 398$

Substitute $y = 398$ into Equation 1: $100(398) - x = 200 \implies x = 39{,}600$

Answer: $(39{,}600, \ 398)$

The lines are not parallel. The scale on the axes must be changed to see the point of intersection.

47. No, it is not possible for a consistent system of linear equations to have exactly two solutions. Either the lines will intersect once or they will coincide and then the system would have infinite solutions.

49. $4x - 8y = -3$ Equation 1

$2x + ky = 16$ Equation 2

Multiply Equation 2 by -2: $-4x - 2ky = -32$

Add this to Equation 1: $-8y - 2ky = -35$

The system in inconsistent if $-8y - 2ky = 0$. This occurs when $k = -4$.

51. Let $x =$ the air speed and $y =$ the wind speed.

$$3.6(x - y) = 1800 \quad \text{Equation 1}$$
$$6(x + y) = 1800 \quad \text{Equation 2}$$

$$
\begin{aligned}
x - y &= 500 \\
x + y &= 600 \\
2x &= 1100 \\
x &= 550 \\
550 + y &= 600 \\
y &= 50
\end{aligned}
$$

Answer: $x = 550$ mph, $y = 50$ mph

53. Let $x =$ the number of liters at 20%

$y =$ the number of liters at 50%.

(a) $x + y = 10$

$0.2x + 0.5y = 0.3(10)$

(c) -2 Equation 1 $-2x - 2y = -20$

10 Equation 2 $2x + 5y = 30$

$$
\begin{aligned}
3y &= 10 \\
y &= \tfrac{10}{3} \\
x + \tfrac{10}{3} &= 10 \\
x &= \tfrac{20}{3}
\end{aligned}
$$

(b)

As x increases, y decreases.

Answer: $x = 6\tfrac{2}{3}$ liters at 20%

$y = 3\tfrac{1}{3}$ liters at 50%

55. Let $x =$ amount invested at 10.5%

$y =$ amount invested at 12%.

$x + y = 12{,}000$ Equation 1

$0.105x + 0.12y = 1350$ Equation 2

$$
\begin{aligned}
-12x - 12y &= -144{,}000 \\
10.5x + 12y &= 135{,}000 \\
-1.5x \quad y &= -9000 \\
x \quad y &= 6000 \\
6000 + y &= 12{,}000 \\
y &= 6000
\end{aligned}
$$

Answer: $y = \$6000$ at 12%

$x = \$6000$ at 10.5%

57. Let x = number of adult tickets sold

      y = number of child tickets sold.

$x + y = 500$	Equation 1
$7.5x + 4y = \$3312.50$	Equation 2

$$-4x - 4y = -2000.00$$
$$7.5x + 4y = 3312.50$$
$$\overline{3.5x = 1312.50}$$
$$x = 375$$
$$375 + y = 500$$
$$y = 125$$

Answer: x = 375 adult tickets

                y = 125 child tickets

59. Let x = distance one person drives

      y = distance other person drives.

$x + y = 300$	Equation 1
$y = 3x$	Equation 2
$x + 3x = 300$	Use substitution
$4x = 300$	
$x = 75$	
$y = 3x = 225$	

Answer: 75 km and 225 km

61.   Demand = Supply

$$50 - 0.5x = 0.125x$$
$$50 = 0.625x$$
$$x = 80 \text{ units}$$
$$p = \$10$$
$$4x = 300$$

Answer: (80, 10)

63. Demand = Supply

$$140 - 0.00002x = 80 + 0.00001x$$
$$60 = 0.00003x$$
$$x = 2,000,000 \text{ units}$$
$$p = \$100.00$$

Answer: (2,000,000, 100)

65. $5b + 10a = 20.2 \implies -10b - 20a = -40.4$

     $10b + 30a = 50.1 \implies 10b + 30a = 50.1$

$$\overline{10a = 9.7}$$
$$a = 0.97$$
$$b = 2.10$$

Least squares regression line:

$$y = 0.97x + 2.10$$

67. $7b + 21a = 35.1 \implies -21b - 63a = -105.3$

     $21b + 91a = 114.2 \implies 21b + 91a = 114.2$

$$\overline{28a = 8.9}$$
$$a = \frac{89}{280}$$
$$b = \frac{1137}{280}$$

Least squares regression line:

$$y = \frac{1}{280}(89x + 1137)$$
$$y \approx 0.318x + 4.061$$

69. $(-2, 0), (0, 1), (2, 3)$

$3b = 4 \implies b = \frac{4}{3}$

$8a = 6 \implies a = \frac{3}{4}$

Least squares regression line:

$y = \frac{3}{4}x + \frac{4}{3}$

71. $(0, 4), (1, 3), (1, 1), (2, 0)$

$$\begin{aligned} 4b + 4a = 8 &\implies \quad 4b + 4a = 8 \\ 4b + 6a = 4 &\implies \underline{-4b - 6a = -4} \\ & \quad\quad\quad -2a = 4 \\ & \quad\quad\quad\quad\, a = -2 \\ & \quad\quad\quad\quad\, b = 4 \end{aligned}$$

Least squares regression line: $y = -2x + 4$

73. $(1.00, 450), (1.25, 375), (1.50, 330)$

$$3b + 3.75a = 1155$$

$$3.75b + 4.8125a = 1413.75$$

By elimination we have $a = -240$ and $b = 685$ and the least squares
regression line is $y = -240x + 685$. When $x = 1.40$, we have $y = -240(1.40) + 685 = 349$ units.

Section 6.3   Multivariable Linear Systems

■   You should know the operations that lead to equivalent systems of equations:

    (a) Interchange any two equations.

    (b) Multiply all terms of an equation by a nonzero constant.

    (c) Replace an equation by the sum of itself and a constant multiple of any other equation in the system.

■   You should be able to use the method of elimination.

Solutions to Odd-Numbered Exercises

1. $2x - y + 5z = 24$ Equation 1

      $y + 2z = 6$ Equation 2

          $z = 4$ Equation 3

Back-substitute $z = 4$ into Equation 2.

    $y + 2(4) = 6$

       $z = -2$

Back-substitute $y = -2$ and $z = 4$ into Equation 1.

    $2x - (-2) + 5(4) = 24$

         $2x + 22 = 24$

             $x = 1$

Answer: $(1, -2, 4)$

3. $2x + y - 3z = 10$ Equation 1

      $y = 2$ Equation 2

      $y - z = 4$ Equation 3

Back-substitute $y = 2$ into Equation 3.

    $2 - z = 4$

       $z = -2$

Back-substitute $y = 2$ and $z = -2$ into Equation 1.

    $2x + 2 - 3(-2) = 10$

           $2x + 8 = 10$

              $x = 1$

Answer: $(1, 2, -2)$

5. $4x - 2y + z = 8$ Equation 1

$\qquad\qquad 2z = 4$ Equation 2

$\qquad -y + z = 4$ Equation 3

From Equation 2 we have $z = 2$. Back-substitute $z = 2$ into Equation 3.

$-y + 2 = 4$

$\qquad y = -2$

Back-substitute $y = -2$ and $z = 2$ into Equation 1.

$4x - 2(-2) + 2 = 8$

$\qquad 4x + 6 = 8$

$\qquad\qquad x = \tfrac{1}{2}$

Answer: $\left(\tfrac{1}{2}, -2, 2\right)$

9. $\qquad x + y + z = \quad 6$ Equation 1

$\quad 2x - y + z = \quad 3$ Equation 2

$\quad 3x \quad - z = \quad 0$ Equation 3

$\quad x + y + z = \quad 6$

$\quad -3y - z = \quad -9$ -2Eq.1 $+$ Eq.2

$\quad -3y - 4z = -18$ -3Eq.1 $+$ Eq.3

$\quad x + y + z = \quad 6$

$\quad -3y - z = \quad -9$

$\qquad\qquad -3z = \quad -9$ $-$Eq.2 $+$ Eq.3

$\qquad -3z = -9 \implies z = 3$

$\quad -3y - 3 = -9 \implies y = 2$

$\quad x + 2 + 3 = \quad 6 \implies x = 1$

Answer: $(1, 2, 3)$

7. $\quad x - 2y + 3z = 5$ Equation 1

$\quad -x + 3y - 5z = 4$ Equation 2

$\quad 2x \qquad - 3z = 0$ Equation 3

Add Equation 1 to Equation 2.

$\quad y - 2z = 9$

This is the first step in putting the system in row-echelon form.

11. $2x \qquad + 2z = 2$ Equation 1

$5x + 3y \qquad = 4$ Equation 2

$3x \qquad - z = 0$ Equation 3

$x + \quad y + \quad z = 6$

$x + 3y - \quad 4z = 0$ -2Eq.1 $+$ Eq.2

$2x \qquad + 2z = 2$ Interchange equations.

$\qquad 3y - 4z = 4$

$x + 3y - \quad 4z = 0$

$\qquad -6y + 10z = 2$ -2Eq.1 $+$ Eq.2

$\qquad 3y - 4z = 4$

$\qquad -6y + 10z = 2$

$\qquad\qquad z = 5$ $\tfrac{1}{2}$ Eq.2 $+$ Eq.3

$\qquad\qquad z = 5$

$-6y + 10(5) = 2 \implies y = 8$

$x + 3(8) - 4(5) = 0 \implies x = -4$

Answer: $(-4, 8, 5)$

13.
$$\begin{aligned} 3x + 3y &= 9 \\ 2x - 3z &= 10 \\ 6y + 4z &= -12 \end{aligned} \qquad \text{Interchange equations.}$$

$$\begin{aligned} x + y &= 3 \qquad \tfrac{1}{3}\text{Eq.1} \\ 2x - 3z &= 10 \\ 6y + 4z &= -12 \end{aligned}$$

$$\begin{aligned} x + y &= 3 \\ -2y - 3z &= 4 \qquad -2\text{Eq.1} + \text{Eq.2} \\ 6y + 4z &= -12 \end{aligned}$$

$$\begin{aligned} x + y &= 3 \\ -2y - 3z &= 4 \\ -5z &= 0 \qquad 3\text{Eq.2} + \text{Eq.3} \end{aligned}$$

$$-5x = 0 \Rightarrow z = 0$$
$$-2y - 3(0) = 4 \Rightarrow y = -2$$
$$x - 2 = 3 \Rightarrow x = 5$$

Answer: $(5, -2, 0)$

17.
$$\begin{aligned} 3x + 3y + 5z &= 1 \\ 3x + 5y + 9z &= 0 \\ 5x + 9y + 17z &= 0 \end{aligned}$$

$$\begin{aligned} 6x + 6y + 10z &= 2 \qquad 2\,\text{Eq.1} \\ 3x + 5y + 9z &= 0 \\ 5x + 9y + 17z &= 0 \end{aligned}$$

$$\begin{aligned} x - 3y - 7z &= 2 \qquad -\text{Eq.3} + \text{Eq.1} \\ 3x + 5y + 9z &= 0 \\ 5x + 9y + 17z &= 0 \end{aligned}$$

$$\begin{aligned} x - 3y - 7z &= 2 \\ 14y + 30z &= -6 \qquad -3\text{Eq.1} + \text{Eq.2} \\ 24y + 52z &= -10 \qquad -5\text{Eq.1} + \text{Eq.3} \end{aligned}$$

15.
$$\begin{aligned} x + y - 2z &= 3 \qquad \text{Interchange equations.} \\ 3x - 2y + 4z &= 1 \\ 2x - 3y + 6z &= 8 \end{aligned}$$

$$\begin{aligned} x + y - 2z &= 3 \\ -5y + 10z &= -8 \qquad -3\text{Eq.1} + \text{Eq.2} \\ -5y + 10z &= 2 \qquad -2\text{Eq.1} + \text{Eq.3} \end{aligned}$$

$$\begin{aligned} x + y - 2z &= 3 \\ -5y + 10z &= -8 \\ 0 &= 10 \qquad \rightarrow \leftarrow -\text{Eq.2} + \text{Eq.3} \end{aligned}$$

No solution, inconsistent

$$\begin{aligned} x - 3y - 7z &= 2 \\ 84y + 180z &= -36 \qquad 6\text{Eq.2} \\ 84y + 182z &= -35 \qquad 3.5\text{Eq.3} \end{aligned}$$

$$\begin{aligned} x - 3y - 7z &= 2 \\ 84y + 180z &= -36 \\ 2z &= 1 \qquad -\text{Eq.2} + \text{Eq.3} \end{aligned}$$

$$2z = 1 \Rightarrow x = \tfrac{1}{2}$$
$$84y + 180\left(\tfrac{1}{2}\right) = -36 \Rightarrow y = \tfrac{1}{2}$$
$$x - 3\left(-\tfrac{3}{2}\right) - 7\left(\tfrac{1}{2}\right) = 2 \Rightarrow x = 1$$

Answer: $\left(1, -\tfrac{3}{2}, \tfrac{1}{2}\right)$

19.
$$x + 2y - 7z = -4$$
$$2x + y + z = 13$$
$$3x + 9y - 36z = -33$$

$$x + 2y - 7z = -4 \qquad -2\text{Eq.2} + \text{Eq.2}$$
$$-3y + 15z = 21 \qquad -3\text{Eq.1} + \text{Eq.3}$$
$$3y - 15z = -21$$

$$x + 2y - 7z = -4$$
$$-3y + 15z = 21$$
$$0 = 0 \qquad \text{Eq.2} + \text{Eq.3}$$

$$x + 2y - 7z = -4$$
$$y - 5z = -7 \qquad \tfrac{1}{3}\text{Eq.2}$$
$$x + 3z = 10 \qquad -2\text{Eq.2} + \text{Eq.1}$$
$$y - 5z = -7$$

Let $z = a$, then:
$$y = 5a - 7$$
$$x = -3a + 10$$
Answer: $(-3a + 10, 5a - 7, a)$

21.
$$3x - 3y + 6z = 6$$
$$x + 2y - z = 5$$
$$5x - 8y + 13z = 7$$

$$x - y + 2z = 2 \qquad \tfrac{1}{3}\text{Eq.1}$$
$$3y - 3z = 3 \qquad -\text{Eq.1} + \text{Eq.2}$$
$$-3y + 3z = -3 \qquad -5\text{Eq.1} + \text{Eq.3}$$

$$x - y + 2z = 2$$
$$y - z = 1 \qquad \tfrac{1}{3}\text{Eq.2}$$
$$0 = 0 \qquad \text{Eq.2} + \text{Eq.3}$$

$$x + z = 3 \qquad \text{Eq.2} + \text{Eq.1}$$
$$y - z - 1$$

Let $z = a$, then:
$$y = a + 1$$
$$x = -a + 3$$
Answer: $(-a + 3, a + 1, a)$

23.
$$x - 2y + 5z = 2$$
$$4x - z = 0$$
Let $z = a$, then $x = \tfrac{1}{4}a$.
$$\tfrac{1}{4}a - 2y + 5a = 2$$
$$a - 8y + 20a = 8$$
$$-8y = -21a + 8$$
$$y = \tfrac{21}{8}a - 1$$
Answer: $\left(\tfrac{1}{4}a, \tfrac{21}{8}a - 1, a\right)$
To avoid fractions, we could go back and let
$z = 8a$, then $4x - 8a = 0 \implies x = 2a$.
$$2a - 2y + 5(8a) = 2$$
$$-2y + 42a = 2$$
$$y = 21a - 1$$
Answer: $(2a, 21a - 1, 8a)$

25.
$$2x - 3y + z = -2$$
$$-4x + 9y = 7$$

$$2x - 3y + z = -2$$
$$3y + 2z = 3 \qquad 2\text{Eq.1} + \text{Eq.2}$$
$$2x + 3z = 1 \qquad \text{Eq.2} + \text{Eq.1}$$
$$3y + 2z = 3$$

Let $x = a$, then:
$$y = -\tfrac{2}{3}a + 1$$
$$x = -\tfrac{3}{2}a + \tfrac{1}{2}$$
Answer: $\left(-\tfrac{3}{2}a + \tfrac{1}{2}, -\tfrac{2}{3}a + 1, a\right)$

27.

$$\begin{aligned} x \qquad\quad + 3w &= 4 \\ 2y - z - w &= 0 \\ 3y \qquad - 2w &= 1 \\ 2x - y + 4z \qquad &= 5 \end{aligned}$$

$$\begin{aligned} x \qquad\quad + 3w &= 4 \\ 2y - z - w &= 0 \\ 3y \qquad - 2w &= 1 \\ - y + 4z - 6w &= -3 \qquad -2\text{Eq.1} + \text{Eq.4} \end{aligned}$$

$$\begin{aligned} x \qquad\quad + 3w &= 4 \\ y - 4z + 6w &= 3 \qquad -\text{Eq.4 and} \\ 2y - z - w &= 0 \qquad \text{interchange} \\ 3y \qquad - 2w &= 1 \qquad \text{the equations.} \end{aligned}$$

$$\begin{aligned} x \qquad\quad + 3w &= 4 \\ y - 4z + 6w &= 3 \\ 7z - 13w &= -6 \qquad -\text{Eq.2} + \text{Eq.3} \\ 12z - 20w &= -8 \qquad -3\text{Eq.2} + \text{Eq.4} \end{aligned}$$

$$\begin{aligned} x \qquad\quad + 3w &= 4 \\ y - 4z + 6w &= 3 \\ z - 3w &= -2 \qquad -\tfrac{1}{2}\text{Eq.4} + \text{Eq.3} \\ 12z - 20w &= -8 \end{aligned}$$

$$\begin{aligned} x \qquad\quad + 3w &= 4 \\ y - 4z + 6w &= 3 \\ z - 3w &= -2 \\ 16w &= 16 \qquad -12\text{Eq.3} + \text{Eq.4} \end{aligned}$$

$$16w = 16 \Rightarrow w = 1$$
$$z - 3(1) = -2 \Rightarrow z = 1$$
$$y - 4(1) + 6(1) = 3 \Rightarrow y = 1$$
$$x + 3(1) = 4 \Rightarrow x = 1$$

Answer: $(1, 1, 1, 1)$

29.

$$\begin{aligned} x \qquad + 4z &= 1 \\ x + y + 10z &= 10 \\ 2x - y + 2z &= -5 \end{aligned}$$

$$\begin{aligned} x \qquad + 4z &= 1 \\ y + 6z &= 9 \qquad -\text{Eq.1} + \text{Eq.2} \\ -y - 6z &= -7 \qquad -2\text{Eq.1} + \text{Eq.3} \end{aligned}$$

$$\begin{aligned} x \qquad + 4z &= 1 \\ y + 6z &= 9 \\ 0 &= 2 \qquad \rightarrow \leftarrow \text{Eq.2} + \text{Eq.3} \end{aligned}$$

No solution, inconsistent

31.

$$\begin{aligned} 2x + 3y \qquad &= 0 \\ 4x + 3y - z &= 0 \\ 8x + 3y + 3z &= 0 \end{aligned}$$

$$\begin{aligned} 2x + 3y \qquad &= 0 \\ -3y - z &= 0 \qquad -2\text{Eq.1} + \text{Eq.2} \\ -9y + 3z &= 0 \qquad -4\text{Eq.1} + \text{Eq.3} \end{aligned}$$

$$\begin{aligned} 2x + 3y \qquad &= 0 \\ -3y - z &= 0 \\ 6z &= 0 \qquad -3\text{Eq.2} + \text{Eq.3} \end{aligned}$$

$$6z = 0 \Rightarrow z = 0$$
$$-3y - 0 = 0 \Rightarrow y = 0$$
$$2x + 3(0) = 0 \Rightarrow x = 0$$

Answer: $(0, 0, 0)$

33.

$$\begin{aligned} 23x + 4y - . z &= 0 \qquad \text{Interchange equations.} \\ 12x + 5y + z &= 0 \end{aligned}$$

$$\begin{aligned} x + 6y + 3z &= 0 \qquad 2\text{Eq.2} - \text{Eq.1} \\ -67y - 35z &= 0 \qquad -12\text{Eq.1} + \text{Eq.2} \end{aligned}$$

To avoid fractions, let $z = 67a$, then:

$$-67y - 35(67a) = 0$$
$$y = -35a$$
$$x + 6(-35a) + 3(67a) = 0$$
$$x = 9a$$

Answer: $(9a, -35a, 67a)$

35. No, they are not equivalent. The constant in the second equation should be -11 and the coefficient of z in the third equation should be 2.

37. There are an infinite number of linear systems that have $(4, -1, 2)$ as their solution. One such system is as follows:

$$3(4) + (-1) - (2) = 9 \Rightarrow 3x + y - z = 9$$
$$(4) + 2(-1) - (2) = 0 \Rightarrow x + 2y - z = 0$$
$$-(4) + (-1) + 3(2) = 1 \Rightarrow -x + y + 3z = 1$$

39. $y = ax^2 + bx + c$ passing through $(0, 0), (2, -2), (4, 0)$

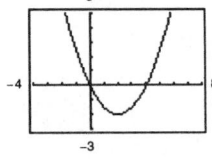

$(0, \ 0)$: $\quad 0 = \qquad\qquad c$

$(2, -2)$: $-2 = \quad 4a + 2b + c \ \Rightarrow \ -1 = 2a + b$

$(4, \ 0)$: $\quad 0 = 16a + 4b + c \ \Rightarrow \quad 0 = 4a + b$

Answer: $a = \frac{1}{2}, b = -2, c = 0$

The equation of the parabola is $y = \frac{1}{2}x^2 - 2x$.

41. $y = ax^2 + bx + c$ passing through $(2, 0), (3, -1), (4, 0)$

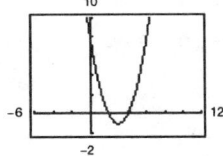

$(2, \ 0)$: $\quad 0 = \ 4a + 2b + c \Rightarrow \quad 0 = -4a - 2b - c$

$(3, -1)$: $-1 = \ 9a + 3b + c \Rightarrow -1 = \quad 5a + b$

$(4, \ 0)$: $\quad 0 = 16a + 4b + c \Rightarrow \quad 0 = 12a + 2b$

Answer: $a = 1, b = -6, c = 8$

The equation of the parabola is $y = x^2 - 6x + 8$.

43. $x^2 + y^2 + Dx + Ey + F = 0$ passing through $(0, 0), (2, 2), (4, 0)$

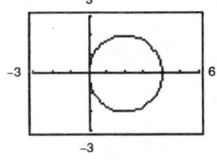

$(0, 0)$: $\qquad\qquad F = 0$

$(2, 2)$: $\ 8 + 2D + 2E + F = 0 \ \Rightarrow \ D + E = -4$

$(4, 0)$: $16 + 4D \qquad + F = 0 \ \Rightarrow \ D = -4$ and $E = 0$

The equation of the circle is $x^2 + y^2 - 4x = 0$.

To graph, let $y_1 = \sqrt{4x - x^2}$ and $y_2 = -\sqrt{4x - x^2}$.

45. $x^2 + y^2 + Dx + Ey + F = 0$ passing through $(-3, -1), (2, 4), (-6, 8)$

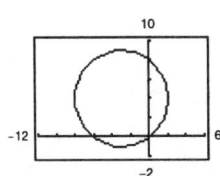

$(-3, -1)$: $\ 10 - 3D - E + F = 0 \ \Rightarrow \quad 10 = \quad 3D + E - F$

$(\ 2, \ \ 4)$: $\ 20 + 2D + 4E + F = 0 \ \Rightarrow \quad 20 = -2D - 4E - F$

$\cdot (-6, \ \ 8)$: $100 - 6D + 8E + F = 0 \ \Rightarrow \ 100 = \quad 6D - 8E - F$

Answer: $D = 6, E = -8, F = 0$

The equation of the circle is $x^2 + y^2 + 6x - 8y = 0$.
To graph, complete the squares first, then solve for y.

$$(x^2 + 6x + 9) + (y^2 - 8y + 16) = 0 + 9 + 16$$
$$(x + 3)^2 + (y - 4)^2 = 25$$
$$(y - 4)^2 = 25 - (x + 3)^2$$
$$y - 4 = \pm\sqrt{25 - (x + 3)^2}$$
$$y = 4 \pm \sqrt{25 - (x + 3)^2}$$

Let $y_1 = 4 + \sqrt{25 - (x + 3)^2}$ and $y_2 = 4 - \sqrt{25 - (x + 3)^2}$.

47. $s = \frac{1}{2}at^2 + v_0t + s_0$

$(1, 128), (2, 80), (3, 0)$

$128 = \frac{1}{2}a + v_0 + s_0 \implies a + 2v_0 + 2s_0 = 256$

$80 = 2a + 2v_0 + s_0 \implies 2a + 2v_0 + s_0 = 80$

$0 = \frac{9}{2}a + 3v_0 + s_0 \implies 9a + 6v_0 + 2s_0 = 0$

Solving this system yields $a = -32$, $v_0 = 0$, $s_0 = 144$.

Thus, $s = \frac{1}{2}(-32)t^2 + (0)t + 144$

$= -16t^2 + 144$.

49. $s = \frac{1}{2}at^2 + v_0t + s_0$

$(1, 452), (2, 372), (3, 260)$

$452 = \frac{1}{2}a + v_0 + s_0 \implies a + 2v_0 + 2s_0 = 904$

$372 = 2a + 2v_0 + s_0 \implies 2a + 2v_0 + s_0 = 372$

$260 = \frac{9}{2}a + 3v_0 + s_0 \implies 9a + 6v_0 + 2s_0 = 520$

Solving this system yields $a = -32$, $v_0 = -32$, $s_0 = 500$.

Thus, $s = \frac{1}{2}(-32)t^2 + (-32)t + 500$

$= -16t^2 - 32t + 500$.

51. Let $x =$ amount at 5%

Let $y =$ amount at 6%

Let $z =$ amount at 7%

$x + y + z = 16{,}000$

$0.05x + 0.06y + 0.07z = 990$

$x + 3000 = z$

$y + 2000 = z$

$(z - 3000) + (z - 2000) + z = 16{,}000$

$3z = 21{,}000$

$z = 7000$

$x = 4000$, $y = 5000$

Check: $0.05(4000) + 0.06(5000) + 0.07(7000) = 990$

Answer: $x = \$4000$ at 5%

$y = \$5000$ at 6%

$z = \$7000$ at 7%

53. Let x = amount at 8%

Let y = amount at 9%

Let z = amount at 10%

$$x + y + z = 775,000$$
$$0.08x + 0.09y + 0.10z = 67,500$$
$$x = 4z$$

$$y + 5z = 775,000$$
$$0.09y + 0.42z = 67,500$$
$$z = 75,000$$
$$y = 775,000 - 5z = 400,000$$
$$x = 4z = 300,000$$

Answer: x = $300,000 at 8%
 y = $400,000 at 9%
 z = $75,000 at 10%

55. Let C = amount in certificates of deposit

Let M = amount in municipal bonds

Let B = amount in blue-chip stocks

Let G = amount in growth or speculative stocks

$$C + M + B + G = 500,000$$
$$0.10C + 0.08M + 0.12B + 0.13G = 0.10(500,000)$$
$$B + G = \tfrac{1}{4}(500,000)$$

This system has infinitely many solutions.

Let $G = s$, then $\quad B = 125,000 - s$

$$M = 125,000 + \tfrac{1}{2}s$$
$$C = 250,000 - \tfrac{1}{2}s$$

Answer:
$(250,000 - \tfrac{1}{2}s, 125,000 + \tfrac{1}{2}s, 125,000 - s, s)$,
where $0 \le s \le 125,000$

One possible solution is to let $s = 50,000$.
$\quad$ Certificates of deposit: $225,000
$\quad$ Municipal bonds: $150,000
$\quad$ Blue-chip stocks: $75,000
$\quad$ Growth or speculative stocks: $50,000

57. Let x = gallons of spray X

Let y = gallons of spray Y

Let z = gallons of spray Z

Chemical A: $\tfrac{1}{5}x + \tfrac{1}{2}z = 12$
Chemical B: $\tfrac{2}{5}x + \tfrac{1}{2}z = 16$ $\Big\} \Rightarrow x = 20, z = 16$

Chemical C: $\tfrac{2}{5}x + y = 26 \Rightarrow y = 18$

Answer: 20 liters of spray X
 18 liters of spray Y
 16 liters of spray Z

59.

	Product	
Truck	A	B
Large	6	3
Medium	4	4
Small	0	3

Possible solutions:

(1) 4 medium trucks

(2) 2 large trucks, 1 medium truck, 2 small trucks

(3) 3 large trucks, 1 medium truck, 1 small truck

(4) 3 large trucks, 3 small trucks

61.
$$t_1 - 2t_2 = 0$$
$$t_1 - 2a = 128 \Rightarrow 2t_2 - 2a = 128$$
$$t_2 + a = 32 \Rightarrow -2t_2 - 2a = -64$$
$$\overline{ -4a = 64}$$
$$a = -16$$
$$t_2 = 48$$
$$t_1 = 96$$

Answer: $t_1 = 96$ lb
 $t_2 = 48$ lb
 $a = -16$ ft/sec^2

63. $\dfrac{1}{x^3 - x} = \dfrac{A}{x} + \dfrac{B}{x - 1} + \dfrac{C}{x + 1}$

$1 = A(x + 1)(x - 1) + Bx(x + 1) + Cx(x - 1)$

$1 = Ax^2 - A + B^2 + Bx + Cx^2 - Cx$

$1 = (A + B + C)x^2 + (B - C)x - A$

By equating coefficients, we have:

$0 = A + B + C$

$0 = B - C$

$1 = -A \quad \Rightarrow \quad A = -1$

$B + C = \ 1$

$\dfrac{B - C = \ 0}{2B = \ 1} \Rightarrow B = \dfrac{1}{2}$

$C = \dfrac{1}{2}$

$\dfrac{A}{x} + \dfrac{B}{x - 1} + \dfrac{C}{x + 1} = \dfrac{-1}{x} + \dfrac{1/2}{x - 1} + \dfrac{1/2}{x + 1} = \dfrac{1}{2}\left(-\dfrac{2}{x} + \dfrac{1}{x - 1} + \dfrac{1}{x + 1}\right)$

65. $\dfrac{x^2 - 3x - 3}{x(x - 2)(x + 3)} = \dfrac{A}{x} + \dfrac{B}{x - 2} + \dfrac{C}{x + B}$

$x^2 - 3x - 3 = A(x - 2)(x + 3) + Bx(x + 3) + Cx(x - 2)$

$x^2 - 3x - 3 = Ax^2 + Ax - 6A + Bx^2 + 3Bx + Cx^2 - 2Cx$

$x^2 - 3x - 3 = (A + B + C)x^2 + (A + 3B - 2C)x - 6A$

By equating coefficients, we have:

$1 = A + B + C$

$-3 = A + 3B - 2C$

$-3 = -6A \quad \Rightarrow \quad A = \tfrac{1}{2}$

$\tfrac{1}{2} = \ B + \ C \ \Rightarrow \ 1 = 2B + 2C$

$-\tfrac{7}{2} = 3B - 2C \ \Rightarrow \ -\tfrac{7}{2} = 3B - 2C$

$\dfrac{}{-\tfrac{5}{2} = 5B}$

$B = -\tfrac{1}{2}$

$C = 1$

$\dfrac{x^2 - 3x - 3}{x(x - 2)(x + 3)} = \dfrac{1/2}{x} - \dfrac{1/2}{x - 2} + \dfrac{1}{x + 3} = \dfrac{1}{2}\left(\dfrac{1}{x} - \dfrac{1}{x - 2} + \dfrac{2}{x + 3}\right)$

67. Least squares regression parabola through $(-4, 5), (-2, 6), (2, 6), (4, 2)$

$$n = 4$$

$$\sum x_i = 0 \qquad\qquad \sum y_i = 19$$

$$\sum x_i^2 = 40 \qquad\qquad \sum x_i^3 = 0$$

$$\sum x_i^4 = 544 \qquad\qquad \sum x_i y_i = -12$$

$$\sum x_i^2 y_i = 160$$

$$\begin{aligned} 4c \quad\;\; + 40a &= 19 \\ 40b \qquad\quad &= -12 \\ 40c \quad\; + 544a &= 160 \end{aligned}$$

Solving this system yields $a = -\frac{5}{24}, b = -\frac{3}{10}$, and $c = \frac{41}{6}$. Thus, $y = -\frac{5}{24}x^2 - \frac{3}{10}x + \frac{41}{6}$.

69. Least squares regression parabola through $(0, 0), (2, 2), (3, 6), (4, 12)$

$$n = 4$$

$$\sum x_i = 9 \qquad\qquad \sum y_i = 20$$

$$\sum x_i^2 = 29 \qquad\qquad \sum x_i^3 = 99$$

$$\sum x_i^4 = 353 \qquad\qquad \sum x_i y_i = 70$$

$$\sum x_i^2 y_i = 254$$

$$\begin{aligned} 4c + 9b + 29a &= 20 \\ 9c + 29b + 99a &= 70 \\ 29c + 99b + 353a &= 254 \end{aligned}$$

Solving this system yields $a = 1, b = -1$, and $c = 0$. Thus, $y = x^2 - x$.

71. (a) Least squares regression parabola through $(20, 25), (30, 55), (40, 105), (50, 188), (60, 300)$

$$n = 5$$

$$\sum x_i = 200 \qquad\qquad \sum y_i = 673$$

$$\sum x_i^2 = 9000 \qquad\qquad \sum x_i^3 = 440{,}000$$

$$\sum x_i^4 = 22{,}740{,}000 \qquad\qquad \sum x_i y_i = 33{,}750$$

$$\sum x_i^2 y_i = 1{,}777{,}500$$

$$\begin{aligned} 5c + 200b + 9000a &= 673 \\ 200c + 9000b + 440{,}000a &= 33{,}750 \\ 9000c + 440{,}000b + 22{,}740{,}000a &= 1{,}777{,}500 \end{aligned}$$

Solving this system yields $a \approx 0.14, b \approx -4.43$, and $c \approx 58.40$.
Thus, $y = 0.14x^2 - 4.43x + 58.40$.

(b)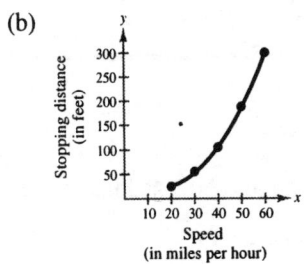

(c) When $x = 70, y \approx 434.3$ feet.

73. $\left.\begin{array}{r} y + \quad \lambda = 0 \\ x + \lambda = 0 \end{array}\right\} \Rightarrow x = y = -\lambda$

$x + y - 10 = 0 \Rightarrow 2x - 10 = 0$

$$x = 5$$

$$y = 5$$

$$\lambda = -5$$

75. $2x - 2x\lambda = 0 \Rightarrow 2x(1 - \lambda) = 0 \Rightarrow \lambda = 1 \text{ or } x = 0$

$-2y + \lambda = 0 \Rightarrow \quad\quad 2y = \lambda \Rightarrow y = \dfrac{1}{2}$

$y - x^2 = 0 \Rightarrow \quad\quad x^2 = y \Rightarrow x = \pm\sqrt{\dfrac{1}{2}} = \pm\dfrac{\sqrt{2}}{2}$

Answer: $\quad x = \pm\dfrac{\sqrt{2}}{2} \text{ or } x = 0$

$\quad\quad\quad y = \dfrac{1}{2} \text{ or } y = 0$

$\quad\quad\quad \lambda = 1 \text{ or } \lambda = 0$

77. The slope represents the average increase in sales per year.

79.$(0.075)(85) = 6.375$

81.$(0.005)(n) = 400$

$n = 80,000$

Section 6.4   Systems of Inequalities

■   You should be able to sketch the graph of an inequality in two variables.

(a) Replace the inequality with an equal sign and graph the equation. Use a dashed line for < or >, a solid line for ≤ or ≥.

(b) Test a point in each region formed by the graph. If the point satisfies the inequality, shade the whole region.

Solutions to Odd-Numbered Exercises

1. $x \geq 2$

Using a solid line, graph the vertical line $x = 2$ and shade to the right of this line.

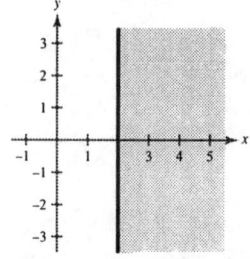

3. $y \geq -1$

Using a solid line, graph the horizontal line $y = -1$ and shade above this line.

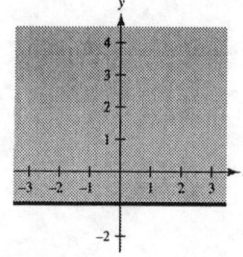

5. $y < 2 - x$

Using a dashed line, graph $y = 2 - x$, and then shade below the line. $\left(\text{Use } (0, 0) \text{ as a test point.}\right)$

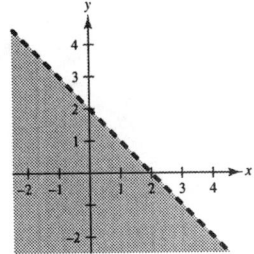

9. $(x + 1)^2 + (y - 2)^2 < 9$

Using a dashed line, sketch the circle $(x + 1)^2 + (y - 2)^2 = 9$.

Center: $(-1, 2)$

Radius: 3

Test point: $(0, 0)$. Shade the inside of the circle.

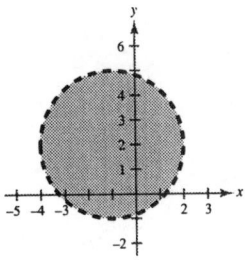

13. $y \geq \dfrac{2}{3}x - 1$

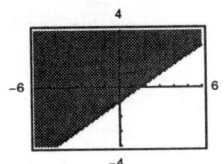

17. The line through $(-4, 0)$ and $(0, 2)$ is $y = \frac{1}{2}x + 2$. For the shaded region below the line, we have $y \leq \frac{1}{2}x + 2$.

7. $2y - x \geq 4$

Using a solid line, graph $2y - x = 4$, and then shade above the line. $\left(\text{Use } (0, 0) \text{ as a test point.}\right)$

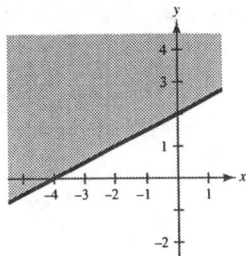

11. $y \leq \dfrac{1}{1 + x^2}$

Using a solid line, graph $y = \dfrac{1}{1 + x^2}$, and then shade below the curve. $\left(\text{Use } (0, 0) \text{ as a test point.}\right)$

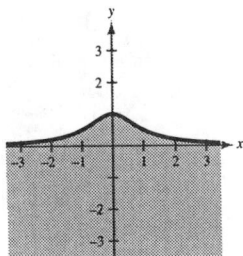

15. $x^2 + 5y - 10 \leq 0$

$$y \leq 2 - \dfrac{x^2}{5}$$

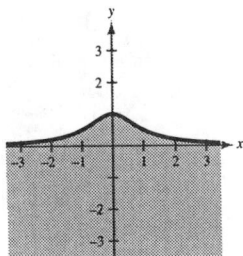

19. The line through $(0, 2)$ and $(3, 0)$ is $y = -\frac{2}{3}x + 2$. For the shaded region above the line, we have

$$y \geq -\dfrac{2}{3}x + 2$$
$$3y \geq -2x + 6$$
$$2x + 3y \geq 6$$
$$\dfrac{x}{3} + \dfrac{y}{2} \geq 1$$

21. $x + y \leq 1$

    $-x + y \leq 1$

        $y \geq 0$

First, find the points of intersection of each pair of equations.

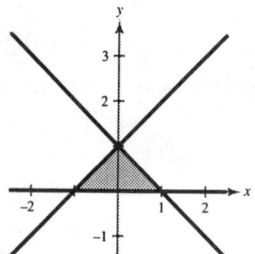

Vertex A	Vertex B	Vertex C
$x + y = 1$	$x + y = 1$	$-x + y = 1$
$-x + y = 1$	$y = 0$	$y = 0$
$(0, 1)$	$(1, 0)$	$(-1, 0)$

23. $x + y \leq 5$

    $x \geq 2$

    $y \geq 0$

First, find the points of intersection of each pair of equations.

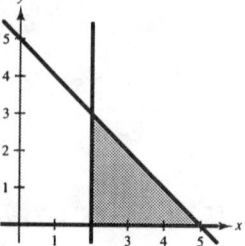

Vertex A	Vertex B	Vertex C
$x + y = 5$	$x + y = 5$	$x = 2$
$x = 2$	$y = 0$	$y = 0$
$(2, 3)$	$(5, 0)$	$(2, 0)$

25. $-3x + 2y < 6$

    $x - 4y > -2$

    $2x + y < 3$

First, find the points of intersection of each pair of equations.

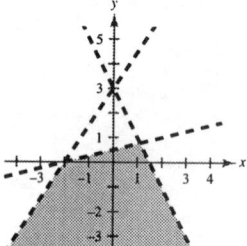

Vertex A	Vertex B	Vertex C
$-3x + 2y = 6$	$-3x + 2y = 6$	$x - 4y = -2$
$x - 4y = -2$	$2x + y = 3$	$2x + y = 3$
$(-2, 0)$	$(0, 3)$	$\left(\frac{10}{9}, \frac{7}{9}\right)$

27. $2x + y > 2$

    $6x + 3y < 2$

The lines are parallel. There are no points of intersection. There is no region common to both inequalities.

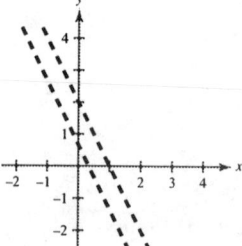

29. $x \quad\quad \geq 1$

$\quad\ x - 2y \leq 3$

$\quad\ 3x + 2y \geq 9$

$\quad\ x + \ y \leq 6$

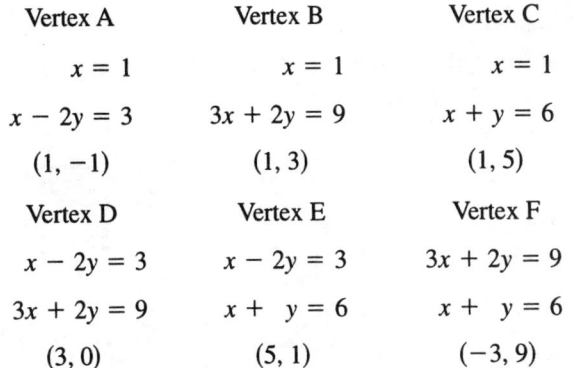

First, find the points of intersection of each pair of equations.

Vertex A	Vertex B	Vertex C
$x = 1$	$x = 1$	$x = 1$
$x - 2y = 3$	$3x + 2y = 9$	$x + y = 6$
$(1, -1)$	$(1, 3)$	$(1, 5)$

Vertex D	Vertex E	Vertex F
$x - 2y = 3$	$x - 2y = 3$	$3x + 2y = 9$
$3x + 2y = 9$	$x + y = 6$	$x + \ y = 6$
$(3, 0)$	$(5, 1)$	$(-3, 9)$

By shading each inequality, we find that the vertices of the
region are $(1, 5)$, $(1, 3)$, $(3, 0)$, and $(5, 1)$.

31. $x^2 + y^2 \leq 9$

$\quad x^2 + y^2 \geq 1$

There are no points of intersection. The region common
to both inequalities is the region between the circles.

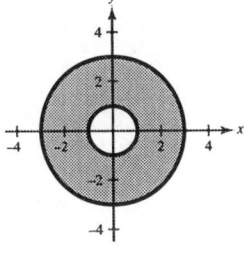

33. $x > y^2$

$\quad x < y + 2$

Points of intersection:

$$y^2 = y + 2$$
$$y^2 - y - 2 = 0$$
$$(y + 1)(y - 2) = 0$$
$$y = -1, 2$$

$(1, -1), (4, 2)$

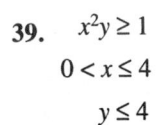

35. $y \leq \sqrt{3x} + 1$

$\quad y \geq x^2 + 1$

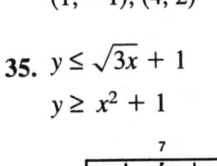

37. $y < x^3 - 2x + 1$

$\quad y > -2x$

$\quad x \leq 1$

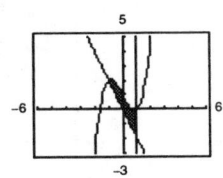

39. $x^2 y \geq 1$

$\quad 0 < x \leq 4$

$\quad\ \ y \leq 4$

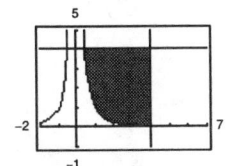

41. $y \le -x + 4 \implies \frac{x}{4} + \frac{y}{4} \le 1$

$\quad x \ge 0 \qquad\qquad\quad x \ge 0$

$\quad y \ge 0 \qquad\qquad\quad y \ge 0$

43. Line through points (0, 4) and (4, 0): $y = 4 - x$

Line through points (0, 2) and (8, 0): $y = 2 - \frac{1}{4}x$

$\quad y \ge 4 - x$

$\quad y \ge 2 - \frac{1}{4}x$

$\quad x \ge 0$

$\quad y \ge 0$

45. $x^2 + y^2 \le 16$

$\quad x \ge 0$

$\quad y \ge 0$

47. Rectangular region with vertices at (2, 1), (5, 1), (5, 7), and (2, 7)

$\quad x \ge 2$

$\quad x \le 5$

$\quad y \ge 1$

$\quad y \le 7$

Thus, $2 \le x \le 5$, $1 \le y \le 7$.

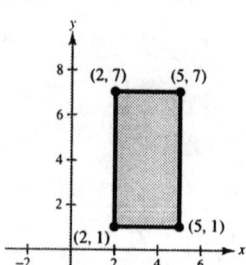

49. Triangle with vertices at (0, 0), (5, 0), (2, 3)

(0, 0), (5, 0) Line: $y \ge 0$

(0, 0), (2, 3) Line: $y \le \frac{3}{2}x$

(2, 3), (5, 0) Line: $y \le -x + 5$

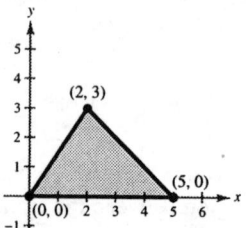

51. Assembly center constraint: $x + \frac{3}{2}y \le 12$

Finishing center constraint: $\frac{4}{3}x + \frac{3}{2}y \le 15$

Point of intersection: (9, 2)

Physical constraints: $x \ge 0$ and $y \ge 0$

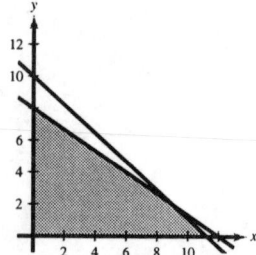

53. Account constraints: $\quad x \ge 5000$

$\quad y \ge 5000$

$\quad 2x \le y$

$\quad x + y \le 20{,}000$

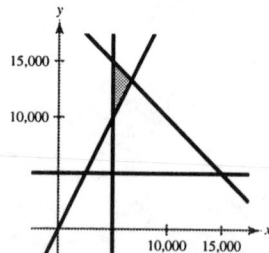

55. x = number of ounces of food X

y = number of ounces of food Y

Calcium: $20x + 10y \ge 280$

Iron: $15x + 10y \ge 160$

Vitamin B: $10x + 20y \ge 180$

$\quad x \ge 0$

$\quad y \ge 0$

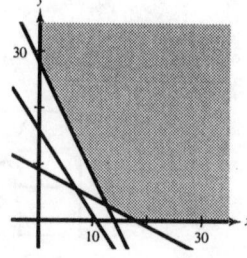

57. $xy \geq 500$ Body-building space

$2x + \pi y \geq 125$ Track (Two semi-circles and two lengths)

$x \geq 0$ Physical constraint

$y \geq 0$ Physical constraint

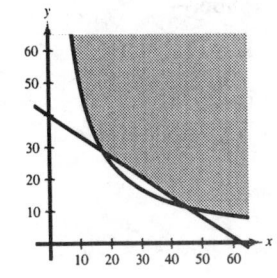

59.   Demand = Supply

$50 - 0.5x = 0.125x$

$50 = 0.625x$

$80 = x$

$10 = p$

Point of equilibrium: $(80, 10)$

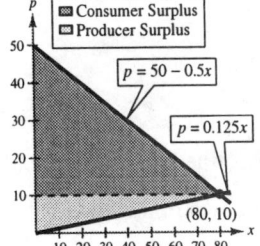

The consumer surplus is the area of the triangle bounded by

$p \leq 50 - 0.5x$

$p \geq 10$

$x \geq 0.$

Consumer surplus $= \frac{1}{2}(\text{base})(\text{height}) = \frac{1}{2}(80)(40) = \$1600$

The producer surplus is the area of the triangle bounded by

$p \geq 0.125x$

$p \leq 10$

$x \geq 0.$

Producer surplus $= \frac{1}{2}(\text{base})(\text{height}) = \frac{1}{2}(80)(10) = \$400$

61.

$$\text{Demand} = \text{Supply}$$

$$140 - 0.00002x = 80 + 0.00001x$$

$$60 = 0.00003x$$

$$2{,}000{,}000 = x$$

$$100 = p$$

Point of equilibrium: $(2{,}000{,}000, 100)$

The consumer surplus is the area of the triangle bounded by

$$p \le 140 - 0.00002x$$

$$p \ge 100$$

$$x \ge 0.$$

Consumer surplus $= \frac{1}{2}(\text{base})(\text{height}) = \frac{1}{2}(2{,}000{,}000)(40) = \$40{,}000{,}000$ or \$40 million

The producer surplus is the area of the triangle bounded by

$$p \ge 80 + 0.00001x$$

$$p \le 100$$

$$x \ge 0.$$

Producer surplus $= \frac{1}{2}(\text{base})(\text{height}) = \frac{1}{2}(2{,}000{,}000)(20) = \$20{,}000{,}000$ or \$20 million

63. Test a point on either side of the boundary.

Section 6.5 Linear Programming

■ To solve a linear programming problem:
1. Sketch the solution set for the system of constraints.
2. Find the vertices of the region.
3. Test the objective function at each of the vertices.

Solutions to Odd-Numbered Exercises

1. $z = 4x + 5y$

At $(0, 6)$: $z = 4(0) + 5(6) = 30$

At $(0, 0)$: $z = 4(0) + 5(0) = 0$

At $(6, 0)$: $z = 4(6) + 5(0) = 24$

The minimum value is 0 at $(0, 0)$.
The maximum value is 30 at $(0, 6)$.

3. $z = 10x + 6y$

At $(0, 6)$: $z = 10(0) + 6(6) = 36$

At $(0, 0)$: $z = 10(0) + 6(0) = 0$

At $(6, 0)$: $z = 10(6) + 6(0) = 60$

The minimum value is 0 at $(0, 0)$.
The maximum value is 60 at $(6, 0)$.

5. $z = 3x + 2y$

At $(0, 5)$: $z = 3(0) + 2(5) = 10$

At $(4, 0)$: $z = 3(4) + 2(0) = 12$

At $(3, 4)$: $z = 3(3) + 2(4) = 17$

At $(0, 0)$: $z = 3(0) + 2(0) = 0$

The minimum value is 0 at $(0, 0)$.
The maximum value is 17 at $(3, 4)$.

7. $z = 5x + 0.5y$

At $(0, 5)$: $z = 5(0) + \frac{5}{2} = \frac{5}{2}$

At $(4, 0)$: $z = 5(4) + \frac{0}{2} = 20$

At $(3, 4)$: $z = 5(3) + \frac{4}{2} = 17$

At $(0, 0)$: $z = 5(0) + \frac{0}{2} = 0$

The minimum value is 0 at $(0, 0)$.
The maximum value is 20 at $(4, 0)$.

9. $z = 10x + 7y$

At $(0, 45)$: $z = 10(0) + 7(45) = 315$

At $(30, 45)$: $z = 10(30) + 7(45) = 615$

At $(60, 20)$: $z = 10(60) + 7(20) = 740$

At $(60, 0)$: $z = 10(60) + 7(0) = 600$

At $(0, 0)$: $z = 10(0) + 7(0) = 0$

The minimum value is 0 at $(0, 0)$.
The maximum value is 740 at $(60, 20)$.

11. $z = 25x + 30y$

At $(0, 45)$: $z = 25(0) + 30(45) = 1350$

At $(30, 45)$: $z = 25(30) + 30(45) = 2100$

At $(60, 20)$: $z = 25(60) + 30(20) = 2100$

At $(60, 0)$: $z = 25(60) + 30(0) = 1500$

At $(0, 0)$: $z = 25(0) + 30(0) = 0$

The minimum value is 0 at $(0, 0)$.
The maximum value is 2100 at any point along the line segment connecting $(30, 45)$ and $(60, 20)$.

13. $z = 6x + 10y$

At $(0, 2)$: $z = 6(0) + 10(2) = 20$

At $(5, 0)$: $z = 6(5) + 10(0) = 30$

At $(0, 0)$: $z = 6(0) + 10(0) = 0$

The minimum value is 0 at $(0, 0)$.
The maximum value is 30 at $(5, 0)$.

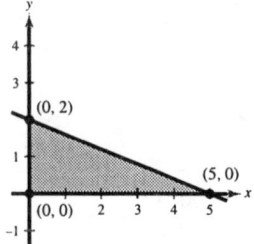

15. $z = 9z + 24y$

At $(0, 2)$: $z = 9(0) + 24(2) = 48$

At $(5, 0)$: $z = 9(5) + 24(0) = 45$

At $(0, 0)$: $z = 9(0) + 24(0) = 0$

The minimum value is 0 at $(0, 0)$.
The maximum value is 48 at $(0, 2)$.

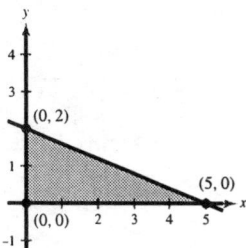

17. $z = 4x + 5y$

At $(10, 0)$: $z = 4(10) + 5(0) = 40$

At $(5, 3)$: $z = 4(5) + 5(3) = 35$

At $(0, 8)$: $z = 4(0) + 5(8) = 40$

The minimum value is 35 at $(5, 3)$.
C is unbounded. Therefore, there is no maximum.

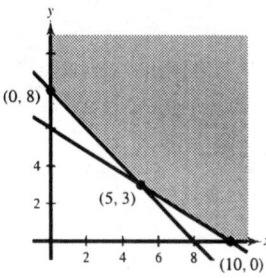

19. $z = 2x + 7y$

At $(10, 0)$: $z = 2(10) + 7(0) = 20$

At $(5, 3)$: $z = 2(5) + 7(3) = 31$

At $(0, 8)$: $z = 2(0) + 7(8) = 56$

The minimum value is 20 at $(10, 0)$.
C is unbounded. Therefore, there is no maximum.

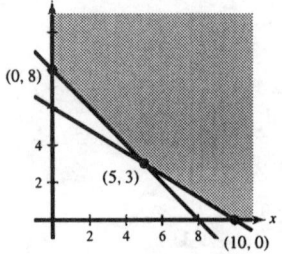

21. $z = 4x + y$

At $(36, 0)$: $z = 4(36) + 0 = 144$

At $(40, 0)$: $z = 4(40) + 0 = 160$

At $(24, 8)$: $z = 4(24) + 8 = 104$

The minimum value is 104 at $(24, 8)$.
The maximum value is 160 at $(40, 0)$.

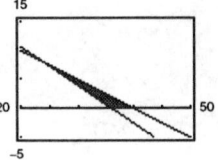

23. $z = x + 4y$

At $(36, 0)$: $z = 36 + 4(0) = 36$

At $(40, 0)$: $z = 40 + 4(0) = 40$

At $(24, 8)$: $z = 24 + 4(8) = 56$

The minimum value is 36 at $(36, 0)$.
The maximum value is 56 at $(24, 8)$.

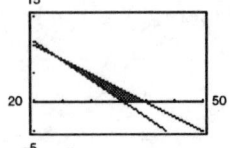

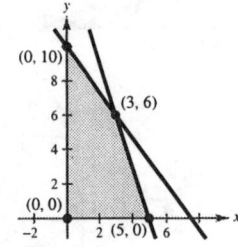

Figure for Exercises 25 and 27

25. $z = 2x + y$

At $(0, 10)$: $z = 2(0) + (10) = 10$

At $(3, 6)$: $z = 2(3) + (6) = 12$

At $(5, 0)$: $z = 2(5) + (0) = 10$

At $(0, 0)$: $z = 2(0) + (0) = 0$

The maximum value is 12 at $(3, 6)$.

27. $z = x + y$

At $(0, 10)$: $z = (0) + (10) = 10$

At $(3, 6)$: $z = (3) + (6) = 9$

At $(5, 0)$: $z = (5) + (0) = 5$

At $(0, 0)$: $z = (0) + (0) = 0$

The maximum value is 10 at $(0, 10)$.

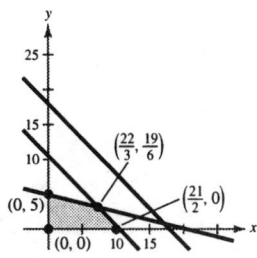

Figure for Exercises 29 and 31

29. $z = x + 5y$

At $(0, 5)$: $z = 0 + 5(5) = 25$

At $\left(\frac{22}{3}, \frac{19}{6}\right)$: $z = \frac{22}{3} + 5\left(\frac{19}{6}\right) = \frac{139}{6}$

At $\left(\frac{21}{2}, 0\right)$: $z = \frac{21}{2} + 5(0) = \frac{21}{2}$

At $(0, 0)$: $z = 0 + 5(0) = 0$

The maximum value is 25 at $(0, 5)$.

31. $z = 4x + 5y$

At $(0, 5)$: $z = 4(0) + 5(5) = 25$

At $\left(\frac{22}{3}, \frac{19}{6}\right)$: $z = 4\left(\frac{22}{3}\right) + 5\left(\frac{19}{6}\right) = \frac{271}{6}$

At $\left(\frac{21}{2}, 0\right)$: $z = 4\left(\frac{21}{2}\right) + 5(0) = 42$

At $(0, 0)$: $z = 4(0) + 5(0) = 0$

The maximum value is $\frac{271}{6}$ at $\left(\frac{22}{3}, \frac{19}{6}\right)$.

33. There are an infinite number of objective functions that would have a maximum at $(0, 4)$. One such objective function is $z = x + 5y$.

35. There are an infinite number of objective functions that would have a maximum at $(5, 0)$. One such objective function is $z = 4x + y$.

37. x = number of Model A

y = number of Model B

Constraints: $2x + 2.5y \le 4000$

$\qquad\qquad 4x + y \le 4800$

$\qquad\qquad x + 0.75y \le 1500$

$\qquad\qquad\qquad x \ge 0$

$\qquad\qquad\qquad y \ge 0$

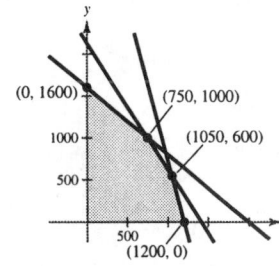

Objective function: $P = 45x + 50y$

Vertices: $(0, 0)$, $(0, 1600)$, $(750, 1000)$, $(1050, 600)$, $(1200, 0)$

At $(0, 0)$: $P = 45(0) + 50(0) = 0$

At $(0, 1600)$: $P = 45(0) + 50(1600) = 80,000$

At $(750, 1000)$: $P = 45(750) + 50(1000) = 83,750$

At $(1050, 600)$: $P = 45(1050) + 50(600) = 77,250$

At $(1200, 0)$: $P = 45(1200) + 50(0) = 54,000$

The maximum profit of \$83,750 occurs when 750 units of Model A and 1000 units of Model B are produced.

39. x = number of $250 models

y = number of $400 models

Constraints: $250x + 400y \leq 70{,}000$

$$x + y \leq 250$$

$$x \geq 0$$

$$y \geq 0$$

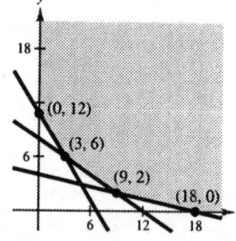

Objective function: $P = 45x + 50y$

Vertices: $(0, 175)$, $(200, 50)$, $(250, 0)$, $(0, 0)$

At $(0, 175)$: $P = 45(0) + 50(175) = 8750$

At $(200, 50)$: $P = 45(200) + 50(50) = 11{,}500$

At $(250, 0)$: $P = 45(250) + 50(0) = 11{,}250$

At $(0, 0)$: $P = 45(0) + 50(0) = 0$

To maximize the profit, the merchant should stock 200 units of the model costing $250 and 50 units of the model costing $400. Then the maximum profit would be $11,500.

41. x = number of bags of Brand X

y = number of bags of Brand Y

Constraints: $2x + y \geq 12$

$$2x + 9y \geq 36$$

$$2x + 3y \geq 24$$

$$x \geq 0$$

$$y \geq 0$$

Objective function: $C = 25x + 20y$

Vertices: $(0, 12)$, $(3, 6)$, $(9, 2)$, $(18, 0)$

At $(0, 12)$: $C = 25(0) + 20(12) = 240$

At $(3, 6)$: $C = 25(3) + 20(6) = 195$

At $(9, 2)$: $C = 25(9) + 20(2) = 265$

At $(18, 0)$: $C = 25(18) + 20(0) = 450$

To minimize cost, use three bags of Brand X and six bags of Brand Y for a total cost of $195.

43. $x =$ the number of audits, and $y =$ the number of tax returns.

Objective function: Maximize $R = 2000x + 300y$.

Constraints: $100x + 12.5y \leq 900$

$$10x + 2.5y \leq 100$$

$$x \geq 0, y \geq 0$$

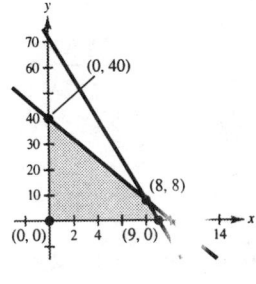

Vertex	Value of $R = 2000x + 300y$
$(0, 0)$	$R = 2000(0) + 300(0) = 0$
$(0, 40)$	$R = 2000(0) + 300(40) = 12{,}000$
$(8, 8)$	$R = 2000(8) + 300(8) = 18{,}400$, maximum value
$(9, 0)$	$R = 2000(9) + 300(0) = 18{,}000$

The revenue will be maximum if the firm does 8 audits and 8 tax returns each week. Maximum revenue is $18,400.

45. Objective function: $z = 2.5x + y$

Constraints: $x \geq 0, y \geq 0, 3x + 5y \leq 15, 5x + 2y \leq 10$

At $(0, 0)$: $z = 0$

At $(2, 0)$: $z = 5$

At $\left(\frac{20}{19}, \frac{45}{19}\right)$: $z = \frac{95}{19} = 5$

At $(0, 3)$: $z = 3$

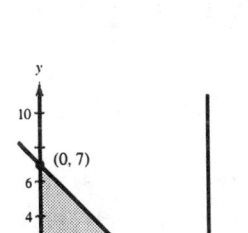

z is the maximum at any point on the line $5x + 2y = 10$ between the points $(2, 0)$ and $\left(\frac{20}{19}, \frac{45}{19}\right)$.

47. Objective function: $z = -x + 2y$

Constraints: $x \geq 0, y \geq 0, x \leq 10, x + y \leq 7$

At $(0, 0)$: $z = -0 + 2(0) = 0$

At $(0, 7)$: $z = -0 + 2(7) = 14$

At $(7, 0)$: $z = -7 + 2(0) = -7$

The constraint $x \leq 10$ is extraneous.

The maximum value of 14 occurs at $(0, 7)$.

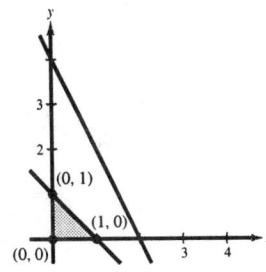

49. Objective function: $z = 3x + 4y$

Constraints: $x \geq 0, y \geq 0, x + y \leq 1, 2x + y \leq 4$

The constraint $2x + y \leq 4$ is extraneous.
The maximum value of $z = 4$ occurs at $(0, 1)$.

51. Constraints: $x \geq 0$, $y \geq 0$, $x + 3y \leq 15$, $4x + y \leq 16$

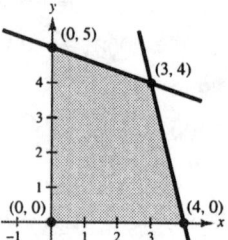

Vertex	Value of $z = 3x + ty$
$(0, 0)$	$z = 0$
$(0, 5)$	$z = 5t$
$(3, 4)$	$z = 9 + 4t$
$(4, 0)$	$z = 12$

(a) For the maximum value to be at $(0, 5)$, $z = 5t$ must be greater than or equal to $z = 9 + 4t$ and $z = 12$.

$$5t \geq 9 + 4t \quad \text{and} \quad 5t \geq 12$$

$$t \geq 9 \qquad\qquad t \geq \tfrac{12}{5}$$

Thus, $t \geq 9$.

(b) For the maximum value to be at $(3, 4)$, $z = 9 + 4t$ must be greater than or equal to $z = 5t$ and $z = 12$.

$$9 + 4t \geq 5t \quad \text{and} \quad 9 + 4t \geq 12$$

$$9 \geq t \qquad\qquad 4t \geq 3$$

$$t \geq \tfrac{3}{4}$$

Thus, $\tfrac{3}{4} \leq t \leq 9$.

53. $\dfrac{\dfrac{9}{x}}{\left(\dfrac{6}{x} + 2\right)} = \dfrac{\dfrac{9}{x}}{\dfrac{6 + 2x}{x}} = \dfrac{9}{x} \cdot \dfrac{x}{2(3 + x)} = \dfrac{9}{2(3 + x)} = \dfrac{9}{2(x + 3)}, x \neq 0$

55. $\dfrac{\left(\dfrac{4}{x^2 - 9} + \dfrac{2}{x - 2}\right)}{\left(\dfrac{1}{x + 3} + \dfrac{1}{x - 3}\right)} = \dfrac{\dfrac{4(x - 2) + 2(x^2 - 9)}{(x - 2)(x^2 - 9)}}{\dfrac{(x - 3) + (x + 3)}{x^2 - 9}}$

$$= \dfrac{2x^2 + 4x - 26}{(x - 2)(x^2 - 9)} \cdot \dfrac{x^2 - 9}{2x}$$

$$= \dfrac{2(x^2 + 2x - 13)}{(x - 2)(2x)}$$

$$= \dfrac{x^2 + 2x - 13}{x(x - 2)}, x \neq \pm 3$$

❑ Review Exercises for Chapter 6

Solutions to Odd-Numbered Exercises

1. $x + y = 2 \implies y = 2 - x$

$x - y = 0 \implies x - (2 - x) = 0$

$\qquad\qquad\qquad 2x - 2 = 0$

$\qquad\qquad\qquad\qquad x = 1$

$\qquad\qquad\qquad\qquad y = 2 - 1 = 1$

Solution: $(1, 1)$

3. $x^2 - y^2 = 9$

$x - y = 1 \implies x = y + 1$

$(y + 1)^2 - y^2 = 9$

$\qquad\qquad 2y + 1 = 9$

$\qquad\qquad\qquad y = 4$

$\qquad\qquad\qquad x = 5$

Solution: $(5, 4)$

5. $y = 2x^2$

$y = x^4 - 2x^2 \implies 2x^2 = x^4 - 2x^2$

$$0 = x^4 - 4x^2$$
$$0 = x^2(x^2 - 4)$$
$$0 = x^2(x + 2)(x - 2)$$
$$x = 0, x = -2, x = 2$$
$$y = 0, y = 8, y = 8$$

Solutions: $(0, 0), (-2, 8), (2, 8)$

7. $y^2 - 2y + x = 0 \implies (y - 1)^2 = 1 - x \implies y = 1 \pm \sqrt{1 - x}$

$\qquad x + y = 0 \implies y = -x$

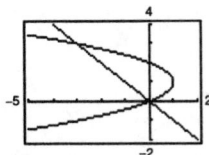

Points of intersection: $(0, 0)$ and $(-3, 3)$

9. $y = 2(6 - x)$

$y = 2^{x-2}$

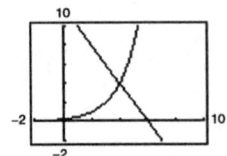

Point of intersection: $(4, 4)$

11. $\qquad 2x - y = \ \ 2 \implies 16x - 8y = 16$

$\qquad 6x + 8y = 39 \implies \underline{6x + 8y = 39}$

$$\qquad\qquad\qquad\qquad 22x \qquad = 55$$
$$x = \tfrac{55}{22} = \tfrac{5}{2}$$
$$y = 3$$

Solution: $\left(\tfrac{5}{2}, 3\right)$

13. $0.2x + 0.3y = 0.14 \implies 20x + 30y = 14 \implies \quad 20x + 30y = \ \ 14$

$\ \ \ \ 0.4x + 0.5y = 0.20 \implies \ 4x + \ \ 5y = \ \ 2 \implies -20x - 25y = -10$

$$\overline{\qquad\qquad\qquad\qquad\qquad 5y = \quad 4}$$
$$y = \quad \tfrac{4}{5}$$
$$x = \quad -\tfrac{1}{2}$$

Solution: $\left(-\tfrac{1}{2}, \tfrac{4}{5}\right) = (-0.5, 0.8)$

15. $\qquad 3x - 2y = 0 \implies 3x - 2y = 0$

$\ \ \ 3x + 2(y + 5) = 10 \implies \underline{3x + 2y = 0}$

$$\qquad\qquad\qquad\qquad 6x \qquad = 0$$
$$x \qquad = 0$$
$$y = 0$$

Solution: $(0, 0)$

17. $1.25x - 2y = 3.5 \implies \quad 5x - 8y = \ \ 14$

$\qquad 5x - 8y = 14 \implies -5x + 8y = -14$

$$\qquad\qquad\qquad\qquad\qquad 0 = \quad 0$$

Infinite solutions

Let $y = a$, then $5x - 8a = 14 \implies x = \tfrac{14}{5} + \tfrac{8}{5}a$.

Solution: $\left(\tfrac{14}{5} + \tfrac{8}{5}a, a\right)$

19. There are infinite linear systems with the solution $\left(\frac{4}{3}, 3\right)$. One possible solution is:

$$3\left(\tfrac{4}{3}\right) + 3 = 7 \implies 3x + y = 7$$
$$-6\left(\tfrac{4}{3}\right) + 3(3) = 1 \implies -6x + 3y = 1$$

21. Revenue $= 4.95x$
Cost $= 2.85x + 10{,}000$
Break even when Revenue $=$ Cost

$$4.95x = 2.85x + 10{,}000$$
$$2.10x = 10{,}000$$
$$x \approx 4762 \text{ units}$$

23. Let $x =$ the amount of 75% solution, and $y =$ amount of 50% solution.

$$x + y = 100 \implies y = 100 - x$$
$$0.75x + 0.50y = 0.60(100)$$
$$0.75x + 0.50(100 - x) = 60$$
$$0.75x + 50 - 0.50x = 60$$
$$0.25x = 10$$
$$x = 40$$
$$y = 100 - x = 60$$

Answer: 40 liters of 75% solution, 60 liters of 60% solution.

25. Let $x =$ speed of the slower plane

Let $y =$ speed of the faster plane

Then, distance of first plane $+$ distance of second plane $= 275$ miles

(rate of first plane)(time) $+$ (rate of second plane)(time) $= 275$ miles

$$x\left(\tfrac{40}{60}\right) + y\left(\tfrac{40}{60}\right) = 275$$
$$y = x + 25$$
$$\tfrac{2}{3}x + \tfrac{2}{3}(x + 25) = 275$$
$$4x + 50 = 825$$
$$4x = 775$$
$$x = 193.75 \text{ mph}$$
$$y = x + 25 = 218.75 \text{ mph}$$

27.   Demand = Supply

$37 - 0.0002x = 22 + 0.00001x$

$15 = 0.00021x$

$x = \dfrac{500,000}{7}, p = \dfrac{159}{7}$

Point of equilibrium: $\left(\dfrac{500,000}{7}, \dfrac{159}{7}\right)$

29.
$$x + 2y + 6z = 4$$
$$-3x + 2y - z = -4$$
$$4x \quad\quad + 2z = 16$$
$$x + 2y + 6z = 4$$
$$8y + 17z = 8 \quad 3\text{Eq.1} + \text{Eq.2}$$
$$-8y - 22z = 0 \quad -4\text{Eq.1} + \text{Eq.3}$$
$$x + 2y + 6z = 4$$
$$8y + 17z = 8$$
$$-5z = 8 \quad \text{Eq.2} + \text{Eq.3}$$
$$z = -\tfrac{8}{5} = -1.6$$
$$8y + 17(-1.6) = 8 \implies y = 4.4$$
$$x + 2(4.4) + 6(-1.6) = 4 \implies x = 4.8$$

Solution: $(4.8, 4.4, -1.6)$

31.
$$x - 2y + z = -6$$
$$2x - 3y = -7$$
$$-x + 3y - 3z = 11$$
$$x - 2y + z = -6$$
$$y - 2z = 5 \quad -2\text{Eq.1} + \text{Eq.2}$$
$$y - 2z = 5 \quad \text{Eq.1} + \text{Eq.3}$$
$$x - 2y + z = -6$$
$$y - 2z = 5$$
$$0 = 0 \quad -\text{Eq.2} + \text{Eq.3}$$

Let $z = a$, then:
$$y = 2a + 5$$
$$x - 2(2a + 5) + a = -6$$
$$x - 3a - 10 = -6$$
$$x = 3a + 4$$

Solution: $(3a + 4, 2a + 5, a)$ where a is any real number.

33. $2x + 5y - 19z = 34 \implies 6x + 15y - 57z = 102$

$3x + 8y - 31z = 54 \implies \underline{-6x - 16y + 62z = -108}$

$\quad\quad\quad\quad\quad\quad\quad\quad -y + 5z = -6$

Let $z = a$. Then:
$$y = 5a + 6$$
$$x = \tfrac{1}{2}[34 - 5(5a + 6) + 19a] = -3a + 2$$

Solution: $(-3a + 2, 5a + 6, a)$ where a is any real number.

35. There are an infinite number of linear systems with the solution $(4, -1, 3)$.
One possible system is as follows:

$$2(4) + (-1) - 2(3) = 1 \implies 2x + y - 2z = 1$$
$$(4) + (-1) - (3) = 0 \implies x + y - z = 0$$
$$2(4) - 3(-1) - 2(3) = 5 \implies 2x - 3y - 2z = 5$$

37. $y = ax^2 + bx + c$ through $(0, -5)$, $(1, -2)$, and $(2, 5)$.

$$(0,-5): -5 = + c \implies c = -5$$
$$(1,-2): -2 = a + b + c \implies a + b = 3$$
$$(2, 5): 5 = 4a + 2b + c \implies 2a + b = 5$$

$$2a + b = 5$$
$$-a - b = -3$$
$$a = 2$$
$$b = 1$$

The equation of the parabola is $y = 2x^2 + x - 5$.

39. $x^2 + y^2 + Dx + Ey + F = 0$ through $(-1, -2)$, $(5, -2)$ and $(2, 1)$.

$$(-1,-2): 5 - D - 2E + F = 0 \implies D + 2E - F = 5$$
$$(5,-2): 29 + 5D - 2E + F = 0 \implies 5D - 2E + F = -29$$
$$(2, 1): 5 + 2D + 2E + F = 0 \implies 2D + E + F = -5$$

From the first two equations we have

$$6D = -24$$
$$D = -4.$$

Substituting $D = -4$ into the second and third equations yields:

$$-20 - 2E + F = -29 \implies -2E + F = -9$$
$$-8 + E + F = -5 \implies -E - F = -3$$
$$\overline{ -3E = -12}$$
$$E = 4$$
$$F = -1$$

The equation of the circle is $x^2 + y^2 - 4x + 4y - 1 = 0$.

41. From the following chart we obtain our system of equations.

	A	B	C
Mixture X	$\frac{1}{5}$	$\frac{2}{5}$	$\frac{2}{5}$
Mixture Y	0	0	1
Mixture Z	$\frac{1}{3}$	$\frac{1}{3}$	$\frac{1}{3}$
Desired Mixture	$\frac{6}{27}$	$\frac{8}{27}$	$\frac{13}{27}$

$$\left.\begin{array}{l} \frac{1}{5}x + \frac{1}{3}z = \frac{6}{27} \\ \frac{2}{5}x + \frac{1}{3}z = \frac{8}{27} \end{array}\right\} x = \frac{10}{27}, z = \frac{12}{27}$$

$$\frac{2}{5}x + y + \frac{1}{3}z = \frac{13}{27} \implies y = \frac{5}{27}$$

To obtain the desired mixture, use 10 gallons of X, 5 gallons of Y, and 12 gallons of Z.

43. $\quad 5b + 10a = 17.8 \quad \Rightarrow \quad -10b - 20a = -35.6$

$\quad\quad 10b + 30a = 45.7 \quad \Rightarrow \quad \underline{\quad 10b + 30a = \quad 45.7}$

$$10a = \quad 10.1$$
$$a = \quad 1.01$$
$$b = \quad 1.54$$

Least squares regression line: $y = 1.01x + 1.54$

45. (a) $\quad 7b + \quad 156.8a = 169.5$

$\quad\quad\quad 156.8b + 3522.78a = 3806.8$

By elimination we have $a \approx 0.956$, $b \approx 2.799$.

The least squares regression line is $y = 0.956x + 2.779$.

(b)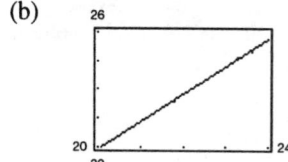

(c) The line is a good model for the data.

(d) A one year change in x (the women's ages) results in y (men's ages) changing 0.956 year.

47. $\quad x + 2y \le 160$

$\quad\quad 3x + \quad y \le 180$

$\quad\quad\quad\quad x \ge 0$

$\quad\quad\quad\quad y \ge 0$

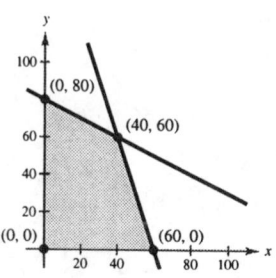

Vertex A	Vertex B	Vertex C	Vertex D	Vertex E	Vertex F
$x + 2y = 160$	$x + 2y = 160$	$3x + y = 180$	$x = 0$	$x + 2y = 160$	$3x + y = 180$
$3x + y = 180$	$x = 0$	$y = 0$	$y = 0$	$y = 0$	$x = 0$
$(40, 60)$	$(0, 80)$	$(60, 0)$	$(0, 0)$	$(160, 0)$	$(0, 180)$
				Outside the region	Outside the region

49. $3x + 2y \geq 24$

$x + 2y \geq 12$

$2 \leq x \leq 15$

$y \leq 15$

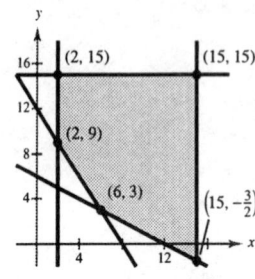

Vertex A	Vertex B	Vertex C	Vertex D	Vertex E
$3x + 2y = 24$	$3x + 2y = 24$	$3x + 2y = 24$	$3x + 2y = 24$	$x + 2y = 12$
$x + 2y = 12$	$x = 2$	$x = 15$	$y = 15$	$x = 2$
$(6, 3)$	$(2, 9)$	$\left(15, -\frac{21}{2}\right)$	$(-2, 15)$	$(2, 5)$
		Outside the region	Outside the region	Outside the region

Vertex F	Vertex G	Vertex H	Vertex I
$x + 2y = 12$	$x + 2y = 12$	$x = 2$	$x = 15$
$x = 15$	$y = 15$	$y = 15$	$y = 15$
$\left(15, -\frac{3}{2}\right)$	$(-18, 15)$	$(2, 15)$	$(15, 15)$
	Outside the region		

51. $y < x + 1$

$y > x^2 - 1$

Vertices:

$x + 1 = x^2 - 1$

$0 = x^2 - x - 2 = (x + 1)(x - 2)$

$x = -1$ or $x = 2$

$y = 0 y = 3$

$(-1, 0) (2, 3)$

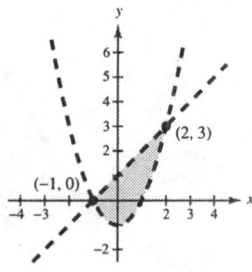

53. $2x - 3y \geq 0$

$2x - y \leq 8$

$y \geq 0$

Vertex A	Vertex B	Vertex C
$2x - 3y = 0$	$2x - 3y = 0$	$2x - y = 8$
$2x - y = 8$	$y = 0$	$y = 0$
$(6, 4)$	$(0, 0)$	$(4, 0)$

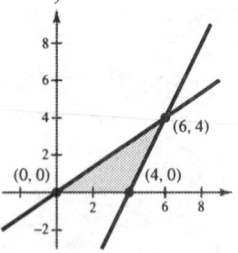

55.

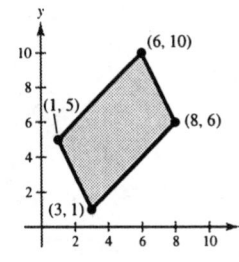

Line through $(1, 5)$, $(3, 1)$: $2x + y = 7$

Line through $(1, 5)$, $(6, 10)$: $-x + y = 4$

Line through $(6, 10)$, $(8, 6)$: $2x + y = 22$

Line through $(8, 6)$, $(3, 1)$: $-x + y = -2$

System of inequalities:

$$-x + y \le 4$$
$$2x + y \le 22$$
$$-x + y \ge -2$$
$$2x + y \ge 7$$

57. Let $x =$ the number of bushels for Harrisburg, and $y =$ the number of bushels for Philadelphia.

$$x \ge 400$$
$$y \ge 600$$
$$x + y \le 1500$$

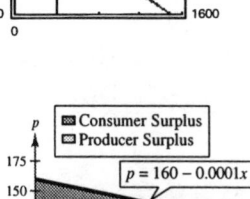

59.

$$\text{Demand} = \text{Supply}$$
$$160 - 0.0001x = 70 + 0.0002x$$
$$90 = 0.0003x$$
$$x = 300{,}000 \text{ units}$$
$$p = \$130$$

Point of equilibrium: $(300{,}000, 130)$

Consumer surplus: $\frac{1}{2}(300{,}000)(30) = \$4{,}500{,}000$

Producer surplus: $\frac{1}{2}(300{,}000)(60) = \$9{,}000{,}000$

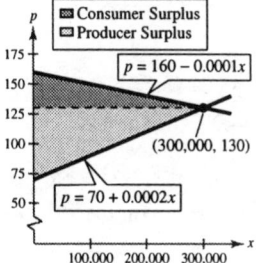

61. Maximize $z = 3x + 4y$ subject to the following constraints.

$$x \ge 0$$
$$y \ge 0$$
$$2x + 5y \le 50$$
$$4x + y \le 28$$

Vertex	Value of $z = 3x + 4y$
$(0, 0)$	$z = 0$
$(0, 10)$	$z = 40$
$(5, 8)$	$z = 47$, maximum value
$(7, 0)$	$z = 21$

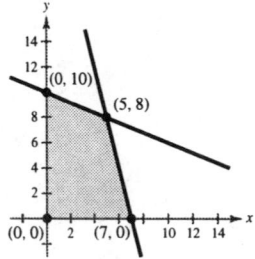

63. Minimize $z = 1.75x + 2.25y$ subject to the following constraints.

$$2x + y \geq 25$$
$$3x + 2y \geq 45$$
$$x \geq 0$$
$$y \geq 0$$

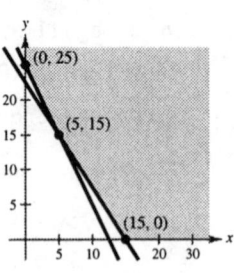

Vertex	Value of $z = 1.75x + 2.25y$
(0, 25)	$z = 56.25$
(5, 15)	$z = 42.5$
(15, 0)	$z = 26.25$, minimum value

65. Let x = number of haircuts
Let y = number of perms
Maximize $R = 17x + 60y$ subject to the following constraints.

$$x \geq 0$$
$$y \geq 0$$
$$\left(\tfrac{20}{60}\right)x + \left(\tfrac{70}{60}\right)y \leq 24 \implies 2x + 7y \leq 144$$

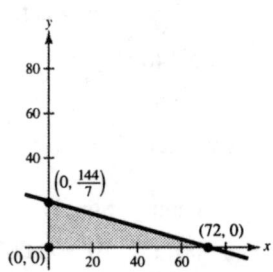

Vertex	Value of $R = 17x + 60y$
(0, 0)	$R = 0$
(72, 0)	$R = 1224$
$\left(0, \tfrac{144}{7}\right)$	$R \approx 1234.29$, maximum value

The revenue is maximum when $y = \tfrac{144}{7} \approx 20$ perms. (Round down since the student cannot work more than 24 hours. Note: Since we rounded down, the student would have enough time left to do 2 haircuts.)

67. Let x = the number of bags of Brand X, and y = the number of bags of Brand Y.
Objective function: Minimize $C = 15x + 30y$
Constraints: $8x + 2y \geq 16$

$$x + y \geq 5$$
$$2x + 7y \geq 20$$
$$x \geq 0, y \geq 0$$

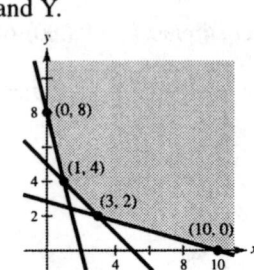

Vertex	Value of $C = 15x + 30y$
(0, 8)	$C = 15(0) + 30(8) = 240$
(1, 4)	$C = 15(1) + 30(4) = 135$
(3, 2)	$C = 15(3) + 30(2) = 105$, minimum value
(10, 0)	$C = 15(10) + 30(0) = 150$

To minimize cost, use three bags of Brand X and two bags of Brand Y.

The cost is $105.

❏ Practice Test for Chapter 6

For Exercises 1–3, solve the given system by the method of substitution.

1. $x + y = 1$

   $3x - y = 15$

2. $x - 3y = -3$

   $x^2 + 6y = 5$

3. $x + y + z = 6$

   $2x - y + 3z = 0$

   $5x + 2y - z = -3$

4. Find the two numbers whose sum is 110 and product is 2800.

5. Find the dimensions of a rectangle if its perimeter is 170 feet and its area is 2800 square feet.

For Exercises 6–8, solve the linear system by elimination.

6. $2x + 15y = 4$

   $x - 3y = 23$

7. $x + y = 2$

   $38x - 19y = 7$

8. $0.4x + 0.5y = 0.112$

   $0.3x - 0.7y = -0.131$

9. Herbert invests $17,000 in two funds that pay 11% and 13% simple interest, respectively. If he receives $2080 in yearly interest, how much is invested in each fund?

10. Find the least squares regression line for the points $(4, 3)$, $(1, 1)$, $(-1, -2)$, and $(-2, -1)$.

For Exercises 11–13, solve the system of equations.

11.
$$x + y \qquad = -2$$
$$2x - y + z = 11$$
$$ 4y - 3z = -20$$

12.
$$4x - y + 5z = 4$$
$$2x + y - z = 0$$
$$2x + 4y + 8z = 0$$

13.
$$3x + 2y - z = 5$$
$$6x - y + 5z = 2$$

14. Find the equation of the parabola $y = ax^2 + bx + c$ passing through the points $(0, -1)$, $(1, 4)$ and $(2, 13)$.

15. Find the position equation $s = \frac{1}{2}at^2 + v_0 t + s_0$ given that $s = 12$ feet after 1 second, $s = 5$ feet after 2 seconds, and $s = 4$ after 3 seconds.

16. Graph $x^2 + y^2 \geq 9$.

17. Graph the solution of the system.
$$x + y \leq 6$$
$$x \geq 2$$
$$y \geq 0$$

18. Derive a set of inequalities to describe the triangle with vertices $(0, 0)$, $(0, 7)$, and $(2, 3)$.

19. Find the maximum value of the objective function, $z = 30z + 26y$, subject to the following constraints.
$$x \geq 0$$
$$y \geq 0$$
$$2x + 3y \leq 21$$
$$5x + 3y \leq 30$$

20. Graph the system of inequalities.
$$x^2 + y^2 \leq 4$$
$$(x - 2)^2 + y^2 \geq 4$$

CHAPTER 7
Matrices and Determinants

CHAPTER 7
Matrices and Determinants

Section 7.1   Matrices and Systems of Equations

■ You should be able to use elementary row operations to produce a row-echelon form (or reduced row-echelon form) of a matrix.

1. Interchange two rows.

2. Multiply a row by a nonzero constant.

3. Add a multiple of one row to another row.

■ You should be able to use either Gaussian elimination with back-substitution or Gauss-Jordan elimination to solve a system of linear equations.

Solutions to Odd-Numbered Exercises

1. Since the matrix has three rows and two columns, its order is 3×2.

3. Since the matrix has three rows and one column, its order is 3×1.

5. Since the matrix has two rows and two columns, its order is 2×2.

7. $4x - 3y = -5$
$-x + 3y = 12$

$$\begin{bmatrix} 4 & -3 & \vdots & -5 \\ -1 & 3 & \vdots & 12 \end{bmatrix}$$

9. $x + 10y - 2z = 2$
$5x - 3y + 4z = 0$
$2x + y = 6$

$$\begin{bmatrix} 1 & 10 & -2 & \vdots & 2 \\ 5 & -3 & 4 & \vdots & 0 \\ 2 & 1 & 0 & \vdots & 6 \end{bmatrix}$$

11. $\begin{bmatrix} 1 & 2 & \vdots & 7 \\ 2 & -3 & \vdots & 4 \end{bmatrix}$

$x + 2y = 7$
$2x - 3y = 4$

13. $\begin{bmatrix} 2 & 0 & 5 & \vdots & -12 \\ 0 & 1 & -2 & \vdots & 7 \\ 6 & 3 & 0 & \vdots & 2 \end{bmatrix}$

$2x + 5z = -12$
$y - 2z = 7$
$6x + 3y = 2$

15. $\begin{bmatrix} 1 & 0 & 0 & 0 \\ 0 & 1 & 1 & 5 \\ 0 & 0 & 0 & 0 \end{bmatrix}$

This matrix is in reduced row-echelon form.

17. $\begin{bmatrix} 2 & 0 & 4 & 0 \\ 0 & -1 & 3 & 6 \\ 0 & 0 & 1 & 5 \end{bmatrix}$

The first nonzero entries in rows one and two are not one. The matrix is not in row-echelon form.

19. $\begin{bmatrix} 1 & 4 & 3 \\ 2 & 10 & 5 \end{bmatrix}$

$-2R_1 + R_2 \rightarrow \begin{bmatrix} 1 & 4 & 3 \\ 0 & \boxed{2} & -1 \end{bmatrix}$

21. $\begin{bmatrix} 1 & 1 & 4 & -1 \\ 3 & 8 & 10 & 3 \\ -2 & 1 & 12 & 6 \end{bmatrix}$

$\begin{aligned} -3R_1 + R_2 &\rightarrow \\ 2R_1 + R_3 &\rightarrow \end{aligned} \begin{bmatrix} 1 & 1 & 4 & -1 \\ 0 & 5 & \boxed{-2} & \boxed{6} \\ 0 & 3 & \boxed{20} & \boxed{4} \end{bmatrix}$

$\tfrac{1}{5}R_2 \rightarrow \begin{bmatrix} 1 & 1 & 4 & -1 \\ 0 & 1 & -\tfrac{2}{5} & \tfrac{6}{5} \\ 0 & 3 & 20 & 4 \end{bmatrix}$

318

23. $\begin{bmatrix} 1 & 2 & 3 \\ 2 & -1 & -4 \\ 3 & 1 & -1 \end{bmatrix}$

(a) $\begin{bmatrix} 1 & 2 & 3 \\ 0 & -5 & -10 \\ 3 & 1 & -1 \end{bmatrix}$

(b) $\begin{bmatrix} 1 & 2 & 3 \\ 0 & -5 & -10 \\ 0 & -5 & -10 \end{bmatrix}$

(c) $\begin{bmatrix} 1 & 2 & 3 \\ 0 & -5 & -10 \\ 0 & 0 & 0 \end{bmatrix}$

(d) $\begin{bmatrix} 1 & 2 & 3 \\ 0 & 1 & 2 \\ 0 & 0 & 0 \end{bmatrix}$

(e) $\begin{bmatrix} 1 & 0 & -1 \\ 0 & 1 & 2 \\ 0 & 0 & 0 \end{bmatrix}$ This matrix is in reduced row-echelon form.

25. $\begin{bmatrix} 1 & 1 & 0 & 5 \\ -2 & -1 & 2 & -10 \\ 3 & 6 & 7 & 14 \end{bmatrix}$

$\begin{matrix} 2R_1 + R_2 \to \\ -3R_1 + R_3 \to \end{matrix} \begin{bmatrix} 1 & 1 & 0 & 5 \\ 0 & 1 & 2 & 0 \\ 0 & 3 & 7 & -1 \end{bmatrix}$

$-3R_2 + R_3 \to \begin{bmatrix} 1 & 1 & 0 & 5 \\ 0 & 1 & 2 & 0 \\ 0 & 0 & 1 & -1 \end{bmatrix}$

27. $\begin{bmatrix} 1 & -1 & -1 & 1 \\ 5 & -4 & 1 & 8 \\ -6 & 8 & 18 & 0 \end{bmatrix}$

$\begin{matrix} -5R_1 + R_2 \to \\ 6R_1 + R_3 \to \end{matrix} \begin{bmatrix} 1 & -1 & -1 & 1 \\ 0 & 1 & 6 & 3 \\ 0 & 2 & 12 & 6 \end{bmatrix}$

$-2R_2 + R_3 \to \begin{bmatrix} 1 & -1 & -1 & 1 \\ 0 & 1 & 6 & 3 \\ 0 & 0 & 0 & 0 \end{bmatrix}$

29. $\begin{bmatrix} 3 & 3 & 3 \\ -1 & 0 & -4 \\ 2 & 4 & -2 \end{bmatrix}$

$\tfrac{1}{3}R_1 \to \begin{bmatrix} 1 & 1 & 1 \\ -1 & 0 & -4 \\ 2 & 4 & -2 \end{bmatrix}$

$\begin{matrix} R_1 + R_2 \to \\ -2R_1 + R_3 \to \end{matrix} \begin{bmatrix} 1 & 1 & 1 \\ 0 & 1 & -3 \\ 0 & 2 & -4 \end{bmatrix}$

$\begin{matrix} -R_2 + R_1 \to \\ -2R_2 + R_3 \to \end{matrix} \begin{bmatrix} 1 & 0 & 4 \\ 0 & 1 & -3 \\ 0 & 0 & 2 \end{bmatrix}$

$\tfrac{1}{2}R_3 \to \begin{bmatrix} 1 & 0 & 4 \\ 0 & 1 & -3 \\ 0 & 0 & 1 \end{bmatrix}$

$\begin{matrix} -4R_3 + R_1 \to \\ 3R_3 + R_2 \to \end{matrix} \begin{bmatrix} 1 & 0 & 0 \\ 0 & 1 & 0 \\ 0 & 0 & 1 \end{bmatrix}$

31. $\begin{bmatrix} 1 & 2 & 3 & -5 \\ 1 & 2 & 4 & -9 \\ -2 & -4 & -4 & 3 \\ 4 & 8 & 11 & -14 \end{bmatrix}$

$\begin{matrix} -R_1 + R_2 \to \\ 2R_1 + R_3 \to \\ -4R_1 + R_4 \to \end{matrix} \begin{bmatrix} 1 & 2 & 3 & -5 \\ 0 & 0 & 1 & -4 \\ 0 & 0 & 2 & -7 \\ 0 & 0 & -1 & 6 \end{bmatrix}$

$\begin{matrix} -3R_2 + R_1 \to \\ \\ -R_2 + R_3 \to \\ R_2 + R_4 \to \end{matrix} \begin{bmatrix} 1 & 2 & 0 & 7 \\ 0 & 0 & 1 & -4 \\ 0 & 0 & 0 & 1 \\ 0 & 0 & 0 & 2 \end{bmatrix}$

$\begin{matrix} -7R_3 + R_1 \to \\ 4R_3 + R_2 \to \\ \\ -2R_3 + R_4 \to \end{matrix} \begin{bmatrix} 1 & 2 & 0 & 0 \\ 0 & 0 & 1 & 0 \\ 0 & 0 & 0 & 1 \\ 0 & 0 & 0 & 0 \end{bmatrix}$

33.
$$x - 2y = 4$$
$$y = -3$$
$$x - 2(-3) = 4$$
$$x = -2$$

Answer: $(-2, -3)$

35.
$$x - y + 2z = 4$$
$$y - z = 2$$
$$z = -2$$
$$y - (-2) = 2$$
$$y = 0$$
$$x - 0 + 2(-2) = 4$$
$$x = 8$$

Answer: $(8, 0, -2)$

37. $\begin{bmatrix} 1 & 0 & \vdots & 7 \\ 0 & 1 & \vdots & -5 \end{bmatrix}$

$$x = 7$$
$$y = -5$$

Answer: $(7, -5)$

39. $\begin{bmatrix} 1 & 0 & 0 & \vdots & -4 \\ 0 & 1 & 0 & \vdots & -8 \\ 0 & 0 & 1 & \vdots & 2 \end{bmatrix}$

$$x = -4$$
$$y = -8$$
$$z = 2$$

Answer: $(-4, -8, 2)$

41.
$$x + 2y = 7$$
$$2x + y = 8$$

$$\begin{bmatrix} 1 & 2 & \vdots & 7 \\ 2 & 1 & \vdots & 8 \end{bmatrix}$$

$$-2R_1 + R_2 \rightarrow \begin{bmatrix} 1 & 2 & \vdots & 7 \\ 0 & -3 & \vdots & -6 \end{bmatrix}$$

$$-\tfrac{1}{3}R_2 \rightarrow \begin{bmatrix} 1 & 2 & \vdots & 7 \\ 0 & 1 & \vdots & 2 \end{bmatrix}$$

$$y = 2$$
$$x + 2(2) = 7 \implies x = 3$$

Answer: $(3, 2)$

43.
$$-3x + 5y = -22$$
$$3x + 4y = 4$$
$$4x - 8y = 32$$

$$\begin{bmatrix} -3 & 5 & \vdots & -22 \\ 3 & 4 & \vdots & 4 \\ 4 & -8 & \vdots & 32 \end{bmatrix}$$

$$R_3 + R_1 \rightarrow \begin{bmatrix} 1 & -3 & \vdots & 10 \\ 3 & 4 & \vdots & 4 \\ 4 & -8 & \vdots & 32 \end{bmatrix}$$

$$\begin{matrix} -3R_1 + R_2 \rightarrow \\ -4R_1 + R_3 \rightarrow \end{matrix} \begin{bmatrix} 1 & -3 & \vdots & 10 \\ 0 & 13 & \vdots & -26 \\ 0 & 4 & \vdots & -8 \end{bmatrix}$$

$$\begin{matrix} \tfrac{1}{13}R_2 \rightarrow \\ -4R_2 + R_3 \rightarrow \end{matrix} \begin{bmatrix} 1 & -3 & \vdots & 10 \\ 0 & 1 & \vdots & -2 \\ 0 & 0 & \vdots & 0 \end{bmatrix}$$

$$y = -2$$
$$x - 3(-2) = 10 \implies x = 4$$

Answer: $(4, -2)$

45.
$$8x - 4y = 7$$
$$5x + 2y = 1$$

$$\begin{bmatrix} 8 & -4 & \vdots & 7 \\ 5 & 2 & \vdots & 1 \end{bmatrix}$$

$$\begin{matrix} 3R_1 \rightarrow \\ 5R_2 \rightarrow \end{matrix} \begin{bmatrix} 24 & -12 & \vdots & 21 \\ 25 & 10 & \vdots & 5 \end{bmatrix}$$

$$-R_2 + R_1 \rightarrow \begin{bmatrix} -1 & -22 & \vdots & 16 \\ 25 & 10 & \vdots & 5 \end{bmatrix}$$

$$25R_1 + R_2 \rightarrow \begin{bmatrix} -1 & -22 & \vdots & 16 \\ 0 & -540 & \vdots & 405 \end{bmatrix}$$

$$\begin{matrix} -R_1 \rightarrow \\ -\tfrac{1}{540}R_2 \rightarrow \end{matrix} \begin{bmatrix} 1 & 22 & \vdots & -16 \\ 0 & 1 & \vdots & -\tfrac{3}{4} \end{bmatrix}$$

$$y = -\tfrac{3}{4}$$
$$x + 22\left(-\tfrac{3}{4}\right) = -16 \implies x = \tfrac{1}{2}$$

Answer: $\left(\tfrac{1}{2}, -\tfrac{3}{4}\right)$

47.
$$-x + 2y = 1.5$$
$$2x - 4y = 3.0$$

$$\begin{bmatrix} -1 & 2 & \vdots & 1.5 \\ 2 & -4 & \vdots & 3.0 \end{bmatrix}$$

$$2R_1 + R_2 \rightarrow \begin{bmatrix} -1 & 2 & \vdots & 1.5 \\ 0 & 0 & \vdots & 6.0 \end{bmatrix}$$

The system is inconsistent and there is no solution.

49.
$$x \qquad - 3z = -2$$
$$3x + y - 2z = 5$$
$$2x + 2y + z = 4$$

$$\begin{bmatrix} 1 & 0 & -3 & \vdots & -2 \\ 3 & 1 & -2 & \vdots & 5 \\ 2 & 2 & 1 & \vdots & 4 \end{bmatrix}$$

$$\begin{matrix} \\ -3R_1 + R_2 \to \\ -2R_1 + R_3 \to \end{matrix} \begin{bmatrix} 1 & 0 & -3 & \vdots & -2 \\ 0 & 1 & 7 & \vdots & 11 \\ 0 & 2 & 7 & \vdots & 8 \end{bmatrix}$$

$$\begin{matrix} \\ \\ -2R_2 + R_3 \to \end{matrix} \begin{bmatrix} 1 & 0 & -3 & \vdots & -2 \\ 0 & 1 & 7 & \vdots & 11 \\ 0 & 0 & -7 & \vdots & -14 \end{bmatrix}$$

$$\begin{matrix} \\ \\ -\tfrac{1}{7}R_3 \to \end{matrix} \begin{bmatrix} 1 & 0 & -3 & \vdots & -2 \\ 0 & 1 & 7 & \vdots & 11 \\ 0 & 0 & 1 & \vdots & 2 \end{bmatrix}$$

$z = 2$

$y + 7(2) = 11 \implies y = -3$

$x - 3(2) = -2 \implies x = 4$

Answer: $(4, -3, 2)$

51.
$$x + y - 5z = 3$$
$$x \qquad - 2z = 1$$
$$2x - y - z = 0$$

$$\begin{bmatrix} 1 & 1 & -5 & \vdots & 3 \\ 1 & 0 & -2 & \vdots & 1 \\ 2 & -1 & -1 & \vdots & 0 \end{bmatrix}$$

$$\begin{matrix} \\ -R_1 + R_2 \to \\ -2R_1 + R_3 \to \end{matrix} \begin{bmatrix} 1 & 1 & -5 & \vdots & 3 \\ 0 & -1 & 3 & \vdots & -2 \\ 0 & -3 & 9 & \vdots & -6 \end{bmatrix}$$

$$\begin{matrix} \\ \\ -3R_2 + R_3 \to \end{matrix} \begin{bmatrix} 1 & 1 & -5 & \vdots & 3 \\ 0 & -1 & 3 & \vdots & -2 \\ 0 & 0 & 0 & \vdots & 0 \end{bmatrix}$$

$$\begin{matrix} R_2 + R_1 \to \\ -R_2 \to \\ \end{matrix} \begin{bmatrix} 1 & 0 & -2 & \vdots & 1 \\ 0 & 1 & -3 & \vdots & 2 \\ 0 & 0 & 0 & \vdots & 0 \end{bmatrix}$$

$z = a$

$y - 3a = 2 \implies y = 3a + 2$

$x - 2a = 1 \implies x = 2a + 1$

Answer: $(2a + 1, 3a + 2, a)$

53.
$$x + 2y + z = 8$$
$$3x + 7y + 6z = 26$$

$$\begin{bmatrix} 1 & 2 & 1 & \vdots & 8 \\ 3 & 7 & 6 & \vdots & 26 \end{bmatrix}$$

$$\begin{matrix} \\ -3R_1 + R_2 \to \end{matrix} \begin{bmatrix} 1 & 2 & 1 & \vdots & 8 \\ 0 & 1 & 3 & \vdots & 2 \end{bmatrix}$$

$$\begin{matrix} -2R_2 + R_1 \to \\ \end{matrix} \begin{bmatrix} 1 & 0 & -5 & \vdots & 4 \\ 0 & 1 & 3 & \vdots & 2 \end{bmatrix}$$

$z = a$

$y + 3a = 2 \implies y = -3a + 2$

$x - 5a = 4 \implies x = 5a + 4$

Answer: $(5a + 4, -3a + 2, a)$

55.
$$x + 2y = 0$$
$$-x - y = 0$$

$$\begin{bmatrix} 1 & 2 & \vdots & 0 \\ -1 & -1 & \vdots & 0 \end{bmatrix}$$

$$\begin{matrix} R_1 + R_2 \to \\ \end{matrix} \begin{bmatrix} 1 & 2 & \vdots & 0 \\ 0 & 1 & \vdots & 0 \end{bmatrix}$$

$y = 0, \ x + 2(0) = 0 \implies x = 0$

Answer: $(0, 0)$

57.
$$3x + 3y + 12z = 6$$
$$x + y + 4z = 2$$
$$2x + 5y + 20z = 10$$
$$-x + 2y + 8z = 4$$

$$\begin{bmatrix} 3 & 3 & 12 & \vdots & 6 \\ 1 & 1 & 4 & \vdots & 2 \\ 2 & 5 & 20 & \vdots & 10 \\ -1 & 2 & 8 & \vdots & 4 \end{bmatrix} \implies \begin{bmatrix} 1 & 0 & 0 & \vdots & 0 \\ 0 & 0 & 0 & \vdots & 0 \\ 0 & 1 & 4 & \vdots & 2 \\ 0 & 0 & 0 & \vdots & 0 \end{bmatrix}$$

$z = a$

$y = -4a + 2$

$x = 0$

Answer: $(0, -4a + 2, a)$

59.
$$\begin{aligned} 2x + y - z + 2w &= -6 \\ 3x + 4y \phantom{{}+{}} + w &= 1 \\ x + 5y + 2z + 6w &= -3 \\ 5x + 2y - z - w &= 3 \end{aligned}$$
$$\begin{bmatrix} 2 & 1 & -1 & 2 & \vdots & -6 \\ 3 & 4 & 0 & 1 & \vdots & 1 \\ 1 & 5 & 2 & 6 & \vdots & -3 \\ 5 & 2 & -1 & -1 & \vdots & 3 \end{bmatrix} \Rightarrow \begin{bmatrix} 1 & 5 & 2 & 6 & \vdots & -3 \\ 0 & 1 & -1 & -3 & \vdots & 2 \\ 0 & 0 & 238 & 629 & \vdots & -306 \\ 0 & 0 & 0 & -71 & \vdots & 142 \end{bmatrix}$$

$x = 1$

$y = 0$

$z = 4$

$w = -2$

Answer: $(1, 0, 4, -2)$

61.
$$\begin{aligned} x + y + z &= 0 \\ 2x + 3y + z &= 0 \\ 3x + 5y + z &= 0 \end{aligned}$$
$$\begin{bmatrix} 1 & 1 & 1 & \vdots & 0 \\ 2 & 3 & 1 & \vdots & 0 \\ 3 & 5 & 1 & \vdots & 0 \end{bmatrix} \Rightarrow \begin{bmatrix} 1 & 0 & 2 & \vdots & 0 \\ 0 & 1 & -1 & \vdots & 0 \\ 0 & 0 & 0 & \vdots & 0 \end{bmatrix}$$

$z = a$

$y = a$

$x = -2a$

Answer: $(-2a, a, a)$

63. $z = a$

$y = -4a + 1$

$x = -3a - 2$

One possible system is:

$$\begin{aligned} x + y + 7z &= (-3a - 2) + (-4a + 1) + 7a = -1 \\ x + 2y + 11z &= (-3a - 2) + 2(-4a + 1) + 11a = -0 \\ 2x + y + 10z &= 2(-3a - 2) + (-4a + 1) + 10a = -3 \end{aligned}$$

65. $x =$ amount at 8%, $y =$ amount at 9%, $z =$ amount at 12%

$$\begin{aligned} x + y + z &= 1{,}500{,}000 \\ 0.08x + 0.09y + 0.12z &= 133{,}000 \\ x \phantom{{}+0.09y} - 4z &= 0 \end{aligned}$$

$$\begin{bmatrix} 1 & 1 & 1 & \vdots & 1{,}500{,}000 \\ 0.08 & 0.09 & 0.12 & \vdots & 133{,}000 \\ 1 & 0 & .-4 & \vdots & 0 \end{bmatrix}$$

$$\begin{matrix} -0.08R_1 + R_2 \to \\ -R_1 + R_3 \to \end{matrix} \begin{bmatrix} 1 & 1 & 1 & \vdots & 1{,}500{,}000 \\ 0 & 0.01 & 0.04 & \vdots & 13{,}000 \\ 0 & -1 & -5 & \vdots & -1{,}500{,}000 \end{bmatrix}$$

$$\begin{matrix} 100R_2 \to \\ R_2 + R_3 \to \end{matrix} \begin{bmatrix} 1 & 1 & 1 & \vdots & 1{,}500{,}000 \\ 0 & 1 & 4 & \vdots & 1{,}300{,}000 \\ 0 & 0 & -1 & \vdots & -200{,}000 \end{bmatrix}$$

$-z = -200{,}000 \implies z = 200{,}000$

$y + 4(200{,}000) = 1{,}300{,}000 \implies y = 500{,}000$

$x + (500{,}000) + (200{,}000) = 1{,}500{,}000 \implies x = 800{,}000$

Answer: \$800,000 at 8%, \$500,000 at 9%, \$200,000 at 12%

67. $\dfrac{4x^2}{(x+1)^2(x-1)} = \dfrac{A}{x-1} + \dfrac{B}{x+1} + \dfrac{C}{(x+1)^2}$

$4x^2 = A(x+1)^2 + B(x+1)(x-1) + C(x-1)$

Let $x = 1$: $4 = \quad 4A \implies A = 1$

Let $x = -1$: $4 = -2C \implies C = -2$

Let $x = 0$: $0 = A - B - C \implies 0 = 1 - B - (-2) \implies B = 3$

Thus, $\dfrac{4x^2}{(x+1)^2(x-1)} = \dfrac{1}{x-1} + \dfrac{3}{x+1} - \dfrac{2}{(x+1)^2}$.

69. $f(x) = ax^2 + bx + c$

$f(1) = a + b + c = 8$

$f(2) = 4a + 2b + c = 13$

$f(3) = 9a + 3b + c = 20$

$$\begin{bmatrix} 1 & 1 & 1 & \vdots & 8 \\ 4 & 2 & 1 & \vdots & 13 \\ 9 & 3 & 1 & \vdots & 20 \end{bmatrix}$$

$$\begin{matrix} \\ -4R_1 + R_2 \rightarrow \\ -9R_1 + R_3 \rightarrow \end{matrix} \begin{bmatrix} 1 & 1 & 1 & \vdots & 8 \\ 0 & -2 & -3 & \vdots & -19 \\ 0 & -6 & -8 & \vdots & -52 \end{bmatrix}$$

$$\begin{matrix} \\ -\frac{1}{2}R_2 \rightarrow \\ -3R_2 + R_3 \rightarrow \end{matrix} \begin{bmatrix} 1 & 1 & 1 & \vdots & 8 \\ 0 & 1 & \frac{3}{2} & \vdots & \frac{19}{2} \\ 0 & 0 & 1 & \vdots & 5 \end{bmatrix}$$

$c = 5$

$b + \frac{3}{2}(5) = \frac{19}{2} \implies b = 2$

$a + 2 + 5 = 8 \implies a = 1$

Answer: $y = x^2 + 2x + 5$

71. (a) $(0, 5.0)$, $(15, 9.6)$, $(30, 12.4)$

$f(x) = ax^2 + bx + c$

$f(0) = c = 5$

$f(15) = 225a + 15b + c = \quad 9.6 \implies 225a + 15b = 4.6$

$f(30) = 900a + 30b + c = 12.4 \implies 900a + 30b = 7.4$

$$\begin{bmatrix} 225 & 15 & \vdots & 4.6 \\ 900 & 30 & \vdots & 7.4 \end{bmatrix}$$

$$\begin{matrix} \frac{1}{225}R_1 \rightarrow \\ -900R_1 + R_2 \rightarrow \end{matrix} \begin{bmatrix} 1 & \frac{1}{15} & \vdots & \frac{23}{1125} \\ 0 & -30 & \vdots & -11 \end{bmatrix}$$

$-30b = -11 \implies b = \dfrac{11}{30} \approx 0.367$

$1 + \dfrac{1}{15}\left(\dfrac{11}{30}\right) = \dfrac{23}{1125} \implies a = -\dfrac{1}{250} = -0.004$

Thus, $y = -0.004x^2 + 0.367x + 5$.

— CONTINUED —

71. — CONTINUED —

(b)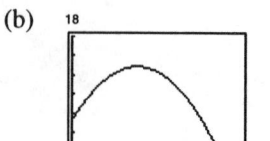

The maximum height is approximately 13 feet and the ball strikes the ground at approximately 104 feet.

(c) The maximum occurs at the vertex.

$$-\frac{b}{2a} = \frac{-0.367}{2(-0.004)} = 45.875$$

$$f(45.875) = -0.004(45.875)^2 + 0.367(45.875) + 5 = 13.418 \text{ feet}$$

The ball strikes the ground when $y = 0$.

$$-0.004x^2 + 0.367x + 5 = 0$$

By the Quadratic Formula and using the positive value for x we have $x \approx 103.793$ feet.

73. (a) $x_1 + x_3 = 600$

$x_1 = x_2 + x_4 \implies x_1 - x_2 - x_4 = 0$

$x_2 + x_5 = 500$

$x_3 + x_6 = 600$

$x_4 + x_7 = x_6 \implies x_4 - x_6 + x_7 = 0$

$x_5 + x_7 = 500$

$$\left[\begin{array}{ccccccc:c}
1 & 0 & 1 & 0 & 0 & 0 & 0 & 600 \\
1 & -1 & 0 & -1 & 0 & 0 & 0 & 0 \\
0 & 1 & 0 & 0 & 1 & 0 & 0 & 500 \\
0 & 0 & 1 & 0 & 0 & 1 & 0 & 600 \\
0 & 0 & 0 & 1 & 0 & -1 & 1 & 0 \\
0 & 0 & 0 & 0 & 1 & 0 & 1 & 500
\end{array}\right]$$

$$\begin{array}{l}
\\
-R_1 + R_2 \rightarrow \\
R_2 + R_3 \rightarrow \\
R_3 + R_4 \rightarrow \\
R_4 + R_5 \rightarrow \\
-R_5 + R_6 \rightarrow
\end{array}
\left[\begin{array}{ccccccc:c}
1 & 0 & 1 & 0 & 0 & 0 & 0 & 600 \\
0 & -1 & -1 & -1 & 0 & 0 & 0 & -600 \\
0 & 0 & -1 & -1 & 1 & 0 & 0 & -100 \\
0 & 0 & 0 & -1 & 1 & 1 & 0 & 500 \\
0 & 0 & 0 & 0 & 1 & 0 & 1 & 500 \\
0 & 0 & 0 & 0 & 0 & 0 & 0 & 0
\end{array}\right]$$

$$\begin{array}{l}
\\
-R_3 + R_2 \rightarrow \\
-R_4 + R_3 \rightarrow \\
-R_4 \rightarrow \\
\\
\\
\end{array}
\left[\begin{array}{ccccccc:c}
1 & 0 & 1 & 0 & 0 & 0 & 0 & 600 \\
0 & -1 & 0 & 0 & -1 & 0 & 0 & -500 \\
0 & 0 & -1 & 0 & 0 & -1 & 0 & -600 \\
0 & 0 & 0 & 1 & -1 & -1 & 0 & -500 \\
0 & 0 & 0 & 0 & 1 & 0 & 1 & 500 \\
0 & 0 & 0 & 0 & 0 & 0 & 0 & 0
\end{array}\right]$$

Let $x_7 = t$ and $x_6 = s$, then $x_5 = 500 - t$,

$x_4 = -500 + s + (500 - t) = s - t$,

$x_3 = 600 - s, x_2 = 500 - (500 - t) = t, x_1 = 600 - (600 - s) = s$.

Answer: $(s, t, 600 - s, s - t, 500 - t, s, t)$

(b) $s = 0, t = 0$: $x_1 = 0, x_2 = 0, x_3 = 600, x_4 = 0, x_5 = 500, x_6 = 0, x_7 = 0$

(c) $s = 0, t = -500$: $x_1 = 0, x_2 = -500, x_3 = 600, x_4 = 500, x_5 = 1000, x_6 = 0, x_7 = -500$

75. $f(x) = 2^{x-1}$

x	-1	0	1	2	3
y	$\frac{1}{4}$	$\frac{1}{2}$	1	2	4

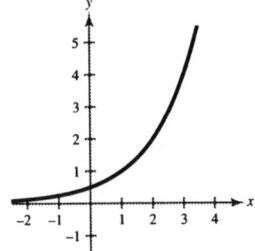

77. $h(x) = \log_2(x-1) \implies 2^y = x - 1 \implies 2^y + 1 = x$

x	$\frac{3}{2}$	2	3	5	9
y	-1	0	1	2	3

Vertical asymptote: $x = 1$

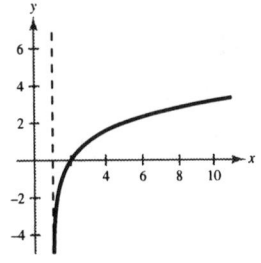

Section 7.2   Operations with Matrices

- $A = B$ if and only if they have the same order and $a_{ij} = b_{ij}$.
- You should be able to perform the operations of matrix addition, scalar multiplication, and matrix multiplication.
- Some properties of matrix addition and scalar multiplication are:
  (a) $A + B = B + A$
  (b) $A + (B + C) = (A + B) + C$
  (c) $(cd)A = c(dA)$
  (d) $1A = A$
  (e) $c(A + B) = cA + cB$
  (f) $(c + d)A = cA + dA$
- You should remember that $AB \neq BA$ in general.

Solutions to Odd-Numbered Exercises

1. $x = -4$, $y = 22$

3. $2x + 1 = 5$, $3y - 5 = 4$

$x = 2$, $y = 3$

5. (a) $A + B = \begin{bmatrix} 1 & -1 \\ 2 & -1 \end{bmatrix} + \begin{bmatrix} 2 & -1 \\ -1 & 8 \end{bmatrix} = \begin{bmatrix} 1+2 & -1-1 \\ 2-1 & -1+8 \end{bmatrix} = \begin{bmatrix} 3 & -2 \\ 1 & 7 \end{bmatrix}$

(b) $A - B = \begin{bmatrix} 1 & -1 \\ 2 & -1 \end{bmatrix} - \begin{bmatrix} 2 & -1 \\ -1 & 8 \end{bmatrix} = \begin{bmatrix} 1-2 & -1+1 \\ 2+1 & -1-8 \end{bmatrix} = \begin{bmatrix} -1 & 0 \\ 3 & -9 \end{bmatrix}$

— CONTINUED —

5. — CONTINUED —

(c) $3A = 3\begin{bmatrix} 1 & -1 \\ 2 & -1 \end{bmatrix} = \begin{bmatrix} 3(1) & 3(-1) \\ 3(2) & 3(-1) \end{bmatrix} = \begin{bmatrix} 3 & -3 \\ 6 & -3 \end{bmatrix}$

(d) $3A - 2B = \begin{bmatrix} 3 & -3 \\ 6 & -3 \end{bmatrix} - 2\begin{bmatrix} 2 & -1 \\ -1 & 8 \end{bmatrix} = \begin{bmatrix} 3 & -3 \\ 6 & -3 \end{bmatrix} + \begin{bmatrix} -4 & 2 \\ 2 & -16 \end{bmatrix} = \begin{bmatrix} -1 & -1 \\ 8 & -19 \end{bmatrix}$

7. $A = \begin{bmatrix} 6 & -1 \\ 2 & 4 \\ -3 & 5 \end{bmatrix}, B = \begin{bmatrix} 1 & 4 \\ -1 & 5 \\ 1 & 10 \end{bmatrix}$

(a) $A + B = \begin{bmatrix} 7 & 3 \\ 1 & 9 \\ -2 & 15 \end{bmatrix}$
    (b) $A - B = \begin{bmatrix} 5 & -5 \\ 3 & -1 \\ -4 & -5 \end{bmatrix}$
    (c) $3A = \begin{bmatrix} 18 & -3 \\ 6 & 12 \\ -9 & 15 \end{bmatrix}$

(d) $3A - 2B = \begin{bmatrix} 18 & -3 \\ 6 & 12 \\ -9 & 15 \end{bmatrix} - \begin{bmatrix} 2 & 8 \\ -2 & 10 \\ 2 & 20 \end{bmatrix} = \begin{bmatrix} 16 & -11 \\ 8 & 2 \\ -11 & -5 \end{bmatrix}$

9. $A = \begin{bmatrix} 2 & 2 & -1 & 0 & 1 \\ 1 & 1 & -2 & 0 & -1 \end{bmatrix}, B = \begin{bmatrix} 1 & 1 & -1 & 1 & 0 \\ -3 & 4 & 9 & -6 & -7 \end{bmatrix}$

(a) $A + B = \begin{bmatrix} 3 & 3 & -2 & 1 & 1 \\ -2 & 5 & 7 & -6 & -8 \end{bmatrix}$

(b) $A - B = \begin{bmatrix} 1 & 1 & 0 & -1 & 1 \\ 4 & -3 & -11 & 6 & 6 \end{bmatrix}$

(c) $3A = \begin{bmatrix} 6 & 6 & -3 & 0 & 3 \\ 3 & 3 & -6 & 0 & -3 \end{bmatrix}$

(d) $3A - 2B = \begin{bmatrix} 6 & 6 & -3 & 0 & 3 \\ 3 & 3 & -6 & 0 & -3 \end{bmatrix} - \begin{bmatrix} 2 & 2 & -2 & 2 & 0 \\ -6 & 8 & 18 & -12 & -14 \end{bmatrix} = \begin{bmatrix} 4 & 4 & -1 & -2 & 3 \\ 9 & -5 & -24 & 12 & 11 \end{bmatrix}$

11. $X = 3\begin{bmatrix} -2 & -1 \\ 1 & 0 \\ 3 & 4 \end{bmatrix} - 2\begin{bmatrix} 0 & 3 \\ 2 & 0 \\ -4 & -1 \end{bmatrix} = \begin{bmatrix} -6 & -3 \\ 3 & 0 \\ 9 & -12 \end{bmatrix} - \begin{bmatrix} 0 & 6 \\ 4 & 0 \\ -8 & -2 \end{bmatrix} = \begin{bmatrix} -6 & -9 \\ -1 & 0 \\ 17 & -10 \end{bmatrix}$

13. $X = -\frac{3}{2}A + \frac{1}{2}B = -\frac{3}{2}\begin{bmatrix} -2 & -1 \\ 1 & 0 \\ 3 & -4 \end{bmatrix} + \frac{1}{2}\begin{bmatrix} 0 & 3 \\ 2 & 0 \\ -4 & -1 \end{bmatrix} = \begin{bmatrix} 3 & 3 \\ -\frac{1}{2} & 0 \\ -\frac{13}{2} & \frac{11}{2} \end{bmatrix}$

15. (a) $AB = \begin{bmatrix} 1 & 2 \\ 4 & 2 \end{bmatrix}\begin{bmatrix} 2 & -1 \\ -1 & 8 \end{bmatrix} = \begin{bmatrix} 2-2 & -1+16 \\ 8-2 & -4+16 \end{bmatrix} = \begin{bmatrix} 0 & 15 \\ 6 & 12 \end{bmatrix}$

(b) $BA = \begin{bmatrix} 2 & -1 \\ -1 & 8 \end{bmatrix}\begin{bmatrix} 1 & 2 \\ 4 & 2 \end{bmatrix} = \begin{bmatrix} 2-4 & 4-2 \\ -1+32 & -2+16 \end{bmatrix} = \begin{bmatrix} -2 & 2 \\ 31 & 14 \end{bmatrix}$

(c) $A^2 = \begin{bmatrix} 1 & 2 \\ 4 & 2 \end{bmatrix}\begin{bmatrix} 1 & 2 \\ 4 & 2 \end{bmatrix} = \begin{bmatrix} 1+8 & 2+4 \\ 4+8 & 8+4 \end{bmatrix} = \begin{bmatrix} 9 & 6 \\ 12 & 12 \end{bmatrix}$

17. (a) $AB = \begin{bmatrix} 3 & -1 \\ 1 & 3 \end{bmatrix} \begin{bmatrix} 1 & -3 \\ 3 & 1 \end{bmatrix} = \begin{bmatrix} 3-3 & -9-1 \\ 1+9 & -3+3 \end{bmatrix} = \begin{bmatrix} 0 & -10 \\ 10 & 0 \end{bmatrix}$

(b) $BA = \begin{bmatrix} 1 & -3 \\ 3 & 1 \end{bmatrix} \begin{bmatrix} 3 & -1 \\ 1 & 3 \end{bmatrix} = \begin{bmatrix} 3-3 & -1-9 \\ 9+1 & -3+3 \end{bmatrix} = \begin{bmatrix} 0 & -10 \\ 10 & 0 \end{bmatrix}$

(c) $A^2 = \begin{bmatrix} 3 & -1 \\ 1 & 3 \end{bmatrix} \begin{bmatrix} 3 & -1 \\ 1 & 3 \end{bmatrix} = \begin{bmatrix} 9-1 & -3-3 \\ 3+3 & -1+9 \end{bmatrix} = \begin{bmatrix} 8 & -6 \\ 6 & 8 \end{bmatrix}$

19. (a) $AB = \begin{bmatrix} 1 & -1 & 7 \\ 2 & -1 & 8 \\ 3 & 1 & -1 \end{bmatrix} \begin{bmatrix} 1 & 1 & 2 \\ 2 & 1 & 1 \\ 1 & -3 & 2 \end{bmatrix} = \begin{bmatrix} 1-2+7 & 1-1-21 & 2-1+14 \\ 2-2+8 & 2-1-24 & 4-1+16 \\ 3+2-1 & 3+1+3 & 6+1-2 \end{bmatrix} = \begin{bmatrix} 6 & -21 & 15 \\ 8 & -23 & 19 \\ 4 & 7 & 5 \end{bmatrix}$

(b) $BA = \begin{bmatrix} 1 & 1 & 2 \\ 2 & 1 & 1 \\ 1 & -3 & 2 \end{bmatrix} \begin{bmatrix} 1 & -1 & 7 \\ 2 & -1 & 8 \\ 3 & 1 & -1 \end{bmatrix} = \begin{bmatrix} 1+2+6 & -1-1+2 & 7+8-2 \\ 2+2+3 & -2-1+1 & 14+8-1 \\ 1-6+6 & -1+3+2 & 7-24-2 \end{bmatrix} = \begin{bmatrix} 9 & 0 & 13 \\ 7 & -2 & 21 \\ 1 & 4 & -19 \end{bmatrix}$

(c) $A^2 = \begin{bmatrix} 1 & -1 & 7 \\ 2 & -1 & 8 \\ 3 & 1 & -1 \end{bmatrix} \begin{bmatrix} 1 & -1 & 7 \\ 2 & -1 & 8 \\ 3 & 1 & -1 \end{bmatrix} = \begin{bmatrix} 1-2+21 & -1+1+7 & 7-8-7 \\ 2-2+24 & -2+1+8 & 14-8-8 \\ 3+2-3 & -3-1-1 & 21+8+1 \end{bmatrix} = \begin{bmatrix} 20 & 7 & -8 \\ 24 & 7 & -2 \\ 2 & -5 & 30 \end{bmatrix}$

21. A is 3×2 and B is $3 \times 3 \implies AB$ is not defined.

23. A is 3×2, B is $2 \times 2 \implies AB$ is 3×2.

$AB = \begin{bmatrix} -1 & 3 \\ 4 & -5 \\ 0 & 2 \end{bmatrix} \begin{bmatrix} 1 & 2 \\ 0 & 7 \end{bmatrix} = \begin{bmatrix} -1 & 19 \\ 4 & -27 \\ 0 & 14 \end{bmatrix}$

25. A is 3×3, B is $3 \times 3 \implies AB$ is 3×3.

$AB = \begin{bmatrix} 5 & 0 & 0 \\ 0 & -8 & 0 \\ 0 & 0 & 7 \end{bmatrix} \begin{bmatrix} \frac{1}{5} & 0 & 0 \\ 0 & -\frac{1}{8} & 0 \\ 0 & 0 & \frac{1}{2} \end{bmatrix} = \begin{bmatrix} 1 & 0 & 0 \\ 0 & 1 & 0 \\ 0 & 0 & \frac{7}{2} \end{bmatrix}$

27. A is 2×1, B is $1 \times 4 \implies AB$ is 2×4.

$\begin{bmatrix} 10 \\ 12 \end{bmatrix} \begin{bmatrix} 6 & -2 & 1 & 6 \end{bmatrix} = \begin{bmatrix} 60 & -20 & 10 & 60 \\ 72 & -24 & 12 & 72 \end{bmatrix}$

29. $\begin{bmatrix} 5 & 6 & -3 \\ -2 & 5 & 1 \\ 10 & -5 & 5 \end{bmatrix} \begin{bmatrix} 1 & -1 & 2 \\ 8 & 1 & 4 \\ 4 & -2 & 9 \end{bmatrix} = \begin{bmatrix} 41 & 7 & 7 \\ 42 & 5 & 25 \\ -10 & -25 & 45 \end{bmatrix}$

31. $\begin{bmatrix} -3 & 8 & -6 & 8 \\ -12 & 15 & 9 & 6 \\ 5 & -1 & 1 & 5 \end{bmatrix} \begin{bmatrix} 3 & 1 & 6 \\ 24 & 15 & 14 \\ 16 & 10 & 21 \\ 8 & -4 & 10 \end{bmatrix} = \begin{bmatrix} 151 & 25 & 48 \\ 516 & 279 & 387 \\ 47 & -20 & 87 \end{bmatrix}$

33. A is 2×4 and B is $2 \times 4 \implies AB$ is not defined.

35. $A = \begin{bmatrix} -1 & 1 \\ -2 & 1 \end{bmatrix}$, $X = \begin{bmatrix} x \\ y \end{bmatrix}$, $B = \begin{bmatrix} 4 \\ 0 \end{bmatrix}$

By Gauss-Jordan elimination on

$$\begin{bmatrix} -1 & 1 & \vdots & 4 \\ -2 & 1 & \vdots & 0 \end{bmatrix}$$

$$\begin{matrix} -R_1 \to \\ 2R_1 + R_2 \to \end{matrix} \begin{bmatrix} 1 & -1 & \vdots & -4 \\ 0 & -1 & \vdots & -8 \end{bmatrix}$$

$$\begin{matrix} R_2 + R_1 \to \\ -R_2 \to \end{matrix} \begin{bmatrix} 1 & 0 & \vdots & 4 \\ 0 & 1 & \vdots & 8 \end{bmatrix}$$

we have $x = 4$ and $y = 8$.

37. $A = \begin{bmatrix} 2 & 3 \\ 1 & 4 \end{bmatrix}$, $X = \begin{bmatrix} x \\ y \end{bmatrix}$, $B = \begin{bmatrix} 5 \\ 10 \end{bmatrix}$

By Gauss-Jordan elimination on

$$\begin{bmatrix} 1 & 4 & \vdots & 10 \\ 2 & 3 & \vdots & 5 \end{bmatrix}$$

$$-2R_1 + R_2 \to \begin{bmatrix} 1 & 4 & \vdots & 10 \\ 0 & -5 & \vdots & -15 \end{bmatrix}$$

$$\begin{matrix} -4R_2 + R_1 \to \\ -\frac{1}{5}R_2 \to \end{matrix} \begin{bmatrix} 1 & 0 & \vdots & -2 \\ 0 & 1 & \vdots & 3 \end{bmatrix}$$

we have $x = -2$ and $y = 3$.

39. $A = \begin{bmatrix} 2 & 0 \\ 4 & 5 \end{bmatrix}$

$$f(A) = A^2 - 5A + 2 = \begin{bmatrix} 2 & 0 \\ 4 & 5 \end{bmatrix}\begin{bmatrix} 2 & 0 \\ 4 & 5 \end{bmatrix} - 5\begin{bmatrix} 2 & 4 \\ 0 & 5 \end{bmatrix} + 2\begin{bmatrix} 1 & 0 \\ 0 & 1 \end{bmatrix} = \begin{bmatrix} -4 & 0 \\ 8 & 2 \end{bmatrix}$$

41. $A = \begin{bmatrix} 3 & 1 & 4 \\ 0 & 2 & 6 \\ 0 & 0 & 5 \end{bmatrix}$

$$f(A) = \begin{bmatrix} 3 & 1 & 4 \\ 0 & 2 & 6 \\ 0 & 0 & 5 \end{bmatrix}^3 - 10\begin{bmatrix} 3 & 1 & 4 \\ 0 & 2 & 6 \\ 0 & 0 & 5 \end{bmatrix}^2 + 31\begin{bmatrix} 3 & 1 & 4 \\ 0 & 2 & 6 \\ 0 & 0 & 5 \end{bmatrix} - 30\begin{bmatrix} 1 & 0 & 0 \\ 0 & 1 & 0 \\ 0 & 0 & 1 \end{bmatrix} = \begin{bmatrix} 0 & 0 & 0 \\ 0 & 0 & 0 \\ 0 & 0 & 0 \end{bmatrix}$$

43. $AC = \begin{bmatrix} 0 & 1 \\ 0 & 1 \end{bmatrix}\begin{bmatrix} 2 & 3 \\ 2 & 3 \end{bmatrix} = \begin{bmatrix} 2 & 3 \\ 2 & 3 \end{bmatrix}$

$BC = \begin{bmatrix} 1 & 0 \\ 1 & 0 \end{bmatrix}\begin{bmatrix} 2 & 3 \\ 2 & 3 \end{bmatrix} = \begin{bmatrix} 2 & 3 \\ 2 & 3 \end{bmatrix}$

Thus, $AC = BC$ even though $A \neq B$.

For 45–53, A is of order 2×3, B is of order 2×3, C is of order 3×2 and D is of order 2×2.

45. $A + 2C$ is not possible. A and C are not of the same order.

47. AB is not possible. The number of columns of A does not equal the number of rows of B.

49. $BC - D$ is possible. The resulting order is 2×2.

51. (CA) is 3×3 so $(CA)D$ is not possible.

53. $D(A - 3B)$ is possible. The resulting order is 2×3.

55. $1.20\begin{bmatrix} 60 & 40 & 20 \\ 30 & 90 & 60 \end{bmatrix} = \begin{bmatrix} 72 & 48 & 24 \\ 36 & 108 & 72 \end{bmatrix}$

57. $BA = \begin{bmatrix} 3.75 & 7.00 \end{bmatrix}\begin{bmatrix} 100 & 75 & 75 \\ 125 & 150 & 100 \end{bmatrix} = \begin{bmatrix} \$1250.00 & \$1331.25 & \$981.25 \end{bmatrix}$

The entries in the last matrix represent the profits for both crops at the three outlets.

59. $ST = \begin{bmatrix} 3 & 2 & 2 & 3 & 0 \\ 0 & 2 & 3 & 4 & 3 \\ 4 & 2 & 1 & 3 & 2 \end{bmatrix} \begin{bmatrix} 840 & 1100 \\ 1200 & 1350 \\ 1450 & 1650 \\ 2650 & 3000 \\ 3050 & 3200 \end{bmatrix} = \begin{bmatrix} \$15{,}770 & \$18{,}300 \\ \$26{,}500 & \$29{,}250 \\ \$21{,}260 & \$24{,}150 \end{bmatrix}$

The entries represent the wholesale and retail inventory values of the inventories at the three outlets.

61. $ST = \begin{bmatrix} 1 & 0.5 & 0.2 \\ 1.6 & 1.0 & 0.2 \\ 2.5 & 2.0 & 0.4 \end{bmatrix} \begin{bmatrix} 12 & 10 \\ 9 & 8 \\ 6 & 5 \end{bmatrix} = \begin{bmatrix} \$17.70 & \$15.00 \\ \$29.40 & \$25.00 \\ \$50.40 & \$43.00 \end{bmatrix}$

This represents the labor cost for each boat size at each plant.

63. $A^2 = \begin{bmatrix} i & 0 \\ 0 & i \end{bmatrix} \begin{bmatrix} i & 0 \\ 0 & i \end{bmatrix} = \begin{bmatrix} -1 & 0 \\ 0 & -1 \end{bmatrix}$ and $i^2 = -1$

$A^3 = A^2 A = \begin{bmatrix} -1 & 0 \\ 0 & -1 \end{bmatrix} \begin{bmatrix} i & 0 \\ 0 & i \end{bmatrix} = \begin{bmatrix} -i & 0 \\ 0 & -i \end{bmatrix}$ and $i^3 = -i$

$A^4 = A^3 A = \begin{bmatrix} -i & 0 \\ 0 & -i \end{bmatrix} \begin{bmatrix} i & 0 \\ 0 & i \end{bmatrix} = \begin{bmatrix} 1 & 0 \\ 0 & 1 \end{bmatrix}$ and $i^4 = 1$

65. The product of two diagonal matrices of the same order is a diagonal matrix whose entries are the products of the corresponding diagonal entries of A and B.

Section 7.3   The Inverse of a Square Matrix

■ You should be able to find the inverse, if it exists, of a square matrix.

(a) Write the $n \times 2n$ matrix that consists of the given matrix A on the left and the $n \times n$ identity matrix I on the right to obtain $[A \;\vdots\; I]$. Note that we separate the matrices A and I by a dotted line. We call this process **adjoining** the matrices A and I.

(b) If possible, row reduce A to I using elementary row operations of the *entire* matrix $[A \;\vdots\; I]$. The result will be the matrix $[I \;\vdots\; A^{-1}]$. If this is not possible, then A is not invertible.

(c) Check your work by multiplying to see that $AA^{-1} = I = A^{-1}A$.

■ You should be able to use inverse matrices to solve systems of equation.

Solutions to Odd-Numbered Exercises

1. $AB = \begin{bmatrix} 2 & 1 \\ 5 & 3 \end{bmatrix} \begin{bmatrix} 3 & -1 \\ -5 & 2 \end{bmatrix} = \begin{bmatrix} 2(3) + 1(-5) & 2(-1) + 1(2) \\ 5(3) + 3(-5) & 5(-1) + 3(2) \end{bmatrix} = \begin{bmatrix} 1 & 0 \\ 0 & 1 \end{bmatrix}$

$BA = \begin{bmatrix} 3 & -1 \\ -5 & 2 \end{bmatrix} \begin{bmatrix} 2 & 1 \\ 5 & 3 \end{bmatrix} = \begin{bmatrix} 3(2) + (-1)(5) & 3(1) + (-1)(3) \\ -5(2) + 2(5) & -5(1) + 2(3) \end{bmatrix} = \begin{bmatrix} 1 & 0 \\ 0 & 1 \end{bmatrix}$

3. $AB = \begin{bmatrix} 1 & 2 \\ 3 & 4 \end{bmatrix}\begin{bmatrix} -2 & 1 \\ \frac{3}{2} & -\frac{1}{2} \end{bmatrix} = \begin{bmatrix} -2+3 & 1-1 \\ -6+6 & 3-2 \end{bmatrix} = \begin{bmatrix} 1 & 0 \\ 0 & 1 \end{bmatrix}$

$BA = \begin{bmatrix} -2 & 1 \\ \frac{3}{2} & -\frac{1}{2} \end{bmatrix}\begin{bmatrix} 1 & 2 \\ 3 & 4 \end{bmatrix} = \begin{bmatrix} -2+3 & -4+4 \\ \frac{3}{2}-\frac{3}{2} & 3-2 \end{bmatrix} = \begin{bmatrix} 1 & 0 \\ 0 & 1 \end{bmatrix}$

5. $AB = \frac{1}{3}\begin{bmatrix} -2 & 2 & 3 \\ 1 & -1 & 0 \\ 0 & 1 & 4 \end{bmatrix}\begin{bmatrix} -4 & -5 & 3 \\ -4 & -8 & 3 \\ 1 & 2 & 0 \end{bmatrix} = \frac{1}{3}\begin{bmatrix} -8+8+3 & 10-16+6 & -6+6 \\ -4+4 & -5+8 & 3-3 \\ -4+4 & -8+8 & 3 \end{bmatrix}$

$= \frac{1}{3}\begin{bmatrix} 3 & 0 & 0 \\ 0 & 3 & 0 \\ 0 & 0 & 3 \end{bmatrix} = \begin{bmatrix} 1 & 0 & 0 \\ 0 & 1 & 0 \\ 0 & 0 & 1 \end{bmatrix}$

$BA = \frac{1}{3}\begin{bmatrix} -4 & -5 & 3 \\ -4 & -8 & 3 \\ 1 & 2 & 0 \end{bmatrix}\begin{bmatrix} -2 & 2 & 3 \\ 1 & -1 & 0 \\ 0 & 1 & 4 \end{bmatrix} = \frac{1}{3}\begin{bmatrix} 8-5 & -8+5+3 & -12+12 \\ 8-8 & -8+8+3 & -12+12 \\ -2+2 & 2-2 & 3 \end{bmatrix} = \begin{bmatrix} 1 & 0 & 0 \\ 0 & 1 & 0 \\ 0 & 0 & 1 \end{bmatrix}$

7. $AB = \begin{bmatrix} 2 & 0 & 1 & 1 \\ 3 & 0 & 0 & 1 \\ -1 & 1 & -2 & 1 \\ 4 & -1 & 1 & 0 \end{bmatrix}\begin{bmatrix} -1 & 2 & -1 & -1 \\ -4 & 9 & -5 & -6 \\ 0 & 1 & -1 & -1 \\ 3 & -5 & 3 & 3 \end{bmatrix} = \begin{bmatrix} 1 & 0 & 0 & 0 \\ 0 & 1 & 0 & 0 \\ 0 & 0 & 1 & 0 \\ 0 & 0 & 0 & 1 \end{bmatrix}$

$BA = \begin{bmatrix} -1 & 2 & -1 & -1 \\ -4 & 9 & -5 & -6 \\ 0 & 1 & -1 & -1 \\ 3 & -5 & 3 & 3 \end{bmatrix}\begin{bmatrix} 2 & 0 & 1 & 1 \\ 3 & 0 & 0 & 1 \\ -1 & 1 & -2 & 1 \\ 4 & -1 & 1 & 0 \end{bmatrix} = \begin{bmatrix} 1 & 0 & 0 & 0 \\ 0 & 1 & 0 & 0 \\ 0 & 0 & 1 & 0 \\ 0 & 0 & 0 & 1 \end{bmatrix}$

9. $[A \ \vdots \ I] = \begin{bmatrix} 2 & 0 & \vdots & 1 & 0 \\ 0 & 3 & \vdots & 0 & 1 \end{bmatrix}$

$\begin{matrix} \frac{1}{2}R_1 \to \\ \frac{1}{3}R_2 \to \end{matrix} \begin{bmatrix} 1 & 0 & \vdots & \frac{1}{2} & 0 \\ 0 & 1 & \vdots & 0 & \frac{1}{3} \end{bmatrix} = [I \ \vdots \ A^{-1}]$

$A^{-1} = \begin{bmatrix} \frac{1}{2} & 0 \\ 0 & \frac{1}{3} \end{bmatrix} = \frac{1}{6}\begin{bmatrix} 3 & 0 \\ 0 & 2 \end{bmatrix}$

11. $[A \ \vdots \ I] = \begin{bmatrix} 1 & -2 & \vdots & 1 & 0 \\ 2 & -3 & \vdots & 0 & 1 \end{bmatrix}$

$-2R_1 + R_2 \to \begin{bmatrix} 1 & -2 & \vdots & 1 & 0 \\ 0 & 1 & \vdots & -2 & 1 \end{bmatrix}$

$2R_2 + R_1 \to \begin{bmatrix} 1 & 0 & \vdots & -3 & 2 \\ 0 & 1 & \vdots & -2 & 1 \end{bmatrix} = [I \ \vdots \ A^{-1}]$

$A^{-1} = \begin{bmatrix} -3 & 2 \\ -2 & 1 \end{bmatrix}$

13. $A \ \vdots \ I] = \begin{bmatrix} -1 & 1 & \vdots & 1 & 0 \\ -2 & 1 & \vdots & 0 & 1 \end{bmatrix}$

$-2R_1 + R_2 \rightarrow \begin{bmatrix} -1 & 1 & \vdots & 1 & 0 \\ 0 & -1 & \vdots & -2 & 1 \end{bmatrix}$

$R_2 + R_1 \rightarrow \begin{bmatrix} -1 & 0 & \vdots & -1 & 1 \\ 0 & -1 & \vdots & -2 & 1 \end{bmatrix}$

$\begin{matrix} -R_1 \rightarrow \\ -R_2 \rightarrow \end{matrix} \begin{bmatrix} 1 & 0 & \vdots & 1 & -1 \\ 0 & 1 & \vdots & 2 & -1 \end{bmatrix} = [I \ \vdots \ A^{-1}]$

$A^{-1} = \begin{bmatrix} 1 & -1 \\ 2 & -1 \end{bmatrix}$

15. $[A \ \vdots \ I] = \begin{bmatrix} 2 & 4 & \vdots & 1 & 0 \\ 4 & 8 & \vdots & 0 & 1 \end{bmatrix}$

$-2R_1 + R_2 \rightarrow \begin{bmatrix} 2 & 4 & \vdots & 1 & 0 \\ 0 & 0 & \vdots & -2 & 1 \end{bmatrix}$

The two zeros in the second row imply that the inverse does not exist.

17. $A = \begin{bmatrix} 2 & 7 & 1 \\ -3 & -9 & 2 \end{bmatrix}$

A has no inverse because it is not square.

19.
$\begin{bmatrix} 1 & 1 & 1 & \vdots & 1 & 0 & 0 \\ 3 & 5 & 4 & \vdots & 0 & 1 & 0 \\ 3 & 6 & 5 & \vdots & 0 & 0 & 1 \end{bmatrix}$

$\begin{matrix} -3R_1 + R_2 \rightarrow \\ -3R_1 + R_3 \rightarrow \end{matrix} \begin{bmatrix} 1 & 1 & 1 & \vdots & 1 & 0 & 0 \\ 0 & 2 & 1 & \vdots & -3 & 1 & 0 \\ 0 & 3 & 2 & \vdots & -3 & 0 & 1 \end{bmatrix}$

$\begin{matrix} -R_2 + R_1 \\ \frac{1}{2}R_2 \rightarrow \\ -3R_2 + R_3 \rightarrow \end{matrix} \begin{bmatrix} 1 & 0 & \frac{1}{2} & \vdots & \frac{5}{2} & -\frac{1}{2} & 0 \\ 0 & 1 & \frac{1}{2} & \vdots & -\frac{3}{2} & \frac{1}{2} & 0 \\ 0 & 0 & \frac{1}{2} & \vdots & \frac{3}{2} & -\frac{3}{2} & 1 \end{bmatrix}$

$\begin{matrix} -R_3 + R_1 \rightarrow \\ -R_3 + R_2 \rightarrow \\ 2R_3 \rightarrow \end{matrix} \begin{bmatrix} 1 & 0 & 0 & \vdots & 1 & 1 & -1 \\ 0 & 1 & 0 & \vdots & -3 & 2 & -1 \\ 0 & 0 & 1 & \vdots & 3 & -3 & 2 \end{bmatrix}$

$A^{-1} = \begin{bmatrix} 1 & 1 & -1 \\ -3 & 2 & -1 \\ 3 & -3 & 2 \end{bmatrix}$

21.
$$[A \ \vdots \ I] = \begin{bmatrix} 1 & 0 & 0 & \vdots & 1 & 0 & 0 \\ 3 & 4 & 0 & \vdots & 0 & 1 & 0 \\ 2 & 5 & 5 & \vdots & 0 & 0 & 1 \end{bmatrix}$$

$$\begin{matrix} \\ -3R_1 + R_2 \rightarrow \\ -2R_1 + R_3 \rightarrow \end{matrix} \begin{bmatrix} 1 & 0 & 0 & \vdots & 1 & 0 & 0 \\ 0 & 4 & 0 & \vdots & -3 & 1 & 0 \\ 0 & 5 & 5 & \vdots & -2 & 0 & 1 \end{bmatrix}$$

$$\begin{matrix} \\ \\ -\frac{5}{4}R_2 + R_3 \rightarrow \end{matrix} \begin{bmatrix} 1 & 0 & 0 & \vdots & 1 & 0 & 0 \\ 0 & 4 & 0 & \vdots & -3 & 1 & 0 \\ 0 & 0 & 5 & \vdots & \frac{7}{4} & -\frac{5}{4} & 1 \end{bmatrix}$$

$$\begin{matrix} \\ \frac{1}{4}R_2 \rightarrow \\ \frac{1}{5}R_3 \rightarrow \end{matrix} \begin{bmatrix} 1 & 0 & 0 & \vdots & 1 & 0 & 0 \\ 0 & 1 & 0 & \vdots & -\frac{3}{4} & \frac{1}{4} & 0 \\ 0 & 0 & 1 & \vdots & \frac{7}{20} & -\frac{1}{4} & \frac{1}{5} \end{bmatrix} = [I \ \vdots \ A^{-1}]$$

$$A^{-1} = \frac{1}{20}\begin{bmatrix} 20 & 0 & 0 \\ -15 & 5 & 0 \\ 7 & -5 & 4 \end{bmatrix} = \begin{bmatrix} 1 & 0 & 0 \\ -0.75 & 0.25 & 0 \\ 0.35 & -0.25 & 0.2 \end{bmatrix}$$

23.
$$[A \ \vdots \ I] = \begin{bmatrix} -8 & 0 & 0 & 0 & \vdots & 1 & 0 & 0 & 0 \\ 0 & 1 & 0 & 0 & \vdots & 0 & 1 & 0 & 0 \\ 0 & 0 & 4 & 0 & \vdots & 0 & 0 & 1 & 0 \\ 0 & 0 & 0 & -5 & \vdots & 0 & 0 & 0 & 1 \end{bmatrix}$$

$$\begin{matrix} -\frac{1}{8}R_1 \rightarrow \\ \\ \frac{1}{4}R_3 \rightarrow \\ -\frac{1}{5}R_4 \rightarrow \end{matrix} \begin{bmatrix} 1 & 0 & 0 & 0 & \vdots & -\frac{1}{8} & 0 & 0 & 0 \\ 0 & 1 & 0 & 0 & \vdots & 0 & 1 & 0 & 0 \\ 0 & 0 & 1 & 0 & \vdots & 0 & 0 & \frac{1}{4} & 0 \\ 0 & 0 & 0 & 1 & \vdots & 0 & 0 & 0 & -\frac{1}{5} \end{bmatrix} = [I \ \vdots \ A^{-1}]$$

$$A^{-1} = \begin{bmatrix} -\frac{1}{8} & 0 & 0 & 0 \\ 0 & 1 & 0 & 0 \\ 0 & 0 & \frac{1}{4} & 0 \\ 0 & 0 & 0 & -\frac{1}{5} \end{bmatrix}$$

25. $A = \begin{bmatrix} 1 & 2 & -1 \\ 3 & 7 & -10 \\ -5 & -7 & -15 \end{bmatrix}$

$A^{-1} = \begin{bmatrix} -175 & 37 & -13 \\ 95 & -20 & 7 \\ 14 & -3 & 1 \end{bmatrix}$

27. $A = \begin{bmatrix} 1 & 1 & 2 \\ 3 & 1 & 0 \\ -2 & 0 & 3 \end{bmatrix}$

$A^{-1} = \frac{1}{2}\begin{bmatrix} -3 & 3 & 2 \\ 9 & -7 & -6 \\ -2 & 2 & 2 \end{bmatrix}$

29. $A = \begin{bmatrix} 0.1 & 0.2 & 0.3 \\ -0.3 & 0.2 & 0.2 \\ 0.5 & 0.4 & 0.4 \end{bmatrix}$

$A^{-1} = \frac{5}{11}\begin{bmatrix} 0 & -4 & 2 \\ -22 & 11 & 11 \\ 22 & -6 & -8 \end{bmatrix}$

31. $A = \begin{bmatrix} 1 & 0 & 3 & 0 \\ 0 & 2 & 0 & 4 \\ 1 & 0 & 3 & 0 \\ 0 & 2 & 0 & 4 \end{bmatrix}$

A^{-1} does not exist.

33. $A = \begin{bmatrix} 1 & -2 & -1 & -2 \\ 3 & -5 & -2 & -3 \\ 2 & -5 & -2 & -5 \\ -1 & 4 & 4 & 11 \end{bmatrix}$

$A^{-1} = \begin{bmatrix} -24 & 7 & 1 & -2 \\ -10 & 3 & 0 & -1 \\ -29 & 7 & 3 & -2 \\ 12 & -3 & -1 & 1 \end{bmatrix}$

35. $AA^{-1} = \begin{bmatrix} a & b \\ c & d \end{bmatrix} \left(\dfrac{1}{ad - bc} \right) \begin{bmatrix} d & -b \\ -c & a \end{bmatrix} = \dfrac{1}{ad - bc} \begin{bmatrix} a & b \\ c & d \end{bmatrix} \begin{bmatrix} d & -b \\ -c & a \end{bmatrix}$

$\qquad = \dfrac{1}{ad - bc} \begin{bmatrix} ad - bc & 0 \\ 0 & ad - bc \end{bmatrix} = \begin{bmatrix} 1 & 0 \\ 0 & 1 \end{bmatrix}$

$\quad A^{-1}A = \dfrac{1}{ad - bc} \begin{bmatrix} d & -b \\ -c & a \end{bmatrix} \begin{bmatrix} a & b \\ c & d \end{bmatrix} = \dfrac{1}{ad - bc} \begin{bmatrix} ad - bc & 0 \\ 0 & ad - bc \end{bmatrix} = \begin{bmatrix} 1 & 0 \\ 0 & 1 \end{bmatrix}$

37. $\begin{bmatrix} x \\ y \end{bmatrix} = \begin{bmatrix} -3 & 2 \\ -2 & 1 \end{bmatrix} \begin{bmatrix} 5 \\ 10 \end{bmatrix} = \begin{bmatrix} 5 \\ 0 \end{bmatrix}$

Answer: $(5, 0)$

39. $\begin{bmatrix} x \\ y \end{bmatrix} = \begin{bmatrix} -3 & 2 \\ -2 & 1 \end{bmatrix} = \begin{bmatrix} 4 \\ 2 \end{bmatrix} = \begin{bmatrix} -8 \\ -6 \end{bmatrix}$

Answer: $(-8, -6)$

41. $\begin{bmatrix} x \\ y \\ z \end{bmatrix} = \begin{bmatrix} 1 & 1 & -1 \\ -3 & 2 & -1 \\ 3 & -3 & 2 \end{bmatrix} \begin{bmatrix} 0 \\ 5 \\ 2 \end{bmatrix} = \begin{bmatrix} 3 \\ 8 \\ -11 \end{bmatrix}$

Answer: $(3, 8, -11)$

43. $\begin{bmatrix} x_1 \\ x_2 \\ x_3 \\ x_4 \end{bmatrix} = \begin{bmatrix} -24 & 7 & 1 & -2 \\ -10 & 3 & 0 & -1 \\ -29 & 7 & 3 & -2 \\ 12 & -3 & -1 & 1 \end{bmatrix} \begin{bmatrix} 0 \\ 1 \\ -1 \\ 2 \end{bmatrix} = \begin{bmatrix} 2 \\ 1 \\ 0 \\ 0 \end{bmatrix}$

Answer: $(2, 1, 0, 0)$

45. $\qquad A = \begin{bmatrix} 3 & 4 \\ 5 & 3 \end{bmatrix}$

$A^{-1} = \dfrac{1}{9 - 20} \begin{bmatrix} 3 & -4 \\ -5 & 3 \end{bmatrix}$

$\begin{bmatrix} x \\ y \end{bmatrix} = -\dfrac{1}{11} \begin{bmatrix} 3 & -4 \\ -5 & 3 \end{bmatrix} \begin{bmatrix} -2 \\ 4 \end{bmatrix} = -\dfrac{1}{11} \begin{bmatrix} -22 \\ 22 \end{bmatrix} = \begin{bmatrix} 2 \\ -2 \end{bmatrix}$

Answer: $(2, \div 2)$

47. $\qquad A = \begin{bmatrix} -0.4 & 0.8 \\ 2 & -4 \end{bmatrix}$

$A^{-1} = \dfrac{1}{1.6 - 1.6} \begin{bmatrix} -4 & -0.8 \\ -2 & -0.4 \end{bmatrix}$

A^{-1} does not exist.

No solution

49. $\qquad A = \begin{bmatrix} 3 & 6 \\ 6 & 14 \end{bmatrix}$

$A^{-1} = \dfrac{1}{42 - 36} \begin{bmatrix} 14 & -6 \\ -6 & 3 \end{bmatrix}$

$\begin{bmatrix} x \\ y \end{bmatrix} = \dfrac{1}{6} \begin{bmatrix} 14 & -6 \\ -6 & 3 \end{bmatrix} \begin{bmatrix} 6 \\ 11 \end{bmatrix} = \dfrac{1}{6} \begin{bmatrix} 18 \\ -3 \end{bmatrix} = \begin{bmatrix} 3 \\ -\frac{1}{2} \end{bmatrix}$

Answer: $\left(3, -\frac{1}{2}\right)$

51. $A = \begin{bmatrix} 4 & -1 & 1 \\ 2 & 2 & 3 \\ 5 & -2 & 6 \end{bmatrix}$

$A^{-1} = \frac{1}{55} \begin{bmatrix} 18 & 4 & -5 \\ 3 & 19 & -10 \\ -14 & 3 & 10 \end{bmatrix}$

$\begin{bmatrix} x \\ y \\ z \end{bmatrix} = \frac{1}{55} \begin{bmatrix} 18 & 4 & -5 \\ 3 & 19 & -10 \\ -14 & 3 & 10 \end{bmatrix} \begin{bmatrix} -5 \\ 10 \\ 1 \end{bmatrix} = \frac{1}{55} \begin{bmatrix} -55 \\ 165 \\ 110 \end{bmatrix} = \begin{bmatrix} -1 \\ 3 \\ 2 \end{bmatrix}$

Answer: $(-1, 3, 2)$

53. $A = \begin{bmatrix} 5 & -3 & 2 \\ 2 & 2 & -3 \\ 1 & -7 & 8 \end{bmatrix}$ A^{-1} does not exist.

No solution

55. $A = \begin{bmatrix} 7 & -3 & 0 & 2 \\ -2 & 1 & 0 & -1 \\ 4 & 0 & 1 & -2 \\ -1 & 1 & 0 & -1 \end{bmatrix}$

$A^{-1} = \begin{bmatrix} 0 & -1 & 0 & 1 \\ -1 & -5 & 0 & 3 \\ -2 & -4 & 1 & -2 \\ -1 & -4 & 0 & 1 \end{bmatrix}$

$\begin{bmatrix} x \\ y \\ z \\ w \end{bmatrix} = \begin{bmatrix} 0 & -1 & 0 & 1 \\ -1 & -5 & 0 & 3 \\ -2 & -4 & 1 & -2 \\ -1 & -4 & 0 & 1 \end{bmatrix} \begin{bmatrix} 41 \\ -13 \\ 12 \\ -8 \end{bmatrix} = \begin{bmatrix} 5 \\ 0 \\ -2 \\ 3 \end{bmatrix}$

Answer: $(5, 0, -2, 3)$

For 57–59 use $A = \begin{bmatrix} 1 & 1 & 1 \\ 0.065 & 0.07 & 0.09 \\ 0 & 2 & -1 \end{bmatrix}$. Using the methods of this section, we have $A^{-1} = \frac{1}{11} \begin{bmatrix} 50 & -600 & -4 \\ -13 & 200 & 5 \\ -26 & 400 & -1 \end{bmatrix}$.

57. $X = A^{-1}B = \frac{1}{11} \begin{bmatrix} 50 & -600 & -4 \\ -13 & 200 & 5 \\ -26 & 400 & -1 \end{bmatrix} \begin{bmatrix} 25,000 \\ 1900 \\ 0 \end{bmatrix} = \begin{bmatrix} 10,000 \\ 5000 \\ 10,000 \end{bmatrix}$

Answer: $10,000 in AAA bonds, $5000 in A bonds, $10,000 in B bonds

59. $X = A^{-1}B = \frac{1}{11} \begin{bmatrix} 50 & -600 & -4 \\ -13 & 200 & 5 \\ -26 & 400 & -1 \end{bmatrix} \begin{bmatrix} 12,000 \\ 835 \\ 0 \end{bmatrix} = \begin{bmatrix} 9000 \\ 1000 \\ 2000 \end{bmatrix}$

Answer: $9000 in AAA bonds, $1000 in A bonds, $2000 in B bonds

61. The inverse matrix remained the same for each system.

63. $A = \begin{bmatrix} 2 & 0 & 4 \\ 0 & 1 & 4 \\ 1 & 1 & -1 \end{bmatrix}$

$A^{-1} = \frac{1}{14} \begin{bmatrix} 5 & -4 & 4 \\ -4 & 6 & 8 \\ 1 & 2 & -2 \end{bmatrix}$

$\begin{bmatrix} I_1 \\ I_2 \\ I_3 \end{bmatrix} = \frac{1}{14} \begin{bmatrix} 5 & -4 & 4 \\ -4 & 6 & 8 \\ 1 & 2 & -2 \end{bmatrix} \begin{bmatrix} 14 \\ 10 \\ 0 \end{bmatrix} = \begin{bmatrix} -3 \\ 8 \\ 5 \end{bmatrix}$

Answer: $I_1 = -3$ amps, $I_2 = 8$ amps, $I_3 = 5$ amps

65. (a) Given $A = \begin{bmatrix} a_{11} & 0 \\ 0 & a_{22} \end{bmatrix}$, $A^{-1} = \begin{bmatrix} \dfrac{1}{a_{11}} & 0 \\ 0 & \dfrac{1}{a_{22}} \end{bmatrix}$.

Given $A = \begin{bmatrix} a_{11} & 0 & 0 \\ 0 & a_{22} & 0 \\ 0 & 0 & a_{33} \end{bmatrix}$, $A^{-1} = \begin{bmatrix} \dfrac{1}{a_{11}} & 0 & 0 \\ 0 & \dfrac{1}{a_{22}} & 0 \\ 0 & 0 & \dfrac{1}{a_{33}} \end{bmatrix}$.

(b) In general, the inverse of the diagonal matrix A is

$$\begin{bmatrix} \dfrac{1}{a_{11}} & 0 & 0 & \cdots & 0 \\ 0 & \dfrac{1}{a_{22}} & 0 & \cdots & 0 \\ 0 & 0 & \dfrac{1}{a_{33}} & \cdots & 0 \\ \vdots & \vdots & \vdots & \cdots & \vdots \\ 0 & 0 & 0 & \cdots & \dfrac{1}{a_{33}} \end{bmatrix}.$$

67. Men: $s = 1.279 - 0.0049(22) \approx 1.171$ minutes

Women: $s = 1.411 - 0.0078(22) \approx 1.239$ minutes

69. $3^{x/2} = 315$

$\ln 3^{x/2} = \ln 315$

$\dfrac{x}{2} \ln 3 = \ln 315$

$x = \dfrac{2 \ln 315}{\ln 3} \approx 10.47$

71. $\log_2 x - 2 = 4.5$

$\log_2 x = 6.5$

$x = 2^{6.5} \approx 90.51$

Section 7.4 The Determinant of a Square Matrix

- You should be able to determine the determinant of a matrix of order 2×2 by using the products of the diagonals.
- You should be able to use expansion by cofactors to find the determinant of a matrix of order 3 or greater.
- The determinant of a triangular matrix equals the product of the entries on the main diagonal.

Solutions to Odd-Numbered Exercises

3. $\begin{vmatrix} 2 & 1 \\ 3 & 4 \end{vmatrix} = 2(4) - 1(3) = 8 - 3 = 5$

5. $\begin{vmatrix} 5 & 2 \\ -6 & 3 \end{vmatrix} = 5(3) - 2(-6) = 15 + 12 = 27$

7. $\begin{vmatrix} -7 & 6 \\ \frac{1}{2} & 3 \end{vmatrix} = -7(3) - 6(\frac{1}{2}) = -21 - 3 = -24$

9. $\begin{vmatrix} 2 & 6 \\ 0 & 3 \end{vmatrix} = 2(3) - 6(0) = 6$

11. $\begin{vmatrix} 2 & -1 & 0 \\ 4 & 2 & 1 \\ 4 & 2 & 1 \end{vmatrix} = 2 \begin{vmatrix} 2 & 1 \\ 2 & 1 \end{vmatrix} - 4 \begin{vmatrix} -1 & 0 \\ 2 & 1 \end{vmatrix} + 4 \begin{vmatrix} -1 & 0 \\ 2 & 1 \end{vmatrix} = 2(0) - 4(-1) + 4(-1) = 0$

13. $\begin{vmatrix} 6 & 3 & -7 \\ 0 & 0 & 0 \\ 4 & -6 & 3 \end{vmatrix} = 0 \begin{vmatrix} 3 & -7 \\ -6 & 3 \end{vmatrix} - 0 \begin{vmatrix} 6 & -7 \\ 4 & 3 \end{vmatrix} + 0 \begin{vmatrix} 6 & 3 \\ 4 & -6 \end{vmatrix} = 0$

15. $\begin{vmatrix} -1 & 2 & 5 \\ 0 & 3 & 4 \\ 0 & 0 & 3 \end{vmatrix} = (-1)(3)(3) = -9$ (Upper Triangular)

17. $\begin{vmatrix} 0.3 & 0.2 & 0.2 \\ 0.2 & 0.2 & 0.2 \\ -0.4 & 0.4 & 0.3 \end{vmatrix} = -0.002$

19. $\begin{vmatrix} 1 & 4 & -2 \\ 3 & 6 & -6 \\ -2 & 1 & 4 \end{vmatrix} = 0$

21. $\begin{bmatrix} 3 & 4 \\ 2 & -5 \end{bmatrix}$

(a) $M_{11} = -5$

$M_{12} = 2$

$M_{21} = 4$

$M_{22} = 3$

(b) $C_{11} = M_{11} = -5$

$C_{12} = -M_{12} = -2$

$C_{21} = -M_{21} = -4$

$C_{22} = M_{22} = 3$

23. $\begin{bmatrix} 3 & -2 & 8 \\ 3 & 2 & -6 \\ -1 & 3 & 6 \end{bmatrix}$

(a) $M_{11} = \begin{vmatrix} 2 & -6 \\ 3 & 6 \end{vmatrix} = 12 + 18 = 30$

(b) $C_{11} = (-1)^2 M_{11} = 30$

$\qquad\quad M_{12} = \begin{vmatrix} 3 & -6 \\ -1 & 6 \end{vmatrix} = 18 - 6 = 12$

$\qquad\qquad C_{12} = (-1)^3 M_{12} = -12$

$\qquad\qquad C_{13} = (-1)^4 M_{13} = 11$

$\qquad\quad M_{13} = \begin{vmatrix} 3 & 2 \\ -1 & 3 \end{vmatrix} = 9 + 2 = 11$

$\qquad\qquad C_{21} = (-1)^3 M_{21} = 36$

$\qquad\qquad C_{22} = (-1)^4 M_{22} = 26$

$\qquad\quad M_{21} = \begin{vmatrix} -2 & 8 \\ 3 & 6 \end{vmatrix} = -12 - 24 = -36$

$\qquad\qquad C_{23} = (-1)^5 M_{23} = -7$

$\qquad\qquad C_{31} = (-1)^4 M_{31} = -4$

$\qquad\quad M_{22} = \begin{vmatrix} 3 & 8 \\ -1 & 6 \end{vmatrix} = 18 + 8 = 26$

$\qquad\qquad C_{32} = (-1)^5 M_{32} = 42$

$\qquad\qquad C_{33} = (-1)^6 M_{33} = 12$

$\qquad\quad M_{23} = \begin{vmatrix} 3 & -2 \\ -1 & 3 \end{vmatrix} = 9 - 2 = 7$

$\qquad\quad M_{31} = \begin{vmatrix} -2 & 8 \\ 2 & -6 \end{vmatrix} = 12 - 16 = -4$

$\qquad\quad M_{32} = \begin{vmatrix} 3 & 8 \\ 3 & -6 \end{vmatrix} = -18 - 24 = -42$

$\qquad\quad M_{33} = \begin{vmatrix} 3 & -2 \\ 3 & 2 \end{vmatrix} = 6 + 6 = 12$

25. (a) $\begin{vmatrix} -3 & 2 & 1 \\ 4 & 5 & 6 \\ 2 & -3 & 1 \end{vmatrix} = -3 \begin{vmatrix} 5 & 6 \\ -3 & 1 \end{vmatrix} - 2 \begin{vmatrix} 4 & 6 \\ 2 & 1 \end{vmatrix} + \begin{vmatrix} 4 & 5 \\ 2 & -3 \end{vmatrix} = -3(23) - 2(-8) - 22 = -75$

(b) $\begin{vmatrix} -3 & 2 & 1 \\ 4 & 5 & 6 \\ 2 & -3 & 1 \end{vmatrix} = -2 \begin{vmatrix} 4 & 6 \\ 2 & 1 \end{vmatrix} + 5 \begin{vmatrix} -3 & 1 \\ 2 & 1 \end{vmatrix} + 3 \begin{vmatrix} -3 & 1 \\ 4 & 6 \end{vmatrix} = -2(-8) + 5(-5) + 3(-22) = -75$

27. (a) $\begin{vmatrix} 5 & 0 & -3 \\ 0 & 12 & 4 \\ 1 & 6 & 3 \end{vmatrix} = 0 \begin{vmatrix} 0 & -3 \\ 6 & 3 \end{vmatrix} + 12 \begin{vmatrix} 5 & -3 \\ 1 & 3 \end{vmatrix} - 4 \begin{vmatrix} 5 & 0 \\ 1 & 6 \end{vmatrix} = 0(18) + 12(18) - 4(30) = 96$

(b) $\begin{vmatrix} 5 & 0 & -3 \\ 0 & 12 & 4 \\ 1 & 6 & 3 \end{vmatrix} = 0 \begin{vmatrix} 0 & 4 \\ 1 & 3 \end{vmatrix} + 12 \begin{vmatrix} 5 & -3 \\ 1 & 3 \end{vmatrix} - 6 \begin{vmatrix} 5 & -3 \\ 0 & 4 \end{vmatrix} = 0(-4) + 12(18) - 6(20) = 96$

29. (a)

$$\begin{vmatrix} 6 & 0 & -3 & 5 \\ 4 & 13 & 6 & -8 \\ -1 & 0 & 7 & 4 \\ 8 & 6 & 0 & 2 \end{vmatrix} = -4\begin{vmatrix} 0 & -3 & 5 \\ 0 & 7 & 4 \\ 6 & 0 & 2 \end{vmatrix} + 13\begin{vmatrix} 6 & -3 & 5 \\ -1 & 7 & 4 \\ 8 & 0 & 2 \end{vmatrix} - 6\begin{vmatrix} 6 & 0 & 5 \\ -1 & 0 & 4 \\ 8 & 6 & 2 \end{vmatrix} - 8\begin{vmatrix} 6 & 0 & -3 \\ -1 & 0 & 7 \\ 8 & 6 & 0 \end{vmatrix}$$

$$= -4(-282) + 13(-298) - 6(-174) - 8(-234) = 170$$

(b)

$$\begin{vmatrix} 6 & 0 & -3 & 5 \\ 4 & 13 & 6 & -8 \\ -1 & 0 & 7 & 4 \\ 8 & 6 & 0 & 2 \end{vmatrix} = 0\begin{vmatrix} 4 & 6 & -8 \\ -1 & 7 & 4 \\ 8 & 0 & 2 \end{vmatrix} + 13\begin{vmatrix} 6 & -3 & 5 \\ -1 & 7 & 4 \\ 8 & 0 & 2 \end{vmatrix} + 0\begin{vmatrix} 6 & -3 & 5 \\ 4 & 6 & -8 \\ 8 & 0 & 2 \end{vmatrix} + 6\begin{vmatrix} 6 & -3 & 5 \\ 4 & 6 & -8 \\ -1 & 7 & 4 \end{vmatrix}$$

$$= 0 + 13(-298) + 0 + 6(674) = 170$$

31. Expand by Column 3.

$$\begin{vmatrix} 1 & 4 & -2 \\ 3 & 2 & 0 \\ -1 & 4 & 3 \end{vmatrix} = -2\begin{vmatrix} 3 & 2 \\ -1 & 4 \end{vmatrix} + 3\begin{vmatrix} 1 & 4 \\ 3 & 2 \end{vmatrix} = -2(14) + 3(-10) = -58$$

33. $\begin{vmatrix} 2 & 4 & 6 \\ 0 & 3 & 1 \\ 0 & 0 & -5 \end{vmatrix} = (2)(3)(-5) = -30$ (Upper Triangular)

35. Expand by Column 3.

$$\begin{vmatrix} 2 & 6 & 6 & 2 \\ 2 & 7 & 3 & 6 \\ 1 & 5 & 0 & 1 \\ 3 & 7 & 0 & 7 \end{vmatrix} = 6\begin{vmatrix} 2 & 7 & 6 \\ 1 & 5 & 1 \\ 3 & 7 & 7 \end{vmatrix} - 3\begin{vmatrix} 2 & 6 & 2 \\ 1 & 5 & 1 \\ 3 & 7 & 7 \end{vmatrix} = 6(-20) - 3(16) = -168$$

37. Expand by Column 1.

$$\begin{vmatrix} 5 & 3 & 0 & 6 \\ 4 & 6 & 4 & 12 \\ 0 & 2 & -3 & 4 \\ 0 & 1 & -2 & 2 \end{vmatrix} = 5\begin{vmatrix} 6 & 4 & 12 \\ 2 & -3 & 4 \\ 1 & -2 & 2 \end{vmatrix} - 4\begin{vmatrix} 3 & 0 & 6 \\ 2 & -3 & 4 \\ 1 & -2 & 2 \end{vmatrix} = 5(0) - 4(0) = 0$$

39. Expand by Column 2, then by Column 4.

$$\begin{vmatrix} 3 & 2 & 4 & -1 & 5 \\ -2 & 0 & 1 & 3 & 2 \\ 1 & 0 & 0 & 4 & 0 \\ 6 & 0 & 2 & -1 & 0 \\ 3 & 0 & 5 & 1 & 0 \end{vmatrix} = -2\begin{vmatrix} -2 & 1 & 3 & 2 \\ 1 & 0 & 4 & 0 \\ 6 & 2 & -1 & 0 \\ 3 & 5 & 1 & 0 \end{vmatrix} = (-2)(-2)\begin{vmatrix} 1 & 0 & 4 \\ 6 & 2 & -1 \\ 3 & 5 & 1 \end{vmatrix} = 4(103) = 412$$

41. $\begin{vmatrix} 3 & 8 & -7 \\ 0 & -5 & 4 \\ 8 & 1 & 6 \end{vmatrix} = -126$

43. $\begin{vmatrix} 7 & 0 & -14 \\ -2 & 5 & 4 \\ -6 & 2 & 12 \end{vmatrix} = 0$

45. $\begin{vmatrix} 1 & -1 & 8 & 4 \\ 2 & 6 & 0 & -4 \\ 2 & 0 & 2 & 6 \\ 0 & 2 & 8 & 0 \end{vmatrix} = -336$

47. $\begin{vmatrix} 3 & -2 & 4 & 3 & 1 \\ -1 & 0 & 2 & 1 & 0 \\ 5 & -1 & 0 & 3 & 2 \\ 4 & 7 & -8 & 0 & 0 \\ 1 & 2 & 3 & 0 & 2 \end{vmatrix} = 410$

49. $\begin{vmatrix} w & x \\ y & z \end{vmatrix} = wz - xy$

$-\begin{vmatrix} y & z \\ w & x \end{vmatrix} = -(xy - wz) = wz - xy$

Thus, $\begin{vmatrix} w & x \\ y & z \end{vmatrix} = -\begin{vmatrix} y & z \\ w & x \end{vmatrix}$.

51. $\begin{vmatrix} w & x \\ y & z \end{vmatrix} = wz - xy$

$\begin{vmatrix} w & x + cw \\ y & z + cy \end{vmatrix} = w(z + cy) - y(x + cw) = wz - xy$

Thus, $\begin{vmatrix} w & x \\ y & z \end{vmatrix} = \begin{vmatrix} w & x + cw \\ y & z + cy \end{vmatrix}$.

53. $\begin{vmatrix} 1 & x & x^2 \\ 1 & y & y^2 \\ 1 & z & z^2 \end{vmatrix} = \begin{vmatrix} y & y^2 \\ z & z^2 \end{vmatrix} - \begin{vmatrix} x & x^2 \\ z & z^2 \end{vmatrix} + \begin{vmatrix} x & x^2 \\ y & y^2 \end{vmatrix}$

$= (yz^2 - y^2z) - (xz^2 - x^2z) + (xy^2 - x^2y)$

$= yz^2 - xz^2 - y^2z + x^2z + xy(y - x)$

$= z^2(y - x) - z(y^2 - x^2) + xy(y - x)$

$= z^2(y - x) - z(y - x)(y + x) + xy(y - x)$

$= (y - x)[z^2 - z(y + x) + xy]$

$= (y - x)[z^2 - zy - zx + xy]$

$= (y - x)[z^2 - zx - zy + xy]$

$= (y - x)[z(z - x) - y(z - x)]$

$= (y - x)(z - x)(z - y)$

55. $= \begin{vmatrix} x - 1 & 2 \\ 3 & x - 2 \end{vmatrix} = 0$

$(x - 1)(x - 2) - 6 = 0$

$x^2 - 3x - 4 = 0$

$(x + 1)(x - 4) = 0$

$x = -1 \text{ or } x = 4$

57. $\begin{vmatrix} 4u & -1 \\ -1 & 2v \end{vmatrix} = 8uv - 1$

59. $\begin{vmatrix} e^{2x} & e^{3x} \\ 2e^{2x} & 3e^{3x} \end{vmatrix} = 3e^{5x} - 2e^{5x} = e^{5x}$

61. $\begin{vmatrix} x & \ln x \\ 1 & \dfrac{1}{x} \end{vmatrix} = 1 - \ln x$

63. (a) $\begin{vmatrix} -1 & 0 \\ 0 & 3 \end{vmatrix} = -3$ (b) $\begin{vmatrix} 2 & 0 \\ 0 & -1 \end{vmatrix} = -2$

(c) $\begin{bmatrix} -1 & 0 \\ 0 & 3 \end{bmatrix} \begin{bmatrix} 2 & 0 \\ 0 & -1 \end{bmatrix} = \begin{bmatrix} -2 & 0 \\ 0 & -3 \end{bmatrix}$ (d) $\begin{vmatrix} -2 & 0 \\ 0 & -3 \end{vmatrix} = 6$

65. (a) $\begin{vmatrix} -1 & 2 & 1 \\ 1 & 0 & 1 \\ 0 & 1 & 0 \end{vmatrix} = 2$ (b) $\begin{vmatrix} -1 & 0 & 0 \\ 0 & 2 & 0 \\ 0 & 0 & 3 \end{vmatrix} = -6$

(c) $\begin{bmatrix} -1 & 2 & 1 \\ 1 & 0 & 1 \\ 0 & 1 & 0 \end{bmatrix} \begin{bmatrix} -1 & 0 & 0 \\ 0 & 2 & 0 \\ 0 & 0 & 3 \end{bmatrix} = \begin{bmatrix} 1 & 4 & 3 \\ -1 & 0 & 3 \\ 0 & 2 & 0 \end{bmatrix}$

(d) $\begin{vmatrix} 1 & 4 & 3 \\ -1 & 0 & 3 \\ 0 & 2 & 0 \end{vmatrix} = -12$

67. Let $A = \begin{bmatrix} 1 & 3 \\ -2 & 4 \end{bmatrix}$ and $B = \begin{bmatrix} -4 & 0 \\ 3 & 5 \end{bmatrix}$.

$|A| = \begin{vmatrix} 1 & 3 \\ -2 & 4 \end{vmatrix} = 10$, $|B| = \begin{vmatrix} -4 & 0 \\ 3 & 5 \end{vmatrix} = -20$, $|A| + |B| = -10$

$A + B = \begin{bmatrix} -3 & 3 \\ 1 & 9 \end{bmatrix}$, $|A + B| = \begin{vmatrix} -3 & 3 \\ 1 & 9 \end{vmatrix} = -30$

Thus, $|A + B| \neq |A| + |B|$. Your answer may differ, depending on how you choose A and B.

69. A square matrix is a square array of numbers. A determinant of a square matrix is a real number.

71. Parabola

Vertex: $(0, 3)$

Focus: $(2, 3)$

Horizontal axis: $(y - k)^2 = 4p(x - h)$

$p = 2$

$(y - 3)^2 = 4(2)(x - 0)$

$(y - 3)^2 = 8x$

73. Ellipse

Vertices: $(\pm 8, 0)$

Foci: $(\pm 6, 0)$

Horizontal major axis

Center: $(0, 0)$

$a = 8, c = 6, b = \sqrt{64 - 36} = \sqrt{28}$

$\dfrac{x^2}{a^2} + \dfrac{y^2}{b^2} = 1$

$\dfrac{x^2}{64} + \dfrac{y^2}{28} = 1$

Section 7.5   Applications of Matrices and Determinants

■ You should be able to use Cramer's Rule to solve a system of linear equations.

■ Now you should be able to solve a system of linear equations by substitution, elimination, elementary row operations on an augmented matrix, using the inverse matrix, or Cramer's Rule.

■ You should be able to find the area of a triangle with vertices (x_1, y_1), (x_2, y_2), and (x_3, y_3).

$$\text{Area} = \pm\frac{1}{2}\begin{vmatrix} x_1 & y_1 & 1 \\ x_2 & y_2 & 1 \\ x_3 & y_3 & 1 \end{vmatrix}$$

The $\pm$ symbol indicates that the appropriate sign should be chosen so that the area is positive.

■ You should be able to test to see if three points, (x_1, y_1), (x_2, y_2), and (x_3, y_3), are collinear.

$$\begin{vmatrix} x_1 & y_1 & 1 \\ x_2 & y_2 & 1 \\ x_3 & y_3 & 1 \end{vmatrix} = 0, \text{ if and only if they are collinear.}$$

■ You should be able to find the equation of the line through (x_1, y_1) and (x_2, y_2) by evaluating.

$$\begin{vmatrix} x & y & 1 \\ x_1 & y_1 & 1 \\ x_2 & y_2 & 1 \end{vmatrix} = 0$$

■ You should be able to encode and decode messages by using an invertible $n \times n$ matrix.

Solutions to Odd-Numbered Exercises

1. $3x + 4y = -2$

$5x + 3y = 4$

$$x = \frac{\begin{vmatrix} -2 & 4 \\ 4 & 3 \end{vmatrix}}{\begin{vmatrix} 3 & 4 \\ 5 & 3 \end{vmatrix}} = \frac{-22}{-11} = 2$$

$$y = \frac{\begin{vmatrix} 3 & -2 \\ 5 & 4 \end{vmatrix}}{\begin{vmatrix} 3 & 4 \\ 5 & 3 \end{vmatrix}} = -\frac{22}{11} = -2$$

Answer: $(2, -2)$

3. $4x - y + z = -5$

$2x + 2y + 3z = 10$

$5x - 2y + 6z = 1$

$$D = \begin{vmatrix} 4 & -1 & 1 \\ 2 & 2 & 3 \\ 5 & -2 & 6 \end{vmatrix} = 55$$

$$x = \frac{\begin{vmatrix} -5 & -1 & 1 \\ 10 & 2 & 3 \\ 1 & -2 & 6 \end{vmatrix}}{55} = \frac{-55}{55} = -1$$

$$y = \frac{\begin{vmatrix} 4 & -5 & 1 \\ 2 & 10 & 3 \\ 5 & 1 & 6 \end{vmatrix}}{55} = \frac{165}{55} = 3$$

$$z = \frac{\begin{vmatrix} 4 & -1 & -5 \\ 2 & 2 & 10 \\ 5 & -2 & 1 \end{vmatrix}}{55} = \frac{110}{55} = 2$$

Answer: $(-1, 3, 2)$

5. $3x + 3y + 5z = 1$

     $3x + 5y + 9z = 2$ $D = \begin{vmatrix} 3 & 3 & 5 \\ 3 & 5 & 9 \\ 5 & 9 & 17 \end{vmatrix} = 4$

     $5x + 9y + 17z = 4$

$$x = \frac{\begin{vmatrix} 1 & 3 & 5 \\ 2 & 5 & 9 \\ 4 & 9 & 17 \end{vmatrix}}{4} = 0, \quad y = \frac{\begin{vmatrix} 3 & 1 & 5 \\ 3 & 2 & 9 \\ 5 & 4 & 17 \end{vmatrix}}{4} = -\frac{1}{2}, \quad z = \frac{\begin{vmatrix} 3 & 3 & 1 \\ 3 & 5 & 2 \\ 5 & 9 & 4 \end{vmatrix}}{4} = \frac{1}{2}$$

Answer: $\left(0, -\frac{1}{2}, \frac{1}{2}\right)$

7. Vertices: $(0, 0)$, $(3, 1)$, $(1, 5)$

$$\text{Area} = \frac{1}{2}\begin{vmatrix} 0 & 0 & 1 \\ 3 & 1 & 1 \\ 1 & 5 & 1 \end{vmatrix} = \frac{1}{2}\begin{vmatrix} 3 & 1 \\ 1 & 5 \end{vmatrix} = 7 \text{ square units}$$

9. Vertices: $(-2, -3)$, $(2, -3)$, $(0, 4)$

$$\text{Area} = \frac{1}{2}\begin{vmatrix} -2 & -3 & 1 \\ 2 & -3 & 1 \\ 0 & 4 & 1 \end{vmatrix} = \frac{1}{2}\left(-2\begin{vmatrix} -3 & 1 \\ 4 & 1 \end{vmatrix} - 2\begin{vmatrix} -3 & 1 \\ 4 & 1 \end{vmatrix}\right) = \frac{1}{2}(14 + 14) = 14 \text{ square units}$$

11. Vertices: $\left(0, \frac{1}{2}\right)$, $\left(\frac{5}{2}, 0\right)$, $(4, 3)$

$$\text{Area} = \frac{1}{2}\begin{vmatrix} 0 & \frac{1}{2} & 1 \\ \frac{5}{2} & 0 & 1 \\ 4 & 3 & 1 \end{vmatrix} = \frac{1}{2}\left(2 + \frac{15}{2} - \frac{5}{4}\right) = \frac{33}{8} \text{ square units}$$

13. Vertices: $(-2, 4)$, $(2, 3)$, $(-1, 5)$

$$\text{Area} = \frac{1}{2}\begin{vmatrix} -2 & 4 & 1 \\ 2 & 3 & 1 \\ -1 & 5 & 1 \end{vmatrix} = \frac{1}{2}\left[\begin{vmatrix} 2 & 3 \\ -1 & 5 \end{vmatrix} - \begin{vmatrix} -2 & 4 \\ -1 & 5 \end{vmatrix} + \begin{vmatrix} -2 & 4 \\ 2 & 3 \end{vmatrix}\right] = \frac{5}{2} \text{ square units}$$

15. Vertices: $(-3, 5)$, $(2, 6)$, $(3, -5)$

$$\text{Area} = -\frac{1}{2}\begin{vmatrix} -3 & 5 & 1 \\ 2 & 6 & 1 \\ 3 & -5 & 1 \end{vmatrix} = -\frac{1}{2}\left[\begin{vmatrix} 2 & 6 \\ 3 & -5 \end{vmatrix} - \begin{vmatrix} -3 & 5 \\ 3 & -5 \end{vmatrix} + \begin{vmatrix} -3 & 5 \\ 2 & 6 \end{vmatrix}\right] = 28 \text{ square units}$$

17. $4 = \pm\dfrac{1}{2}\begin{vmatrix} -5 & 1 & 1 \\ 0 & 2 & 1 \\ -2 & x & 1 \end{vmatrix}$

$\pm 8 = -5\begin{vmatrix} 2 & 1 \\ x & 1 \end{vmatrix} - 2\begin{vmatrix} 1 & 1 \\ 2 & 1 \end{vmatrix}$

$\pm 8 = -5(2 - x) - 2(-1)$

$\pm 8 = 5x - 8$

$x = \dfrac{8 \pm 8}{5}$

$x = \dfrac{16}{5}$ OR $x = 0$

19. Vertices: $(0, 25),\ (10, 0),\ (28, 5)$

Area $= \dfrac{1}{2}\begin{vmatrix} 0 & 25 & 1 \\ 10 & 0 & 1 \\ 28 & 5 & 1 \end{vmatrix} = 250$ square miles

21. Points: $(3, -1),\ (0, -3),\ (12, 5)$

$\begin{vmatrix} 3 & -1 & 1 \\ 0 & -3 & 1 \\ 12 & 5 & 1 \end{vmatrix} = 3\begin{vmatrix} -3 & 1 \\ 5 & 1 \end{vmatrix} + 12\begin{vmatrix} -1 & 1 \\ -3 & 1 \end{vmatrix} = 0$

The points are collinear.

23. Points: $\left(2, -\tfrac{1}{2}\right),\ (-4, 4),\ (6, -3)$

$\begin{vmatrix} 2 & -\frac{1}{2} & 1 \\ -4 & 4 & 1 \\ 6 & -3 & 1 \end{vmatrix} = \begin{vmatrix} -4 & 4 \\ 6 & -3 \end{vmatrix} - \begin{vmatrix} 2 & -\frac{1}{2} \\ 6 & -3 \end{vmatrix} + \begin{vmatrix} 2 & -\frac{1}{2} \\ -4 & 4 \end{vmatrix} = -3 \neq 0$

The points are not collinear.

25. Points: $(0, 2),\ (1, 2.4),\ (-1, 1.6)$

$\begin{vmatrix} 0 & 2 & 1 \\ 1 & 2.4 & 1 \\ -1 & 1.6 & 1 \end{vmatrix} = -2\begin{vmatrix} 1 & 1 \\ -1 & 1 \end{vmatrix} + \begin{vmatrix} 1 & 2.4 \\ -1 & 1.6 \end{vmatrix} = 0$

The points are collinear.

27. Points: $(0, 0),\ (5, 3)$

Equation: $\begin{vmatrix} x & y & 1 \\ 0 & 0 & 1 \\ 5 & 3 & 1 \end{vmatrix} = 5y - 3x = 0 \Longrightarrow 3x - 5y = 0$

29. Points: $(-4, 3),\ (2, 1)$

Equation: $\begin{vmatrix} x & y & 1 \\ -4 & 3 & 1 \\ 2 & 1 & 1 \end{vmatrix} = 2x - 6y - 10 = 0 \Longrightarrow x + 3y - 5 = 0$

31. Points: $\left(-\frac{1}{2}, 3\right)$, $\left(\frac{5}{2}, 1\right)$

Equation: $\begin{vmatrix} x & y & 1 \\ -\frac{1}{2} & 3 & 1 \\ \frac{5}{2} & 1 & 1 \end{vmatrix} = 2x + 3y - 8 = 0$

33. $\begin{vmatrix} 2 & -5 & 1 \\ 4 & x & 1 \\ 5 & -2 & 1 \end{vmatrix} = 0$

$2\begin{vmatrix} x & 1 \\ -2 & 1 \end{vmatrix} + 5\begin{vmatrix} 4 & 1 \\ 5 & 1 \end{vmatrix} + \begin{vmatrix} 4 & x \\ 5 & -2 \end{vmatrix} = 0$

$2(x + 2) + 5(-1) + (-8 - 5x) = 0$

$$-3x - 9 = 0$$

$$x = -3$$

35. The uncoded row matrices are the rows of the 7×3 matrix on the left.

$$
\begin{array}{ccc}
T & R & 0 \\
U & B & L \\
E & & I \\
N & & R \\
I & V & E \\
R & & C \\
I & T & Y
\end{array}
\begin{bmatrix}
20 & 18 & 15 \\
21 & 2 & 12 \\
5 & 0 & 9 \\
14 & 0 & 18 \\
9 & 22 & 5 \\
18 & 0 & 3 \\
9 & 20 & 25
\end{bmatrix}
\begin{bmatrix}
1 & -1 & 0 \\
1 & 0 & -1 \\
-6 & 2 & 3
\end{bmatrix} =
\begin{bmatrix}
-52 & 10 & 27 \\
-49 & 3 & 34 \\
-49 & 13 & 27 \\
-94 & 22 & 54 \\
1 & 1 & -7 \\
0 & -12 & 9 \\
-121 & 41 & 55
\end{bmatrix}
$$

Answer: $[-52, 10, 27], [-49, 3, 34], [-49, 13, 27], [-94, 22, 54], [1, 1, -7], [0, -12, 9], [-121, 41, 55]$

In Exercises 37–39, use the matrix $A = \begin{bmatrix} 1 & 2 & 2 \\ 3 & 7 & 9 \\ -1 & -4 & -7 \end{bmatrix}$.

37. L A N   D I N   G _ S   U C C   E S S   F U L

$[12 \ 1 \ 14]$ $[4 \ 9 \ 14]$ $[7 \ 0 \ 19]$ $[21 \ 3 \ 3]$ $[5 \ 19 \ 19]$ $[6 \ 21 \ 12]$

$[12 \quad 1 \quad 14] A = [\quad 1 \quad -25 \quad -65]$

$[4 \quad 9 \quad 14] A = [\quad 17 \quad 15 \quad -9]$

$[7 \quad 0 \quad 19] A = [-12 \quad -62 \quad -119]$

$[21 \quad 3 \quad 3] A = [\quad 27 \quad 51 \quad 48]$

$[5 \quad 19 \quad 19] A = [\quad 43 \quad 67 \quad 48]$

$[6 \quad 21 \quad 12] A = [\quad 57 \quad 111 \quad 117]$

Cryptogram: 1 -25 -65 17 15 -9 -12 -62 -119 27 51 48 43 67 48 57 111 117

39.   H   A   P   P   Y   _   B   I   R   T   H   D   A   Y   _

[8   1   16]   [16   25   0]   [2   9   18]   [20   8   4]   [1   25   0]

$[\,8 \quad 1 \quad 16\,]\,A = [\,5 \quad -41 \quad -87\,]$

$[\,16 \quad 25 \quad 0\,]\,A = [\,91 \quad 207 \quad 257\,]$

$[\,2 \quad 9 \quad 18\,]\,A = [\,11 \quad -5 \quad -41\,]$

$[\,20 \quad 8 \quad 4\,]\,A = [\,40 \quad 80 \quad 84\,]$

$[\,1 \quad 25 \quad 0\,]\,A = [\,76 \quad 177 \quad 227\,]$

Cryptogram: $-5 \quad -41 \quad -87 \quad 91 \quad 207 \quad 257 \quad 11 \quad -5 \quad -41 \quad 40 \quad 80 \quad 84 \quad 76 \quad 177 \quad 227$

41. $A^{-1} = \begin{bmatrix} 1 & 2 \\ 3 & 5 \end{bmatrix}^{-1} = \begin{bmatrix} -5 & 2 \\ 3 & -1 \end{bmatrix}$

$$\begin{bmatrix} 11 & 21 \\ 64 & 112 \\ 25 & 50 \\ 29 & 53 \\ 23 & 46 \\ 40 & 75 \\ 55 & 92 \end{bmatrix} \begin{bmatrix} -5 & 2 \\ 3 & -1 \end{bmatrix} = \begin{bmatrix} 8 & 1 \\ 16 & 16 \\ 25 & 0 \\ 14 & 5 \\ 23 & 0 \\ 25 & 5 \\ 1 & 18 \end{bmatrix} \begin{matrix} \text{H} & \text{A} \\ \text{P} & \text{P} \\ \text{Y} & \\ \text{N} & \text{E} \\ \text{W} & \\ \text{Y} & \text{E} \\ \text{A} & \text{R} \end{matrix}$$

Message: HAPPY NEW YEAR

43. $A^{-1} = \begin{bmatrix} 1 & 2 & 2 \\ 3 & 7 & 9 \\ -1 & -4 & -7 \end{bmatrix}^{-1} = \begin{bmatrix} -13 & 6 & 4 \\ 12 & -5 & -3 \\ -5 & 2 & 1 \end{bmatrix}$

$$\begin{bmatrix} 20 & 17 & -15 \\ -12 & -56 & -104 \\ 1 & -25 & -65 \\ 62 & 143 & 181 \end{bmatrix} \begin{bmatrix} -13 & 6 & 4 \\ 12 & -5 & -3 \\ -5 & 2 & 1 \end{bmatrix} = \begin{bmatrix} 19 & 5 & 14 \\ 4 & 0 & 16 \\ 12 & 1 & 14 \\ 5 & 19 & 0 \end{bmatrix} \begin{matrix} \text{S} & \text{E} & \text{N} \\ \text{D} & & \text{P} \\ \text{L} & \text{A} & \text{N} \\ \text{E} & \text{S} & \end{matrix}$$

Message: SEND PLANES

45. Let A be the 2×2 matrix needed to decode the message.

$$\begin{bmatrix} -18 & -18 \\ 1 & 16 \end{bmatrix} A = \begin{bmatrix} 0 & 18 \\ 15 & 14 \end{bmatrix} \begin{matrix} \text{R} \\ \text{O} \quad \text{N} \end{matrix}$$

$$A = \begin{bmatrix} -18 & -18 \\ 1 & 16 \end{bmatrix}^{-1} \begin{bmatrix} 0 & 18 \\ 15 & 14 \end{bmatrix} = \begin{bmatrix} -\frac{8}{135} & -\frac{1}{15} \\ \frac{1}{270} & \frac{1}{15} \end{bmatrix} \begin{bmatrix} 0 & 18 \\ 15 & 14 \end{bmatrix} = \begin{bmatrix} -1 & -2 \\ 1 & 1 \end{bmatrix}$$

$$\begin{bmatrix} 8 & 21 \\ -15 & -10 \\ -13 & -13 \\ 5 & 10 \\ 5 & 25 \\ 5 & 19 \\ -1 & 6 \\ 20 & 40 \\ -18 & -18 \\ 1 & 16 \end{bmatrix} \begin{bmatrix} -1 & -2 \\ 1 & 1 \end{bmatrix} = \begin{bmatrix} 13 & 5 \\ 5 & 20 \\ 0 & 13 \\ 5 & 0 \\ 20 & 15 \\ 14 & 9 \\ 7 & 8 \\ 20 & 0 \\ 0 & 18 \\ 15 & 14 \end{bmatrix} \begin{matrix} \text{M} & \text{E} \\ \text{E} & \text{T} \\ & \text{M} \\ \text{E} & \\ \text{T} & \text{O} \\ \text{N} & \text{I} \\ \text{G} & \text{H} \\ \text{T} & \\ & \text{R} \\ \text{O} & \text{N} \end{matrix}$$

Message: MEET ME TONIGHT RON

❑ Review Exercises for Chapter 7

Solutions to Odd-Numbered Exercises

1. $\begin{bmatrix} 3 & -10 & \vdots & 15 \\ 5 & 4 & \vdots & 22 \end{bmatrix}$

3. $\begin{bmatrix} 5 & 1 & 7 & \vdots & -9 \\ 4 & 2 & 0 & \vdots & 10 \\ 9 & 4 & 2 & \vdots & 3 \end{bmatrix}$ $\begin{array}{l} 5x + y + 7z = -9 \\ 4x + 2y = 10 \\ 9x + 4y + 2z = 3 \end{array}$

5.

$\begin{bmatrix} 0 & 1 & 1 \\ 1 & 2 & 3 \\ 2 & 2 & 2 \end{bmatrix}$

$\begin{array}{l} R_1 + R_2 \rightarrow \\ -R_1 + R_2 \rightarrow \\ -2R_1 + R_3 \rightarrow \end{array} \begin{bmatrix} 1 & 3 & 4 \\ 0 & -1 & -1 \\ 0 & -4 & -6 \end{bmatrix}$

$\begin{array}{l} 3R_2 + R_1 \rightarrow \\ -R_2 \rightarrow \\ -4R_2 + R_3 \rightarrow \end{array} \begin{bmatrix} 1 & 0 & 1 \\ 0 & 1 & 1 \\ 0 & 0 & -2 \end{bmatrix}$

$\begin{array}{l} -R_3 + R_1 \rightarrow \\ -R_3 + R_2 \rightarrow \\ -\frac{1}{2}R_3 \rightarrow \end{array} \begin{bmatrix} 1 & 0 & 0 \\ 0 & 1 & 0 \\ 0 & 0 & 1 \end{bmatrix}$

7.

$\begin{bmatrix} 5 & 4 & \vdots & 2 \\ -1 & 1 & \vdots & -22 \end{bmatrix}$

$\begin{array}{l} 4R_2 + R_1 \rightarrow \\ R_1 + R_2 \rightarrow \end{array} \begin{bmatrix} 1 & 8 & \vdots & -86 \\ 0 & 9 & \vdots & -108 \end{bmatrix}$

$\begin{array}{l} -8R_2 + R_1 \rightarrow \\ \frac{1}{9}R_2 \rightarrow \end{array} \begin{bmatrix} 1 & 0 & \vdots & 10 \\ 0 & 1 & \vdots & -12 \end{bmatrix}$

$x = 10, \ y = -12$

Answer: $(10, -12)$

9.

$\begin{bmatrix} 2 & 1 & \vdots & 0.3 \\ 3 & -1 & \vdots & -1.3 \end{bmatrix}$

$\begin{array}{l} R_2 - R_1 \rightarrow \\ -3R_1 + R_2 \rightarrow \end{array} \begin{bmatrix} 1 & -2 & \vdots & -1.6 \\ 0 & 5 & \vdots & 3.5 \end{bmatrix}$

$\begin{array}{l} 2R_2 + R_1 \rightarrow \\ \frac{1}{5}R_2 \rightarrow \end{array} \begin{bmatrix} 1 & 0 & \vdots & -0.2 \\ 0 & 1 & \vdots & 0.7 \end{bmatrix}$

$x = -0.2, \ y = 0.7$

Answer: $(-0.2, 0.7)$

11.

$\begin{bmatrix} -1 & 1 & 2 & \vdots & 1 \\ 2 & 3 & 1 & \vdots & -2 \\ 5 & 4 & 2 & \vdots & 4 \end{bmatrix}$

$\begin{array}{l} -R_1 \rightarrow \\ 2R_1 + R_2 \rightarrow \\ 5R_1 + R_3 \rightarrow \end{array} \begin{bmatrix} 1 & -1 & -2 & \vdots & -1 \\ 0 & 5 & 5 & \vdots & 0 \\ 0 & 9 & 12 & \vdots & 9 \end{bmatrix}$

$\begin{array}{l} R_2 + R_1 \rightarrow \\ \frac{1}{5}R_2 \rightarrow \\ -9R_2 + R_3 \rightarrow \end{array} \begin{bmatrix} 1 & 0 & -1 & \vdots & -1 \\ 0 & 1 & 1 & \vdots & 0 \\ 0 & 0 & 3 & \vdots & 9 \end{bmatrix}$

$\begin{array}{l} R_3 + R_1 \rightarrow \\ -R_3 + R_2 \rightarrow \\ \frac{1}{3}R_3 \rightarrow \end{array} \begin{bmatrix} 1 & 0 & 0 & \vdots & 2 \\ 0 & 1 & 0 & \vdots & -3 \\ 0 & 0 & 1 & \vdots & 3 \end{bmatrix}$

$x = 2, y = -3, z = 3$

Answer: $(2, -3, 3)$

13.

$\begin{bmatrix} 4 & 4 & 4 & \vdots & 5 \\ 4 & -2 & -8 & \vdots & 1 \\ 5 & 3 & 8 & \vdots & 6 \end{bmatrix}$

$\begin{array}{l} R_3 - R_1 \rightarrow \\ -4R_1 + R_2 \rightarrow \\ -5R_1 + R_3 \rightarrow \end{array} \begin{bmatrix} 1 & -1 & 4 & \vdots & 1 \\ 0 & 2 & -24 & \vdots & -3 \\ 0 & 8 & -12 & \vdots & 1 \end{bmatrix}$

$\begin{array}{l} R_2 + R_1 \rightarrow \\ \frac{1}{2}R_2 \rightarrow \\ -8R_2 + R_3 \rightarrow \end{array} \begin{bmatrix} 1 & 0 & -8 & \vdots & -\frac{1}{2} \\ 0 & 1 & -12 & \vdots & -\frac{3}{2} \\ 0 & 0 & 84 & \vdots & 13 \end{bmatrix}$

$\begin{array}{l} 8R_3 + R_1 \rightarrow \\ 12R_3 + R_2 \rightarrow \\ \frac{1}{84}R_3 \rightarrow \end{array} \begin{bmatrix} 1 & 0 & 0 & \vdots & \frac{31}{42} \\ 0 & 1 & 0 & \vdots & \frac{5}{14} \\ 0 & 0 & 1 & \vdots & \frac{13}{84} \end{bmatrix}$

$x = \frac{31}{42}, y = \frac{5}{14}, z = \frac{13}{84}$

Answer: $\left(\frac{31}{42}, \frac{5}{14}, \frac{13}{84}\right)$

15.
$$\begin{bmatrix} 2 & 1 & 2 & \vdots & 4 \\ 2 & 2 & 0 & \vdots & 5 \\ 2 & -1 & 6 & \vdots & 2 \end{bmatrix}$$

$$\begin{matrix} \\ -R_1 + R_2 \rightarrow \\ -R_1 + R_3 \rightarrow \end{matrix} \begin{bmatrix} 2 & 1 & 2 & \vdots & 4 \\ 0 & 1 & -2 & \vdots & 1 \\ 0 & -2 & 4 & \vdots & -2 \end{bmatrix}$$

$$\begin{matrix} -R_2 + R_1 \rightarrow \\ \\ 2R_2 + R_3 \rightarrow \end{matrix} \begin{bmatrix} 2 & 0 & 4 & \vdots & 3 \\ 0 & 1 & -2 & \vdots & 1 \\ 0 & 0 & 0 & \vdots & 0 \end{bmatrix}$$

Let $z = a$, then:

$y - 2a = 1 \implies y = 2a + 1$

$2x + 4a = 3 \implies x = -2a + \frac{3}{2}$

Answer: $\left(-2a + \frac{3}{2}, 2a + 1, a\right)$

17.
$$\begin{bmatrix} 1 & 2 & 6 & \vdots & 1 \\ 2 & 5 & 15 & \vdots & 4 \\ 3 & 1 & 3 & \vdots & -6 \end{bmatrix}$$

$$\begin{matrix} \\ -2R_1 + R_2 \rightarrow \\ -3R_1 + R_3 \rightarrow \end{matrix} \begin{bmatrix} 1 & 2 & 6 & \vdots & 1 \\ 0 & 1 & 3 & \vdots & 2 \\ 0 & -5 & -15 & \vdots & -9 \end{bmatrix}$$

$$\begin{matrix} -2R_2 + R_1 \rightarrow \\ \\ 5R_2 + R_3 \rightarrow \end{matrix} \begin{bmatrix} 1 & 0 & 0 & \vdots & -3 \\ 0 & 1 & 3 & \vdots & 2 \\ 0 & 0 & 0 & \vdots & 1 \end{bmatrix}$$

$x = -3$

$y + 3z = 2$

$0 = 1$

Inconsistent, no solution

19. If a system of linear equations has a unique solution, the augmented matrix reduces to a form in which the number of rows with nonzero entries on the coefficient side of the matrix equals the number of variables.

21.
$$\begin{bmatrix} 2 & 1 & 0 \\ 0 & 5 & -4 \end{bmatrix} - 3\begin{bmatrix} 5 & 3 & -6 \\ 0 & -2 & 5 \end{bmatrix} = \begin{bmatrix} 2 & 1 & 0 \\ 0 & 5 & -4 \end{bmatrix} - \begin{bmatrix} 15 & 9 & -18 \\ 0 & -6 & 15 \end{bmatrix}$$

$$= \begin{bmatrix} -13 & -8 & 18 \\ 0 & 11 & -19 \end{bmatrix}$$

23.
$$\begin{bmatrix} 1 & 2 \\ 5 & -4 \\ 6 & 0 \end{bmatrix}\begin{bmatrix} 6 & -2 & 8 \\ 4 & 0 & 0 \end{bmatrix} = \begin{bmatrix} 1(6) + 2(4) & 1(-2) + 2(0) & 1(8) + 2(0) \\ 5(6) + (-4)(4) & 5(-2) + (-4)(0) & 5(8) + (-4)(0) \\ 6(6) + (0)(4) & 6(-2) + (0)(0) & 6(8) + (0)(0) \end{bmatrix}$$

$$= \begin{bmatrix} 14 & -2 & 8 \\ 14 & -10 & 40 \\ 36 & -12 & 48 \end{bmatrix}$$

25.
$$\begin{bmatrix} 1 & 5 & 6 \\ 2 & -4 & 0 \end{bmatrix}\begin{bmatrix} 6 & 4 \\ -2 & 0 \\ 8 & 0 \end{bmatrix} = \begin{bmatrix} 1(6) + 5(-2) + 6(8) & 1(4) + 5(0) + 6(0) \\ 2(6) - 4(-2) + 0(8) & 2(4) - 4(0) + 0(0) \end{bmatrix}$$

$$= \begin{bmatrix} 44 & 4 \\ 20 & 8 \end{bmatrix}$$

27.
$$\begin{bmatrix} 1 & 3 & 2 \\ 0 & 2 & -4 \\ 0 & 0 & 3 \end{bmatrix}\begin{bmatrix} 4 & -3 & 2 \\ 0 & 3 & -1 \\ 0 & 0 & 2 \end{bmatrix} = \begin{bmatrix} 1(4) & 1(-3) + 3(3) & 1(2) + 3(-1) + 2(2) \\ 0 & 2(3) & 2(-1) + (-4)(2) \\ 0 & 0 & 3(2) \end{bmatrix}$$

$$= \begin{bmatrix} 4 & 6 & 3 \\ 0 & 6 & -10 \\ 0 & 0 & 6 \end{bmatrix}$$

29. $3\begin{bmatrix} 8 & -2 & 5 \\ 1 & 3 & -1 \end{bmatrix} + 6\begin{bmatrix} 4 & -2 & -3 \\ 2 & 7 & 6 \end{bmatrix} = \begin{bmatrix} 48 & -18 & -3 \\ 15 & 51 & 33 \end{bmatrix}$

31. $\begin{bmatrix} 4 & 1 \\ 11 & -7 \\ 12 & 3 \end{bmatrix}\begin{bmatrix} 3 & -5 & 6 \\ 2 & -2 & -2 \end{bmatrix} = \begin{bmatrix} 14 & -22 & 22 \\ 19 & -41 & 80 \\ 42 & -66 & 66 \end{bmatrix}$

33. $X = 3A - 2B = 3\begin{bmatrix} -4 & 0 \\ 1 & -5 \\ -3 & 2 \end{bmatrix} - 2\begin{bmatrix} 1 & 2 \\ -2 & 1 \\ 4 & 4 \end{bmatrix} = \begin{bmatrix} -14 & -4 \\ 7 & -17 \\ -17 & -2 \end{bmatrix}$

35. $X = \frac{1}{3}[B - 2A] = \frac{1}{3}\left(\begin{bmatrix} 1 & 2 \\ -2 & 1 \\ 4 & 4 \end{bmatrix} - 2\begin{bmatrix} -4 & 0 \\ 1 & -5 \\ -3 & 2 \end{bmatrix}\right) = \frac{1}{3}\begin{bmatrix} 9 & 2 \\ -4 & 11 \\ 10 & 0 \end{bmatrix}$

37. $\begin{bmatrix} 5 & 4 \\ -1 & 1 \end{bmatrix}\begin{bmatrix} x \\ y \end{bmatrix} = \begin{bmatrix} 2 \\ -22 \end{bmatrix}$

$\begin{bmatrix} 5x + 4y \\ -x + y \end{bmatrix} = \begin{bmatrix} 2 \\ -22 \end{bmatrix}$

$5x + 4y = 2$

$-x + 4y = -22$

39. $\begin{bmatrix} 2 & 6 \\ 3 & -6 \end{bmatrix}^{-1} = \begin{bmatrix} \frac{1}{5} & \frac{1}{5} \\ \frac{1}{10} & -\frac{1}{15} \end{bmatrix}$

41. $\begin{bmatrix} 2 & 0 & 3 \\ -1 & 1 & 1 \\ 2 & -2 & 1 \end{bmatrix}^{-1} = \begin{bmatrix} \frac{1}{2} & -1 & -\frac{1}{2} \\ \frac{1}{2} & -\frac{2}{3} & -\frac{5}{6} \\ 0 & \frac{2}{3} & \frac{1}{3} \end{bmatrix}$

43. $\begin{vmatrix} 50 & -30 \\ 10 & 5 \end{vmatrix} = 50(5) - (-30)(10) = 550$

45. $\begin{vmatrix} 3 & 0 & -4 & 0 \\ 0 & 8 & 1 & 2 \\ 6 & 1 & 8 & 2 \\ 0 & 3 & -4 & 1 \end{vmatrix} = 3\begin{vmatrix} 8 & 1 & 2 \\ 1 & 8 & 2 \\ 3 & -4 & 1 \end{vmatrix} + (-4)\begin{vmatrix} 0 & 8 & 2 \\ 6 & 1 & 2 \\ 0 & 3 & 1 \end{vmatrix}$ (Expansion along Row 1)

$= 3[8(8 - (-8)) - 1(1 - 6) + 2(-4 - 24)] - 4[0 - 6(8 - 6) + 0]$

$= 3[128 + 5 - 56] - 4[-12]$

$= 279$

47. $x + 2y = -1$

$3x + 4y = -5$

$\begin{bmatrix} 1 & 2 \\ 3 & 4 \end{bmatrix}^{-1} = \begin{bmatrix} -2 & 1 \\ \frac{3}{2} & -\frac{1}{2} \end{bmatrix}$

$\begin{bmatrix} x \\ y \end{bmatrix} = \begin{bmatrix} -2 & 1 \\ \frac{3}{2} & -\frac{1}{2} \end{bmatrix}\begin{bmatrix} -1 \\ -5 \end{bmatrix} = \begin{bmatrix} -3 \\ 1 \end{bmatrix}$

$x = -3, y = 1$

Answer: $(-3, 1)$

49. $-3x - 3y - 4z = 2$

$y + z = -1$

$4x + 3y + 4z = -1$

$\begin{bmatrix} -3 & -3 & -4 \\ 0 & 1 & 1 \\ 4 & 3 & 4 \end{bmatrix}^{-1} = \begin{bmatrix} 1 & 0 & 1 \\ 4 & 4 & 3 \\ -4 & -3 & -3 \end{bmatrix}$

$\begin{bmatrix} x \\ y \\ z \end{bmatrix} = \begin{bmatrix} 1 & 0 & 1 \\ 4 & 4 & 3 \\ -4 & -3 & -3 \end{bmatrix}\begin{bmatrix} 2 \\ -1 \\ -1 \end{bmatrix} = \begin{bmatrix} 1 \\ 1 \\ -2 \end{bmatrix}$

$x = 1, y = 1, z = -2$

Answer: $(1, 1, -2)$

51.
$$x + 3y + 2z = 2$$
$$-2x - 5y - z = 10$$
$$2x + 4y = -12$$

$$\begin{bmatrix} 1 & 3 & 2 \\ -2 & -5 & -1 \\ 2 & 4 & 0 \end{bmatrix}^{-1} = \begin{bmatrix} 2 & 4 & \frac{7}{2} \\ -1 & -2 & -\frac{3}{2} \\ 1 & 1 & \frac{1}{2} \end{bmatrix}$$

$$\begin{bmatrix} x \\ y \\ z \end{bmatrix} = \begin{bmatrix} 1 & 0 & 1 \\ 4 & 4 & 3 \\ -4 & -3 & -3 \end{bmatrix}\begin{bmatrix} 2 \\ -1 \\ -1 \end{bmatrix} = \begin{bmatrix} 1 \\ 1 \\ -2 \end{bmatrix}$$

$$x = 2, y = -4, z = 6$$
Answer: $(2, -4, 6)$

53.
$$-x + y + z = 6$$
$$4x - 3y + z = 20$$
$$2x - y + 3z = 8$$

$$\begin{bmatrix} -1 & 1 & 1 \\ 4 & -3 & 1 \\ 2 & -1 & 3 \end{bmatrix}^{-1} \text{ does not exist}$$

The system is inconsistent and has no solution.

55. x = number of carnations, y = number of roses
$$x + y = 12$$
$$0.75x + 1.50y = 12.00$$

$$x = \frac{\begin{vmatrix} 12 & 1 \\ 12 & 1.50 \end{vmatrix}}{\begin{vmatrix} 1 & 1 \\ 0.75 & 1.50 \end{vmatrix}} = \frac{6}{0.75} = 8$$

$$y = \frac{\begin{vmatrix} 1 & 12 \\ 0.75 & 12 \end{vmatrix}}{\begin{vmatrix} 1 & 1 \\ 0.75 & 1.50 \end{vmatrix}} = \frac{3}{0.75} = 4$$

Answer: 8 carnations, 4 roses

57. $(-1, 2), (0, 3), (1, 6)$
$$f(x) = ax^2 + bx + c$$
$$f(-1) = a - b + c = 2$$
$$f(0) = c = 3$$
$$f(1) = a + b + c = 6$$

$$D = \begin{vmatrix} 1 & -1 & 1 \\ 0 & 0 & 1 \\ 1 & 1 & 1 \end{vmatrix} = -2$$

$$a = \frac{\begin{vmatrix} 2 & -1 & 1 \\ 3 & 0 & 1 \\ 6 & 1 & 1 \end{vmatrix}}{-2} = \frac{-2}{-2} = 1$$

$$b = \frac{\begin{vmatrix} 1 & 2 & 1 \\ 0 & 3 & 1 \\ 1 & 6 & 1 \end{vmatrix}}{-2} = \frac{-4}{-2} = 2; c = 3$$

Thus, $y = x^2 + 2x + 3$.

59.
$$13a + 91b = 1107$$
$$91a + 819b = 8404.7$$

(a) $a \approx 59.9, b \approx 3.6$
$$y = 59.9 + 3.6t$$

(b)
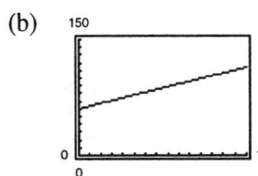

(c) The median price of one-family homes sold in the United States has been increasing by an average of 3.6 thousand dollars ($3600) each year.

(d) $y(15) = 59.9 + 3.6(15) = 113.9$ which corresponds to a price of $113,900.

61. $(1, 0)$, $(5, 0)$, $(5, 8)$

$$\text{Area} = \frac{1}{2}\begin{vmatrix} 1 & 0 & 1 \\ 5 & 0 & 1 \\ 5 & 8 & 1 \end{vmatrix} = \frac{1}{2}(32)$$

$$= 16 \text{ square units}$$

63. $(1, 2)$, $(4, -5)$, $(3, 2)$

$$\text{Area} = \frac{1}{2}\begin{vmatrix} 1 & 2 & 1 \\ 4 & -5 & 1 \\ 3 & 1 & 1 \end{vmatrix} = \frac{1}{2}(14)$$

$$= 7 \text{ square units}$$

65. $(-4, 0)$, $(4, 4)$

$$\begin{vmatrix} x & y & 1 \\ -4 & 0 & 1 \\ 4 & 4 & 1 \end{vmatrix} = 0$$

$$-4x + 8y - 16 = 0$$

$$x - 2y + 4 = 0$$

67. $\left(-\frac{5}{2}, 3\right)$, $\left(\frac{7}{2}, 1\right)$

$$\begin{vmatrix} x & y & 1 \\ -\frac{5}{2} & 3 & 1 \\ \frac{7}{2} & 1 & 1 \end{vmatrix} = 0$$

$$2x + 6y - 13 = 0$$

69. Expansion by Row 3

$$\begin{vmatrix} a_{11} & a_{12} & a_{13} \\ a_{21} & a_{22} & a_{23} \\ a_{31} + c_1 & a_{32} + c_2 & a_{33} + c_3 \end{vmatrix} = (a_{31} + c_1)\begin{vmatrix} a_{12} & a_{13} \\ a_{22} & a_{23} \end{vmatrix} - (a_{32} + c_2)\begin{vmatrix} a_{11} & a_{13} \\ a_{21} & a_{23} \end{vmatrix} + (a_{33} + c_3)\begin{vmatrix} a_{11} & a_{12} \\ a_{21} & a_{22} \end{vmatrix}$$

$$= a_{31}\begin{vmatrix} a_{12} & a_{13} \\ a_{22} & a_{23} \end{vmatrix} - a_{32}\begin{vmatrix} a_{11} & a_{13} \\ a_{21} & a_{23} \end{vmatrix} + a_{33}\begin{vmatrix} a_{11} & a_{12} \\ a_{21} & a_{22} \end{vmatrix} + c_1\begin{vmatrix} a_{12} & a_{13} \\ a_{22} & a_{23} \end{vmatrix}$$

$$- c_2\begin{vmatrix} a_{11} & a_{13} \\ a_{21} & a_{23} \end{vmatrix} + c_3\begin{vmatrix} a_{11} & a_{12} \\ a_{21} & a_{22} \end{vmatrix}$$

$$= \begin{vmatrix} a_{11} & a_{12} & a_{13} \\ a_{21} & a_{22} & a_{23} \\ a_{31} & a_{32} & a_{33} \end{vmatrix} + \begin{vmatrix} a_{11} & a_{12} & a_{13} \\ a_{21} & a_{22} & a_{23} \\ c_1 & c_2 & c_3 \end{vmatrix}$$

Note: Expand each of these matrices by Row 3 to see the previous step.

❏ Practice Test for Chapter 7

1. Put the matrix in reduced echelon form.

$$\begin{bmatrix} 1 & -2 & 4 \\ 3 & -5 & 9 \end{bmatrix}$$

For Exercises 2–4, use matrices to solve the system of equations.

2. $3x + 5y = 3$
   $2x - y = -11$

3. $2x + 3y = -3$
   $3x + 2y = 8$
   $x + y = 1$

4. $x + 3z = -5$
   $2x + y = 0$
   $3x + y - z = 3$

5. Multiply $\begin{bmatrix} 1 & 4 & 5 \\ 2 & 0 & -3 \end{bmatrix} \begin{bmatrix} 1 & 6 \\ 0 & -7 \\ -1 & 2 \end{bmatrix}$.

6. Given $A = \begin{bmatrix} 9 & 1 \\ -4 & 8 \end{bmatrix}$ and $B = \begin{bmatrix} 6 & -2 \\ 3 & 5 \end{bmatrix}$, find $3A - 5B$.

7. Find $f(A)$:

   $$f(x) = x^2 - 7x + 8, \quad A = \begin{bmatrix} 3 & 0 \\ 7 & 1 \end{bmatrix}.$$

8. True or false:

   $(A + B)(A + 3B) = A^2 + 4AB + 3B^2$ where A and B are matrices.

   (Assume that A^2, AB, and B^2 exist.)

For Exercises 9–10, find the inverse of the matrix, if it exists.

9. $\begin{bmatrix} 1 & 2 \\ 3 & 5 \end{bmatrix}$

10. $\begin{bmatrix} 1 & 1 & 1 \\ 3 & 6 & 5 \\ 6 & 10 & 8 \end{bmatrix}$

11. Use an inverse matrix to solve the systems.

   (a) $x + 2y = 4$
      $3x + 5y = 1$

   (b) $x + 2y = 3$
      $3x + 5y = -2$

For Exercises 12–14, find the determinant of the matrix.

12. $\begin{bmatrix} 6 & -1 \\ 3 & 4 \end{bmatrix}$

13. $\begin{bmatrix} 1 & 3 & -1 \\ 5 & 9 & 0 \\ 6 & 2 & -5 \end{bmatrix}$

14. $\begin{bmatrix} 1 & 4 & 2 & 3 \\ 0 & 1 & -2 & 0 \\ 3 & 5 & -1 & 1 \\ 2 & 0 & 6 & 1 \end{bmatrix}$

15. Evaluate $\begin{vmatrix} 6 & 4 & 3 & 0 & 6 \\ 0 & 5 & 1 & 4 & 8 \\ 0 & 0 & 2 & 7 & 3 \\ 0 & 0 & 0 & 9 & 2 \\ 0 & 0 & 0 & 0 & 1 \end{vmatrix}$

16. Use a determinant to find the area of the triangle with vertices $(0, 7)$, $(5, 0)$, and $(3, 9)$.

17. Find the equation of the line through $(2, 7)$ and $(-1, 4)$.

For Exercises 18–20, use Cramer's Rule to find the indicated value.

18. Find x.

$$6x - 7y = 4$$
$$2x + 5y = 11$$

19. Find z.

$$3x + z = 1$$
$$ y + 4z = 3$$
$$x - y = 2$$

20. Find y.

$$721.4x - 29.1y = 33.77$$
$$45.9x + 105.6y = 19.85$$

CHAPTER 8
Sequences and Probability

CHAPTER 8
Sequences and Probability

Section 8.1   Sequences and Summation Notation

- ■ Given the general nth term in a sequence, you should be able to find, or list, some of the terms.
- ■ You should be able to find an expression for the nth term of a sequence.
- ■ You should be able to use and evaluate factorials.
- ■ You should be able to use sigma notation for a sum.

Solutions to Odd-Numbered Exercises

1. $a_n = 2n + 1$
$a_1 = 2(1) + 1 = 3$
$a_2 = 2(2) + 1 = 5$
$a_3 = 2(3) + 1 = 7$
$a_4 = 2(4) + 1 = 9$
$a_5 = 2(5) + 1 = 11$

3. $a_n = 2^n$
$a_1 = 2^1 = 2$
$a_2 = 2^2 = 4$
$a_3 = 2^3 = 8$
$a_4 = 2^4 = 16$
$a_5 = 2^5 = 32$

5. $a_n = (-2)^n$
$a_1 = (-2)^1 = -2$
$a_2 = (-2)^2 = 4$
$a_3 = (-2)^3 = -8$
$a_4 = (-2)^4 = 16$
$a_5 = (-2)^5 = -32$

7. $a_n = \dfrac{n+1}{n}$
$a_1 = \dfrac{1+1}{1} = 2$
$a_2 = \dfrac{3}{2}$
$a_3 = \dfrac{4}{3}$
$a_4 = \dfrac{5}{4}$
$a_5 = \dfrac{6}{5}$

9. $a_n = \dfrac{6n}{3n^2 - 1}$
$a_1 = \dfrac{6(1)}{3(1)^2 - 1} = 3$
$a_2 = \dfrac{6(2)}{3(2)^2 - 1} = \dfrac{12}{11}$
$a_3 = \dfrac{6(3)}{3(3)^2 - 1} = \dfrac{9}{13}$
$a_4 = \dfrac{6(4)}{3(4)^2 - 1} = \dfrac{24}{47}$
$a_5 = \dfrac{6(5)}{3(5)^2 - 1} = \dfrac{15}{37}$

11. $a_n = \dfrac{1 + (-1)^n}{n}$
$a_1 = 0$
$a_2 = \dfrac{2}{2} = 1$
$a_3 = 0$
$a_4 = \dfrac{2}{4} = \dfrac{1}{2}$
$a_5 = 0$

13. $a_n = 3 - \dfrac{1}{2^n}$

$a_1 = 3 - \dfrac{1}{2} = \dfrac{5}{2}$

$a_2 = 3 - \dfrac{1}{4} = \dfrac{11}{4}$

$a_3 = 3 - \dfrac{1}{8} = \dfrac{23}{8}$

$a_4 = 3 - \dfrac{1}{16} = \dfrac{47}{16}$

$a_5 = 3 - \dfrac{1}{32} = \dfrac{95}{32}$

15. $a_n = \dfrac{1}{n^{3/2}}$

$a_1 = \dfrac{1}{1} = 1$

$a_2 = \dfrac{1}{2^{3/2}}$

$a_3 = \dfrac{1}{3^{3/2}}$

$a_4 = \dfrac{1}{4^{3/2}} = \dfrac{1}{8}$

$a_5 = \dfrac{1}{5^{3/2}}$

17. $a_n = \dfrac{3^n}{n!}$

$a_1 = \dfrac{3^1}{1!} = \dfrac{3}{1} = 3$

$a_2 = \dfrac{3^2}{2!} = \dfrac{9}{2}$

$a_3 = \dfrac{27}{6} = \dfrac{9}{2}$

$a_4 = \dfrac{81}{24} = \dfrac{27}{8}$

$a_5 = \dfrac{243}{120} = \dfrac{81}{40}$

19. $a_n = \dfrac{(-1)^n}{n^2}$

$a_1 = -\dfrac{1}{1} = -1$

$a_2 = \dfrac{1}{4}$

$a_3 = -\dfrac{1}{9}$

$a_4 = \dfrac{1}{16}$

$a_5 = -\dfrac{1}{25}$

21. $a_n = \dfrac{2}{3}$

$a_1 = \dfrac{2}{3}$

$a_2 = \dfrac{2}{3}$

$a_3 = \dfrac{2}{3}$

$a_4 = \dfrac{2}{3}$

$a_5 = \dfrac{2}{3}$

23. $a_{25} = (-1)^{25}(3(25) - 2) = -73$

25. $a_1 = 28$ and $a_{k+1} = a_k - 4$

$a_1 = 28$

$a_2 = a_1 - 4 = 28 - 4 = 24$

$a_3 = a_2 - 4 = 24 - 4 = 20$

$a_4 = a_3 - 4 = 20 - 4 = 16$

$a_5 = a_4 - 4 = 16 - 4 = 12$

27. $a_1 = 3$ and $a_{k+1} = 2(a_k - 1)$

$a_1 = 3$

$a_2 = 2(a_1 - 1) = 2(3 - 1) = 4$

$a_3 = 2(a_2 - 1) = 2(4 - 1) = 6$

$a_4 = 2(a_3 - 1) = 2(6 - 1) = 10$

$a_5 = 2(a_4 - 1) = 2(10 - 1) = 18$

29. $a_n = \dfrac{2}{3}n$

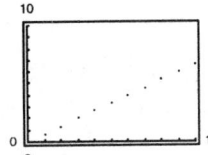

31. $a_n = 16(-0.5)^{n-1}$

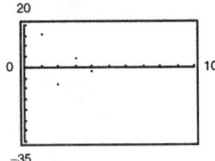

33. $a_n = \dfrac{2n}{n + 1}$

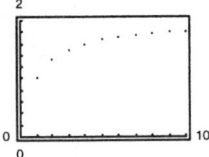

35. $a_n = \dfrac{8}{n + 1}$

$a_n \to 0$ as $n \to \infty$

$a_1 = 4, \ a_{10} = \dfrac{8}{11}$

Matches graph (c).

37. $a_n = 4(0.5)^{n-1}$

$a_n \to 0$ as $n \to \infty$

$a_1 = 4, \ a_{10} \approx 0.008$

Matches graph (d).

39. $\dfrac{4!}{6!} = \dfrac{4!}{6 \cdot 5 \cdot 4!} = \dfrac{1}{30}$

41. $\dfrac{10!}{8!} = \dfrac{10 \cdot 9 \cdot 8!}{8!} = 90$

43. $\dfrac{(n + 1)!}{n!} = \dfrac{(n + 1)n!}{n!} = n + 1$

45. $\dfrac{(2n - 1)!}{(2n + 1)!} = \dfrac{(2n - 1)!}{(2n + 1)(2n)(2n - 1)!}$

$= \dfrac{1}{2n(2n + 1)}$

47. $1, 4, 7, 10, 13, \ldots$

$a_n = 1 + (n - 1)3 = 3n - 2$

49. $0, 3, 8, 15, 24, \ldots$

$a_n = n^2 - 1$

51. $\dfrac{2}{3}, \dfrac{3}{4}, \dfrac{4}{5}, \dfrac{5}{6}, \dfrac{6}{7}, \ldots$

$a_n = \dfrac{n + 1}{n + 2}$

53. $\dfrac{1}{2}, \dfrac{-1}{4}, \dfrac{1}{8}, \dfrac{-1}{16}, \ldots$

$a_n = \dfrac{(-1)^{n+1}}{2^n}$

55. $1 + \dfrac{1}{1}, 1 + \dfrac{1}{2}, 1 + \dfrac{1}{3}, 1 + \dfrac{1}{4}, 1 + \dfrac{1}{5}, \ldots$

$a_n = 1 + \dfrac{1}{n}$

57. $1, \dfrac{1}{2}, \dfrac{1}{6}, \dfrac{1}{24}, \dfrac{1}{120}, \ldots$

$a_n = \dfrac{1}{n!}$

59. $1, -1, 1, -1, 1, \ldots$

$a_n = (-1)^{n+1}$

61. $a_1 = 6$ and $a_{k+1} = a_k + 2$

$a_1 = 6$

$a_2 = a_1 + 2 = 6 + 2 = 8$

$a_3 = a_2 + 2 = 8 + 2 = 10$

$a_4 = a_3 + 2 = 10 + 2 = 12$

$a_5 = a_4 + 2 = 12 + 2 = 14$

In general, $a_n = 2n + 4$.

63. $a_1 = 81$ and $a_{k+1} = \dfrac{1}{3}a_k$

$a_1 = 81$

$a_2 = \dfrac{1}{3}a_1 = \dfrac{1}{3}(81) = 27$

$a_3 = \dfrac{1}{3}a_2 = \dfrac{1}{3}(27) = 9$

$a_4 = \dfrac{1}{3}a_3 = \dfrac{1}{3}(9) = 3$

$a_5 = \dfrac{1}{3}a_4 = \dfrac{1}{3}(3) = 1$

In general, $a_n = 81\left(\dfrac{1}{3}\right)^{n-1} = 81(3)\left(\dfrac{1}{3}\right)^n = \dfrac{243}{3^n}$.

65. $\displaystyle\sum_{i=1}^{5} (2i + 1) = (2 + 1) + (4 + 1) + (6 + 1) + (8 + 1) + (10 + 1) = 35$

67. $\displaystyle\sum_{k=1}^{4} 10 = 10 + 10 + 10 + 10 = 40$

69. $\displaystyle\sum_{i=0}^{4} i^2 = 0^2 + 1^2 + 2^2 + 3^2 + 4^2 = 30$

71. $\displaystyle\sum_{k=0}^{3} \frac{1}{k^2 + 1} = \frac{1}{1} + \frac{1}{1 + 1} + \frac{1}{4 + 1} + \frac{1}{9 + 1} = \frac{9}{5}$

73. $\displaystyle\sum_{i=1}^{4} [(i - 1)^2 + (i + 1)^3] = [(0)^2 + (2)^3] + [(1)^2 + (3)^3] + [(2)^2 + (4)^3] + [(3)^2 + (5)^3] = 238$

75. $\displaystyle\sum_{i=1}^{4} 2^i = 2^1 + 2^2 + 2^3 + 2^4 = 30$ **77.** $\displaystyle\sum_{j=1}^{6} (24 - 3j) = 81$

79. $\displaystyle\sum_{k=0}^{4} \frac{(-1)^k}{k + 1} = 1 - \frac{1}{2} + \frac{1}{3} - \frac{1}{4} + \frac{1}{5} = \frac{47}{60}$

81. $\displaystyle\frac{1}{3(1)} + \frac{1}{3(2)} + \frac{1}{3(3)} + \cdots + \frac{1}{3(9)} = \sum_{i=1}^{9} \frac{1}{3i}$

83. $\displaystyle\left[2\left(\frac{1}{8}\right) + 3\right] + \left[2\left(\frac{2}{8}\right) + 3\right] + \left[2\left(\frac{3}{8}\right) + 3\right] + \cdots + \left[2\left(\frac{8}{8}\right) + 3\right] = \sum_{i=1}^{8} \left[2\left(\frac{i}{8}\right) + 3\right]$

85. $\displaystyle 3 - 9 + 27 - 81 + 243 - 729 = \sum_{i=1}^{6} (-1)^{i+1} 3^i$

87. $\displaystyle\frac{1}{1^2} - \frac{1}{2^2} + \frac{1}{3^2} - \frac{1}{4^2} + \cdots - \frac{1}{20^2} = \sum_{i=1}^{20} \frac{(-1)^{i+1}}{i^2}$

89. $\displaystyle\frac{1}{4} + \frac{3}{8} + \frac{7}{16} + \frac{15}{32} + \frac{31}{64} = \sum_{i=1}^{5} \frac{2i - 1}{2^{i+1}}$

91. $A_n = 5000\left(1 + \dfrac{0.08}{4}\right)^n, \ n = 1, 2, 3, \ldots$

    (a) $A_1 = \$5100.00$

        $A_2 = \$5202.00$

        $A_3 = \$5306.04$

        $A_4 = \$5412.16$

        $A_5 = \$5520.40$

        $A_6 = \$5630.81$

        $A_7 = \$5743.43$

        $A_8 = \$5858.30$

    (b) $A_{40} = \$11,040.20$

93. $a_n = 510.13 + 16.37n + 3.23n^2, \ n = 1, \ldots, 11$

    $a_1 = 529.73$

    $a_2 = 555.79$

    $a_3 = 588.31$

    $a_4 = 627.29$

    $a_5 = 672.73$

    $a_6 = 724.63$

    $a_7 = 782.99$

    $a_8 = 847.81$

    $a_9 = 919.09$

    $a_{10} = 996.83$

    $a_{11} = 1081.03$

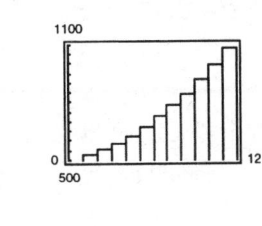

95. $\displaystyle\sum_{n=5}^{14} (129.9 + 0.9n^3) = \$11,131.5 \text{ million}$

97. $a_1 = 1, a_2 = 1, a_{k+2} = a_{k+1} + a_k$

$a_1 = 1$	$b_1 = \frac{1}{1} = 1$
$a_2 = 1$	$b_2 = \frac{2}{1} = 2$
$a_3 = 1 + 1 = 2$	$b_3 = \frac{3}{2}$
$a_4 = 2 + 1 = 3$	$b_4 = \frac{5}{3}$
$a_5 = 3 + 2 = 5$	$b_5 = \frac{8}{5}$
$a_6 = 5 + 3 = 8$	$b_6 = \frac{13}{8}$
$a_7 = 8 + 5 = 13$	$b_7 = \frac{21}{13}$
$a_8 = 13 + 8 = 21$	$b_8 = \frac{34}{21}$
$a_9 = 21 + 13 = 34$	$b_9 = \frac{55}{34}$
$a_{10} = 34 + 21 = 55$	$b_{10} = \frac{89}{55}$
$a_{11} = 55 + 34 = 89$	
$a_{12} = 89 + 55 = 144$	

99. $\dfrac{327.15 + 785.69 + 433.04 + 265.38 + 604.12 + 590.30}{6} \approx \$500.95$

101. $\displaystyle\sum_{i=1}^{n}(x_i - \bar{x}) = \sum_{i=1}^{n} x_i - \sum_{i=1}^{n} \bar{x}$

$\displaystyle = \sum_{i=1}^{n} x_i - n\bar{x}$

$\displaystyle = \sum_{i=1}^{n} x_i - n\left(\frac{1}{n}\sum_{i=1}^{n} x_i\right)$

$= 0$

103. True, $\displaystyle\sum_{i=1}^{4}(i^2 + 2i) = \sum_{i=1}^{4} i^2 + 2\sum_{i=1}^{4} i$

by the properties of sums.

Section 8.2 Arithmetic Sequences

> ■ You should be able to recognize an arithmetic sequence, find its common difference, and find its nth term.
>
> ■ You should be able to find the nth partial sum of an arithmetic sequence by using the formula
>
> $$S_n = \frac{n}{2}(a_1 + a_n).$$

Solutions to Odd-Numbered Exercises

1. $10, 8, 6, 4, 2, \ldots$

Arithmetic sequence, $d = -2$

3. $1, 2, 4, 8, 16, 32, \ldots$

Not an arithmetic sequence

5. $\frac{9}{4}, 2, \frac{7}{4}, \frac{3}{2}, \frac{5}{4}, 1, \ldots$

Arithmetic sequence, $d = -\frac{1}{4}$

7. $-12, -8, -4, 0, 4, \ldots$

Arithmetic sequence, $d = 4$

9. $5.3, 5.7, 6.1, 6.5, 6.9, \ldots$

Arithmetic sequence, $d = 0.4$

11. $a_n = 5 + 3n$

$8, 11, 14, 17, 20$

Arithmetic sequence, $d = 3$

13. $a_n = \dfrac{1}{n+1}$

$\dfrac{1}{2}, \dfrac{1}{3}, \dfrac{1}{4}, \dfrac{1}{5}, \dfrac{1}{6}$

Not an arithmetic sequence

15. $a_n = 100 - 3n$

97, 94, 91, 88, 85

Arithmetic sequence, $d = -3$

17. $a_n = 3 + \dfrac{(-1)^n 2}{n}$

$1, 4, \dfrac{7}{3}, \dfrac{7}{2}, \dfrac{13}{5}$

Not an arithmetic sequence

19. $a_1 = 15, \; a_{k+1} = a_k + 4$

$a_2 = 15 + 4 = 19$

$a_3 = 19 + 4 = 23$

$a_4 = 23 + 4 = 27$

$a_5 = 27 + 4 = 31$

$a_n = 11 + 4n$

21. $a_1 = 200, \; a_{k+1} = a_k - 10$

$a_2 = 200 - 10 = 190$

$a_3 = 190 - 10 = 180$

$a_4 = 180 - 10 = 170$

$a_5 = 170 - 10 = 160$

$a_n = 210 - 10n$

23. $a_1 = \frac{3}{2}, \; a_{k+1} = a_k - \frac{1}{4}$

$a_2 = \frac{3}{2} - \frac{1}{4} = \frac{5}{4}$

$a_3 = \frac{5}{4} - \frac{1}{4} = 1$

$a_4 = 1 - \frac{1}{4} = \frac{3}{4}$

$a_5 = \frac{3}{4} - \frac{1}{4} = \frac{1}{2}$

$a_n = \frac{7}{4} - \frac{1}{4}n$

25. $a_1 = 5, \; d = 6$

$a_1 = 5$

$a_2 = 5 + 6 = 11$

$a_3 = 11 + 6 = 17$

$a_4 = 17 + 6 = 23$

$a_5 = 23 + 6 = 29$

27. $a_1 = -2.6, \; d = -0.4$

$a_1 = -2.6$

$a_2 = -2.6 + (-0.4) = -3.0$

$a_3 = -3.0 + (-0.4) = -3.4$

$a_4 = -3.4 + (-0.4) = -3.8$

$a_5 = -3.8 + (-0.4) = -4.2$

29. $a_1 = 2, \; a_{12} = 46$

$46 = 2 + (12 - 1)d$

$44 = 11d$

$4 = d$

$a_1 = 2$

$a_2 = 2 + 4 = 6$

$a_3 = 6 + 4 = 10$

$a_4 = 10 + 4 = 14$

$a_5 = 14 + 4 = 18$

31. $a_8 = 26, \; a_{12} = 42$

$26 = a_8 = a_1 + (n-1)d = a_1 + 7d$

$42 = a_{12} = a_1 + (n-1)d = a_1 + 11d$

Answer: $d = 4, \; a_1 = -2$

$a_1 = -2$

$a_2 = -2 + 4 = 2$

$a_3 = 2 + 4 = 6$

$a_4 = 6 + 4 = 10$

$a_5 = 10 + 4 = 14$

33. $a_1 = 1, \; d = 3$

$a_n = a_1 + (n-1)d = 1 + (n-1)(3)$

35. $a_1 = 100, \; d = -8$

$a_n = a_1 + (n-1)d = 100 + (n-1)(-8)$

37. $a_1 = x, \; d = 2x$

$a_n = a_1 + (n-1)d = x + (n-1)(2x)$

39. $4, \frac{3}{2}, -1, -\frac{7}{2}, \ldots$

$d = -\frac{5}{2}$

$a_n = a_1 + (n-1)d = 4 + (n-1)\left(-\frac{5}{2}\right)$

41. $a_1 = 5, \; a_4 = 15$

$a_4 = a_1 + 3d \;\Rightarrow\; 15 = 5 + 3d \;\Rightarrow\; d = \frac{10}{3}$

$a_n = a_1 + (n-1)d = 5 + (n-1)\left(\frac{10}{3}\right)$

43. $a_3 = 94, \; a_6 = 85$

$a_6 = a_3 + 3d \;\Rightarrow\; 85 = 94 + 3d \;\Rightarrow\; d = -3$

$a_1 = a_3 - 2d \;\Rightarrow\; a_1 = 94 - 2(-3) = 100$

$a_n = a_1 + (n-1)d = 100 + (n-1)(-3)$

45. $a_n = -\frac{2}{3}n + 6$

$d = -\frac{2}{3}$ so the sequence is decreasing, and $a_1 = 5\frac{1}{3}$. Matches (b).

47. $a_n = 2 + \frac{3}{4}n$

$d = \frac{3}{4}$ so the sequence is increasing, and $a_1 = 2\frac{3}{4}$. Matches (c).

49. $a_n = 15 - \frac{3}{2}n$

51. $a_n = 0.2n + 3$

53. Since $a_n = dn + c$, its geometric pattern is linear.

55. 8, 20, 32, 44, . . .

$a_1 = 8$, $d = 12$, $n = 10$

$a_{10} = 8 + 9(12) = 116$

$S_{10} = \frac{10}{2}(8 + 116) = 620$

57. $-6, -2, 2, 6, \ldots$

$a_1 = -6$, $d = 4$, $n = 50$

$a_{50} = -6 + 49(4) = 190$

$S_{50} = \frac{50}{2}(-6 + 190) = 4600$

59. 40, 37, 34, 31, . . .

$a_1 = 40$, $d = -3$, $n = 10$

$a_{10} = 40 + 9(-3) = 13$

$S_{10} = \frac{10}{2}(40 + 13) = 265$

61. $a_1 = 100$, $a_{25} = 220$, $n = 25$

$S_n = \frac{n}{2}[a_1 + a_n]$

$S_{25} = \frac{25}{2}(100 + 220) = 4000$

63. $a_1 = 1$, $a_{50} = 50$, $n = 50$

$$\sum_{n=1}^{50} n = \frac{50}{2}(1 + 50) = 1275$$

65. $a_1 = 5$, $a_{100} = 500$, $n = 100$

$$\sum_{n=1}^{100} 5n = \frac{100}{2}(5 + 500) = 25{,}250$$

67. $\displaystyle\sum_{n=11}^{30} n - \sum_{n=1}^{10} n = \frac{20}{2}(11 + 30) - \frac{10}{2}(1 + 10) = 355$

69. $a_1 = 4$, $a_{500} = 503$, $n = 500$

$$\sum_{n=1}^{500} (n + 3) = \frac{500}{2}(4 + 503) = 126{,}750$$

71. $a_1 = 7$, $a_{20} = 45$, $n = 20$

$$\sum_{n=1}^{20} (2n + 5) = \frac{20}{2}(7 + 45) = 520$$

73. $a_0 = 1000$, $a_{50} = 750$, $n = 51$

$$\sum_{n=0}^{50} (100 - 5n) = \frac{51}{2}(1000 + 750) = 44{,}625$$

75. $a_1 = \frac{742}{3}$, $a_{60} = 90$, $n = 60$

$$\sum_{i=1}^{60} \left(250 - \frac{8}{3}i\right) = \frac{60}{2}\left(\frac{742}{3} + 90\right) = 10{,}120$$

77. $a_1 = 1$, $a_{100} = 199$, $n = 100$

$$\sum_{n=1}^{100} (2n - 1) = \frac{100}{2}(1 + 199) = 10{,}000$$

79. (a) $a_1 = 32{,}500$, $d = 1500$

$a_6 = a_1 + 5d = 32{,}500 + 5(1500) = \$40{,}000$

(b) $S_6 = \frac{6}{2}[32{,}500 + 40{,}000] = \$217{,}500$

81. $a_1 = 20$, $d = 4$, $n = 30$

$a_{30} = 20 + 29(4) = 136$

$S_{30} = \frac{30}{2}(20 + 136) = 2340$ seats

83. $a_1 = 14$, $a_{18} = 31$

$S_{18} = \frac{18}{2}(14 + 31) = 405$ bricks

85. (a) $1 + 3 = 4$

$1 + 3 + 5 = 9$

$1 + 3 + 5 + 7 = 16$

$1 + 3 + 5 + 7 + 9 = 25$

$1 + 3 + 5 + 7 + 9 + 11 = 36$

(b) $S_n = n^2$

$S_7 = 1 + 3 + 5 + 7 + 9 + 11 + 13 = 49 = 7^2$

(c) $S_n = \dfrac{n}{2}[1 + (2n - 1)] = \dfrac{n}{2}(2n) = n^2$

87. $S_{20} = \dfrac{20}{2}\{a_1 + [a_1 + (20 - 1)(3)]\} = 650$

$10(2a_1 + 57) = 650$

$2a_1 + 57 = 65$

$2a_1 = 8$

$a_1 = 4$

Section 8.3   Geometric Sequences

- You should be able to identify a geometric sequence, find its common ratio, and find the nth term.
- You should be able to find the nth partial sum of a geometric sequence with common ratio r using the formula.

$$S_n = a_1\left(\frac{1 - r^n}{1 - r}\right)$$

- You should know that if $|r| < 1$, then

$$\sum_{n=1}^{\infty} a_1 r^{n-1} = \frac{a_1}{1 - r}.$$

Solutions to Odd-Numbered Exercises

1. $5,\ 15,\ 45,\ 135, \ldots$

Geometric sequence, $r = 3$

3. $3,\ 12,\ 21,\ 30, \ldots$

Not a geometric sequence

Note: It is an arithmetic sequence with $d = 9$.

5. $1,\ -\frac{1}{2},\ \frac{1}{4},\ -\frac{1}{8}, \ldots$

Geometric sequence, $r = -\frac{1}{2}$

7. $\frac{1}{2},\ \frac{2}{3},\ \frac{3}{4},\ \frac{4}{5}, \ldots$

Not a geometric sequence

9. $1,\ \frac{1}{2},\ \frac{1}{3},\ \frac{1}{4}, \ldots$

Not a geometric sequence

11. $a_1 = 2,\ r = 3$

$a_1 = 2$

$a_2 = 2(3) = 6$

$a_3 = 6(3) = 18$

$a_4 = 18(3) = 54$

$a_5 = 54(3) = 162$

13. $a_1 = 1,\ r = \frac{1}{2}$

$a_1 = 1$

$a_2 = 1\left(\frac{1}{2}\right) = \frac{1}{2}$

$a_3 = \frac{1}{2}\left(\frac{1}{2}\right) = \frac{1}{4}$

$a_4 = \frac{1}{4}\left(\frac{1}{2}\right) = \frac{1}{8}$

$a_5 = \frac{1}{8}\left(\frac{1}{2}\right) = \frac{1}{16}$

15. $a_1 = 5,\ r = -\frac{1}{10}$

$a_1 = 5$

$a_2 = 5\left(-\frac{1}{10}\right) = -\frac{1}{2}$

$a_3 = \left(-\frac{1}{2}\right)\left(-\frac{1}{10}\right) = \frac{1}{20}$

$a_4 = \frac{1}{20}\left(-\frac{1}{10}\right) = -\frac{1}{200}$

$a_5 = \left(-\frac{1}{200}\right)\left(-\frac{1}{10}\right) = \frac{1}{2000}$

17. $a_1 = 1,\ r = e$

$a_1 = 1$

$a_2 = 1(e) = e$

$a_3 = (e)(e) = e^2$

$a_4 = (e^2)(e) = e^3$

$a_5 = (e^3)(e) = e^4$

19. $a_1 = 3,\ r = \dfrac{x}{2}$

$a_1 = 3$

$a_2 = 3\left(\dfrac{x}{2}\right) = \dfrac{3x}{2}$

$a_3 = \left(\dfrac{3x}{2}\right)\left(\dfrac{x}{2}\right) = \dfrac{3x^2}{4}$

$a_4 = \left(\dfrac{3x^2}{4}\right)\left(\dfrac{x}{2}\right) = \dfrac{3x^3}{8}$

$a_5 = \left(\dfrac{3x^3}{8}\right)\left(\dfrac{x}{2}\right) = \dfrac{3x^4}{16}$

21. $a_1 = 64,\ a_{k+1} = \dfrac{1}{2}a_k$

$a_1 = 64$

$a_2 = \dfrac{1}{2}(64) = 32$

$a_3 = \dfrac{1}{2}(32) = 16$

$a_4 = \dfrac{1}{2}(16) = 8$

$a_5 = \dfrac{1}{2}(8) = 4$

$a_n = 64\left(\dfrac{1}{2}\right)^{n-1} = 128\left(\dfrac{1}{2}\right)^n$

23. $a_1 = 4,\ a_{k+1} = 3a_k$

$a_1 = 4$

$a_2 = 3(4) = 12$

$a_3 = 3(12) = 36$

$a_4 = 3(36) = 108$

$a_5 = 3(108) = 324$

$a_n = 4(3)^{n-1} = \dfrac{4}{3}(3)^n$

25. $a_k = 6,\ a_{k+1} = -\dfrac{3}{2}a_k$

$a_1 = 6$

$a_2 = -\dfrac{3}{2}(6) = -9$

$a_3 = -\dfrac{3}{2}(-9) = \dfrac{27}{2}$

$a_4 = -\dfrac{3}{2}\left(\dfrac{27}{2}\right) = -\dfrac{81}{4}$

$a_5 = -\dfrac{3}{2}\left(-\dfrac{81}{4}\right) = \dfrac{243}{8}$

$a_n = 6\left(-\dfrac{3}{2}\right)^{n-1}$ or $a_n = -4\left(-\dfrac{3}{2}\right)^n$

27. $a_1 = 4,\ r = \dfrac{1}{2},\ n = 10$

$a_n = a_1 r^{n-1}$

$a_{10} = 4\left(\dfrac{1}{2}\right)^9 = \left(\dfrac{1}{2}\right)^7 = \dfrac{1}{128}$

29. $a_1 = 6,\ r = -\dfrac{1}{3},\ n = 12$

$a_n = a_1 r^{n-1}$

$a_{12} = 6\left(-\dfrac{1}{3}\right)^{11} = \dfrac{-2}{3^{10}}$

31. $a_1 = 100,\ r = e^x,\ n = 9$

$a_n = a_1 r^{n-1}$

$a_9 = 100(e^x)^8 = 100e^{8x}$

33. $a_1 = 500,\ r = 1.02,\ n = 40$

$a_n = a_1 r^{n-1}$

$a_{40} = 500(1.02)^{39} \approx 1082.37$

35. $a_1 = 16,\ a_4 = \dfrac{27}{4},\ n = 3$

$\dfrac{27}{4} = 16r^3 \implies r = \dfrac{3}{4}$

$a_n = a_1 r^{n-1}$

$a_3 = 16\left(\dfrac{3}{4}\right)^2 = 9$

37. $a_2 = a_1 r = -18 \implies a_1 = \dfrac{-18}{r}$

$a_5 = a_1 r^4 = (a_1 r)r^3 = -18r^3 = \dfrac{2}{3} \implies r = -\dfrac{1}{3}$

$a_1 = \dfrac{-18}{r} = \dfrac{-18}{-1/3} = 54$

$a_6 = a_1 r^5 = 54\left(\dfrac{-1}{3}\right)^5 = -\dfrac{54}{243} = -\dfrac{2}{9}$

39. $a_n = 18\left(\frac{2}{3}\right)^{n-1}$

$r = \frac{2}{3} < 1$, so the sequence is decreasing.

Matches (a).

41. $a_n = 18\left(\frac{3}{2}\right)^{n-1}$

$r = \frac{3}{2} > 1$, so the sequence is increasing.

Matches (b).

43. $a_n = 12(-0.75)^{n-1}$

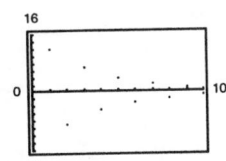

45. $a_n = 2(1.3)^{n-1}$

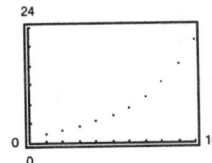

47. Given real numbers r between -1 and 1, as the exponent increases, r^n approaches zero.

49. $A = P\left(1 + \dfrac{r}{n}\right)^{nt} = 1000\left(1 + \dfrac{0.10}{n}\right)^{n(10)}$

    (a) $n = 1$, $A = 1000(1 + 0.10)^{10} \approx \$2593.74$

    (b) $n = 2$, $A = 1000\left(1 + \dfrac{0.10}{2}\right)^{2(10)} \approx \$2653.30$

    (c) $n = 4$, $A = 1000\left(1 + \dfrac{0.10}{4}\right)^{4(10)} \approx \$2685.06$

    (d) $n = 12$, $A = 1000\left(1 + \dfrac{0.10}{12}\right)^{12(10)} \approx \$2707.04$

    (e) $n = 365$, $A = 1000\left(1 + \dfrac{0.10}{365}\right)^{365(10)} \approx \$2717.91$

51. $V_5 = 135{,}000(0.70)^5 = \$22{,}689.45$

53. $8, -4, 2, -1, \frac{1}{2}, \ldots$

$S_1 = 8$

$S_2 = 8 + (-4) = 4$

$S_3 = 8 + (-4) + 2 = 6$

$S_4 = 8 + (-4) + 2 + (-1) = 5$

55. $\displaystyle\sum_{n=1}^{9} 2^{n-1} \implies a_1 = 1, \ r = 2$

$S_9 = \dfrac{1(1 - 2^9)}{1 - 2} = 511$

57. $\displaystyle\sum_{i=1}^{7} 64\left(-\frac{1}{2}\right)^{i-1} \implies a_1 = 64, \ r = -\frac{1}{2}$

$S_7 = 64\left[\dfrac{1 - \left(-\frac{1}{2}\right)^7}{1 - \left(-\frac{1}{2}\right)}\right] = \dfrac{128}{3}\left[1 - \left(-\frac{1}{2}\right)^7\right] = 43$

59. $\displaystyle\sum_{n=0}^{20} 3\left(\frac{3}{2}\right)^{n} = \sum_{n=1}^{21} 3\left(\frac{3}{2}\right)^{n-1} \implies a_1 = 3, \ r = \frac{3}{2}$

$S_{21} = 3\left[\dfrac{1 - \left(\frac{3}{2}\right)^{21}}{1 - \frac{3}{2}}\right] = -6\left[1 - \left(\frac{3}{2}\right)^{21}\right] \approx 29{,}921.31$

61. $\sum\limits_{i=1}^{10} 8\left(-\frac{1}{4}\right)^{i-1} \implies a_1 = 8, \; r = -\frac{1}{4}$

$S_{10} = 8\left[\dfrac{1-\left(-\frac{1}{4}\right)^{10}}{1-\left(-\frac{1}{4}\right)}\right] = \dfrac{32}{5}\left[1-\left(-\frac{1}{4}\right)^{10}\right] \approx 6.4$

63. $\sum\limits_{n=0}^{5} 300(1.06)^n = \sum\limits_{n=1}^{6} 300(1.06)^{n-1} \implies a_1 = 300, \; r = 1.06$

$S_6 = 300\left[\dfrac{1-(1.06)^6}{1-1.06}\right] \approx 2092.60$

65. $5 + 15 + 45 + \cdots + 3645$

$r = 3$ and $3645 = 5(3)^{n-1} \implies n = 7$

Thus, the sum can be written as $\sum\limits_{n=1}^{7} 5(3)^{n-1}$.

67. $A = \sum\limits_{n=1}^{60} 100\left(1+\dfrac{0.10}{12}\right)^n = 100\left(1+\dfrac{0.10}{12}\right)\cdot\dfrac{\left[1-\left(1+\frac{0.10}{12}\right)^{60}\right]}{\left[1-\left(1+\frac{0.10}{12}\right)\right]} \approx \$7808.24$

69. Let $N = 12t$ be the total number of deposits.

$A = P\left(1+\dfrac{r}{12}\right) + P\left(1+\dfrac{r}{12}\right)^2 + \cdots + P\left(1+\dfrac{r}{12}\right)^N$

$= \left(1+\dfrac{r}{12}\right)\left[P + P\left(r+\dfrac{r}{12}\right) + \cdots + P\left(1+\dfrac{r}{12}\right)^{N-1}\right]$

$= P\left(1+\dfrac{r}{12}\right)\sum\limits_{n=1}^{N}\left(1+\dfrac{r}{12}\right)^{n-1}$

$= P\left(1+\dfrac{r}{12}\right)\dfrac{1-\left(1+\frac{r}{12}\right)^N}{1-\left(1+\frac{r}{12}\right)}$

$= P\left(1+\dfrac{r}{12}\right)\left(-\dfrac{12}{r}\right)\left[1-\left(1+\dfrac{r}{12}\right)^N\right]$

$= P\left(\dfrac{12}{r}+1\right)\left[-1+\left(1+\dfrac{r}{12}\right)^N\right]$

$= P\left[\left(1+\dfrac{r}{12}\right)^N - 1\right]\left(1+\dfrac{12}{r}\right)$

$= P\left[\left(1+\dfrac{r}{12}\right)^{12t} - 1\right]\left(1+\dfrac{12}{r}\right)$

71. $P = \$50, \; r = 7\%, \; t = 20$ years

(a) Compounded monthly: $A = 50\left[\left(1+\dfrac{0.07}{12}\right)^{12(20)} - 1\right]\left(1+\dfrac{12}{0.07}\right) \approx \$26,198.27$

(b) Compounded continuously: $A = \dfrac{50e^{0.07/12}(e^{0.07(20)}-1)}{e^{0.07/12}-1} \approx \$26,263.88$

73. $P = \$100$, $r = 10\%$, $t = 40$ years

(a) Compounded monthly: $A = 100\left[\left(1 + \dfrac{0.10}{12}\right)^{12(40)} - 1\right]\left(1 + \dfrac{12}{0.10}\right) \approx \$637,678.02$

(b) Compounded continuously: $A = \dfrac{100e^{0.10/12}(e^{(0.10)(40)} - 1)}{e^{0.10/12} - 1} \approx \$645,861.43$

75. $P = W\displaystyle\sum_{n=1}^{12t}\left[\left(1 + \dfrac{r}{12}\right)^{-1}\right]^{n}$

$= W\left(1 + \dfrac{r}{12}\right)^{-1}\left[\dfrac{1 - \left(1 + \dfrac{r}{12}\right)^{-12t}}{1 - \left(1 - \dfrac{r}{12}\right)^{-1}}\right]$

$= W\left(\dfrac{1}{1 + \dfrac{r}{12}}\right)\dfrac{\left[1 - \left(1 + \dfrac{r}{12}\right)^{-12t}\right]}{1 - \dfrac{1}{\left(1 + \dfrac{r}{12}\right)}}$

$= W\dfrac{\left[1 - \left(1 + \dfrac{r}{12}\right)^{-12t}\right]}{\left(1 + \dfrac{r}{12}\right) - 1}$

$= W\left(\dfrac{12}{r}\right)\left[1 - \left(1 - \dfrac{r}{12}\right)^{-12t}\right]$

77. $64 + 32 + 16 + 8 + 4 + 2 = 126$
Total area of shaded region is
approximately 126 square inches.

79. $S_n = \displaystyle\sum_{i=1}^{n} 0.01(2)^{i-1}$
$S_{29} = \$5,368,709.11$
$S_{30} = \$10,737,418.23$
$S_{31} = \$21,474,836.47$

81. $a_1 = 1$, $r = \dfrac{1}{2}$

$\displaystyle\sum_{n=0}^{\infty}\left(\dfrac{1}{2}\right)^n = \dfrac{a_1}{1 - r} = \dfrac{1}{1 - (1/2)} = 2$

83. $a_1 = 1$, $r = -\dfrac{1}{2}$

$\displaystyle\sum_{n=1}^{\infty}\left(-\dfrac{1}{2}\right)^{n-1} = \dfrac{a_1}{1 - r} = \dfrac{1}{1 - (-1/2)} = \dfrac{2}{3}$

85. $a_1 = 4$, $r = \dfrac{1}{4}$

$\displaystyle\sum_{n=0}^{\infty}4\left(\dfrac{1}{4}\right)^n = \dfrac{a_1}{1 - r} = \dfrac{4}{1 - (1/4)} = \dfrac{16}{3}$

87. $8 + 6 + \dfrac{9}{2} + \dfrac{27}{8} + \cdots = \displaystyle\sum_{n=0}^{\infty}8\left(\dfrac{3}{4}\right)^n = \dfrac{8}{1 - 3/4} = 32$

89. $0.\overline{36} = \displaystyle\sum_{n=0}^{\infty}0.36(0.01)^n = \dfrac{0.36}{1 - 0.01} = \dfrac{0.36}{0.99} = \dfrac{36}{99} = \dfrac{4}{11}$

91. $0.3\overline{18} = 0.3 + \sum_{n=0}^{\infty} 0.018(0.01)^n = \dfrac{3}{10} + \dfrac{0.018}{1-0.01}$

$= \dfrac{3}{10} + \dfrac{0.018}{0.99} = \dfrac{3}{10} + \dfrac{18}{990} = \dfrac{3}{10} + \dfrac{2}{110}$

$= \dfrac{35}{110} = \dfrac{7}{22}$

93. $f(x) = 6\left[\dfrac{1-(0.5)^x}{1-(0.5)}\right], \; \sum_{n=0}^{\infty} 6\left(\dfrac{1}{2}\right)^n = \dfrac{6}{1-1/2} = 12$

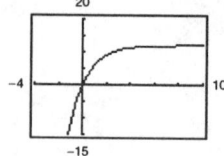

The horizontal asymptote of $f(x)$ is $y = 12$. This corresponds to the sum of the series.

95. (a) Total distance $= \left[\sum_{n=0}^{\infty} 32(0.81)^n\right] - 16 = \dfrac{32}{1-0.81} - 16 \approx 152.42$ feet

(b) $t = 1 + 2\sum_{n=1}^{\infty}(0.9)^n = 1 + 2\left[\dfrac{0.9}{1-0.9}\right] = 19$ seconds

Section 8.4   Mathematical Induction

- You should be sure that you understand the principle of mathematical induction. If P_n is a statement involving the positive integer n, where P_1 is true and the truth of P_k implies the truth of P_{k+1}, then P_n is true for all positive integers n.

- You should be able to verify (by induction) the formulas for the sums of powers of integers and be able to use these formulas.

Solutions to Odd-Numbered Exercises

1. $P_k = \dfrac{5}{k(k+1)}$

$P_{k+1} = \dfrac{5}{(k+1)((k+1)+1)} = \dfrac{5}{(k+1)(k+2)}$

3. $P_k = \dfrac{k^2(k+1)^2}{4}$

$P_{k+1} = \dfrac{(k+1)^2((k+1)+1)^2}{4} = \dfrac{(k+1)^2(k+2)^2}{4}$

5. 1. When $n = 1$, $S_1 = 2 = 1(1+1)$.

2. Assume that

$S_k = 2 + 4 + 6 + 8 + \cdots + 2k = k(k+1)$.

Then,

$S_{k+1} = 2 + 4 + 6 + 8 + \cdots + 2k + 2(k+1)$

$= S_k + 2(k+1) = k(k+1) + 2(k+1) = (k+1)(k+2)$.

We conclude by mathematical induction that the formula is valid for all positive integer values of n.

7. 1. When $n = 1$, $S_1 = 2 = \dfrac{1}{2}(5(1) - 1)$.

   2. Assume that
   $$S_k = 2 + 7 + 12 + 17 + \cdots + (5k - 3) = \frac{k}{2}(5k - 1).$$
   Then,
   $$S_{k+1} = 2 + 7 + 12 + 17 + \cdots + (5k - 3) + [5(k + 1) - 3]$$
   $$= S_k + (5k + 5 - 3) = \frac{k}{2}(5k - 1) + 5k + 2$$
   $$= \frac{5k^2 - k + 10k + 4}{2} = \frac{5k^2 + 9k + 4}{2}$$
   $$= \frac{(k + 1)(5k + 4)}{2} = \frac{(k + 1)}{2}[5(k + 1) - 1].$$
   We conclude by mathematical induction that the formula is valid for all positive integer values of n.

9. 1. When $n = 1$, $S_1 = 1 = 2^1 - 1$.

   2. Assume that
   $$S_k = 1 + 2 + 2^2 + 2^3 + \cdots + 2^{k-1} = 2^k - 1.$$
   Then,
   $$S_{k+1} = 1 + 2 + 2^2 + 2^3 + \cdots + 2^{k-1} + 2^k$$
   $$= S_k + 2^k = 2^k - 1 + 2^k = 2(2^k) - 1 = 2^{k+1} - 1.$$
   Therefore, by mathematical induction, the formula is valid for all positive integer values of n.

11. 1. When $n = 1$, $S_1 = 1 = \dfrac{1(1 + 1)}{2}$.

   2. Assume that
   $$S_k = 1 + 2 + 3 + 4 + \cdots + k = \frac{k(k + 1)}{2}.$$
   Then,
   $$S_{k+1} = 1 + 2 + 3 + 4 + \cdots + k + (k + 1)$$
   $$= S_k + (k + 1) = \frac{k(k + 1)}{2} + \frac{2(k + 1)}{2} = \frac{(k + 1)(k + 2)}{2}.$$
   Therefore, we conclude that this formula holds for all positive integer values of n.

13. 1. When $n = 1$, $S_1 = 1^3 = 1 = \dfrac{1(1 + 1)^2}{4}$.

   2. Assume that
   $$S_k = 1^3 + 2^3 + 3^3 + 4^3 + \cdots + k^3 = \frac{k^2(k + 1)^2}{4}.$$
   Then,
   $$S_{k+1} = 1^3 + 2^3 + 3^3 + 4^3 + \cdots + k^3 + (k + 1)^3$$
   $$= S_k + (k + 1)^3 = \frac{k^2(k + 1)^2}{4} + (k + 1)^3 = \frac{k^2(k + 1)^2 + 4(k + 1)^3}{4}$$
   $$= \frac{(k + 1)^2[k^2 + 4(k + 1)]}{4} = \frac{(k + 1)^2(k^2 + 4k + 4)}{4} = \frac{(k + 1)^2(k + 2)^2}{4}.$$
   Therefore, we conclude that this formula holds for all positive integer values of n.

15. 1. When $n = 1$, $S_1 = 1 = \dfrac{(1)^2(1 + 1)^2(2(1)^2 + 2(1) - 1)}{12}$.

2. Assume that

$$S_k = \sum_{i=1}^{k} i^5 = \frac{k^2(k + 1)^2(2k^2 + 2k - 1)}{12}.$$

Then,

$$S_{k+1} = \sum_{i=1}^{k+1} i^5 = \sum_{i=1}^{k} i^5 + (k + 1)^5$$

$$= \frac{k^2(k + 1)^2(2k^2 + 2k - 1)}{12} + \frac{12(k + 1)^5}{12}$$

$$= \frac{(k + 1)^2[k^2(2k^2 + 2k - 1) + 12(k + 1)^3]}{12}$$

$$= \frac{(k + 1)^2[2k^4 + 2k^3 - k^2 + 12(k^3 + 3k^2 + 3k + 1)]}{12}$$

$$= \frac{(k + 1)^2[2k^4 + 14k^3 + 35k^2 + 36k + 12]}{12}$$

$$= \frac{(k + 1)^2(k^2 + 4k + 4)(2k^2 + 6k + 3)}{12}$$

$$= \frac{(k + 1)^2(k + 2)^2[2(k + 1)^2 + 2(k + 1) - 1]}{12}.$$

Therefore, we conclude that this formula holds for all positive integer values of n.

17. 1. When $n = 1$, $S_1 = 2 = \dfrac{1(2)(3)}{3}$.

2. Assume that,

$$S_k = 1(2) + 2(3) + 3(4) + \cdots + k(k + 1) = \frac{k(k + 1)(k + 2)}{3}.$$

Then,

$$S_{k+1} = 1(2) + 2(3) + 3(4) + \cdots + k(k + 1) + (k + 1)(k + 2)$$

$$= S_k + (k + 1)(k + 2) = \frac{k(k + 1)(k + 2)}{3} + \frac{3(k + 1)(k + 2)}{3}$$

$$= \frac{(k + 1)(k + 2)(k + 3)}{3}.$$

Thus, this formula is valid for all positive integer values of n.

19. $\displaystyle\sum_{n=1}^{20} n = \frac{20(20 + 1)}{2} = 210$

21. $\displaystyle\sum_{n=1}^{6} n^2 = \frac{6(6 + 1)(2(6) + 1)}{6} = 91$

23. $\displaystyle\sum_{n=1}^{5} n^4 = \frac{5(5 + 1)(2(5) + 1)(3(5)^2 + 3(5) - 1)}{30} = 979$

25. $\displaystyle\sum_{n=1}^{6} (n^2 - n) = \sum_{n=1}^{6} n^2 - \sum_{n=1}^{6} n = \frac{6(6 + 1)(2(6) + 1)}{6} - \frac{6(6 + 1)}{2} = 91 - 21 = 70$

27. $\displaystyle\sum_{i=1}^{6}(6i - 8i^3) = 6\sum_{i=1}^{6}i - 8\sum_{i=1}^{6}i^3 = 6\left[\frac{6(6+1)}{2}\right] - 8\left[\frac{(6)^2(6+1)^2}{4}\right] = 6(21) - 8(441) = -3402$

29. $1 + 5 + 9 + 13 + \cdots = n(2n - 1)$ **31.** $1 + \frac{9}{10} + \frac{81}{100} + \frac{729}{1000} + \cdots = 10 - 10\left(\frac{9}{10}\right)^n$

33. $\dfrac{1}{4} + \dfrac{1}{12} + \dfrac{1}{24} + \dfrac{1}{40} + \cdots + \dfrac{1}{2n(n-1)} + \cdots = \dfrac{n}{2(n+1)}$

35. 1. When $n = 4$, $4! = 24$ and $2^4 = 16$, thus $4! > 2^4$.

  2. Assume

   $k! > 2^k$, $k > 4$.

   Then,

   $(k+1)! = k!(k+1) > 2^k(2)$ since $k + 1 > 2$.

   Thus, $(k+1)! > 2^{k+1}$.

  Therefore, by mathematical induction, the formula is valid for all integers n such that $n \geq 4$.

37. 1. When $n = 2$, $\dfrac{1}{\sqrt{1}} + \dfrac{1}{\sqrt{2}} \approx 1.707$ and $\sqrt{2} \approx 1.414$, thus $\dfrac{1}{\sqrt{1}} + \dfrac{1}{\sqrt{2}} > \sqrt{2}$.

  2. Assume

   $\dfrac{1}{\sqrt{1}} + \dfrac{1}{\sqrt{2}} + \dfrac{1}{\sqrt{3}} + \cdots + \dfrac{1}{\sqrt{k}} > \sqrt{k}, k > 2$.

   Then,

   $\dfrac{1}{\sqrt{1}} + \dfrac{1}{\sqrt{2}} + \dfrac{1}{\sqrt{3}} + \cdots + \dfrac{1}{\sqrt{k}} + \dfrac{1}{\sqrt{k+1}} > \sqrt{k} + \dfrac{1}{\sqrt{k+1}}$.

   Now we need to show that

   $\sqrt{k} + \dfrac{1}{\sqrt{k+1}} > \sqrt{k+1}, k > 2$.

   This is true since

   $\sqrt{k(k+1)} > k$

   $\sqrt{k(k+1)} + 1 > k + 1$

   $\dfrac{\sqrt{k(k+1)} + 1}{\sqrt{k+1}} > \dfrac{k+1}{\sqrt{k+1}}$

   $\sqrt{k} + \dfrac{1}{\sqrt{k+1}} > \sqrt{k+1}$.

   Therefore,

   $\dfrac{1}{\sqrt{1}} + \dfrac{1}{\sqrt{2}} + \dfrac{1}{\sqrt{3}} + \ldots + \dfrac{1}{\sqrt{k}} + \dfrac{1}{\sqrt{k+1}} > \sqrt{k+1}$.

  Therefore, by mathematical induction, the formula is valid for all integers n such that $n \geq 2$.

39. 1. When $n = 1$, $(ab)^1 = a^1b^1 = ab$.

2. Assume that $(ab)^k = a^kb^k$.

Then, $(ab)^{k+1} = (ab)^k(ab)$

$= a^kb^kab$

$= a^{k+1}b^{k+1}$.

Thus, $(ab)^n = a^nb^n$.

41. 1. When $n = 1$, $(x_1)^{-1} = x_1^{-1}$.

2. Assume that

$(x_1x_2x_3 \cdots x_k)^{-1} = x_1^{-1}x_2^{-1}x_3^{-1} \cdots x_k^{-1}$.

Then,

$(x_1x_2x_3 \cdots x_kx_{k+1})^{-1} = [(x_1x_2x_3 \cdots x_k)x_{k+1}]^{-1}$

$= (x_1x_2x_3 \ldots x_k)^{-1}x_{k+1}^{-1}$

$= x_1^{-1}x_2^{-1}x_3^{-1} \cdots x_k^{-1}x_{k+1}^{-1}$.

Thus, the formula is valid.

43. 1. When $n = 1$, $x(y_1) = xy_1$.

2. Assume that

$x(y_1 + y_2 + \cdots + y_k) = xy_1 + xy_2 + \cdots + xy_k$.

Then,

$xy_1 + xy_2 + \cdots + xy_k + xy_{k+1} = x(y_1 + y_2 + \cdots + y_k) + xy_{k+1}$

$= x[(y_1 + y_2 + \cdots + y_k) + y_{k+1}]$

$= x(y_1 + y_2 + \cdots + y_k + y_{k+1})$.

Hence, the formula holds.

45. 1. When $n = 1$, $\sin(x + n) = (-1)^1 \sin x$

2. Assume that $\sin(x + k\pi) = (-1)^k \sin x$.

Then, $\sin[x + (k+1)\pi] = \sin[(x + k\pi) + \pi] = (-1)^1 \sin(x + k\pi)$

$= (-1)^1(-1)^k \sin x = (-1)^{k+1} \sin x$.

Thus, the formula is valid for all positive interger values of n.

47. 1. When $n = 1$, $(1^3 + 3(1)^2 + 2(1)) = 6$ and 3 is a factor.

2. Assume that 3 is a factor of $(k^3 + 3k^2 + 2k)$.

Then,

$[(k + 1)^3 + 3(k + 1)^2 + 2(k + 1)]$

$= k^3 + 3k^2 + 3k + 1 + 3k^2 + 6k + 3 + 2k + 2$

$= (k^3 + 3k^2 + 2k) + (3k^2 + 9k + 6)$

$= (k^3 + 3k^2 + 2k) + 3(k^2 + 3k + 2)$.

Since 3 is a factor of $(k^3 + 3k^2 + 2k)$, our assumption, and 3 is a factor of $3(k^2 + 3k + 2)$, then 3 is a factor of the whole sum.

Thus, 3 is a factor of $(n^3 + 3n^2 + 2n)$ for every positive integer n.

49. See page 622 in the textbook.

51. $a_0 = 1, a_n = a_{n-1} + 2$

$a_0 = 1$

$a_1 = a_0 + 2 = 1 + 2 = 3$

$a_2 = a_1 + 2 = 3 + 2 = 5$

$a_3 = a_2 + 2 = 5 + 2 = 7$

$a_4 = a_3 + 2 = 7 + 2 = 9$

53. $a_0 = 4, a_1 = 2, a_n = a_{n-1} - a_{n-2}$

$a_0 = 4$

$a_1 = 2$

$a_2 = a_1 - a_0 = 2 - 4 = -2$

$a_3 = a_2 - a_1 = -2 - 2 = -4$

$a_4 = a_3 - a_2 = -4 - (-2) = -2$

55. $f(1) = 0, a_n = a_{n-1} + 3$

$a_1 = f(1) = 0$

$a_2 = a_1 + 3 = 0 + 3 = 3$

$a_3 = a_2 + 3 = 3 + 3 = 6$

$a_4 = a_3 + 3 = 6 + 3 = 9$

$a_5 = a_4 + 3 = 9 + 3 = 12$

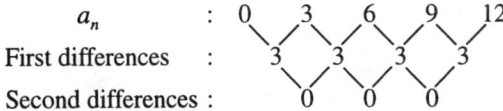

Since the first differences are equal, the sequence has a linear model.

57. $f(1) = 3, a_n = a_{n-1} - n$

$a_1 = f(1) = 3$

$a_2 = a_1 - 2 = 3 - 2 = 1$

$a_3 = a_2 - 3 = 1 - 3 = -2$

$a_4 = a_3 - 4 = -2 - 4 = -6$

$a_5 = a_4 - 5 = -6 - 5 = -11$

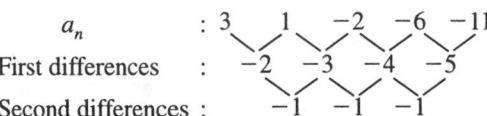

Since the second differences are all the same, the sequence has a quadratic model.

59. $a_0 = 0, a_n = a_{n-1} + n$

$a_0 = 0$

$a_1 = a_0 + 1 = 0 + 1 = 1$

$a_2 = a_1 + 2 = 1 + 2 = 3$

$a_3 = a_2 + 3 = 3 + 3 = 6$

$a_4 = a_3 + 4 = 6 + 4 = 10$

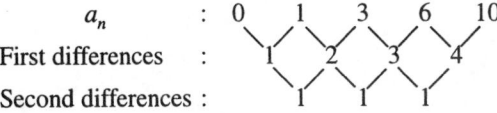

Since the second differences are equal, the sequence has a quadratic model.

61. $f(1) = 2, a_n = a_{n-1} + 2$

$a_1 = f(1) = 2$

$a_2 = a_1 + 2 = 2 + 2 = 4$

$a_3 = a_2 + 2 = 4 + 2 = 6$

$a_4 = a_3 + 2 = 6 + 2 = 8$

$a_5 = a_4 + 2 = 8 + 2 = 10$

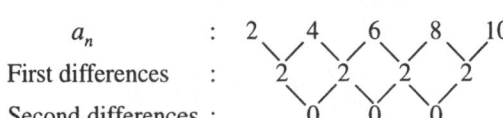

Since the first differences are equal, the sequence has a linear model.

63. $a_0 = 1, a_n = a_{n-1} + n^2$

$a_0 = 1$

$a_1 = 1 + 1^2 = 2$

$a_2 = 2 + 2^2 = 6$

$a_3 = 6 + 3^2 = 15$

$a_4 = 15 + 4^2 = 31$

a_n : 1   2   6   15   31

First differences : 1   4   9   16

Second differences : 3   5   7

Since neither the first differences nor the second differences are equal, the sequence does not have a linear or a quadratic model.

65. $a_0 = 3, a_1 = 3, a_4 = 15$

Let $a_n = an^2 + bn + c$.

Thus: $a_0 = a(0)^2 + b(0) + c = \ \ 3 \implies \ c = 3$

$\quad\ \ a_1 = a(1)^2 + b(1) + c = \ \ 3 \implies a + b + c = 3$

$\qquad\qquad\qquad\qquad\qquad\qquad\quad a + b = 0$

$\quad\ \ a_4 = a(4)^2 + b(4) + c = 15 \implies 16a + 4b + c = 15$

$\qquad\qquad\qquad\qquad\qquad\qquad\quad 16a + 4b = 12$

$\qquad\qquad\qquad\qquad\qquad\qquad\quad\ 4a + \ b = 3$

By elimination: $\ -a - b = 0$

$\qquad\qquad\qquad\ \ \underline{4a + b = 3}$

$\qquad\qquad\qquad\ \ 3a \quad\ \ = 3$

$\qquad\qquad\qquad\qquad a = 1 \implies b = -1$

Thus, $a_n = n^2 - n + 3$.

67. $a_0 = -3, a_2 = 1, a_4 = 9$

Let $a_n = an^2 + bn + c$.

Then: $a_0 = a(0)^2 + b(0) + c = -3 \implies \ c = -3$

$\quad\ \ a_2 = a(2)^2 + b(2) + c = \ \ \ 1 \implies 4a + 2b + c = 1$

$\qquad\qquad\qquad\qquad\qquad\qquad\quad 4a + 2b = 4$

$\qquad\qquad\qquad\qquad\qquad\qquad\quad 2a + \ b = 2$

$\quad\ \ a_4 = a(4)^2 + b(4) + c = \ \ \ 9 \implies 16a + 4b + c = 9$

$\qquad\qquad\qquad\qquad\qquad\qquad\quad 16a + 4b = 12$

$\qquad\qquad\qquad\qquad\qquad\qquad\quad\ 4a + \ b = 3$

By elimination: $\ -2a - b = -2$

$\qquad\qquad\qquad\ \ \underline{4a + b = \ \ \ 3}$

$\qquad\qquad\qquad\ \ 2a \quad\ \ = -1$

$\qquad\qquad\qquad\qquad a = \ \tfrac{1}{2} \implies b = 1$

Thus, $a_n = \tfrac{1}{2}n^2 + n - 3$.

69. $y = x^2$

$\quad -3x + 2y = 2 \quad \implies \quad -3x + 2x^2 = 2$

$\qquad\qquad\qquad\qquad\qquad\ 2x^2 - 3x - 2 = 0$

$\qquad\qquad\qquad\qquad\quad (2x + 1)(x - 2) = 0$

$\qquad\qquad\qquad\qquad\qquad x = -\tfrac{1}{2} \ \text{ or } \ x = 2$

$\qquad\qquad\qquad\qquad\qquad\ \ y = \tfrac{1}{4} \qquad\ \ y = 4$

Points of intersection: $\left(-\tfrac{1}{2}, \tfrac{1}{4}\right), (2, 4)$

71.
$$
\begin{aligned}
x - y &= -1 \\
x + 2y - 2z &= 3 \\
3x - y + 2z &= 3
\end{aligned}
$$

Using an augmented matrix, we have:

$$
\begin{bmatrix}
1 & -1 & 0 & \vdots & -1 \\
1 & 2 & -2 & \vdots & 3 \\
3 & -1 & 2 & \vdots & 3
\end{bmatrix}
$$

$$
\begin{matrix}
\\
-R_1 + R_2 \rightarrow \\
-3R_1 + R_3 \rightarrow
\end{matrix}
\begin{bmatrix}
1 & -1 & 0 & \vdots & -1 \\
0 & 3 & -2 & \vdots & 4 \\
0 & 2 & 2 & \vdots & 6
\end{bmatrix}
$$

$$
\begin{matrix}
\\
-R_3 + R_2 \rightarrow \\
\tfrac{1}{2}R_3 \rightarrow
\end{matrix}
\begin{bmatrix}
1 & -1 & 0 & \vdots & -1 \\
0 & 1 & -4 & \vdots & -2 \\
0 & 1 & 1 & \vdots & 3
\end{bmatrix}
$$

$$
\begin{matrix}
R_2 + R_1 \rightarrow \\
\\
-R_2 + R_3 \rightarrow
\end{matrix}
\begin{bmatrix}
1 & 0 & -4 & \vdots & -3 \\
0 & 1 & -4 & \vdots & -2 \\
0 & 0 & 5 & \vdots & 5
\end{bmatrix}
$$

$$
\begin{matrix}
4R_3 + R_1 \rightarrow \\
4R_3 + R_2 \rightarrow \\
\tfrac{1}{5}R_3 \rightarrow
\end{matrix}
\begin{bmatrix}
1 & 0 & 0 & \vdots & 1 \\
0 & 1 & 0 & \vdots & 2 \\
0 & 0 & 1 & \vdots & 1
\end{bmatrix}
$$

Thus, $x = 1, y = 2, z = 1$.

Answer: $(1, 2, 1)$

Section 8.5   The Binomial Theorem

- You should be able to use the formula
$$
(x + y)^n = x^n + nx^{n-1}y + \frac{n(n-1)}{2!}x^{n-2}y^2 + \cdots + {}_nC_r x^{n-r}y^r + \cdots + y^n
$$
where ${}_nC_r = \dfrac{n!}{(n-r)!r!}$, to expand $(x + y)^n$.

- You should be able to use Pascal's Triangle in binomial expansion.

Solutions to Odd-Numbered Exercises

1. ${}_5C_3 = \dfrac{5!}{3!2!} = \dfrac{5 \cdot 4}{2 \cdot 1} = 10$

3. ${}_{12}C_0 = \dfrac{12!}{0!12!} = 1$

5. ${}_{20}C_{15} = \dfrac{20!}{15!5!} = \dfrac{20 \cdot 19 \cdot 18 \cdot 17 \cdot 16}{5 \cdot 4 \cdot 3 \cdot 2 \cdot 1} = 15{,}504$

7. ${}_{100}C_{98} = \dfrac{100!}{98!2!} = \dfrac{100 \cdot 99}{2 \cdot 1} = 4950$

9. ${}_{100}C_2 = \dfrac{100!}{2!98!} = \dfrac{100 \cdot 99}{2 \cdot 1} = 4950$

11. The first and last number in each row is 1. Every other number is found by adding the two numbers immediately above it.

13.
```
          1
        1   1
       1  2  1
      1  3  3  1
     1  4  6  4  1
    1  5 10 10  5  1
   1  6 15 20 15  6  1
  1  7 21 35 (35) 21  7  1
```

$_7C_4 = 35$, the 5th entry in the 8th row.

15.
```
          1
        1   1
       1  2  1
      1  3  3  1
     1  4  6  4  1
    1  5 10 10  5  1
   1  6 15 20 15  6  1
  1  7 21 35 35 21  7  1
 1  8 28 56 70 (56) 28  8  1
```

$_8C_5 = 56$, the 6th entry in the 9th row.

17. $(x + 1)^4 = {}_4C_0x^4 + {}_4C_1x^3(1) + {}_4C_2x^2(1)^2 + {}_4C_3x(1)^3 + {}_4C_4(1)^4$

$\qquad = x^4 + 4x^3 + 6x^2 + 4x + 1$

19. $(a + 2)^3 = {}_3C_0a^3 + {}_3C_1a^2(2) + {}_3C_2a(2)^2 + {}_3C_3(2)^3$

$\qquad\qquad = a^3 + 3a^2(2) + 3a(2)^2 + (2)^3$

$\qquad\qquad = a^3 + 6a^2 + 12a + 8$

21. $(y - 2)^4 = {}_4C_0y^4 - {}_4C_1y^3(2) + {}_4C_2y^2(2)^2 - {}_4C_3y(2)^3 + {}_4C_4(2)^4$

$\qquad\qquad = y^4 - 4y^3(2) + 6y^2(4) - 4y(8) + 16$

$\qquad\qquad = y^4 - 8y^3 + 24y^2 - 32y + 16$

23. $(x + y)^5 = {}_5C_0x^5 + {}_5C_1x^4y + {}_5C_2x^3y^2 + {}_5C_3x^2y^3 + {}_5C_4xy^4 + {}_5C_5y^5$

$\qquad\qquad = x^5 + 5x^4y + 10x^3y^2 + 10x^2y^3 + 5xy^4 + y^5$

25. $(r + 3s)^6 = {}_6C_0r^6 + {}_6C_1r^5(3s) + {}_6C_2r^4(3s)^2 + {}_6C_3r^3(3s)^3 + {}_6C_4r^2(3s)^4$

$\qquad\qquad\quad + {}_6C_5r(3s)^5 + {}_6C_6(3s)^6$

$\qquad\qquad = r^6 + 18r^5s + 135r^4s^2 + 540r^3s^3 + 1215r^2s^4 + 1458rs^5 + 729s^6$

27. $(x - y)^5 = {}_5C_0x^5 - {}_5C_1x^4y + {}_5C_2x^3y^2 - {}_5C_3x^2y^3 + {}_5C_4xy^4 - {}_5C_5y^5$

$\qquad\qquad = x^5 - 5x^4y + 10x^3y^2 - 10x^2y^3 + 5xy^4 - y^5$

29. $(1 - 2x)^3 = {}_3C_01^3 - {}_3C_11^2(2x) + {}_3C_21(2x)^2 - {}_3C_3(2x)^3$

$\qquad\qquad = 1 - 3(2x) + 3(2x)^2 - (2x)^3$

$\qquad\qquad = 1 - 6x + 12x^2 - 8x^3$

31. $(x^2 + 5)^4 = {}_4C_0(x^2)^4 + {}_4C_1(x^2)^3(5) + {}_4C_2(x^2)^2(5)^2 + {}_4C_3(x^2)(5)^3 + {}_4C_4(5)^4$

$\qquad\qquad = x^8 + 4x^6(5) + 6x^4(25) + 4x^2(125) + 625$

$\qquad\qquad = x^8 + 20x^6 + 150x^4 + 500x^2 + 625$

33. $\left(\dfrac{1}{x} + y\right)^5 = {}_5C_0\left(\dfrac{1}{x}\right)^5 + {}_5C_1\left(\dfrac{1}{x}\right)^4y + {}_5C_2\left(\dfrac{1}{x}\right)^3y^2 + {}_5C_3\left(\dfrac{1}{x}\right)^2y^3 + {}_5C_4\left(\dfrac{1}{x}\right)y^4 + {}_5C_5y^5$

$\qquad\qquad = \dfrac{1}{x^5} + \dfrac{5y}{x^4} + \dfrac{10y^2}{x^3} + \dfrac{10y^3}{x^2} + \dfrac{5y^4}{x} + y^5$

35. $2(x - 3)^4 + 5(x - 3)^2 = 2[x^4 - 4(x^3)(3) + 6(x^2)(3^2) - 4(x)(3^3) + 3^4] + 5[x^2 - 2(x)(3) + 3^2]$

$\qquad\qquad\qquad = 2(x^4 - 12x^3 + 54x^2 - 108x + 81) + 5(x^2 - 6x + 9)$

$\qquad\qquad\qquad = 2x^4 - 24x^3 + 113x^2 - 246x + 207$

37. 5th Row of Pascal's Triangle: 1   5   10   10   5   1

$\quad (2t - s)^5 = 1(2t)^5 + 5(2t)^4(-s) + 10(2t)^3(-s)^2 + 10(2t)^2(-s)^3 + 5(2t)(-s)^4 + 1(-s)^5$

$\qquad\qquad = 32t^5 - 80t^4s + 80t^3s^2 - 40t^2s^3 + 10ts^4 - s^5$

39. 4th Row of Pascal's Triangle: 1   4   6   4   1

$\quad (3 - 2z)^4 = 3^4 - 4(3)^3(2z) + 6(3)^2(2z)^2 - 4(3)(2z)^3 + (2z)^4$

$\qquad\qquad = 81 - 216z + 216z^2 - 96z^3 + 16z^4$

41. The term involving x^5 in the expansion of $(x + 3)^{12}$ is

$\quad {}_{12}C_7 x^5 (3)^7 = \dfrac{12!}{7!5!} \cdot 3^7 x^5 = 1{,}732{,}104 x^5$. The coefficient is $1{,}732{,}104$.

43. The term involving $x^8 y^2$ in the expansion of $(x - 2y)^{10}$ is

$\quad {}_{10}C_2 x^8 (-2y)^2 = \dfrac{10!}{2!8!} \cdot 4x^8 y^2 = 180 x^8 y^2$. The coefficient is 180.

45. The coefficient of $x^4 y^5$ in the expansion of $(3x - 2y)^9$ is

$\quad {}_9C_5 (3)^4 (-2)^5 = \dfrac{9!}{5!4!}(81)(-32) = -326{,}592$.

47. The coefficient of $x^8 y^6 = (x^2)^4 y^6$ in the expansion of $(x^2 + y)^{10}$ is ${}_{10}C_6 = 210$.

49. There are $n + 1$ terms in the expansion of $(x + y)^n$.

51. $\left(\sqrt{x} + 3\right)^4 = \left(\sqrt{x}\right)^4 + 4\left(\sqrt{x}\right)^3(3) + 6\left(\sqrt{x}\right)^2(3)^2 + 4\left(\sqrt{x}\right)(3)^3 + (3)^4$

$\qquad\qquad = x^2 + 12x\sqrt{x} + 54x + 108\sqrt{x} + 81$

$\qquad\qquad = x^2 + 12x^{3/2} + 54x + 108x^{1/2} + 81$

53. $(x^{2/3} - y^{1/3})^3 = (x^{2/3})^3 - 3(x^{2/3})^2 (y^{1/3}) + 3(x^{2/3})(y^{1/3})^2 - (y^{1/3})^3$

$\qquad\qquad = x^2 - 3x^{4/3}y^{1/3} + 3x^{2/3}y^{2/3} - y$

55. $\dfrac{f(x + h) - f(x)}{h} = \dfrac{(x + .h)^3 - x^3}{h}$

$\qquad\qquad = \dfrac{x^3 + 3x^2 h + 3xh^2 + h^3 - x^3}{h}$

$\qquad\qquad = \dfrac{h(3x^2 + 3xh + h^2)}{h}$

$\qquad\qquad = 3x^2 + 3xh + h^2$

57. $\dfrac{f(x + h) - f(x)}{h} = \dfrac{\sqrt{x + h} \div \sqrt{x}}{h}$

$\qquad\qquad = \dfrac{\sqrt{x + h} - \sqrt{x}}{h} \cdot \dfrac{\sqrt{x + h} + \sqrt{x}}{\sqrt{x + h} + \sqrt{x}}$

$\qquad\qquad = \dfrac{(x + h) - x}{h\left(\sqrt{x + h} + \sqrt{x}\right)}$

$\qquad\qquad = \dfrac{1}{\sqrt{x + h} + \sqrt{x}}$

59. $(1 + i)^4 = {}_4C_0 1^4 + {}_4C_1 (1)^3 i + {}_4C_2 (1)^2 i^2 + {}_4C_3 1 \cdot i^3 + {}_4C_4 i^4$

$\qquad\qquad = 1 + 4i - 6 - 4i + 1$

$\qquad\qquad = -4$

61. $(2 - 3i)^6 = {}_6C_0 2^6 - {}_6C_1 2^5(3i) + {}_6C_2 2^4(3i)^2 - {}_6C_3 2^3(3i)^3 + {}_6C_4 2^2(3i)^4 - {}_6C_5 2(3i)^5 + {}_6C_6(3i)^6$

$= 64 - 576i - 2160 + 4320i + 4860 - 2916i - 729$

$= 2035 + 828i$

63. $\left(-\dfrac{1}{2} + \dfrac{\sqrt{3}}{2}i\right)^3 = \dfrac{1}{8}\left(-1 + \sqrt{3}i\right)^3$

$= \dfrac{1}{8}\left[(-1)^3 + 3(-1)^2(\sqrt{3}i) + 3(-1)(\sqrt{3}i)^2 + (\sqrt{3}i)^3\right]$

$= \dfrac{1}{8}\left[-1 + 3\sqrt{3}i + 9 - 3\sqrt{3}i\right]$

$= 1$

65. ${}_7C_4\left(\dfrac{1}{2}\right)^4\left(\dfrac{1}{2}\right)^3 = 35\left(\dfrac{1}{16}\right)\left(\dfrac{1}{8}\right) \approx 0.273$ **67.** ${}_8C_4\left(\dfrac{1}{3}\right)^4\left(\dfrac{2}{3}\right)^4 = 70\left(\dfrac{1}{81}\right)\left(\dfrac{16}{81}\right) \approx 0.171$

69. $(1.02)^8 = (1 + 0.02)^8 = 1 + 8(0.02) + 28(0.02)^2 + 56(0.02)^3 + 70(0.02)^4 + 56(0.02)^5$

$+ 28(0.02)^6 + 8(0.02)^7 + (0.02)^8$

$= 1 + 0.16 + 0.0112 + 0.000448 + \cdots \approx 1.172$

71. $(2.99)^{12} = (3 - 0.01)^{12}$

$= 3^{12} - 12(3)^{11}(0.01) + 66(3)^{10}(0.01)^2 - 220(3)^9(0.01)^3 + 495(3)^8(0.01)^4$

$- 792(3)^7(0.01)^5 + 924(3)^6(0.01)^6 - 792(3)^5(0.01)^7 + 495(3)^4(0.01)^8$

$- 220(3)^3(0.01)^9 + 66(3)^2(0.01)^{10} - 12(3)(0.01)^{11} + (0.01)^{12} \approx 510{,}568.785$

73. $f(x) = x^3 - 4x$

$g(x) = f(x + 4)$

$= (x + 4)^3 - 4(x + 4)$

$= x^3 + 3x^2(4) + 3x(4)^2 + (4)^3 - 4x - 16$

$= x^3 + 12x^2 + 48x + 64 - 4x - 16$

$= x^3 + 12x^2 + 44x + 48$

The graph of g is the same as the graph of f shifted 4 units to the left.

75. ${}_nC_{n-r} = \dfrac{n!}{(n - (n - r))!(n - r)!}$

$= \dfrac{n!}{r!(n - r)!}$

$= \dfrac{n!}{(n - r)!r!}$

$= {}_nC_r$

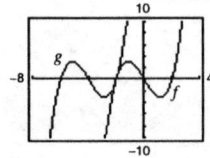

77. $_nC_r + {_nC_{r-1}} = \dfrac{n!}{(n-r)!r!} + \dfrac{n!}{(n-r+1)!(r-1)!}$

$$= \dfrac{n!(n-r+1)!(r-1)! + n!(n-r)!r!}{(n-r)!r!(n-r+1)!(r-1)!}$$

$$= \dfrac{n![(n-r+1)!(r-1)! + r!(n-r)!]}{(n-r)!r!(n-r+1)!(r-1)!}$$

$$= \dfrac{n!(r-1)![(n-r+1)! + r(n-r)!]}{(n-r)!r!(n-r+1)!(r-1)!}$$

$$= \dfrac{n!(n-r)![(n-r+1) + r]}{(n-r)!r!(n-r+1)!}$$

$$= \dfrac{n![n+1]}{r!(n-r+1)!}$$

$$= \dfrac{(n+1)!}{[(n+1)-r]!r!}$$

$$= {_{n+1}C_r}$$

79. $f(t) = 0.1506t^2 + 0.7361t + 21.1374, \; 0 \le t \le 22$

(a) $g(t) = f(t+10)$

$\qquad = 0.1506(t+10)^2 + 0.7361(t+10) + 21.1374$

$\qquad = 0.1506(t^2 + 20t + 100) + 0.7361t + 7.361 + 21.1374$

$\qquad = 0.1506t^2 + 3.7481t + 43.5584$

(b)

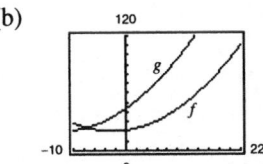

81. $f(x) = (1-x)^3$

$g(x) = 1 - 3x$

$h(x) = 1 - 3x + 3x^2$

$p(x) = 1 - 3x + 3x^2 - x^3$

Since $p(x)$ is the expansion of $f(x)$, they have the same graph.

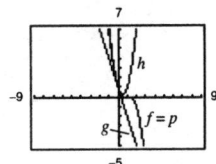

83. $g(x) = f(x-3)$

$g(x)$ is shifted 3 units to the right of $f(x)$.

85. $g(x) = -f(x)$

$g(x)$ is the reflection of $f(x)$ in the x-axis.

Section 8.6   Counting Principles

- You should know The Fundamental Principle of Counting.

- $_nP_r = \dfrac{n!}{(n-r)!}$ is the number of permutations of n elements taken r at a time.

- Given a set of n objects that has n_1 of one kind, n_2 of a second kind, and so on, the number of distinguishable permutations is

$$\frac{n!}{n_1!n_2!\ldots n_k!}.$$

- $_nC_r = \dfrac{n!}{(n-r)!r!}$ is the number of combinations of n elements taken r at a time.

Solutions to Odd-Numbered Exercises

1. Odd integers: 1, 3, 5, 7, 9, 11
6 ways

3. Prime integers: 2, 3, 5, 7, 11
5 ways

5. Divisible by 4: 4, 8, 12
3 ways

7. Sum is 8: $1 + 7$, $2 + 6$, $3 + 5$, $4 + 4$
4 ways

9. Amplifiers: 2 choices
Compact disc players: 4 choices
Speakers: 6 choices
Total: $2 \cdot 4 \cdot 6 = 48$ ways

11. Chemist: 3 choices
Statistician: 4 choices
Total: $3 \cdot 4 = 12$ ways

13. $2^6 = 64$

15. 1st Position: 2 choices
2nd Position: 3 choices
3rd Position: 2 choices
4th Position: 1 choices
Total: $2 \cdot 3 \cdot 2 \cdot 1 = 12$ ways

Label the four people A, B, C, and D and suppose that A and B are willing to take the first position. The twelve combinations are as follows

ABCD	BACD
ABDC	BADC
ACBD	BCAD
ACDB	BCDA
ADBC	BDAC
ADCB	BDCA

17. $26 \cdot 26 \cdot 10 \cdot 10 \cdot 10 \cdot 10 = 6{,}760{,}000$

19. (a) $9 \cdot 10 \cdot 10 = 900$

    (b) $9 \cdot 9 \cdot 8 = 648$

    (c) $9 \cdot 10 \cdot 2 = 180$

    (d) $10 \cdot 10 \cdot 10 - 400 = 600$

21. $40^3 = 64,000$

23. (a) $6 \cdot 5 \cdot 4 \cdot 3 \cdot 2 \cdot 1 = 720$

    (b) $6 \cdot 1 \cdot 4 \cdot 1 \cdot 2 \cdot 1 = 48$

25. $_nP_r = \dfrac{n!}{(n-r)!}$

    So, $_4P_4 = \dfrac{4!}{0!} = 4! = 24.$

27. $_8P_3 = \dfrac{8!}{5!} = 8 \cdot 7 \cdot 6 = 336$

29. $_5P_4 = \dfrac{5!}{1!} = 120$

31. $14 \cdot {}_nP_3 = {}_{n+2}P_4$ Note: $n \geq 3$ for this to be defined.

    $14\left(\dfrac{n!}{(n-3)!}\right) = \dfrac{(n+2)!}{(n-2)!}$

    $14n(n-1)(n-2) = (n+2)(n+1)n(n-1)$ (We can divide here by $n(n-1)$ since $n \neq 0, n \neq 1$.)

        $14n(n-2) = (n+2)(n+1)$

        $14n - 28 = n^2 + 3n + 2$

           $0 = n^2 - 11n + 30$

           $0 = (n-5)(n-6)$

           $n = 5$ or $n = 6$

33. $_{20}P_5 = 1,860,480$

35. $_{100}P_3 = 970,200$

37. $_{20}C_5 = 15,504$

39. $_{100}P_{80} \approx 3.836 \times 10^{139}$

This number is too large for some calculators to evaluate.

41.

ABCD	BACD	CABD	DABC
ABDC	BADC	CADB	DACB
ACBD	BCAD	CBAD	DBAC
ACDB	BCDA	CBDA	DBCA
ADBC	BDAC	CDAB	DCAB
ADCB	BDCA	CDBA	DCBA

43. $5! = 120$ ways

45. $_{12}P_4 = \dfrac{12!}{8!} = 12 \cdot 11 \cdot 10 \cdot 9 = 11,880$ ways

47. $\dfrac{7!}{2!1!3!1!} = \dfrac{7!}{2!3!} = 420$

49. $\dfrac{7!}{2!1!1!1!1!1!} = \dfrac{7!}{2!} = 7 \cdot 6 \cdot 5 \cdot 4 \cdot 3 = 2520$

51. $_6C_2 = 15$

The 15 ways are listed below.

AB, AC, AD, AE, AF,

BC, BD, BE, BF, CD,

CE, CF, DE, DF, EF

53. $_{20}C_4 = 4845$ groups

55. $_{40}C_6 = 3,838,380$ ways

57. $_{100}C_4 = 3,921,225$ subsets

59. $_7C_2 = 21$ lines

61. (a) $_8C_4 = \dfrac{8!}{(8-4)!4!} = \dfrac{8!}{4!4!} = \dfrac{8 \cdot 7 \cdot 6 \cdot 5}{4 \cdot 3 \cdot 2} = 70$ ways

(b) $_3C_2 \cdot \, _5C_2 = \dfrac{3!}{(3-2)!2!} \cdot \dfrac{5!}{(5-2)!2!} = 3 \cdot 10 = 30$ ways

63. (a) $_8C_4 = \dfrac{8!}{4!4!} = 70$ ways

(b) There are 10 ways that a group of four can be formed without any couples in the group. Therefore, if at least one couple is to be in the group, there are $70 - 10 = 60$ ways that could occur.

(c) $2 \cdot 2 \cdot 2 \cdot 2 = 16$ ways

65. $_5C_2 - 5 = 10 - 5 = 5$ diagonals

67. $_8C_2 - 8 = 28 - 8 = 20$ diagonals

69. $_nP_{n-1} = \dfrac{n!}{(n-(n-1))!} = \dfrac{n!}{1!} = \dfrac{n!}{0!} = \, _nP_n$

71. $_nC_{n-1} = \dfrac{n!}{(n-(n-1))!(n-1)!} = \dfrac{n!}{(1)!(n-1)!} = \dfrac{n!}{(n-1)!1!} = \, _nC_1$

Section 8.7   Probability

You should know the following basic principles of probability.

■ If an event E has $n(E)$ equally likely outcomes and its sample space has $n(S)$ equally likely outcomes, then the probability of event E is

$$P(E) = \frac{n(E)}{n(S)}, \text{ where } 0 \le P(E) \le 1.$$

■ If A and B are mutually exclusive events, then $P(A \cup B) = P(A) + P(B)$.

If A and B are not mutually exclusive events, then $P(A \cup B) = P(A) + P(B) - P(A \cap B)$.

■ If A and B are independent events, then the probability that both A and B will occur is $P(A)P(B)$.

■ The complement of an event E is $P(E') = 1 - P(E)$.

Solutions to Odd-Numbered Exercises

1. $\{(h, 1), (h, 2), (h, 3), (h, 4), (h, 5), (h, 6),$
$(t, 1), (t, 2), (t, 3), (t, 4), (t, 5), (t, 6)\}$

3. $\{ABC, ACB, BAC, BCA, CAB, CBA\}$

5. $\{(A, B), (A, C), (A, D), (A, E), (B, C), (B, D), (B, E), (C, D), (C, E), (D, E)\}$

7. $E = \{HHT, HTH, THH\}$

$$P(E) = \frac{n(E)}{n(S)} = \frac{3}{8}$$

9. $E = \{HHH, HHT, HTH, HTT, THH, THT, TTH\}$

$$P(E) = \frac{n(E)}{n(S)} = \frac{7}{8}$$

11. $E = \{K, K, K, K, Q, Q, Q, Q, J, J, J, J\}$

$$P(E) = \frac{n(E)}{n(S)} = \frac{12}{52} = \frac{3}{13}$$

13. $E = \{K, K, Q, Q, J, J\}$

$$P(E) = \frac{n(E)}{n(S)} = \frac{6}{52} = \frac{3}{26}$$

15. $E = \{(1, 3), (2, 2), (3, 1)\}$

$$P(E) = \frac{n(E)}{n(S)} = \frac{3}{36} = \frac{1}{12}$$

17. not $E = \{(5, 6), (6, 5), (6, 6)\}$

$$n(E) = n(S) - n(\text{not } E) = 36 - 3 = 33$$

$$P(E) = \frac{n(E)}{n(S)} = \frac{33}{36} = \frac{11}{12}$$

19. $E_3 = \{(1, 2), (2, 1)\}, \ n(E_3) = 2$

$E_5 = \{(1, 4), (2, 3), (3, 2), (4, 1)\}, \ n(E_5) = 4$

$E_7 = \{(1, 6), (2, 5), (3, 4), (4, 3), (5, 2), (6, 1)\}, \ n(E_7) = 6$

$E = E_3 \cup E_5 \cup E_7$

$n(E) = 2 + 4 + 6 = 12$

$$P(E) = \frac{n(E)}{n(S)} = \frac{12}{36} = \frac{1}{3}$$

21. $P(E) = \dfrac{{}_3C_2}{{}_6C_2} = \dfrac{3}{15} = \dfrac{1}{5}$

23. $P(E) = \dfrac{{}_4C_2}{{}_6C_2} = \dfrac{6}{15} = \dfrac{2}{5}$

25. $1 - p = 1 - 0.7 = 0.3$

27. $1 - p = 1 - 0.15 = 0.85$

29. (a) $0.37(2.5) = 0.925$ million $= 925,000$

(b) 18%

(c) $11\% + 6\% = 17\%$

31. (a) $\frac{290}{500} = 0.58 = 58\%$

(b) $\frac{478}{500} = 0.956 = 95.6\%$

(c) $\frac{2}{500} = 0.004 = 0.4\%$

33. (a) $\dfrac{672}{1254}$

(b) $\dfrac{582}{1254}$

(c) $\dfrac{672 - 124}{1254} = \dfrac{548}{1254}$

35. $p + p + 2p = 1$

$\qquad\qquad p = 0.25$

Taylor: $0.50 = \frac{1}{2}$

Moore: $0.25 = \frac{1}{4}$

Jenkins: $0.25 = \frac{1}{4}$

37. (a) $\dfrac{{}_{15}C_{10}}{{}_{20}C_{10}} = \dfrac{3003}{184,756} = \dfrac{21}{1292} \approx 0.016$

(b) $\dfrac{{}_{15}C_8 \cdot {}_5C_2}{{}_{20}C_{10}} = \dfrac{64,350}{184,756} = \dfrac{225}{646} \approx 0.348$

(c) $\dfrac{{}_{15}C_9 \cdot {}_5C_1}{{}_{20}C_{10}} + \dfrac{{}_{15}C_{10}}{{}_{20}C_{10}} = \dfrac{25,025 + 3003}{184,756} = \dfrac{28,028}{184,756} = \dfrac{49}{323} \approx 0.152$

39. Total ways to insert letters: $4! = 24$ ways

    4 correct: 1 way

    3 correct: not possible

    2 correct: 6 ways

    1 correct: 8 ways

    0 correct: 9 ways

(a) $\dfrac{8}{24} = \dfrac{1}{3}$

(b) $\dfrac{8 + 6 + 1}{24} = \dfrac{15}{24} = \dfrac{5}{8}$

41. (a) $\dfrac{1}{{}_5P_5} = \dfrac{1}{120}$

    (b) $\dfrac{1}{{}_4P_4} = \dfrac{1}{24}$

43. (a) $\dfrac{4}{52} \cdot \dfrac{4}{52} = \dfrac{1}{169}$

    (b) $\dfrac{4}{52} \cdot \dfrac{3}{51} = \dfrac{1}{221}$

45. (a) $\dfrac{{}_9C_4}{{}_{12}C_4} = \dfrac{126}{495} = \dfrac{14}{55}$ (4 good units)

    (c) $\dfrac{({}_9C_3)({}_3C_1)}{{}_{12}C_4} = \dfrac{252}{495} = \dfrac{28}{55}$ (3 good units)

(b) $\dfrac{({}_9C_2)({}_3C_2)}{{}_{12}C_4} = \dfrac{108}{495} = \dfrac{12}{55}$ (2 good units)

    At least 2 good units: $\dfrac{12}{55} + \dfrac{28}{55} + \dfrac{14}{55} = \dfrac{54}{55}$

47. (a) $P(EE) = \dfrac{15}{30} \cdot \dfrac{15}{30} = \dfrac{1}{4}$

    (c) $P(N_1 < 10, N_2 < 10) = \dfrac{9}{30} \cdot \dfrac{9}{30} = \dfrac{9}{100}$

(b) $P(EO \text{ or } OE) = 2\left(\dfrac{15}{30}\right)\left(\dfrac{15}{30}\right) = \dfrac{1}{2}$

(d) $P(N_1 N_1) = \dfrac{30}{30} \cdot \dfrac{1}{30} = \dfrac{1}{30}$

49. (a) $P(SS) = (0.985)^2 \approx 0.9702$

    (b) $P(S) = 1 - P(FF) = 1 - (0.015)^2 \approx 0.9998$

    (c) $P(FF) = (0.015)^2 \approx 0.0002$

51. (a) $\left(\dfrac{1}{4}\right)^5 = \dfrac{1}{1024}$

    (b) $\left(\dfrac{3}{4}\right)^5 = \dfrac{243}{1024}$

    (c) $1 - \dfrac{243}{1024} = \dfrac{781}{1024}$

53. $(0.32)^2 = 0.1024$

55. $1 - \dfrac{(45)^2}{(60)^2} = 1 - \left(\dfrac{45}{60}\right)^2 = 1 - \left(\dfrac{3}{4}\right)^2 = 1 - \dfrac{9}{16} = \dfrac{7}{16}$

57. (a) As you consider successive people with distinct birthdays, the probabilities must decrease to take into account the birth dates already used. Because the birth dates of people are independent events, multiply the respective probabilities of distinct birthdays.

(b) $\dfrac{365}{365} \cdot \dfrac{364}{365} \cdot \dfrac{363}{365} \cdot \dfrac{362}{365}$

(c) $P_1 = \dfrac{365}{365} = 1$

$P_2 = \dfrac{365}{365} \cdot \dfrac{364}{365} = \dfrac{364}{365} P_1 = \dfrac{365 - (2 - 1)}{365} P_1$

$P_3 = \dfrac{365}{365} \cdot \dfrac{364}{365} \cdot \dfrac{363}{365} = \dfrac{363}{365} P_2 = \dfrac{365 - (3 - 1)}{365} P_2$

$P_n = \dfrac{365}{365} \cdot \dfrac{364}{365} \cdot \dfrac{363}{365} \cdot \ldots \dfrac{365 - (n - 1)}{365} = \dfrac{365 - (n - 1)}{365} P_{n-1}$

(d) Q_n is the probability that the birthdays are *not* distinct which is equivalent to at least 2 people having the same birthday.

(e)

n	10	15	20	23	30	40	50
P_n	0.88	0.75	0.59	0.49	0.29	0.11	0.03
Q_n	0.12	0.25	0.41	(0.51)	0.71	0.89	0.97

(f) 23,   See the chart above.

59. $1 - 0.546 = 0.454$

❑ **Review Exercises for Chapter 8**

Solutions to Odd-Numbered Exercises

1. $a_n = 2 + \dfrac{6}{n}$

$a_1 = 2 + \dfrac{6}{1} = 8$

$a_2 = 2 + \dfrac{6}{2} = 5$

$a_3 = 2 + \dfrac{6}{3} = 4$

$a_4 = 2 + \dfrac{6}{4} = \dfrac{7}{2}$

$a_5 = 2 + \dfrac{6}{5} = \dfrac{16}{5}$

3. $a_n = \dfrac{72}{n!}$

$a_1 = \dfrac{72}{1!} = 72$

$a_2 = \dfrac{72}{2!} = 36$

$a_3 = \dfrac{72}{3!} = 12$

$a_4 = \dfrac{72}{4!} = 3$

$a_5 = \dfrac{72}{5!} = \dfrac{3}{5}$

5. $a_n = \dfrac{3}{2}n$

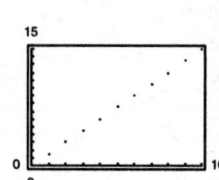

7. $a_n = \dfrac{3n}{n + 2}$

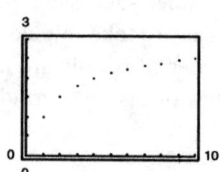

9. $\dfrac{1}{2(1)} + \dfrac{1}{2(2)} + \dfrac{1}{2(3)} + \cdots + \dfrac{1}{2(20)} = \displaystyle\sum_{k=1}^{20} \dfrac{1}{2k}$

11. $\dfrac{1}{2} + \dfrac{2}{3} + \dfrac{3}{4} + \cdots + \dfrac{9}{10} = \displaystyle\sum_{k=1}^{9} \dfrac{k}{k + 1}$

13. $\displaystyle\sum_{i=1}^{6} 5 = 6(5) = 30$

15. $\displaystyle\sum_{j=1}^{4} \dfrac{6}{j^2} = \dfrac{6}{1^2} + \dfrac{6}{2^2} + \dfrac{6}{3^2} + \dfrac{6}{4^2} = 6 + \dfrac{3}{2} + \dfrac{2}{3} + \dfrac{3}{8} = \dfrac{205}{24}$

17. $\displaystyle\sum_{k=1}^{10} 2k^3 = 2(1)^3 + 2(2)^3 + 2(3)^3 + \cdots + 2(10)^3 = 6050$

19. $\displaystyle\sum_{n=0}^{10} (n^2 + 3) = \displaystyle\sum_{n=0}^{10} n^2 + \displaystyle\sum_{n=0}^{10} 3 = \dfrac{10(11)(21)}{6} + 11(3) = 418$

21. $a_1 = 3, d = 4$

$a_1 = 3$

$a_2 = 3 + 4 = 7$

$a_3 = 7 + 4 = 11$

$a_4 = 11 + 4 = 15$

$a_5 = 15 + 4 = 19$

23. $a_4 = 10 \quad a_{10} = 28$

$a_{10} = a_4 + 6d$

$28 = 10 + 6d$

$18 = 6d$

$3 = d$

$a_1 = a_4 - 3d$

$a_1 = 10 - 3(3)$

$a_1 = 1$

$a_2 = 1 + 3 = 4$

$a_3 = 4 + 3 = 7$

$a_4 = 7 + 3 = 10$

$a_5 = 10 + 3 = 13$

25. $a_1 = 35, a_{k+1} = a_k - 3$

$a_1 = 35$

$a_2 = a_1 - 3 = 35 - 3 = 32$

$a_3 = a_2 - 3 = 32 - 3 = 29$

$a_4 = a_3 - 3 = 29 - 3 = 26$

$a_5 = a_4 - 3 = 26 - 3 = 23$

$a_n = 35 + (n - 1)(-3) = 38 - 3n$

27. $a_1 = 9, a_{k+1} = a_k + 7$

$a_1 = 9$

$a_2 = a_1 + 7 = 9 + 7 = 16$

$a_3 = a_2 + 7 = 16 + 7 = 23$

$a_4 = a_3 + 7 = 23 + 7 = 30$

$a_5 = a_4 + 7 = 30 + 7 = 37$

$a_n = 9 + (n - 1)(7) = 2 + 7n$

29. $a_n = 100 + (n - 1)(-3) = 103 - 3n, a_1 = 100, a_{20} = 43, S_{20} = \frac{20}{2}(100 + 43) = 1430$

31. $\displaystyle\sum_{j=1}^{10}(2j - 3)$ is arithmetic. Therefore, $a_1 = -1, a_{10} = 17, S_{10} = \frac{10}{2}[-1 + 17] = 80.$

33. $\displaystyle\sum_{k=1}^{11}\left(\frac{2}{3}k + 4\right)$ is arithmetic. Therefore, $a_1 = \frac{14}{3}, a_{11} = \frac{34}{3}, s_{11} = \frac{11}{2}\left[\frac{14}{3} + \frac{34}{3}\right] = 88.$

35. $\displaystyle\sum_{k=1}^{100} 5k$ is arithmetic. Therefore, $a_1 = 5, a_{100} = 500, s_{500} = \frac{100}{2}(5 + 500) = 25,250.$

37. (a) $34,000 + 4(2250) = \$43,000$

   (b) $\displaystyle\sum_{k=1}^{5}[34,000 + (k - 1)(2250)] = \sum_{k=1}^{5}(31,750 + 2250k) = \$192,500$

39. $a_1 = 4,\ r = -\frac{1}{4}$

$a_1 = 4$

$a_2 = 4\left(-\frac{1}{4}\right) = -1$

$a_3 = -1\left(-\frac{1}{4}\right) = \frac{1}{4}$

$a_4 = \frac{1}{4}\left(-\frac{1}{4}\right) = -\frac{1}{16}$

$a_5 = -\frac{1}{16}\left(-\frac{1}{4}\right) = \frac{1}{64}$

41. $a_1 = 9,\ a_3 = 4$

$a_3 = a_1 r^2$

$4 = 9r^2$

$\frac{4}{9} = r^2 \implies r = \pm\frac{2}{3}$

$a_1 = 9$	$a_1 = 9$
$a_2 = 9\left(\frac{2}{3}\right) = 6$	$a_2 = 9\left(-\frac{2}{3}\right) = -6$
$a_3 = 6\left(\frac{2}{3}\right) = 4$ OR	$a_3 = -6\left(-\frac{2}{3}\right) = 4$
$a_4 = 4\left(\frac{2}{3}\right) = \frac{8}{3}$	$a_4 = 4\left(-\frac{2}{3}\right) = -\frac{8}{3}$
$a_5 = \frac{8}{3}\left(\frac{2}{3}\right) = \frac{16}{9}$	$a_5 = -\frac{8}{3}\left(-\frac{2}{3}\right) = \frac{16}{9}$

43. $a_1 = 120, a_{k+1} = \frac{1}{3}a_k$

$a_1 = 120$

$a_2 = \frac{1}{3}(120) = 40$

$a_3 = \frac{1}{3}(40) = \frac{40}{3}$

$a_4 = \frac{1}{3}\left(\frac{40}{3}\right) = \frac{40}{9}$

$a_5 = \frac{1}{3}\left(\frac{40}{9}\right) = \frac{40}{27}$

$a_n = 120\left(\frac{1}{3}\right)^{n-1}$

45. $a_1 = 25,\ a_{k+1} = -\frac{3}{5}a_k$

$a_1 = 25$

$a_2 = -\frac{3}{5}(25) = -15$

$a_3 = -\frac{3}{5}(-15) = 9$

$a_4 = -\frac{3}{5}(9) = -\frac{27}{5}$

$a_5 = -\frac{3}{5}\left(-\frac{27}{5}\right) = \frac{81}{25}$

$a_n = 25\left(-\frac{3}{5}\right)^{n-1}$

47. $a_2 = a_1 r$

$-8 = 16r$

$-\frac{1}{2} = r$

$a_n = 16\left(-\frac{1}{2}\right)^{n-1}$

$\displaystyle\sum_{n=1}^{20} 16\left(-\frac{1}{2}\right)^{n-1} = 16\left[\frac{1-\left(-\frac{1}{2}\right)^{20}}{1-\left(-\frac{1}{2}\right)}\right] \approx 10.67$

49. $\displaystyle\sum_{i=1}^{7} 2^{i-1} = \frac{1-2^7}{1-2} = 127$

51. $\displaystyle\sum_{i=1}^{\infty}\left(\frac{7}{8}\right)^{i-1} = \frac{1}{1-\frac{7}{8}} = 8$

53. $\displaystyle\sum_{k=1}^{\infty} 4\left(\frac{2}{3}\right)^{k-1} = \frac{4}{1-\frac{2}{3}} = 12$

55. $\displaystyle\sum_{i=1}^{10} 10\left(\frac{3}{5}\right)^{i-1} \approx 24.849$

57. (a) $a_t = 120{,}000(0.7)^t$

(b) $a_5 = 120{,}000(0.7)^5 = \$20{,}168.40$

59. $A = \displaystyle\sum_{i=1}^{24} 200\left(1+\frac{0.06}{12}\right)^t \approx \$5111.82$

61. 1. When $n = 1$, $1 = \frac{1}{2}(3(1) - 1)$.

2. Assume that

$S_k = 1 + 4 + \cdots + (3k-2) = \frac{k}{2}(3k-1).$

Then,

$S_{k+1} = 1 + 4 + \cdots + (3k-2) + (3(k+1)-2) = S_k + (3k+1)$

$= \frac{k}{2}(3k-1) + (3k+1) = \frac{k(3k-1)+2(3k+1)}{2}$

$= \frac{3k^2 + 5k + 2}{2} = \frac{(k+1)(3k+2)}{2} = \frac{(k+1)}{2}(3(k+1)-1).$

Therefore, by mathematical induction, the formula is valid for all positive integer values of n.

63. 1. When $n = 1$, $a = a\left(\dfrac{1 - r}{1 - r}\right)$.

   2. Assume that

   $$S_k = \sum_{i=0}^{k-1} ar^i = \frac{a(1 - r^k)}{1 - r}$$

   Then,

   $$S_{k+1} = \sum_{i=0}^{k} ar^i = \sum_{i=0}^{k-1} ar^i + ar^k = \frac{a(1 - r^k)}{1 - r} + ar^k$$

   $$= \frac{a(1 - r^k + r^k - r^{k+1})}{1 - r} = \frac{a(1 - r^{k+1})}{1 - r}.$$

   Therefore, by mathematical induction, the formula is valid for all positive integer values of n.

65. $_6C_4 = \dfrac{6!}{2!4!} = 15$ **67.** $_8P_5 = \dfrac{8!}{3!} = 6720$

69. $\left(\dfrac{x}{2} + y\right)^4 = \left(\dfrac{x}{2}\right)^4 + 4\left(\dfrac{x}{2}\right)^3 y + 6\left(\dfrac{x}{2}\right)^2 y^2 + 4\left(\dfrac{x}{2}\right) y^3 + y^4$

   $$= \frac{x^4}{16} + \frac{x^3 y}{2} + \frac{3x^2 y^2}{2} + 2xy^3 + y^4$$

71. $\left(\dfrac{2}{x} - 3x\right)^6 = \left(\dfrac{2}{x}\right)^6 + 6\left(\dfrac{2}{x}\right)^5 (-3x) + 15\left(\dfrac{2}{x}\right)^4 (-3x)^2 + 20\left(\dfrac{2}{x}\right)^3 (-3x)^3$

   $$+ 15\left(\frac{2}{x}\right)^2 (-3x)^4 + 6\left(\frac{2}{x}\right)(-3x)^5 + (-3x)^6$$

   $$= \frac{64}{x^6} - \frac{576}{x^4} + \frac{2160}{x^2} - 4320 + 4860x^2 - 2916x^4 + 729x^6$$

73. $(5 + 2i)^4 = (5)^4 + 4(5)^3(2i) + 6(5)^2(2i)^2 + 4(5)(2i)^3 + (2i)^4$

   $$= 625 + 1000i + 600i^2 + 160i^3 + 16i^4$$

   $$= 625 + 1000i - 600 - 160i + 16 = 41 + 840i$$

75. $(26)(26)(10)(26)(26)(26) = 118,813,760$ **77.** $\dfrac{10}{10} \cdot \dfrac{1}{9} = \dfrac{1}{9}$

79. Chance of rolling a 3 with one die is $\dfrac{1}{6}$. With two dice

   $E = \{(1, 5),\ (2, 4),\ (3, 3),\ (4, 2),\ (5, 1)\}$ and $P(E) = \dfrac{5}{36}$.

   The probability of rolling a 3 with one die is higher.

81. $1 - P(HHHHH) = 1 - \left(\dfrac{1}{2}\right)^5 = \dfrac{31}{32}$ **83.** $P(2 \text{ pairs}) = \dfrac{(_{13}C_2)(_4C_2)(_4C_2)(_{44}C_1)}{(_{52}C_5)} = 0.0475$

❑ Practice Test for Chapter 8

1. Write out the first five terms of the sequence $a_n = \dfrac{2n}{(n+2)!}$.

2. Write an expression for the nth term of the sequence $\left\{ \dfrac{4}{3}, \dfrac{5}{9}, \dfrac{6}{27}, \dfrac{7}{81}, \dfrac{8}{243}, \ldots \right\}$.

3. Find the sum $\displaystyle\sum_{i=1}^{6}(2i-1)$.

4. Write out the first five terms of the arithmetic sequence where $a_1 = 23$ and $d = -2$.

5. Find a_n for the arithmetic sequence with $a_1 = 12$, $d = 3$, and $n = 50$.

6. Find the sum of the first 200 positive integers.

7. Write out the first five terms of the geometric sequence with $a_1 = 7$ and $r = 2$.

8. Evaluate $\displaystyle\sum_{n=0}^{9} 6\left(\dfrac{2}{3}\right)^n$.

9. Evaluate $\displaystyle\sum_{n=0}^{\infty}(0.03)^n$.

10. Use mathematical induction to prove that $1 + 2 + 3 + 4 + \cdots + n = \dfrac{n(n+1)}{2}$.

11. Use mathematical induction to prove that $n! > 2^n$, $n \geq 4$.

12. Evaluate $_{13}C_4$.

13. Expand $(x+3)^5$.

14. Find the term involving x^7 in $(x-2)^{12}$.

15. Evaluate $_{30}P_4$.

16. How many ways can six people sit at a table with six chairs?

17. Twelve cars run in a race. How many different ways can they come in first, second, and third place? (Assume that there are no ties.)

18. Two six-sided dice are tossed. Find the probability that the total of the two dice is less than 5.

19. Two cards are selected at random form a deck of 52 playing cards without replacement. Find the probability that the first card is a King and the second card is a black ten.

20. A manufacturer has determined that for every 1000 units it produces, 3 will be faulty. What is the probability that an order of 50 units will have one or more faulty units?

❏ Chapter P Practice Test Solutions

1. $\dfrac{|-42| - 20}{15 - |-4|} = \dfrac{42 - 20}{15 - 4} = \dfrac{22}{11} = 2$

2. $\dfrac{x}{z} - \dfrac{z}{y} = \dfrac{x}{z} \cdot \dfrac{y}{y} - \dfrac{z}{y} \cdot \dfrac{z}{z} = \dfrac{xy - z^2}{yz}$

3. $|x - 7| \le 4$

4. $10(-5)^3 = 10(-125) = -1250$

5. $(-4x^3)(-2x^{-5})\left(\dfrac{1}{16}x\right) = (-4)(-2)\left(\dfrac{1}{16}\right)x^{3+(-5)+1} = \dfrac{8}{16}x^{-1} = \dfrac{1}{2x}$

6. $0.0000412 = 4.12 \times 10^{-5}$

7. $125^{2/3} = \left(\sqrt[3]{125}\right)^2 = (5)^2 = 25$

8. $\sqrt[4]{64x^7y^9} = \sqrt[4]{16 \cdot 4x^4x^3y^8y} = 2xy^2\sqrt[4]{4x^3y}$

9. $\dfrac{6}{\sqrt{12}} = \dfrac{6}{2\sqrt{3}} \cdot \dfrac{\sqrt{3}}{\sqrt{3}} = \dfrac{6\sqrt{3}}{6} = \sqrt{3}$

10. $3\sqrt{80} - 7\sqrt{500} = 3(4\sqrt{5}) - 7(10\sqrt{5}) = 12\sqrt{5} - 70\sqrt{5} = -58\sqrt{5}$

11. $(8x^4 - 9x^2 + 2x - 1) - (3x^3 + 5x + 4) = 8x^4 - 3x^3 - 9x^2 - 3x - 5$

12. $(x - 3)(x^2 + x - 7) = x^3 + x^2 - 7x - 3x^2 - 3x + 21 = x^3 - 2x^2 - 10x + 21$

13. $[(x - 2) - y]^2 = (x - 2)^2 - 2y(x - 2) + y^2$
$$= x^2 - 4x + 4 - 2xy + 4y + y^2 = x^2 + y^2 - 2xy - 4x + 4y + 4$$

14. $16x^4 - 1 = (4x^2 + 1)(4x^2 - 1) = (4x^2 + 1)(2x + 1)(2x - 1)$

15. $6x^2 + 5x - 4 = (2x - 1)(3x + 4)$

16. $x^3 - 64 = x^3 - 4^3 = (x - 4)(x^2 + 4x + 16)$

17. $-\dfrac{3}{x} + \dfrac{x}{x^2 + 2} = \dfrac{-3(x^2 + 2) + x^2}{x(x^2 + 2)} = \dfrac{-2x^2 - 6}{x(x^2 + 2)} = -\dfrac{2(x^2 + 3)}{x(x^2 + 2)}$

18. $\dfrac{x - 3}{4x} \div \dfrac{x^2 - 9}{x^2} = \dfrac{x - 3}{4x} \cdot \dfrac{x^2}{(x + 3)(x - 3)} = \dfrac{x}{4(x + 3)}$

19. $\dfrac{1 - \dfrac{1}{x}}{1 - \dfrac{1}{1 - (1/x)}} = \dfrac{\dfrac{x - 1}{x}}{1 - \dfrac{1}{(x - 1)/x}} = \dfrac{\dfrac{x - 1}{x}}{1 - \dfrac{x}{x - 1}} = \dfrac{\dfrac{x - 1}{x}}{\dfrac{-1}{x - 1}} = \dfrac{x - 1}{x} \cdot \dfrac{x - 1}{-1} = \dfrac{-(x - 1)^2}{x}$

20. (a)

(b) $d = \sqrt{[5 - (-3)]^2 + (-1 - 7)^2}$
$= \sqrt{(8)^2 + (-8)^2}$
$= \sqrt{64 + 64}$
$= \sqrt{128}$
$= 8\sqrt{2}$

(c) $\left(\dfrac{-3 + 5}{2}, \dfrac{7 + (-1)}{2}\right)$
$= (1, 3)$

❑ Chapter 1 Practice Test Solutions

1. $3x - 5y = 15$

Line
x-intercept: $(5, 0)$
y-intercept: $(0, -3)$

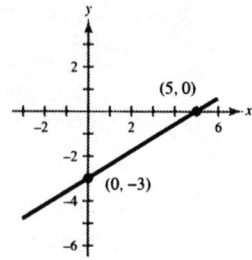

2. $y = \sqrt{9 - x}$

Domain: $(-\infty, 9]$
x-intercept: $(9, 0)$
y-intercept: $(0, 3)$

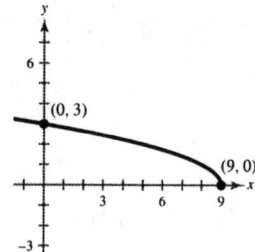

3. $5x + 4 = 7x - 8$

$4 + 8 = 7x - 5x$

$12 = 2x$

$x = 6$

4. $\dfrac{x}{3} - 5 = \dfrac{x}{5} + 1$

$15\left(\dfrac{x}{3} - 5\right) = 15\left(\dfrac{x}{5} + 1\right)$

$5x - 75 = 3x + 15$

$2x = 90$

$x = 45$

5. $\dfrac{3x + 1}{6x - 7} = \dfrac{2}{5}$

$5(3x + 1) = 2(6x - 7)$

$15x + 5 = 12x - 14$

$3x = -19$

$x = -\dfrac{19}{3}$

6. $(x - 3)^2 + 4 = (x + 1)^2$

$x^2 - 6x + 9 + 4 = x^2 + 2x + 1$

$-8x = -12$

$x = \dfrac{-12}{-8}$

$x = \dfrac{3}{2}$

7. $A = \dfrac{1}{2}(a + b)h$

$2A = ah + bh$

$2A - bh = ah$

$\dfrac{2A - bh}{h} = a$

8. Percent $= \dfrac{301}{4300} = 0.07 = 7\%$

9. Let x = number of quarters.

Then $53 - x$ = number of nickels.

$$25x + 5(53 - x) = 605$$

$$20x + 265 = 605$$

$$20x = 340$$

$$x = 17 \text{ quarters}$$

$$53 - x = 36 \text{ nickels}$$

10. Let x = amount in $9\frac{1}{2}\%$ fund.

Then $15{,}000 - x$ = amount in 11% fund.

$$0.095x + 0.11(15{,}000 - x) = 1582.50$$

$$-0.015x + 1650 = 1582.50$$

$$-0.015x = -67.5$$

$$x = \$4500 \text{ at } 9\tfrac{1}{2}\%$$

$$15{,}000 - x = \$10{,}500 \text{ at } 11\%$$

11. $28 + 5x - 3x^2 = 0$

$$(4 - x)(7 + 3x) = 0$$

$$4 - x = 0 \implies x = 4$$

$$7 + 3x = 0 \implies x = -\tfrac{7}{3}$$

12. $(x - 2)^2 = 24$

$$x - 2 = \pm\sqrt{24}$$

$$x - 2 = \pm 2\sqrt{6}$$

$$x = 2 \pm 2\sqrt{6}$$

13. $x^2 - 4x - 9 = 0$

$$x^2 - 4x + 2^2 = 9 + 2^2$$

$$(x - 2)^2 = 13$$

$$x - 2 = \pm\sqrt{13}$$

$$x = 2 \pm \sqrt{13}$$

14. $x^2 + 5x - 1 = 0$

$$a = 1, \ b = 5, \ c = -1$$

$$x = \frac{-5 \pm \sqrt{(5)^2 - 4(1)(-1)}}{2(1)}$$

$$= \frac{-5 \pm \sqrt{25 + 4}}{2} = \frac{-5 \pm \sqrt{29}}{2}$$

15. $3x^2 - 2x + 4 = 0$

$$a = 3, \ b = -2, \ c = 4$$

$$x = \frac{-(-2) \pm \sqrt{(-2)^2 - 4(3)(4)}}{2(3)}$$

$$= \frac{2 \pm \sqrt{4 - 48}}{6}$$

$$= \frac{2 \pm \sqrt{-44}}{6}$$

$$= \frac{2 \pm 2i\sqrt{11}}{6}$$

$$= \frac{1 \pm i\sqrt{11}}{3} = \frac{1}{3} \pm \frac{\sqrt{11}}{3}i$$

16.

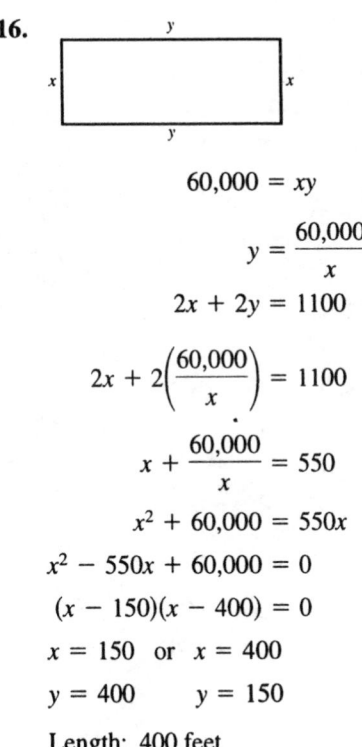

$$60{,}000 = xy$$

$$y = \frac{60{,}000}{x}$$

$$2x + 2y = 1100$$

$$2x + 2\left(\frac{60{,}000}{x}\right) = 1100$$

$$x + \frac{60{,}000}{x} = 550$$

$$x^2 + 60{,}000 = 550x$$

$$x^2 - 550x + 60{,}000 = 0$$

$$(x - 150)(x - 400) = 0$$

$$x = 150 \quad \text{or} \quad x = 400$$

$$y = 400 \qquad y = 150$$

Length: 400 feet

Width: 150 feet

17.
$$x(x + 2) = 624$$
$$x^2 + 2x - 624 = 0$$
$$(x - 24)(x + 26) = 0$$
$$x = 24 \quad \text{or} \quad x = -26, \text{ (extraneous solution)}$$
$$x + 2 = 26$$

18.
$$x^3 - 10x^2 + 24x = 0$$
$$x(x^2 - 10x + 24) = 0$$
$$x(x - 4)(x - 6) = 0$$
$$x = 0, \; x = 4, \; x = 6$$

19.
$$\sqrt[3]{6 - x} = 4$$
$$6 - x = 64$$
$$-x = 58$$
$$x = -58$$

20.
$$(x^2 - 8)^{2/5} = 4$$
$$x^2 - 8 = \pm 4^{5/2}$$
$$x^2 - 8 = 32 \quad \text{or} \quad x^2 - 8 = -32$$
$$x^2 = 40 \qquad x^2 = -24$$
$$x = \pm\sqrt{40} \qquad x = \pm\sqrt{-24}$$
$$x = \pm 2\sqrt{10} \qquad x = \pm 2\sqrt{6}i$$

21.
$$x^4 - x^2 - 12 = 0$$
$$(x^2 - 4)(x^2 + 3) = 0$$
$$x^2 = 4 \quad \text{or} \quad x^2 = -3$$
$$x^2 = \pm 2 \qquad x = \pm\sqrt{3}i$$

22.
$$4 - 3x > 16$$
$$-3x > 12$$
$$x < -4$$

23.
$$\left|\frac{x - 3}{2}\right| < 5$$
$$-5 < \frac{x - 3}{2} < 5$$
$$-10 < x - 3 < 10$$
$$-7 < x < 13$$

24.
$$\frac{x + 1}{x - 3} < 2$$
$$\frac{x + 1}{x - 3} - 2 < 0$$
$$\frac{x + 1 - 2(x - 3)}{x - 3} < 0$$
$$\frac{7 - x}{x - 3} < 0$$

Critical numbers: $x = 7$ and $x = 3$

Test intervals: $(-\infty, 3), (3, 7), (7, \infty)$

Solution intervals: $(-\infty, 3) \cup (7, \infty)$

25. $|3x - 4| \geq 9$
$$3x - 4 \leq -9 \quad \text{or} \quad 3x - 4 \geq 9$$
$$3x \leq -5 \qquad 3x \geq 13$$
$$x \leq -\frac{5}{3} \qquad x \geq \frac{13}{3}$$

❑ Chapter 2 Practice Test Solutions

1. $m = \dfrac{-1 - 4}{3 - 2} = -5$

$y - 4 = -5(x - 2)$

$y - 4 = -5x + 10$

$y = -5x + 14$

2. $y = \dfrac{4}{3}x - 3$

3. $2x + 3y = 0$

$y = -\dfrac{2}{3}x$

$m_1 = -\dfrac{2}{3}$

$\perp m_2 = \dfrac{3}{2}$ through $(4, 1)$

$y - 1 = \dfrac{3}{2}(x - 4)$

$y - 1 = \dfrac{3}{2}x - 6$

$y = \dfrac{3}{2}x - 5$

4. $(5, 32)$ and $(9, 44)$

$m = \dfrac{44 - 32}{9 - 5} = \dfrac{12}{4} = 3$

$y - 32 = 3(x - 5)$

$y - 32 = 3x - 15$

$y = 3x + 17$

When $x = 20$, $y = 3(20) + 17$

$y = \$77.$

5. $f(x - 3) = (x - 3)^2 - 2(x - 3) + 1$

$= x^2 - 6x + 9 - 2x + 6 + 1$

$= x^2 - 8x + 16$

6. $f(3) = 12 - 11 = 1$

$\dfrac{f(x) - f(3)}{x - 3} = \dfrac{(4x - 11) - 1}{x - 3}$

$= \dfrac{4x - 12}{x - 3}$

$= \dfrac{4(x - 3)}{x - 3} = 4, x \neq 3$

7. $f(x) = \sqrt{36 - x^2} = \sqrt{(6 + x)(6 - x)}$

Domain: $[-6, 6]$

Range: $[0, 6]$, because

$(6 + x)(6 - x) \geq 0$ on this interval

8. (a) $6x - 5y + 4 = 0$

$y = \dfrac{6x + 4}{5}$ is a function of x.

(b) $x^2 + y^2 = 9$

$y = \pm\sqrt{9 - x^2}$ is not a function of x.

(c) $y^3 = x^2 + 6$

$y = \sqrt[3]{x^2 + 6}$ is a function of x.

9. Parabola

Vertex: $(0, -5)$

Intercepts: $(0, -5)$, $\left(\pm\sqrt{5}, 0\right)$

y-axis symmetry

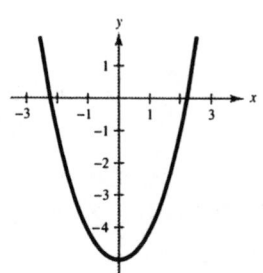

10. Intercepts: $(0, 3)$, $(-3, 0)$

x	0	1	-1	2	-2	-3	-4
y	3	4	2	5	1	0	1

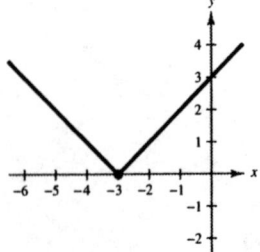

11.

x	0	1	2	3
y	1	3	5	7

x	-1	-2	-3
y	2	6	12

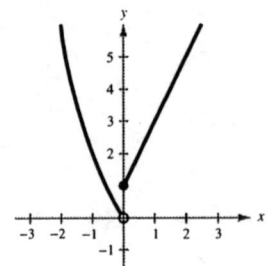

12. (a) $f(x + 2)$

Horizontal shift two units to the left

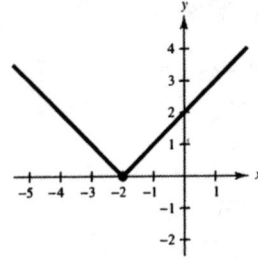

(b) $-f(x) + 2$

Reflection in the x-axis and a vertical shift two units upward

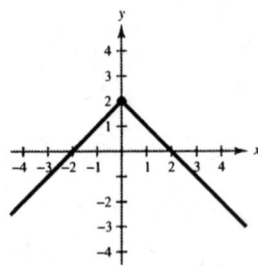

13. (a) $(g - f)(x) = g(x) - f(x)$
$$= (2x^2 - 5) - (3x + 7)$$
$$= 2x^2 - 3x - 12$$

(b) $(fg)(x) = f(x)g(x)$
$$= (3x + 7)(2x^2 - 5)$$
$$= 6x^3 + 14x^2 - 15x - 35$$

14. $f(g(x)) = f(2x + 3)$
$$= (2x + 3)^2 - 2(2x + 3) + 16$$
$$= 4x^2 + 12x + 9 - 4x - 6 + 16$$
$$= 4x^2 + 8x + 19$$

15. $f(x) = x^3 + 7$
$$y = x^3 + 7$$
$$x = y^3 + 7$$
$$x - 7 = y^3$$
$$\sqrt[3]{x - 7} = y$$
$$f^{-1}(x) = \sqrt[3]{x - 7}$$

16. (a) $f(x) = |x - 6|$ does not have an inverse. Its graph does not pass the horizontal line test.

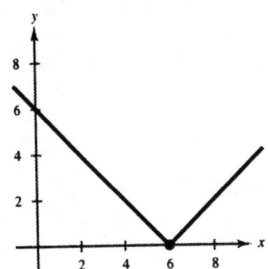

(b) $f(x) = ax + b, a \neq 0$ does have an inverse.

$$y = ax + b$$
$$x = ay + b$$
$$\frac{x - b}{a} = y$$
$$f^{-1}(x) = \frac{x - b}{a}$$

(c) $f(x) = x^3 - 19$ does have an inverse.

$$y = x^3 - 19$$
$$x = y^3 - 19$$
$$x + 19 = y^3$$
$$\sqrt[3]{x + 19} = y$$
$$f^{-1}(x) = \sqrt[3]{x + 19}$$

17.
$$f(x) = \sqrt{\frac{3 - x}{x}}, \ 0 < x \leq 3, y \geq 0$$
$$y = \sqrt{\frac{3 - x}{x}}$$
$$x = \sqrt{\frac{3 - y}{y}}$$
$$x^2 = \frac{3 - y}{y}$$
$$x^2 y = 3 - y$$
$$x^2 y + y = 3$$
$$y(x^2 + 1) = 3$$
$$y = \frac{3}{x^2 + 1}$$
$$f^{-1}(x) = \frac{3}{x^2 + 1}, \ x \geq 0$$

18. False. The slopes of 3 and $\frac{1}{3}$ are not **negative** reciprocals.

19. True. Let $y = (f \circ g)(x)$. Then $x = (f \circ g)^{-1}(y)$. Also,

$$(f \circ g)(x) = y$$
$$f(g(x)) = y$$
$$g(x) = f^{-1}(y)$$
$$x = g^{-1}(f^{-1}(y))$$
$$x = (g^{-1} \circ f^{-1})(y)$$

Since $x = x$, we have $(f \circ g)^{-1}(y) = (g^{-1} \circ f^{-1})(y)$.

20. True. It must pass the vertical line test to be a function and it must pass the horizontal line test to have an inverse.

❏ Chapter 3 Practice Test Solutions

1. *x*-intercepts: $(1, 0)$, $(5, 0)$

y-intercept: $(0, 5)$

Vertex: $(3, -4)$

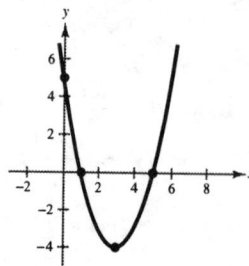

2. $a = 0.01$, $b = -90$

$$\frac{-b}{2a} = \frac{90}{2(0.01)} = 4500 \text{ units}$$

3. Vertex $(1, 7)$ opening downward through $(2, 5)$

$y = a(x - 1)^2 + 7$ Standard form

$5 = a(2 - 1)^2 + 7$

$5 = a + 7$

$a = -2$

$y = -2(x - 1)^2 + 7$

$\quad = -2(x^2 - 2x + 1) + 7$

$\quad = -2x^2 + 4x + 5$

4. $y = \pm a(x - 2)(3x - 4)$ where a is any real nonzero number.

$y = \pm(3x^2 - 10x + 8)$

5. Leading coefficient: -3

Degree: 5

Moves down to the right and up to the left.

6. $0 = x^5 - 5x^3 + 4x$

$\quad = x(x^4 - 5x^2 + 4)$

$\quad = x(x^2 - 1)(x^2 - 4)$

$\quad = x(x + 1)(x - 1)(x + 2)(x - 2)$

$x = 0, x = \pm 1, x = \pm 2$

7. $f(x) = x(x - 3)(x + 2)$

$\quad = x(x^2 - x - 6)$

$\quad = x^3 - x^2 - 6x$

8. Intercepts: $(0, 0)$, $\left(\pm 2\sqrt{3}, 0\right)$

Moves up to the right.

Moves down to the left.

x	-2	-1	0	1	2
y	16	11	0	-11	-16

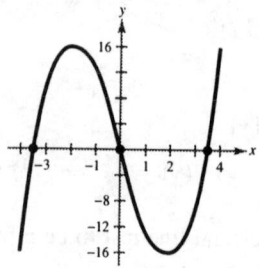

9.

$$3x^3 + 9x^2 + 20x + 62 + \frac{176}{x - 3}$$

$$x - 3\overline{)3x^4 + 0x^3 - 7x^2 + 2x - 10}$$

$$\underline{3x^4 - 9x^3}$$
$$9x^3 - 7x^2$$
$$\underline{9x^3 - 27x^2}$$
$$20x^2 + 2x$$
$$\underline{20x^2 - 60x}$$
$$62x - 10$$
$$\underline{62x - 186}$$
$$176$$

10.

$$x - 2 + \frac{5x - 13}{x^2 + 2x - 1}$$

$$x^2 + 2x - 1\overline{)x^3 + 0x^2 + 0x - 11}$$

$$\underline{x^3 + 2x^2 - x}$$
$$-2x^2 + x - 11$$
$$\underline{-2x^2 - 4x + 2}$$
$$5x - 13$$

11.

$$\begin{array}{r|rrrrrr} -5 & 3 & 13 & 0 & 0 & 12 & -1 \\ & & -15 & 10 & -50 & 250 & -1310 \\ \hline & 3 & -2 & 10 & -50 & 262 & -1311 \end{array}$$

$$\frac{3x^5 + 13x^4 + 12x - 1}{x + 5} = 3x^4 - 2x^3 + 10x^2 - 50x + 262 - \frac{1311}{x + 5}$$

12.

$$\begin{array}{r|rrrr} -6 & 7 & 40 & -12 & 15 \\ & & -42 & 12 & 0 \\ \hline & 7 & -2 & 0 & 15 \end{array}$$

$$f(-6) = 15$$

13. $0 = x^3 - 19x - 30$

Possible rational roots: $\pm 1, \pm 2, \pm 3, \pm 5, \pm 6, \pm 10, \pm 15, \pm 30$

$$\begin{array}{r|rrrr} -2 & 1 & 0 & -19 & -30 \\ & & -2 & 4 & 30 \\ \hline & 1 & -2 & -15 & 0 \end{array}$$

$$0 = (x + 2)(x^2 - 2x - 15)$$
$$0 = (x + 2)(x + 3)(x - 5)$$

Zeros: $x = -2, x = -3, x = 5$

14. $0 = x^4 + x^3 - 8x^2 - 9x - 9$

Possible rational roots: ± 1, ± 3, ± 9

$$
\begin{array}{r|rrrr}
3 & 1 & 1 & -8 & -9 & -9 \\
  &   & 3 & 12 & 12 & 9 \\
\hline
  & 1 & 4 & 4 & 3 & 0
\end{array}
$$

$0 = (x - 3)(x^3 + 4x^2 + 4x + 3)$

Possible rational roots of $x^3 + 4x^2 + 4x + 3$: $\pm 1, \pm 3$

$$
\begin{array}{r|rrrr}
-3 & 1 & 4 & 4 & 3 \\
   &   & -3 & -3 & -3 \\
\hline
   & 1 & 1 & 1 & 0
\end{array}
$$

$0 = (x - 3)(x + 3)(x^2 + x + 1)$

The zeros of $x^2 + x + 1$ are $x = \dfrac{-1 \pm \sqrt{3}\,i}{2}$.

Zeros: $x = 3, x = -3, x = -\dfrac{1}{2} + \dfrac{\sqrt{3}}{2}i, x = -\dfrac{1}{2} - \dfrac{\sqrt{3}}{2}i$

15. $0 = 6x^3 - 5x^2 + 4x - 15$

Possible rational roots: ± 1, ± 3, ± 5, ± 15, $\pm\frac{1}{2}$, $\pm\frac{3}{2}$, $\pm\frac{5}{2}$, $\pm\frac{15}{2}$, $\pm\frac{1}{3}$, $\pm\frac{5}{3}$, $\pm\frac{1}{6}$, $\pm\frac{5}{6}$

16. $0 = x^3 - \frac{20}{3}x^2 + 9x - \frac{10}{3}$

$0 = 3x^3 - 20x^2 + 27x - 10$

Possible rational roots: ± 1, ± 2, ± 5, ± 10, $\pm\frac{1}{3}$, $\pm\frac{2}{3}$, $\pm\frac{5}{3}$, $\pm\frac{10}{3}$

$$
\begin{array}{r|rrrr}
1 & 3 & -20 & 27 & -10 \\
  &   & 3 & -17 & 10 \\
\hline
  & 3 & -17 & 10 & 0
\end{array}
$$

$0 = (x - 1)(3x^2 - 17x + 10)$

$0 = (x - 1)(3x - 2)(x - 5)$

Zeros: $x = 1, x = \frac{2}{3}, x = 5$

17. Possible rational roots: $\pm 1,\ \pm 2,\ \pm 5,\ \pm 10$

$$
\begin{array}{r|rrrrr}
1 & 1 & 1 & 3 & 5 & -10 \\
  &   & 1 & 2 & 5 & 10 \\
\hline
  & 1 & 2 & 5 & 10 & 0
\end{array}
$$

$$
\begin{array}{r|rrrr}
-2 & 1 & 2 & 5 & 10 \\
   &   & -2 & 0 & -10 \\
\hline
   & 1 & 0 & 5 & 0
\end{array}
$$

$$
\begin{aligned}
f(x) &= (x - 1)(x + 2)(x^2 + 5) \\
     &= (x - 1)(x + 2)(x + 5i)(x - 5i)
\end{aligned}
$$

18.
$$
\begin{aligned}
f(x) &= (x - 2)[x - (3 + i)][x - (3 - i)] \\
     &= (x - 2)[(x - 3) - i][(x - 3) + i)] \\
     &= (x - 2)[(x - 3)^2 - (i)^2] \\
     &= (x - 2)[x^2 - 6x + 10] \\
     &= x^3 - 8x^2 + 22x - 20
\end{aligned}
$$

19.
$$
\begin{array}{r|rrrr}
3i & 1 & 4 & 9 & 36 \\
   &   & 3i & 12i - 9 & -36 \\
\hline
   & 1 & 4 + 3i & 12i & 0
\end{array}
$$

20. $z = \dfrac{kx^2}{\sqrt{y}}$

❏ Chapter 4 Practice Test Solutions

1. Vertical asymptote: $x = 0$

Horizontal asymptote: $y = \frac{1}{2}$

x-intercept: $(1, 0)$

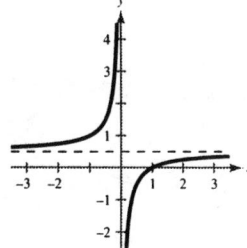

2. Vertical asymptote: $x = 0$

Horizontal asymptote: $y = 3x$

x-intercept: $\left(\pm\dfrac{2}{\sqrt{3}}, 0\right)$

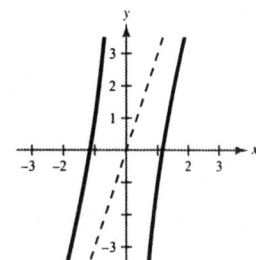

3. $y = 8$ is a horizontal asymptote since the degree on the numerator equals the degree of the denominator. There are no vertical asymptotes.

4. $x = 1$ is a vertical asymptote.

$$\frac{4x^2 - 2x + 7}{x - 1} = 4x + 2 + \frac{9}{x - 1}$$

so $y = 4x + 2$ is a slant asymptote.

5. $f(x) = \dfrac{x - 5}{(x - 5)^2} = \dfrac{1}{x - 5}$

Vertical asymptote: $x = 5$

Horizontal asymptote: $y = 0$

y-intercept: $\left(0, -\dfrac{1}{5}\right)$

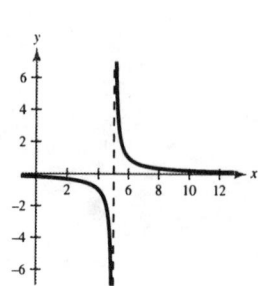

6. $\dfrac{1 - 2x}{x^2 + x} = \dfrac{1 - 2x}{x(x + 1)} = \dfrac{A}{x} + \dfrac{B}{x + 1}$

$1 - 2x = A(x + 1) + Bx$

When $x = 0,\ 1 = A.$

When $x = -1,\ 3 = -B \implies B = -3.$

$\dfrac{1 - 2x}{x^2 + x} = \dfrac{1}{x} - \dfrac{3}{x + 1}$

7. $\dfrac{6x}{x^2 - x - 2} = \dfrac{6x}{(x + 1)(x - 2)} = \dfrac{A}{x + 1} + \dfrac{B}{x - 2}$

$6x = A(x - 2) + B(x + 1)$

When $x = -1,\ -6 = -3A \implies A = 2.$

When $x = 2,\ 12 = 3B \implies B = 4.$

$\dfrac{6x}{x^2 - x - 2} = \dfrac{2}{x + 1} + \dfrac{4}{x - 2}$

8. $\dfrac{6x - 17}{(x - 3)^2} = \dfrac{A}{x - 3} + \dfrac{B}{(x - 3)^2}$

$6x - 17 = A(x - 3) + B$

When $x = 3,\ 1 = B.$

When $x = 0,\ -17 = -3A + B \implies A = 6.$

$\dfrac{6x - 17}{(x - 3)^2} = \dfrac{6}{x - 3} + \dfrac{1}{(x - 3)^2}$

9. $\dfrac{3x^2 - x + 8}{x^3 + 2x} = \dfrac{3x^2 - x + 8}{x(x^2 + 2)} = \dfrac{A}{x} + \dfrac{Bx + C}{x^2 + 2}$

$3x^2 - x + 8 = A(x^2 + 2) + (Bx + C)x$

When $x = 0,\ 8 = 2A \implies A = 4.$

When $x = 1,\ 10 = 3A + B + C \implies \quad -2 = B + C.$

When $x = -1,\ 12 = 3A + B - C \implies \quad \underline{\ \ 0 = B - C\ \ }$

$\qquad\qquad\qquad\qquad\qquad\qquad\quad -2 = 2B \qquad \implies B = -1$

$\qquad\qquad\qquad\qquad\qquad\qquad\qquad\qquad\qquad\qquad C = -1$

$\dfrac{3x^2 - x + 8}{x^3 + 2x} = \dfrac{4}{x} - \dfrac{x + 1}{x^2 + 2}$

10. $(x - 0)^2 = 4(5)(y - 0)$

Vertex: $(0, 0)$

Focus: $(0, 5)$

Directrix: $y = -5$

11. $(y - 0)^2 = 4(7)(x - 0)$

$y^2 = 28x$

12. $a = 12, b = 5, h = k = 0,$

$c = \sqrt{144 - 25} = \sqrt{119}$

Center: $(0, 0)$

Foci: $\left(\pm\sqrt{119}, 0\right)$

Vertices: $(\pm 12, 0)$

13. Center: $(0, 0)$

$c = 4, 2b = 6 \implies b = 3,$

$a = \sqrt{16 + 9} = 5$

$\dfrac{x^2}{25} + \dfrac{y^2}{9} = 1$

14. $a = 12, b = 13, c = \sqrt{144 + 169} = \sqrt{313}$

Center: $(0, 0)$

Foci: $\left(0, \pm\sqrt{313}\right)$

Vertices: $(0, \pm 12)$

Asymptotes: $y = \pm\dfrac{12}{13}x$

15. Center: $(0, 0)$

$a = 4, \pm\dfrac{1}{2} = \pm\dfrac{b}{4} \implies b = 2$

$\dfrac{x^2}{16} - \dfrac{y^2}{4} = 1$

16. $p = 4$

$(x - 6)^2 = 4(4)(y + 1)$

$(x - 6)^2 = 16(y + 1)$

17.

$$16x^2 - 96x + 9y^2 + 36y = -36$$

$$16(x^2 - 6x + 9) + 9(y^2 + 4y + 4) = -36 + 144 + 36$$

$$16(x - 3)^2 + 9(y + 2)^2 = 144$$

$$\frac{(x - 3)^2}{9} + \frac{(y + 2)^2}{16} = 1$$

$a = 4, b = 3, c = \sqrt{16 - 9} = \sqrt{7}$

Center: $(3, -2)$

Foci: $\left(3, -2 \pm \sqrt{7}\right)$

Vertices: $(3, -2 \pm 4)$ OR $(3, 2)$ and $(3, -6)$

18. Center: $(3, 1)$

$a = 4, 2b = 2 \implies b = 1$

$\dfrac{(x - 3)^2}{16} + \dfrac{(y - 1)^2}{1} = 1$

19. $a = \dfrac{1}{2}, b = \dfrac{1}{3}, c = \sqrt{\dfrac{1}{4} + \dfrac{1}{9}} = \dfrac{\sqrt{13}}{6}$

$$\dfrac{(x + 3)^2}{1/4} - \dfrac{(y - 1)^2}{1/9} = 1$$

Center: $(-3, 1)$

Vertices: $\left(-3 \pm \dfrac{1}{2}, 1\right)$ OR $\left(-\dfrac{5}{2}, 1\right)$ and $\left(-\dfrac{7}{2}, 1\right)$

Foci: $\left(-3 \pm \dfrac{\sqrt{13}}{6}, 1\right)$

Asymptotes: $y = \pm\dfrac{1/3}{1/2}(x + 3) + 1 = \pm\dfrac{2}{3}(x + 3) + 1$

20. Center: $(3, 0)$

$a = 4, c = 7, b = \sqrt{49 - 16} = \sqrt{33}$

$$\dfrac{y^2}{16} - \dfrac{(x - 3)^2}{33} = 1$$

❑ Chapter 5 Practice Test Solutions

1. $x^{3/5} = 8$

$x = 8^{5/3} = \left(\sqrt[3]{8}\right)^5 = 2^5 = 32$

2. $3^{x-1} = \frac{1}{81}$

$3^{x-1} = 3^{-4}$

$x - 1 = -4$

$x = -3$

3. $f(x) = 2^{-x} = \left(\frac{1}{2}\right)^x$

x	-2	-1	0	1	2
$f(x)$	4	2	1	$\frac{1}{2}$	$\frac{1}{4}$

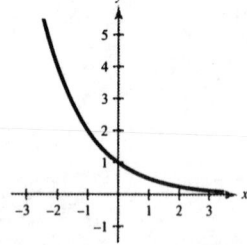

4. $g(x) = e^x + 1$

x	-2	-1	0	1	2
$g(x)$	1.14	1.37	2	3.72	8.39

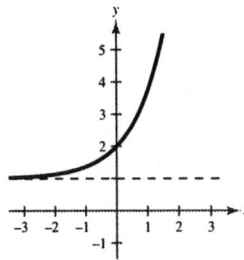

5. $A = P\left(1 + \dfrac{r}{n}\right)^{nt}$ OR $A = Pe^{rt}$

(a) $A = 5000\left(1 + \dfrac{0.09}{12}\right)^{12(3)} \approx \$6543.23$

(b) $A = 5000\left(1 + \dfrac{0.09}{4}\right)^{4(3)} \approx \$6530.25$

(c) $A = 5000e^{(0.09)(3)} \approx \$6549.82$

6. $\quad 7^{-2} = \dfrac{1}{49}$

$\log_7 \dfrac{1}{49} = -2$

7. $x - 4 = \log_2 \dfrac{1}{64}$

$2^{x-4} = \dfrac{1}{64}$

$2^{x-4} = 2^{-6}$

$x - 4 = -6$

$x = -2$

8. $\log_b \sqrt[4]{\dfrac{8}{25}} = \dfrac{1}{4} \log_b \dfrac{8}{25}$

$= \dfrac{1}{4}[\log_b 8 - \log_b 25]$

$= \dfrac{1}{4}[\log_b 2^3 - \log_b 5^2]$

$= \dfrac{1}{4}[3 \log_b 2 - 2 \log_b 5]$

$= \dfrac{1}{4}[3(0.3562) - 2(0.8271)]$

$= -0.1464$

9. $5 \ln x - \dfrac{1}{2} \ln y + 6 \ln z = \ln x^5 - \ln \sqrt{y} + \ln z^6 = \ln\left(\dfrac{x^5 z^6}{\sqrt{y}}\right)$

10. $\log_9 28 = \dfrac{\log 28}{\log 9} \approx 1.5166$

11. $\log N = 0.6646$

$N = 10^{0.6646} \approx 4.62$

12.

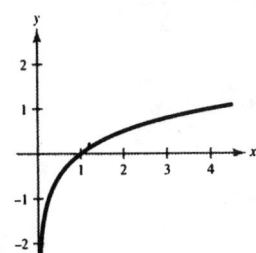

13. Domain:

$x^2 - 9 > 0$

$(x + 3)(x - 3) > 0$

$x < -3 \text{ or } x > 3$

14.

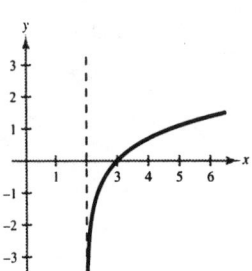

15. $\dfrac{\ln x}{\ln y} \neq \ln(x - y)$ since $\dfrac{\ln x}{\ln y} = \log_y x$.

16. $5^3 = 41$

$x = \log_5 41 = \dfrac{\ln 41}{\ln 5} \approx 2.3074$

17. $x - x^2 = \log_5 \frac{1}{25}$

$5^{x-x^2} = \frac{1}{25}$

$5^{x-x^2} = 5^{-2}$

$x - x^2 = -2$

$0 = x^2 - x - 2$

$0 = (x + 1)(x - 2)$

$x = -1$ or $x = 2$

18. $\log_2 x + \log_2(x - 3) = 2$

$\log_2[x(x - 3)] = 2$

$x(x - 3) = 2^2$

$x^2 - 3x = 4$

$x^2 - 3x - 4 = 0$

$(x + 1)(x - 4) = 0$

$x = 4$

$x = -1$ (extraneous)

$x = 4$ is the only solution.

19. $\dfrac{e^x + e^{-x}}{3} = 4$

$e^x(e^x + e^{-x}) = 12e^x$

$e^{2x} + 1 = 12e^x$

$e^{2x} - 12e^x + 1 = 0$

$e^x = \dfrac{12 \pm \sqrt{144 - 4}}{2}$

$e^x \approx 11.9161$ or $e^x \approx 0.0839$

$x = \ln 11.9161$ $x = \ln 0.0839$

$x \approx 2.478$ $x \approx -2.478$

20. $A = Pe^{et}$

$12{,}000 = 6000e^{0.13t}$

$2 = e^{0.13t}$

$0.13t = \ln 2$

$t = \dfrac{\ln 2}{0.13}$

$t \approx 5.3319$ years or 5 years 4 months

❑ Chapter 6 Practice Test Solutions

1. $x + y = 1$

$3x - y = 15 \implies y = 3x - 15$

$x + (3x - 15) = 1$

$4x = 16$

$x = 4$

$y = -3$

Answer: $(4, -3)$

2. $x - 3y = -3 \implies x = 3y - 3$

$x^2 + 5y = 5$

$(3y - 3)^2 + 6y = 5$

$9y^2 - 18y + 9 + 6y = 5$

$9y^2 - 12y + 4 = 0$

$(3y - 2)^2 = 0$

$y = \frac{2}{3}$

$x = -1$

Answer: $\left(-1, \frac{2}{3}\right)$

3. $x + y + z = 6 \implies z = 6 - x - y$

$2x - y + 3z = 0 \implies 2x - y + 3(6 - x - y) = 0 \implies -x - 4y = -18 \implies x = 18 - 4y$

$5x + 2y - z = -3 \implies 5x + 2y - (6 - x - y) = -3 \implies 6x + 3y = 3$

$6(18 - 4y) + 3y = 3$

$$-21y = -105$$

$$y = 5$$

$$x = 18 - 4y = -2$$

$$z = 6 - x - y = 3$$

Answer: $(-2, 5, 3)$

4. $x + y = 110 \implies y = 110 - x$

$xy = 2800$

$x(110 - x) = 2800$

$$0 = x^2 - 110x + 2800$$

$$0 = (x - 40)(x - 70)$$

$$x = 40 \quad \text{or} \quad x = 70$$

$$y = 70 \qquad y = 40$$

Answer: The two numbers are 40 and 70.

5. $2x + 2y = 170 \implies y = \dfrac{170 - 2x}{2} = 85 - x$

$xy = 2800$

$x(85 - x) = 2800$

$$0 = x^2 - 85x + 2800$$

$$0 = (x - 25)(x - 60)$$

$$x = 25 \quad \text{or} \quad x = 60$$

$$y = 60 \qquad y = 25$$

Dimensions: $60' \times 25'$

6. $2x + 15y = 4 \implies 2x + 15y = 4$

$x - 3y = 23 \implies \underline{5x - 15y = 115}$

$$7x \qquad\quad = 119$$

$$x = 17$$

$$y = \frac{x - 23}{3}$$

$$= -2$$

Answer: $(17, -2)$

7. $x + y = 2 \implies 19x + 19y = 38$

$38x - 19y = 7 \implies \underline{38x - 19y = 7}$

$$57x \qquad\quad = 45$$

$$x = \frac{45}{57} = \frac{15}{19}$$

$$y = 2 - x = \frac{38}{19} - \frac{15}{19} = \frac{23}{19}$$

Answer: $\left(\dfrac{15}{19}, \dfrac{23}{19} \right)$

8. $0.4x + 0.5y = 0.112 \implies 0.28x + 0.35y = 0.0784$

$0.3x - 0.7y = -0.131 \implies \underline{0.15x - 0.35y = -0.0655}$

$$0.43x \qquad\quad = 0.0129$$

$$x = \frac{0.0129}{0.43} = 0.03$$

$$y = \frac{0.112 - 0.4x}{0.5} = 0.20$$

Answer: $(0.03, 0.20)$

9. Let $x = $ amount in 11% fund and $y = $ amount in 13% fund.

$$x + y = 17000 \implies y = 17000 - x$$

$$0.11x + 0.13y = 2080$$

$$0.11x + 0.13(17000 - x) = 2080$$

$$-0.02x = -130$$

$$x = \$6500 \quad \text{at } 11\%$$

$$y = \$10,500 \text{ at } 13\%$$

10. $(4, 3), (1, 1), (-1, -2), (-2, -1)$

$$n = 4, \sum_{i=1}^{4} x_i = 2, \sum_{i=1}^{4} y_i = 1, \sum_{i=1}^{4} x_i^2 = 22, \sum_{i=1}^{4} x_i y_i = 17$$

$$4b + 2a = 1 \implies 4b + 2a = 1$$

$$2b + 22a = 17 \implies \underline{-4b - 44a = -34}$$

$$-42a = -33$$

$$a = \tfrac{33}{42} = \tfrac{11}{14}$$

$$b = \tfrac{1}{4}\left(1 - 2\left(\tfrac{33}{42}\right)\right) = -\tfrac{1}{7}$$

$$y = ax + b = \tfrac{11}{14} x - \tfrac{1}{7}$$

11.

$$\begin{aligned}
x + y &= -2 \implies -2x - 2y = 4 & -9y + 3z &= 45 \\
2x - y + z &= 11 \qquad\quad 2x - y + z = 11 & 4y - 3z &= -20 \\
4y - 3z &= -20 \qquad\quad \overline{ -3y + z = 15} & \overline{-5y} &= 25 \\
& & y &= -5 \\
& & x &= 3 \\
& & z &= 0
\end{aligned}$$

Answer: $(3, -5, 0)$

12.

$$\begin{aligned}
4x - y + 5z &= 4 \implies 4x - y + 5z = 4 \\
2x + y - z &= 0 \implies -4x - 2y + 2z = 0 \\
2x + 4y + 8z &= 0 \qquad\quad \overline{ -3y + 7z = 4} \\
& \qquad\qquad\quad\; 2x + 4y + 8z = 0 \\
& \qquad\qquad\quad -2x - y + z = 0 \\
& \qquad\qquad\quad \overline{ \; 3y + 9z = 0} \\
& \qquad\qquad\quad -3y + 7z = 4 \\
& \qquad\qquad\quad \overline{ \; 16z = 4} \\
& \qquad\qquad\qquad\quad\; z = \tfrac{1}{4} \\
& \qquad\qquad\qquad\quad\; y = -\tfrac{3}{4} \\
& \qquad\qquad\qquad\quad\; x = \tfrac{1}{2}
\end{aligned}$$

Answer: $\left(\tfrac{1}{2}, -\tfrac{3}{4}, \tfrac{1}{4}\right)$

13. $3x + 2y - z = 5 \implies 6x + 4y - 2z = 10$

   $6x - y + 5z = 2 \implies -6x + y - 5z = -2$

$$\overline{ 5y - 7z = 8}$$

$$y = \frac{8 + 7z}{5}$$

$3x + 2y - z = 5$

$12x - 2y + 10z = 4$

$$\overline{15x + 9z = 9}$$

$$x = \frac{9 - 9z}{15} = \frac{3 - 3z}{5}$$

Let $z = a$, then $x = \dfrac{3 - 3a}{5}$ and $y = \dfrac{8 + 7a}{5}$.

Answer: $\left(\dfrac{3 - 3a}{5}, \dfrac{8 + 7a}{5}, a\right)$ where a is any real number.

14. $y = ax^2 + bx + c$ passes through $(0, -1)$, $(1, 4)$, and $(2, 13)$.

At $(0, -1)$: $-1 = a(0)^2 + b(0) + c \implies c = -1$

At $(1, 4)$: $4 = a(1)^2 + b(1) - 1 \implies 5 = a + b \implies 5 = a + b$

At $(2, 13)$: $13 = a(2)^2 + b(2) - 1 \implies 14 = 4a + 2b \implies -7 = -2a - b$

$$-2 = -a$$
$$a = 2$$
$$b = 3$$

Thus, $y = 2x^2 + 3x - 1$.

15. $s = \frac{1}{2}at^2 + v_0 t + s_0$ passes through $(1, 12)$, $(2, 5)$, and $(3, 4)$.

At $(1, 12)$: $12 = \frac{1}{2}a + v_0 + s_0 \implies 24 = a + 2v_0 + 2s_0$

At $(2, 5)$: $5 = 2a + 2v_0 + s_0 \implies -5 = -2a - 2v_0 - s_0$

At $(3, 4)$: $4 = \frac{9}{2}a + 3v_0 + s_0 \implies \overline{19 = -a + s_0}$

$$15 = 6a + 6v_0 + 3s_0$$
$$\cdot -8 = -9a - 6v_0 - 2s_0$$
$$\overline{7 = -3a + s_0}$$
$$-19 = a - s_0$$
$$\overline{-12 = -2a}$$
$$a = 6$$
$$s_0 = 25$$
$$v_0 = -16$$

Thus, $s = \frac{1}{2}(6)t^2 - 16t + 25 = 3t^2 - 16t + 25$.

16. $x^2 + y^2 \geq 9$

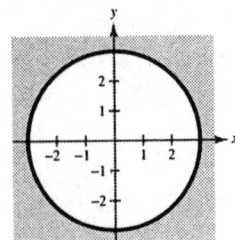

17. $x + y \leq 6$

$x \geq 2$

$y \geq 0$

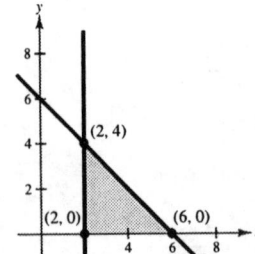

18. Line through $(0, 0)$ and $(0, 7)$:

$x = 0$

Line through $(0, 0)$ and $(2, 3)$:

$y = \frac{3}{2}x$ or $3x - 2y = 0$

Line through $(0, 7)$ and $(2, 3)$:

$y = -2x + 7$ or $2x + y = 7$

Inequalities: $x \geq 0$

$3x - 2y \leq 0$

$2x + y \leq 7$

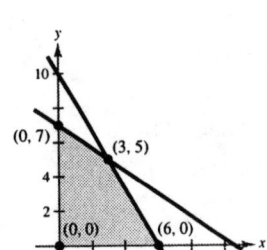

19. Vertices: $(0, 0)$, $(0, 7)$, $(6, 0)$, $(3, 5)$

$z = 30x + 26y$

At $(0, 0)$: $z = 0$

At $(0, 7)$: $z = 182$

At $(6, 0)$: $z = 180$

At $(3, 5)$: $z = 220$

The maximum value of z is 220.

20. $x^2 + y^2 \leq 4$

$(x - 2)^2 + y^2 \geq 4$

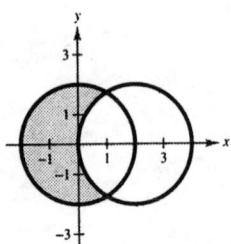

❑ **Chapter 7 Practice Test Solutions**

1.
$$\begin{bmatrix} 1 & -2 & 4 \\ 3 & -5 & 9 \end{bmatrix}$$

$-3R_1 + R_2 \rightarrow \begin{bmatrix} 1 & -2 & 4 \\ 0 & 1 & -3 \end{bmatrix}$

$2R_2 + R_1 \rightarrow \begin{bmatrix} 1 & 0 & -2 \\ 0 & 1 & -3 \end{bmatrix}$

2. $3x + 5y = 3$
$\quad 2x - y = -11$

$$\begin{bmatrix} 3 & 5 & \vdots & 3 \\ 2 & -1 & \vdots & -11 \end{bmatrix}$$

$-R_2 + R_1 \rightarrow \begin{bmatrix} 1 & 6 & \vdots & 14 \\ 2 & -1 & \vdots & -11 \end{bmatrix}$

$-2R_1 + R_2 \rightarrow \begin{bmatrix} 1 & 6 & \vdots & 14 \\ 0 & -13 & \vdots & -39 \end{bmatrix}$

$-\frac{1}{13}R_2 \rightarrow \begin{bmatrix} 1 & 6 & \vdots & 14 \\ 0 & 1 & \vdots & 3 \end{bmatrix}$

$-6R_2 + R_1 \rightarrow \begin{bmatrix} 1 & 0 & \vdots & -4 \\ 0 & 1 & \vdots & 3 \end{bmatrix}$

$x = -4, y = 3$

Answer: $(-4, 3)$

3. $2x + 3y = -3$
$\quad 3x - 2y = 8$
$\quad\ x + y = 1$

$$\begin{bmatrix} 2 & 3 & \vdots & -3 \\ 3 & 2 & \vdots & 8 \\ 1 & 1 & \vdots & 1 \end{bmatrix}$$

$\begin{matrix} R_3 \rightarrow \\ \\ R_1 \rightarrow \end{matrix} \begin{bmatrix} 1 & 1 & \vdots & 1 \\ 3 & 2 & \vdots & 8 \\ 2 & 3 & \vdots & -3 \end{bmatrix}$

$\begin{matrix} \\ -3R_1 + R_2 \rightarrow \\ -2R_1 + R_3 \rightarrow \end{matrix} \begin{bmatrix} 1 & 1 & \vdots & 1 \\ 0 & -1 & \vdots & 5 \\ 0 & 1 & \vdots & -5 \end{bmatrix}$

$\begin{matrix} R_2 + R_1 \rightarrow \\ -R_2 \rightarrow \\ -R_2 + R_3 \rightarrow \end{matrix} \begin{bmatrix} 1 & 0 & \vdots & 6 \\ 0 & 1 & \vdots & -5 \\ 0 & 0 & \vdots & 0 \end{bmatrix}$

$x = 6, y = -5$

Answer: $(6, -5)$

4. $x \quad\ + 3z = -5$
$\quad 2x + y \quad\ = 0$
$\quad 3x + y - z = -3$

$$\begin{bmatrix} 1 & 0 & 3 & \vdots & -5 \\ 2 & 1 & 0 & \vdots & 0 \\ 3 & 1 & -1 & \vdots & 3 \end{bmatrix}$$

$\begin{matrix} \\ -2R_1 + R_2 \rightarrow \\ -3R_1 + R_3 \rightarrow \end{matrix} \begin{bmatrix} 1 & 0 & 3 & \vdots & -5 \\ 0 & 1 & -6 & \vdots & 10 \\ 0 & 1 & -10 & \vdots & 18 \end{bmatrix}$

$\begin{matrix} \\ \\ -R_2 + R_3 \rightarrow \end{matrix} \begin{bmatrix} 1 & 0 & 3 & \vdots & -5 \\ 0 & 1 & -6 & \vdots & 10 \\ 0 & 0 & -4 & \vdots & 8 \end{bmatrix}$

$\begin{matrix} -3R_3 + R_1 \rightarrow \\ 6R_3 + R_2 \rightarrow \\ -\frac{1}{4}R_4 \rightarrow \end{matrix} \begin{bmatrix} 1 & 0 & 0 & \vdots & 1 \\ 0 & 1 & 0 & \vdots & -2 \\ 0 & 0 & 1 & \vdots & -2 \end{bmatrix}$

$x = 1, y = -2, z = -2$

Answer: $(1, -2, -2)$

5. $\begin{bmatrix} 1 & 4 & 5 \\ 2 & 0 & -3 \end{bmatrix} \begin{bmatrix} 1 & 6 \\ 0 & -7 \\ -1 & 2 \end{bmatrix} = \begin{bmatrix} -4 & -12 \\ 5 & 6 \end{bmatrix}$

6. $3A - 5B = 3 \begin{bmatrix} 9 & 1 \\ -4 & 8 \end{bmatrix} - 5 \begin{bmatrix} 6 & -2 \\ 3 & 5 \end{bmatrix}$

$= \begin{bmatrix} 27 & 3 \\ -12 & 24 \end{bmatrix} - \begin{bmatrix} 30 & -10 \\ 15 & 25 \end{bmatrix}$

$= \begin{bmatrix} -3 & 13 \\ -27 & -1 \end{bmatrix}$

7. $f(A) = \begin{bmatrix} 3 & 0 \\ 7 & 1 \end{bmatrix}^2 - 7 \begin{bmatrix} 3 & 0 \\ 7 & 1 \end{bmatrix} + 8 \begin{bmatrix} 1 & 0 \\ 0 & 1 \end{bmatrix}$

$= \begin{bmatrix} 3 & 0 \\ 7 & 1 \end{bmatrix} \begin{bmatrix} 3 & 0 \\ 7 & 1 \end{bmatrix} - \begin{bmatrix} 21 & 0 \\ 49 & 7 \end{bmatrix} + \begin{bmatrix} 8 & 0 \\ 0 & 8 \end{bmatrix}$

$= \begin{bmatrix} 9 & 0 \\ 28 & 1 \end{bmatrix} - \begin{bmatrix} 21 & 0 \\ 49 & 7 \end{bmatrix} + \begin{bmatrix} 8 & 0 \\ 0 & 8 \end{bmatrix}$

$= \begin{bmatrix} -4 & 0 \\ -21 & 2 \end{bmatrix}$

8. False since

$(A + B)(A + 3B) = A(A + 3B) + B(A + 3B)$

$\qquad\qquad = A^2 + 3AB + BA + 3B^2$ and, in general, $AB \neq BA$.

9. $\begin{bmatrix} 1 & 2 & \vdots & 1 & 0 \\ 3 & 5 & \vdots & 0 & 1 \end{bmatrix}$

$-3R_1 + R_2 \rightarrow \begin{bmatrix} 1 & 2 & \vdots & 1 & 0 \\ 0 & -1 & \vdots & -3 & 1 \end{bmatrix}$

$\begin{matrix} 2R_2 + R_1 \rightarrow \\ -R_2 \rightarrow \end{matrix} \begin{bmatrix} 1 & 0 & \vdots & -5 & 2 \\ 0 & 1 & \vdots & 3 & -1 \end{bmatrix}$

$A^{-1} = \begin{bmatrix} -5 & 2 \\ 3 & -1 \end{bmatrix}$

10. $\begin{bmatrix} 1 & 1 & 1 & \vdots & 1 & 0 & 0 \\ 3 & 6 & 5 & \vdots & 0 & 1 & 0 \\ 6 & 10 & 8 & \vdots & 0 & 0 & 1 \end{bmatrix}$

$\begin{matrix} -3R_1 + R_2 \rightarrow \\ -6R_1 + R_3 \rightarrow \end{matrix} \begin{bmatrix} 1 & 1 & 1 & \vdots & 1 & 0 & 0 \\ 0 & 3 & 2 & \vdots & -3 & 1 & 0 \\ 0 & 4 & 2 & \vdots & -6 & 0 & 1 \end{bmatrix}$

$\begin{matrix} -R_2 + R_1 \rightarrow \\ \frac{1}{3}R_2 \rightarrow \\ -4R_2 + R_3 \rightarrow \end{matrix} \begin{bmatrix} 1 & 0 & \frac{1}{3} & \vdots & 2 & -\frac{1}{3} & 0 \\ 0 & 1 & \frac{2}{3} & \vdots & -1 & \frac{1}{3} & 0 \\ 0 & 0 & -\frac{2}{3} & \vdots & -2 & -\frac{4}{3} & 1 \end{bmatrix}$

$\begin{matrix} \frac{1}{2}R_3 + R_1 \rightarrow \\ R_3 + R_2 \rightarrow \\ -\frac{3}{2}R_3 \rightarrow \end{matrix} \begin{bmatrix} 1 & 0 & 0 & \vdots & 1 & -1 & \frac{1}{2} \\ 0 & 1 & 0 & \vdots & -3 & -1 & 1 \\ 0 & 0 & 1 & \vdots & 3 & 2 & -\frac{3}{2} \end{bmatrix}$

$A^{-1} = \begin{bmatrix} 1 & -1 & \frac{1}{2} \\ -3 & -1 & 1 \\ 3 & 2 & -\frac{3}{2} \end{bmatrix}$

11. (a) $x + 2y = 4$

$3x + 5y = 1$

$$\begin{bmatrix} 1 & 2 & \vdots & 1 & 0 \\ 3 & 5 & \vdots & 0 & 1 \end{bmatrix}$$

$-3R_1 + R_2 \rightarrow \begin{bmatrix} 1 & 2 & \vdots & 1 & 0 \\ 0 & -1 & \vdots & -3 & 1 \end{bmatrix}$

$\begin{matrix} -2R_2 + R_1 \rightarrow \\ -R_2 \rightarrow \end{matrix} \begin{bmatrix} 1 & 0 & \vdots & -5 & 2 \\ 0 & 1 & \vdots & 3 & -1 \end{bmatrix}$

$X = A^{-1}B = \begin{bmatrix} -5 & 2 \\ 3 & -1 \end{bmatrix}\begin{bmatrix} 4 \\ 1 \end{bmatrix} = \begin{bmatrix} -18 \\ 11 \end{bmatrix}$

$x = -18, y = 11$

Answer: $(-18, 11)$

(b) $x + 2y = 3$

$3x + 5y = -2$

$X = A^{-1}B = \begin{bmatrix} -5 & 2 \\ 3 & -1 \end{bmatrix}\begin{bmatrix} 3 \\ -2 \end{bmatrix} = \begin{bmatrix} -19 \\ 11 \end{bmatrix}$

$x = -19, y = 11$

Answer: $(-19, 11)$

12. $\begin{vmatrix} 6 & -1 \\ 3 & 4 \end{vmatrix} = 24 - (-3) = 27$

13. $\begin{vmatrix} 1 & 3 & -1 \\ 5 & 9 & 0 \\ 6 & 2 & -5 \end{vmatrix} -1\begin{vmatrix} 5 & 9 \\ 6 & 2 \end{vmatrix} - 5\begin{vmatrix} 1 & 3 \\ 5 & 9 \end{vmatrix} = 74$

14. $\begin{vmatrix} 1 & 4 & 2 & 3 \\ 0 & 1 & -2 & 0 \\ 3 & 5 & -2 & 1 \\ 2 & 0 & 6 & 1 \end{vmatrix} = \begin{vmatrix} 1 & 2 & 3 \\ 3 & -1 & 1 \\ 2 & 6 & 1 \end{vmatrix} + 2\begin{vmatrix} 1 & 4 & 3 \\ 3 & 5 & 1 \\ 2 & 0 & 1 \end{vmatrix}$

$= 51 + 2(-29) = -7$ (Expansion along Row 2.)

15. $\begin{vmatrix} 6 & 4 & 3 & 0 & 6 \\ 0 & 5 & 1 & 4 & 8 \\ 0 & 0 & 2 & 7 & 3 \\ 0 & 0 & 0 & 9 & 2 \\ 0 & 0 & 0 & 0 & 1 \end{vmatrix} = 6(5)(2)(9)(1) = 540$ (Upper triangular)

16. Area $= \frac{1}{2}\begin{vmatrix} 0 & 7 & 1 \\ 5 & 0 & 1 \\ 3 & 9 & 1 \end{vmatrix} = \frac{1}{2}(31)$

17. $\begin{vmatrix} x & y & 1 \\ 2 & 7 & 1 \\ -1 & 4 & 1 \end{vmatrix} = 3x - 3y + 15 = 0 \text{ OR} = x - y + 5 = 0$

18. $x = \dfrac{\begin{vmatrix} 4 & -7 \\ 11 & 5 \end{vmatrix}}{\begin{vmatrix} 6 & -7 \\ 2 & 5 \end{vmatrix}} = \dfrac{97}{44}$

19. $z = \dfrac{\begin{vmatrix} 3 & 0 & 1 \\ 0 & 1 & 3 \\ 1 & -1 & 2 \end{vmatrix}}{\begin{vmatrix} 3 & 0 & 1 \\ 0 & 1 & 4 \\ 1 & -1 & 0 \end{vmatrix}} = \dfrac{14}{11}$

20. $y = \dfrac{\begin{vmatrix} 721.4 & 33.77 \\ 45.9 & 19.85 \end{vmatrix}}{\begin{vmatrix} 721.4 & -29.1 \\ 45.9 & 105.6 \end{vmatrix}} = \dfrac{12{,}769.747}{77{,}515.530} \approx 0.1647$

❏ Chapter 8 Practice Test Solutions

1. $a_n = \dfrac{2n}{(n+2)!}$

$a_1 = \dfrac{2(1)}{3!} = \dfrac{2}{6} = \dfrac{1}{3}$

$a_2 = \dfrac{2(2)}{4!} = \dfrac{4}{24} = \dfrac{1}{6}$

$a_3 = \dfrac{2(3)}{5!} = \dfrac{6}{120} = \dfrac{1}{20}$

$a_4 = \dfrac{2(4)}{6!} = \dfrac{8}{720} = \dfrac{1}{90}$

$a_5 = \dfrac{2(5)}{7!} = \dfrac{10}{5040} = \dfrac{1}{504}$

Terms: $\dfrac{1}{3}, \dfrac{1}{6}, \dfrac{1}{20}, \dfrac{1}{90}, \dfrac{1}{504}$

2. $a_n = \dfrac{n+3}{3^n}$

3. $\displaystyle\sum_{i=1}^{6}(2i - 1) = 1 + 3 + 5 + 7 + 9 + 11 = 36$

4. $a_1 = 23, d = -2$

$a_2 = 23 + (-2) = 21$

$a_3 = 21 + (-2) = 19$

$a_4 = 19 + (-2) = 17$

$a_5 = 17 + (-2) = 15$

Terms: 23, 21, 19, 17, 15

5. $a_1 = 12, d = 3, n = 50$

$a_n = a_1 + (n-1)d$

$a_{50} = 12 + (50 - 1)3 = 159$

6. $a_1 = 1$

$a_{200} = 200$

$S_n = \dfrac{n}{2}(a_1 + a_n)$

$S_{200} = \dfrac{200}{2}(1 + 200) = 20{,}100$

7. $a_1 = 7,\ r = 2$

$a_2 = 7(2) = 14$

$a_3 = 7(2)^2 = 28$

$a_4 = 7(2)^3 = 56$

$a_5 = 7(2)^4 = 112$

Terms: 7, 14, 28, 56, 112

8. $\displaystyle\sum_{n=0}^{9} 6\left(\dfrac{2}{3}\right)^n,\ a_1 = 6,\ r = \dfrac{2}{3},\ n = 10$

$S_n = \dfrac{a_1(1 - r^n)}{1 - r} = \dfrac{6\left(1 - \left(\frac{2}{3}\right)^{10}\right)}{1 - \frac{2}{3}} \approx 17.6879$

9. $\displaystyle\sum_{n=0}^{\infty} (0.03)^n,\ a_1 = 1,\ r = 0.03$

$S = \dfrac{a_1}{1 - r} = \dfrac{1}{1 - 0.03} = \dfrac{1}{0.97} = \dfrac{100}{97} \approx 1.0309$

10. For $n = 1$, $1 = \dfrac{1(1 + 1)}{2}$.

Assume that $1 + 2 + 3 + 4 + \cdots + k = \dfrac{k(k + 1)}{2}$.

Now for $n = k + 1$,

$1 + 2 + 3 + 4 + \cdots + k + (k + 1) = \dfrac{k(k + 1)}{2} + k + 1$

$= \dfrac{k(k + 1)}{2} + \dfrac{2(k + 1)}{2}$

$= \dfrac{(k + 1)(k + 2)}{2}$.

Thus, $1 + 2 + 3 + 4 + \cdots + n = \dfrac{n(n + 1)}{2}$ for all integers $n \geq 1$.

11. For $n = 4$, $4! > 2^4$. Assume that $k! > 2^k$.

Then $(k + 1)! = (k + 1)(k!) > (k + 1)2^k > 2 \cdot 2^k = 2^{k+1}$.

Thus, $n! > 2^n$ for all integers $n \geq 4$.

12. $_{13}C_4 = \dfrac{13!}{(13 - 4)!\,4!} = 715$

13. $(x + 3)^5 = x^5 + 5x^4(3) + 10x^3(3)^2 + 10x^2(3)^3 + 5x(3)^4 + (3)^5$

$= x^5 + 15x^4 + 90x^3 + 270x^2 + 405x + 243$

14. $_{12}C_5 x^7(-2)^5 = -25{,}344x^7$

15. $_{30}P_4 = \dfrac{30!}{(30 - 4)!} = 657{,}720$

16. $6! = 720$ ways

17. $_{12}P_3 = 1320$

18. $P(2) + P(3) + P(4) = \dfrac{1}{36} + \dfrac{2}{36} + \dfrac{3}{36}$

$$= \dfrac{6}{36} = \dfrac{1}{6}$$

19. $P(K, B10) = \dfrac{4}{52} \cdot \dfrac{2}{51} = \dfrac{2}{663}$

20. Let A = probability of no faulty units.

$$P(A) = \left(\dfrac{997}{1000}\right)^{50} \approx 0.8605$$

$$P(A') = 1 - P(A) \approx 0.1395$$

PART II

❑ Chapter Test Solutions for Chapter P

1. $-\frac{10}{3} = -3\frac{1}{3}$

$-|-4| = -4$

$-\frac{10}{3} > -|-4|$

2. $\left|-5.4 - 3\frac{3}{4}\right| = 9.15$

3. (a) $27\left(-\frac{2}{3}\right) = -18$

(b) $\dfrac{5}{18} \div \dfrac{15}{8} = \dfrac{15}{8} \cdot \dfrac{8}{15} = \dfrac{4}{27}$

4. (a) $\left(-\dfrac{3}{5}\right)^3 = -\dfrac{27}{125}$

(b) $\left(\dfrac{3^2}{2}\right)^{-3} = \left(\dfrac{2}{9}\right)^3 = \dfrac{8}{729}$

5. (a) $\sqrt{5} \cdot \sqrt{125} = \sqrt{625} = 25$

(b) $\dfrac{\sqrt{72}}{\sqrt{2}} = \sqrt{36} = 6$

6. (a) $\dfrac{5.4 \times 10^8}{3 \times 10^3} = \dfrac{5.4}{3} \times 10^{8-3} = 1.8 \times 10^5$

(b) $(3 \times 10^4)^3 = 27 \times 10^{12} = 2.7 \times 10^{13}$

7. (a) $3z^2(2z^3)^2 = 3z^2(4z^6) = 12z^8$

(b) $(u-2)^{-4}(u-2)^{-3} = (u-2)^{-7} = \dfrac{1}{(u-2)^7}$

8. (a) $\left(\dfrac{x^{-2}y^2}{3}\right)^{-1} = \dfrac{x^2y^{-2}}{3^{-1}} = \dfrac{3x^2}{y^2}$

(b) $\sqrt[3]{\dfrac{16}{v^5}} = \dfrac{2\sqrt[3]{2}}{v\sqrt[3]{v^2}} = \dfrac{2}{v}\sqrt[3]{\dfrac{2}{v^2}}$

9. (a) $9z\sqrt{8z} - 3\sqrt{2z^3} = 18z\sqrt{2z} - 3z\sqrt{2z} = 15z\sqrt{2z}$

(b) $-5\sqrt{16y} + 10\sqrt{y} = -20\sqrt{y} + 10\sqrt{y} = -10\sqrt{y}$

10. $(x^2+3) - [3x + (8 - x^2)] = x^2 + 3 - 3x - 8 + x^2$

$\qquad\qquad\qquad\qquad\qquad = 2x^2 - 3x - 5$

11. $\left(x + \sqrt{5}\right)\left(x - \sqrt{5}\right) = x^2 - \left(\sqrt{5}\right)^2 = x^2 - 5$

12. $\dfrac{8x}{x-3} + \dfrac{24}{3-x} = \dfrac{8x}{x-3} - \dfrac{24}{x-3} = \dfrac{8x-24}{x-3}$

$\qquad\qquad\qquad = \dfrac{8(x-3)}{x-3} = 8, \quad x \neq 3$

13. $\dfrac{\left(\dfrac{2}{x} - \dfrac{2}{x+1}\right)}{\left(\dfrac{4}{x^2-1}\right)} = \dfrac{2(x+1) - 2x}{x(x+1)} \cdot \dfrac{x^2-1}{4}$

$\qquad\qquad = \dfrac{2}{x(x+1)} \cdot \dfrac{(x+1)(x-1)}{4}$

$\qquad\qquad = \dfrac{x-1}{2x}, \, x \neq \pm 1$

14. (a) $2x^4 - 3x^3 - 2x^2 = x^2(2x^2 - 3x - 2)$

$$= x^2(2x + 1)(x - 2)$$

(b) $x^3 + 2x^2 - 4x - 8 = x^2(x + 2) - 4(x + 2)$

$$= (x + 2)(x^2 - 4)$$

$$= (x + 2)(x + 2)(x - 2)$$

$$= (x + 2)^2(x - 2)$$

15. (a) $\dfrac{16}{\sqrt[3]{16}} = \dfrac{16}{\sqrt[3]{16}} \cdot \dfrac{\sqrt[3]{4}}{\sqrt[3]{4}} = \dfrac{16\sqrt[3]{4}}{\sqrt[3]{64}} = \dfrac{16\sqrt[3]{4}}{4} = 4\sqrt[3]{4}$

(b) $\dfrac{6}{1 - \sqrt{3}} = \dfrac{6}{1 - \sqrt{3}} \cdot \dfrac{1 + \sqrt{3}}{1 + \sqrt{3}} = \dfrac{6(1 + \sqrt{3})}{1 - 3} = -3(1 + \sqrt{3})$

16.

Midpoint: $\left(\dfrac{-2 + 6}{2}, \dfrac{5 + 0}{2}\right) = \left(2, \dfrac{5}{2}\right)$

Distance: $d = \sqrt{(-2 - 6)^2 + (5 - 0)^2}$

$$= \sqrt{64 + 25}$$

$$= \sqrt{89}$$

17. Area = Area of large triangle − Area of small triangle

$$A = \dfrac{1}{2}(3x)(\sqrt{3x}) - \dfrac{1}{2}(2x)\left(\dfrac{2}{3}\sqrt{3x}\right)$$

$$= \dfrac{3\sqrt{3x^2}}{2} - \dfrac{2\sqrt{3x^2}}{3}$$

$$= \dfrac{9\sqrt{3x^2} = 4\sqrt{3x^2}}{6}$$

$$= \dfrac{5\sqrt{3x^2}}{6}$$

$$= \dfrac{5}{6}\sqrt{3x^2}$$

❏ Chapter Test Solutions for Chapter 1

1. $y = 4 - \dfrac{3}{4}x$

No symmetry

x-intercept: $\left(\dfrac{16}{3}, 0\right)$

y-intercept: $(0, 4)$

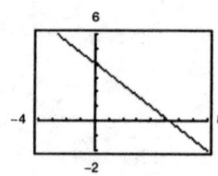

2. $y = 4 - \frac{3}{4}|x|$

y-axis symmetry

x-intercept: $\left(\pm \frac{16}{3}, 0\right)$

y-intercept: $(0, 4)$

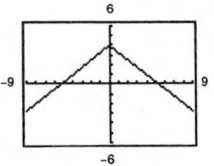

3. $y = 4 - (x - 2)^2$

Parabola; vertex: $(2, 4)$
No symmetry
x-intercepts: $(0, 0)$ and $(4, 0)$

$$0 = 4 - (x - 2)^2$$
$$(x - 2)^2 = 4$$
$$x - 2 = \pm 2$$
$$x = 2 \pm 2$$
$$x = 4 \quad \text{or} \quad x = 0$$

y-intercept: $(0, 0)$

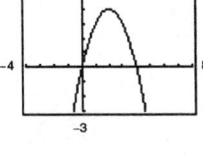

4. $y = x - x^3$

Origin symmetry

x-intercepts: $(0, 0), (1, 0), (-1, 0)$

$$0 = x - x^3$$
$$0 = x(1 + x)(1 - x)$$
$$x = 0, x = \pm 1$$

y-intercept: $(0, 0)$

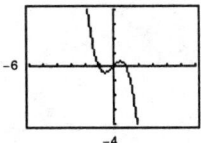

5. $y = \sqrt{3 - x}$

Domain: $x \leq 3$
No symmetry
x-intercept: $(3, 0)$
y-intercept: $\left(0, \sqrt{3}\right)$

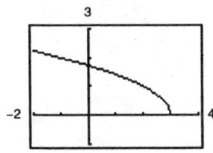

6. $(x - 3)^2 + y^2 = 9$

Circle
Center: $(3, 0)$
Radius: 3
x-axis symmetry
x-intercepts: $(0, 0)$ and $(6, 0)$
y-intercepts: $(0, 0)$

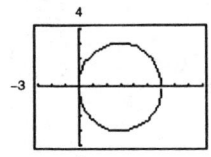

7. $\frac{2}{3}(x-1)+\frac{1}{4}x=10$

$12\left[\frac{2}{3}(x-1)+\frac{1}{4}x\right]=12(10)$

$8(x-1)+3x=120$

$8x-8+3x=120$

$11x=128$

$x=\frac{128}{11}$

8. $(x-3)(x+2)=14$

$x^2-x-6=14$

$x^2-x-20=0$

$(x+4)(x-5)=0$

$x=-4 \quad \text{or} \quad x=5$

9. $\dfrac{x-2}{x+2}+\dfrac{4}{x+2}+4=0$

$\dfrac{x+2}{x+2}=-4$

$1 \neq -4 \implies \text{No solution}$

10. $x^4+x^2-6=0$

$(x^2-2)(x^2+3)=0$

$x^2=2 \implies x=\pm\sqrt{2}$

$x^2=-3 \implies x=\pm\sqrt{3}i$

11. $2\sqrt{3}-\sqrt{2x+1}=1$

$-\sqrt{2x+1}=1-2\sqrt{x}$

$\left(-\sqrt{2x+1}\right)^2=\left(1-2\sqrt{x}\right)^2$

$2x+1=1-4\sqrt{x}+4x$

$-2x=-4\sqrt{x}$

$x=2\sqrt{x}$

$x^2=4x$

$x^2-4x=0$

$x(x-4)=0$

$x=0 \quad \text{or} \quad x=4$

Only $x=4$ is a solution to the original equation.
$x=0$ is extraneous

12. $|3x-1|=7$

$3x-1=7 \quad \text{or} \quad 3x-1=-7$

$3x=8 \qquad\qquad 3x=-6$

$x=\frac{8}{3} \qquad\qquad x=-2$

13. $-3\leq 2(x+4)<14$

$-3\leq 2x+8<14$

$-11\leq 2x<6$

$-\frac{11}{2}\leq x<3$

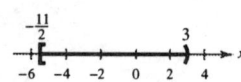

14.
$$\frac{2}{x} > \frac{5}{x+6}$$

$$\frac{2}{x} - \frac{5}{x+6} > 0$$

$$\frac{2(x+6) - 5x}{x(x+6)} > 0$$

$$\frac{-3x+12}{x(x+6)} > 0$$

$$\frac{-3(x-4)}{x(x+6)} > 0$$

Critical numbers: $x = 4, x = 0, x = -6$

Test intervals: $(-\infty, -6), (-6, 0), (0, 4), (4, \infty)$

Test: Is $\dfrac{-3(x-4)}{x(x+6)} > 0$?

Solution set: $(-\infty, -6) \cup (0, 4)$

In inequality notation: $x < -6$ or $0 < x < 4$

15. $(100 \text{ km/hr})(2\frac{1}{4} \text{ hr}) + (x \text{ km/hr})(1\frac{1}{3} \text{ hr}) = 350 \text{ km}$

$$225 + \tfrac{4}{3}x = 350$$

$$\tfrac{4}{3}x = 125$$

$$x = \tfrac{375}{4} = 93\tfrac{3}{4} \text{ km/hr}$$

16. $a + b = 100 \Longrightarrow b = 100 - a$

Area or ellipse = Area of circle

$$\pi ab = \pi(40)^2$$

$$a(100 - a) = 1600$$

$$0 = a^2 - 100a + 1600$$

$$0 = (a - 80)(a - 20)$$

$$a = 80 \Longrightarrow b = 20$$

OR

$$a = 20 \Longrightarrow b = 80$$

Since $a > b$, we choose $a = 80$ and $b = 20$.

17. (a) $10i - \left(3 + \sqrt{-25}\right) = 10i - (3 + 5i) = -3 + 5i$

(b) $\left(2 + \sqrt{3}i\right)\left(2 - \sqrt{3}i\right) = 4 - 3i^2 = 4 + 3 = 7$

(c) $\dfrac{5}{2+i} = \dfrac{5}{2+i} \cdot \dfrac{2-i}{2-i} = \dfrac{5(2-i)}{4+1} = 2 - i$

❏ Cumulative Test Solutions for Chapters P–2

1. $\dfrac{8x^2y^{-3}}{30x^{-1}y^2} = \dfrac{8x^2x}{30y^2y^3} = \dfrac{4x^3}{15y^5}, x \neq 0$

2. $\sqrt{24x^4y^3} = \sqrt{4x^4y^26y} = 2x^2y\sqrt{6y}$

3. $4x - [2x + 3(2 - x)] = 4x - [2x + 6 - 3x]$

$$= 4x - [-x + 6]$$

$$= 5x - 6$$

4. $(x - 2)(x^2 + x - 3) = x^3 + x^2 - 3x - 2x^2 - 2x + 6$

$$= x^3 - x^2 - 5x + 6$$

5. $\dfrac{2}{s + 3} - \dfrac{1}{s + 1} = \dfrac{2(s + 1) - (s + 3)}{(s + 3)(s + 1)}$

$$= \dfrac{2s + 2 - s - 3}{(s + 3)(s + 1)}$$

$$= \dfrac{s - 1}{(s + 3)(s + 1)}$$

6. $25 - (x - 2)^2 = [5 + (x - 2)][5 - (x - 2)]$

$$= (3 + x)(7 - x)$$

7. $x - 5x^2 - 6x^3 = x(1 - 5x - 6x^2)$

$$= x(1 + x)(1 - 6x)$$

8. $54 - 16x^3 = 2(27 - 8x^3)$

$$= 2(3 - 2x)(9 + 6x + 4x^2)$$

9. $x - 3y + 12 = 0$
Line
x-intercept: $(-12, 0)$
y-intercept: $(0, 4)$

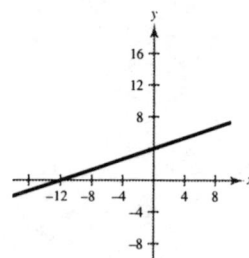

10. $y = x^2 - 9$

Parabola
x-intercepts: $(\pm 3, 0)$
y-intercepts: $(0, -9)$

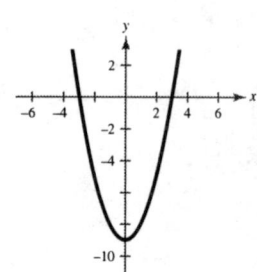

11. $y = \sqrt{4 - x}$

Domain: $x \le 4$
x-intercept: $(4, 0)$
y-intercept: $(0, 2)$

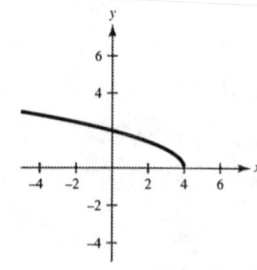

12. $2x - 3(x - 4) = 5$

$\quad 2x - 3x + 12 = 5$

$\qquad\qquad -x = -7$

$\qquad\qquad\ x = 7$

14. $\sqrt{x + 10} = x - 2$

$\quad x + 10 = x^2 - 4x + 4$

$\qquad 0 = x^2 - 5x - 6$

$\qquad 0 = (x - 6)(x + 1)$

$\qquad x = 6 \ \text{ or } \ x = -1$

Only $x = 6$ is a solution to the original equation.

$x = -1$ is extraneous.

13. $3y^2 + 6y + 2 = 0$

$\qquad 3y^2 + 6y = -2$

$\qquad\quad y^2 + 2y = -\dfrac{2}{3}$

$\quad y^2 + 2y + 1 = -\dfrac{2}{3} + 1$

$\qquad\quad (y + 1)^2 = \dfrac{1}{3}$

$\qquad\quad\ y + 1 = \pm\sqrt{\dfrac{1}{3}}$

$\qquad\qquad\ \ y = -1 \pm \dfrac{\sqrt{3}}{3}$

$\qquad\qquad\quad = -1 \pm \dfrac{1}{3}\sqrt{3}$

15. $|4(x - 2)| < 28$

$\quad -28 < 4(x - 2) < 28$

$\qquad -7 < x - 2 < 7$

$\qquad -5 < x < 9$

16. $\quad m = \dfrac{8 - 1}{3 - (-1/2)} = \dfrac{7}{7/2} = 2$

$\quad y - 8 = 2(x - 3)$

$\quad y - 8 = 2x - 6$

$\qquad\ 0 = 2x - y + 2$

17. It fails the vertical line test. For some values of x there corresponds two values of y.

18. $f(x) = \dfrac{x}{x - 2}$

$\quad$(a) $f(6) = \dfrac{6}{4} = \dfrac{3}{2}$

$\quad$(b) $f(2) = \dfrac{2}{0}$ is undefined.

$\quad$(c) $f(s + 2) = \dfrac{s + 2}{(s + 2) - 2} = \dfrac{s + 2}{s}$

19. $y = \sqrt[3]{x}$

$\quad$(a) $r(x) = \frac{1}{2}\sqrt[3]{x}$ is a vertical shrink by $\frac{1}{2}$.

$\quad$(b) $h(x) = \sqrt[3]{x} + 2$ is a vertical shift two units upward.

$\quad$(c) $g(x) = \sqrt[3]{x + 2}$ is a horizontal shift two units to the left.

20.

$$h(x) = 5x - 2$$

$$y = 5x - 2$$

$$x = 5y - 2$$

$$x + 2 = 5y$$

$$\frac{1}{5}(x + 2) = y$$

$$h^{-x}(x) = \frac{1}{5}(x + 2)$$

21. Cost per person: $\dfrac{36{,}000}{n}$

If three additional people join the group, the cost per person is $\dfrac{36{,}000}{n + 3}$.

$$\frac{36{,}000}{n} = \frac{36{,}000}{n + 3} + 1000$$

$$36{,}000\,(n + 3) = 36{,}000n + 1000n(n + 3)$$

$$36(n + 3) = 36n + n(n + 3)$$

$$36n + 108 = 36n + n^2 + 3n$$

$$0 = n^2 + 3n - 108$$

$$0 = (n + 12)(n - 9)$$

Choosing the positive value, we have $n = 9$ people.

❑ Chapter Test Solutions for Chapter 3

1. $f(x) = x^2$

(a) $g(x) = 2 - x^2$
Reflection in the x-axis followed by a vertical shift two units upward

(b) $g(x) = \left(x - \frac{3}{2}\right)^2$
Horizontal shift $\frac{3}{2}$ units to the right

2. $y = x^2 + 4x + 3$

$$= x^2 + 4x + 4 - 4 + 3$$

$$= (x + 2)^2 - 1$$

Vertex: $(-2, -1)$

x-intercepts: $0 = x^2 + 4x + 3$

$$0 = (x + 3)(x + 1)$$

$$x = -3 \ \text{ or } \ x = -1$$

$$(-3, 0) \ \text{ or } \ (-1, 0)$$

y-intercept: $(0, 3)$

3. Vertex: $(3, -6)$

$y = a(x - 3)^2 - 6$

Point on the graph: $(0, 3)$

$3 = a(0 - 3)^2 - 6$

$9 = 9a \Longrightarrow a = 1$

Thus, $y = (x - 3)^2 - 6$.

4. (a) $y = -\frac{1}{20}x^2 + 3x + 5$

$$= -\frac{1}{20}\left(x^2 - 60x + 900 - 900\right) + 5$$

$$= -\frac{1}{20}\left[(x - 30)^2 - 900\right] + 5$$

$$= -\frac{1}{20}(x - 30)^2 + 50$$

Vertex: $(30, 50)$
The maximum height is 50 feet.

(b) The constant term, $c = 5$, determines the height at which the ball was thrown. Changing this constant results in a vertical translation of the graph, so it *does* change the maximum height.

5. $h(t) = -\dfrac{3}{4}t^5 + 2t^2$

The degree is odd and the leading coefficient is negative. The graph rises to the left and falls to the right.

6.

$$x^2 + 0x + 1 \overline{\smash{\big)}\ 3x^3 + 0x^2 + 4x - 1}$$

$$\phantom{x^2 + 0x + 1 \overline{\big)}} 3x + \dfrac{x-1}{x^2+1}$$

$$-(3x^3 + 0x^2 + 3x) -$$

$$x - 1$$

Thus, $\dfrac{3x^3 + 4x - 1}{x^2 + 1} = 3x + \dfrac{x - 1}{x^2 + 1}$.

7.

$$
\begin{array}{r|rrrrr}
2 & 2 & 0 & -5 & 0 & -3 \\
  &   & 4 & 8 & 6 & 12 \\
\hline
  & 2 & 4 & 3 & 6 & 9
\end{array}
$$

Thus, $\dfrac{2x^4 - 5x^2 - 3}{x - 2} = 2x^3 + 4x^2 + 3x + 6 + \dfrac{9}{x - 2}$.

8.

$$
\begin{array}{r|rrrr}
\sqrt{3} & 4 & -1 & -12 & 3 \\
         &   & 4\sqrt{3} & 12 - \sqrt{3} & -3 \\
\hline
\end{array}
$$

$$
\begin{array}{r|rrr}
-\sqrt{3} & 4 & 4\sqrt{3} - 1 & -\sqrt{3} & 0 \\
          &   & -4\sqrt{3} & \sqrt{3} \\
\hline
          & 4 & -1 & 0
\end{array}
$$

$4x^3 - x^2 - 12x + 3 = \left(x - \sqrt{3}\right)\left(x + \sqrt{3}\right)(4x - 1)$

The real solutions are $x = \pm\sqrt{3}$ and $x = \dfrac{1}{4}$.

9. $g(t) = 2t^4 - 3t^3 + 16t - 24$

Possible rational zeros: $\pm 1, \pm 2, \pm 3, \pm 4, \pm 6, \pm 8, \pm 12, \pm 24, \pm\frac{1}{2}, \pm\frac{3}{2}$

From the graph, we have $x = -2$ and $x = \frac{3}{2}$.

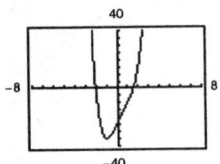

10. $h(x) = 3x^5 + 2x^4 - 3x - 2$

Possible rational zeros: $\pm 1, \pm 2, \pm\frac{1}{3}, \pm\frac{2}{3}$

From the graph, we have $x = \pm 1$ and $x = -\frac{2}{3}$.

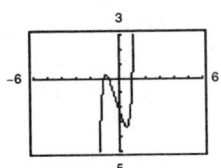

11. $f(x) = x^4 - x^3 - 1$

   $x \approx 1.380$ and $x \approx -0.819$

12. $f(x) = 3x^5 + 2x^4 - 12x - 8$

   $x \approx \pm 1.414$ and $x \approx -0.667$

13. $f(x) = x(x - 3)[x - (3 + i)][x - (3 - i)]$

$\quad = (x^2 - 3x)[(x - 3) - i][(x - 3) + i]$

$\quad = (x^2 - 3x)[(x - 3)^2 - i^2]$

$\quad = (x^2 - 3x)(x^2 - 6x + 10)$

$\quad = x^4 - 9x^3 + 28x^2 - 30x$

14. $f(x) = \left[x - \left(1 + \sqrt{3}i\right)\right]\left[x - \left(1 - \sqrt{3}i\right)\right](x - 2)(x - 2)$

$\quad = \left[(x - 1) - \sqrt{3}i\right]\left[(x - 1) + \sqrt{3}i\right](x^2 - 4x + 4)$

$\quad = \left[(x - 1)^2 - 3i^2\right](x^2 - 4x + 4)$

$\quad = (x^2 - 2x + 4)(x^2 - 4x + 4)$

$\quad = x^4 - 6x^3 + 16x^2 - 24x + 16$

❏ Chapter Test Solutions for Chapter 4

1. $y = \dfrac{3}{4 - x}$

   Domain: all real numbers except $x = 4$

   Vertical asymptote: $x = 4$

   Horizontal asymptote: $y = 0$

2. $f(x) = \dfrac{2 - x^2}{2 + x^2} = \dfrac{-x^2 + 2}{x^2 + 2}$

   Domain: all real numbers

   Vertical asymptote: None

   Horizontal asymptote: $y = \dfrac{-1}{1} = -1$

3. $g(x) = \dfrac{x^2 + 2x + 3}{x - 2} = x + 4 + \dfrac{5}{x - 2}$

   Domain: all real numbers except $x = 2$

   Vertical asymptote: $x = 2$

   Slant asymptote: $y = x + 4$

4. $h(x) = \dfrac{4}{x^2} - 1$

   Vertical asymptote: $x = 0$

   Horizontal asymptote: $y = -1$

   x-intercepts: $(\pm 2, 0)$

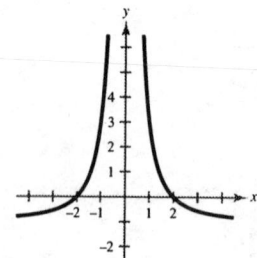

5. $g(x) = \dfrac{x^2 + 2}{x - 1} = x + 1 + \dfrac{3}{x - 1}$

   Vertical asymptote: $x = 1$

   Slant asymptote: $y = x + 1$

   y-intercepts: $(0, -2)$

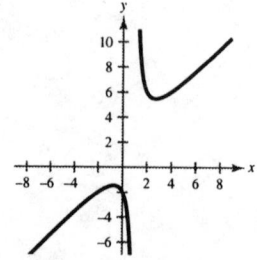

6. $x = \pm 3$ must make the denominator zero. The numerator must have the same degree as the denominator and a leading coefficient of 4 (assuming that the leading coefficient of the denominator is one). One possible function is

$$f(x) = \frac{4x^2}{x^2 - 9}.$$

7. (a) Equate the slopes.

$$\frac{y - 1}{0 - 2} = \frac{1 - 0}{2 - x}$$

$$\frac{y - 1}{-2} = \frac{1}{2 - x}$$

$$y - 1 = -2\left(\frac{1}{2 - x}\right)$$

$$y = 1 + \frac{2}{x - 2}$$

(b) $A = \frac{1}{2}xy = \frac{1}{2}x\left[1 + \frac{2}{x - 2}\right] = \frac{x}{2} + \frac{x}{x - 2} = \frac{x^2}{2(x - 2)}$

In context, we have $x > 2$ for the domain.

(c)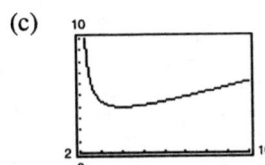

The minimum area occurs at $x = 4$ and is $A = 4$.

8. $\dfrac{2x + 5}{(x - 2)(x + 1)} = \dfrac{A}{x - 2} + \dfrac{B}{x + 1}$

$2x + 5 = A(x + 1) + B(x - 2)$

Let $x = 2$: $9 = 3A \Longrightarrow A = 3$

Let $x = -1$: $3 = -3B \Longrightarrow B = -1$

Thus, $\dfrac{2x + 5}{x^2 - x - 2} = \dfrac{3}{x - 2} - \dfrac{1}{x + 1}$.

9. $\dfrac{3x^2 - 2x + 4}{x^2(2 - x)} = \dfrac{A}{x} + \dfrac{B}{x^2} + \dfrac{C}{2 - x}$

$3x^2 - 2x + 4 = Ax(2 - x) + B(2 - x) + Cx^2$

Let $x = 0$: $4 = 2B \Longrightarrow B = 2$

Let $x = 2$: $12 = 4C \Longrightarrow C = 3$

Let $x = 1$: $5 = A + B + C \Longrightarrow A = 0$

Thus, $\dfrac{3x^2 - 2x + 4}{x^2(2 - x)} = \dfrac{2}{x^2} + \dfrac{3}{2 - x} = \dfrac{2}{x^2} - \dfrac{3}{x - 2}$.

10. $\dfrac{x^2 + 1}{x(x - 1)(x + 1)} = \dfrac{A}{x} + \dfrac{B}{x - 1} + \dfrac{C}{x + 1}$

$x^2 + 1 = A(x - 1)(x + 1) + Bx(x + 1) + Cx(x - 1)$

Let $x = 0$: $1 = -A \Longrightarrow A = -1$

Let $x = 1$: $2 = 2B \Longrightarrow B = 1$

Let $x = -1$: $2 = 2C \Longrightarrow C = 1$

Thus, $\dfrac{x^2 + 1}{x^3 - x} = -\dfrac{1}{x} + \dfrac{1}{x - 1} + \dfrac{1}{x + 1}$.

11. $\dfrac{x^2 - 1}{x(x^2 + 1)} = \dfrac{A}{x} + \dfrac{Bx + C}{x^2 + 1}$

$x^2 - 1 = A(x^2 + 1) + (Bx + C)(x)$

$= (A + B)x^2 + Cx + A$

By equating coefficients we have:

$1 = A + B$

$0 = C$

$-1 = A \Longrightarrow B = 2$

Thus, $\dfrac{x^2 - 1}{x^3 + x} = -\dfrac{1}{x} + \dfrac{2x}{x^2 + 1}$.

12. $y^2 - 4x + 4 = 0$

$$y^2 = 4(x - 1)$$

Parabola, opens to the right.
Vertex: $(1, 0)$
Focus: $(2, 0)$

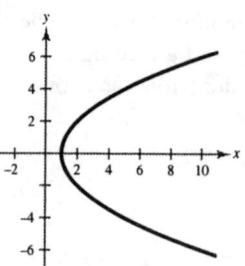

13. $\dfrac{x^2}{1} - \dfrac{y^2}{4} = 1$

Hyperbola
Center: $(0, 0)$
$a = 1, b = 2, c = \sqrt{5}$
Horizontal transverse axis
Vertices: $(\pm 1, 0)$
Foci: $\left(\pm \sqrt{5}, 0\right)$
Asymptotes: $y = \pm 2x$

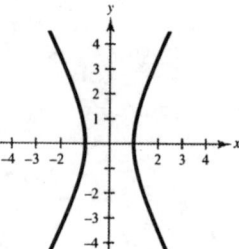

14. $x^2 - 4y^2 - 4x = 0$

$$(x^2 - 4x + 4) - 4y^2 = 0 + 4$$

$$\frac{(x - 2)^2}{4} - \frac{y^2}{1} = 1$$

Hyperbola
Center: $(2, 0)$
$a = 2, b = 1, c = \sqrt{5}$
Horizontal transverse axis
Vertices: $(0, 0)$ and $(4, 0)$
Foci: $\left(2 \pm \sqrt{5}, 0\right)$

Asymptotes: $y = \pm \dfrac{1}{2}(x - 2)$

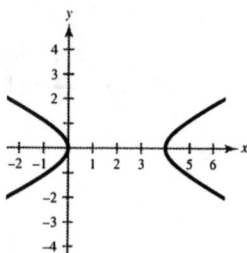

15. Ellipse
Vertices: $(0, 2)$ and $(8, 2)$
Center: $(4, 2)$
Horizontal major axis
$a = 4$

Minor axis of length 4: $2b = 4 \Longrightarrow b = 2$

$$\frac{(x - h)^2}{a^2} + \frac{(y - k)^2}{b^2} = 1$$

$$\frac{(x - 4)^2}{16} + \frac{(y - 2)^2}{4} = 1$$

16. Hyperbola
Vertices: $(0, \pm 3)$
Center: $(0, 0)$
Vertical transverse axis
$a = 3$

Asymptotes: $y = \pm \dfrac{3}{2}x$

$$\pm \frac{a}{b} = \pm \frac{3}{2} \Longrightarrow b = 2$$

$$\frac{(y - k)^2}{a^2} - \frac{(x - h)^2}{b^2} = 1$$

$$\frac{y^2}{9} - \frac{x^2}{4} = 1$$

17. $x^2 + y^2 = 36 \Longrightarrow y = \pm\sqrt{36 - x^2}$

$$x^2 - \frac{y^2}{4} = 1 \Longrightarrow y = \pm 2\sqrt{x^2 - 1}$$

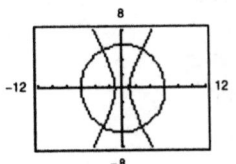

There are four points of intersection.
Solving the system algebraically yields.
$\left(\pm 2\sqrt{2}, \pm 2\sqrt{7}\right)$

❏ Cumulative Test Solutions for Chapters 3–5

1. $h(x) = -(x^2 + 4x)$

$= -(x^2 + 4x + 4 - 4)$

$= -(x + 2)^2 + 4$

Parabola

Vertex: $(-2, 4)$

Intercepts: $(-4, 0), (0, 0)$

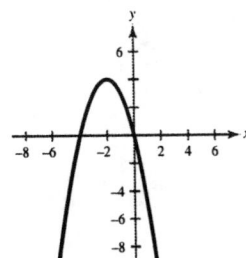

2. $f(t) = \frac{1}{4} t(t - 2)^2$

Cubic

Falls to the left

Rises to the right

Intercepts: $(0, 0), (2, 0)$

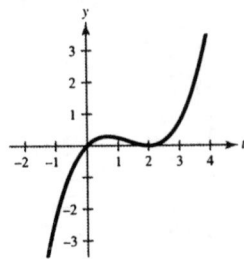

3. $g(s) = \dfrac{2s}{s - 3}$

Vertical asymptote: $s = 3$

Horizontal asymptote: $y = 2$

Intercept: $(0, 0)$

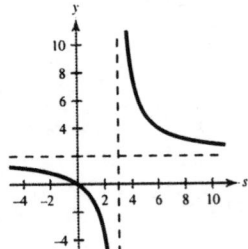

4. $g(s) = \dfrac{2s^2}{s - 3} = 2s + 6 + \dfrac{18}{s - 3}$

Vertical asymptote: $s = 3$

Slant asymptote: $y = 2s + 6$

Intercept: $(0, 0)$

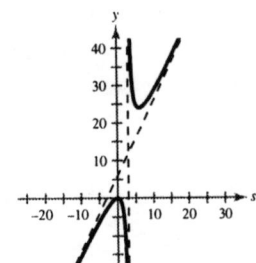

5. $f(x) = 6(2^{-x}) = 6\left(\frac{1}{2}\right)^x$

Exponential decay

Horizontal asymptote: $y = 0$

Intercept: $(0, 6)$

6. $g(x) = \log_3 x \Longrightarrow 3^y = x$

Vertical asymptote: $x = 0$

Intercept: $(1, 0)$

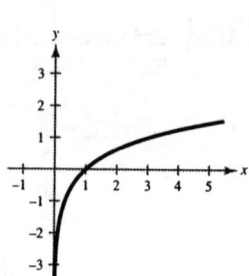

7.

$$2x^2 + 0x + 1 \overline{\smash{\big)}\, 6x^3 - 4x^2 + 0x + 0} \quad \left[3x - 2 + \frac{-3x + 2}{2x^2 + 1} \right]$$

$$\underline{-(6x^3 + 0x^2 + 3x)}$$
$$-4x^2 - 3x + 0$$
$$\underline{-(-4x^2 + 0x - 2)}$$
$$-3x + 2$$

Thus, $\dfrac{6x^3 - 4x^2}{2x^2 + 1} = 3x - 2 - \dfrac{3x - 2}{2x^2 + 1}$.

8. $f(x) = x^3 + 2x^2 + 4x + 8$

$\qquad = x^2(x + 2) + 4(x + 2)$

$\qquad = (x + 2)(x^2 + 4)$

$\quad x + 2 = 0 \Longrightarrow x = -2$

$\quad x^2 + 4 = 0 \Longrightarrow x = \pm 2i$

The zeros of $f(x)$ are -2 and $\pm 2i$.

9. $g(x) = x^3 + 3x^2 - 6$

$\qquad x \approx 1.20$

10. $6x - y^2 = 0$

$\qquad y^2 = 6x$

Parabola opening to the right

Vertex: $(0, 0)$

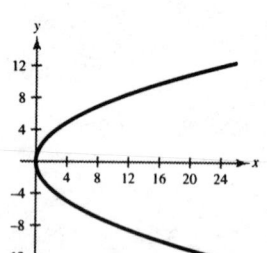

11. $\dfrac{(x - 2)^2}{4} + \dfrac{(y + 1)^2}{9} = 1$

Ellipse

Center: $(2, -1)$

Vertices: $(2, -4)$ and $(2, 2)$

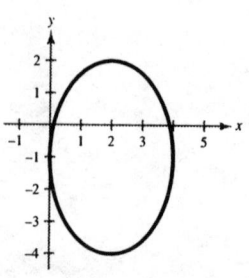

12. Parabola

Vertex: $(3, -2) \Longrightarrow y = a(x - 3)^2 - 2$

Point: $(0, 4) \Longrightarrow 4 = a(0 - 3)^2 - 2$

$$6 = 9a \Longrightarrow a = \tfrac{2}{3}$$

Equation: $y = \tfrac{2}{3}(x - 3)^2 - 2$

$$3y = 2(x^2 - 6x + 9) - 6$$

$$0 = 2x^2 - 12x - 3y + 12$$

13. Hyperbola

Foci: $(0, 0)$ and $(0, 4) \Longrightarrow$ Center: $(0, 2)$ and vertical transverse axis

Asymptotes: $y = \pm\dfrac{1}{2}x + 2 \Longrightarrow \dfrac{a}{b} = \dfrac{1}{2} \Longrightarrow 2a = b$

$c^2 = a^2 + b^2 \Longrightarrow 4 = a^2 + 4a^2 \Longrightarrow a^2 = \dfrac{4}{5}$ and $b^2 = \dfrac{16}{5}$

Equation: $\dfrac{(y - 2)^2}{4/5} - \dfrac{x^2}{16/5} = 1$

$$20(y - 2)^2 - 5x^2 = 16$$

$$20y^2 - 80y - 5x^2 + 80 = 16$$

$$0 = 5x^2 - 20y^2 + 80y - 64$$

14. $2 \ln x - \dfrac{1}{2} \ln(x + 5) = \ln x^2 - \ln\sqrt{x + 5}$

$$= \ln \dfrac{x^2}{\sqrt{x + 5}}, x > 0$$

15. $f(x) = \dfrac{1000}{1 + 4e^{-0.2x}}$

Horizontal asymptotes: $y = 0$ and $y = 1000$

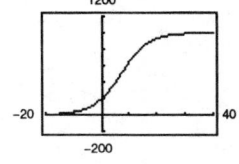

16. $6e^{2x} = 72$

$e^{2x} = 12$

$2x = \ln 12$

$x = \tfrac{1}{2} \ln 12 \approx 1.2425$

17. $\log_2 x + \log_2 5 = 6$

$\log_2 5x = 6$

$5x = 2^6$

$x = \dfrac{64}{5}$

18. $P = -\tfrac{1}{2}x^2 + 20x + 230$

$$= -\tfrac{1}{2}(x^2 - 40x + 400 - 400) + 230$$

$$= -\tfrac{1}{2}(x - 20)^2 + 430$$

The maximum occurs at the vertex when $x = 20$, which corresponds to spending $(20)(100) = \$2000$ on advertising.

19. $A = 2500e^{(0.075)(25)} \approx \$16,302.05$

❏ Chapter Test Solutions for Chapter 6

1. $x - y = 4 \implies y = x - 4$

$3x + 2y = 2 \implies 3x + 2(x - 4) = 2$

$$5x - 8 = 2$$

$$5x = 10$$

$$x = 2 \implies y = -2$$

Solution: $(2, -2)$

2. $y = x - 1$

$y = (x - 1)^3$

$x - 1 = (x - 1)^3$

$x - 1 = x^3 - 3x^2 + 3x - 1$

$0 = x^3 - 3x^2 + 2x$

$0 = x(x - 1)(x - 2)$

$x = 0, \quad x = 1, \quad x = 2$

$y = -1, \quad y = 0, \quad y = 1$

Solutions: $(0, -1), (1, 0), (2, 1)$

3. $x - y = 4 \implies x = y + 4$

$2x - y^2 = 0 \implies 2(y + 4) - y^2 = 0$

$$0 = y^2 - 2y - 8$$

$$0 = (y + 2)(y - 4)$$

$$y = -2 \quad \text{or} \quad y = 4$$

$$x = 2 \qquad x = 8$$

Solutions: $(2, -2), (8, 4)$

4. $2x - 3y = 0$

$2x + 3y = 12$

Solution: $(3, 2)$

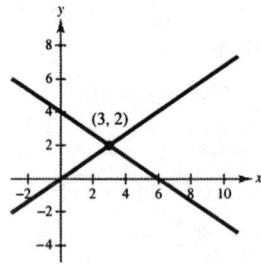

5. $y = 9 - x^2$

$y = x + 3$

Solutions: $(-3, 0), (2, 5)$

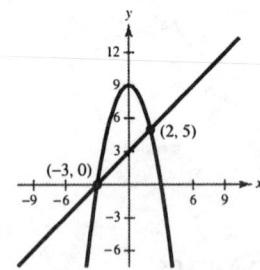

6. $y = \log_3 x \implies 3^y = x$

$y = -\frac{1}{3}x + 2$

Solutions: $(3, 1)$

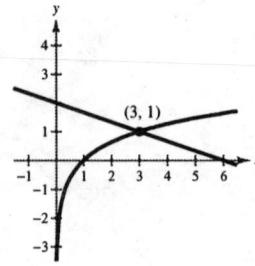

7. $2x + 3y = 17 \implies 8x + 12y = 68$

$5x - 4y = -15 \implies 15x - 12y = -45$

$$23x = 23$$

$$x = 1 \implies y = 5$$

Solution: $(1, 5)$

8.
$$x - 2y + 3z = 11$$
$$2x \quad\quad - z = 3$$
$$3y + z = -8$$

$$x - 2y + 3z = 11$$
$$4y - 7z = -19 \quad -2\,\text{Eq.1} + \text{Eq. 2}$$
$$3y + z = -8$$

$$x - 2y + 3z = 11$$
$$y - 8z = -11 \quad -\text{Eq. 3} + \text{Eq. 2}$$
$$25z = 25 \Longrightarrow z = 1$$
$$y - 8 = -11 \Longrightarrow y = -3$$
$$x - 2(-3) + 3(1) = 11 \Longrightarrow x = 2$$

Solution: $(2, -3, 1)$

9. There are infinitely many systems with the solution $\left(\frac{4}{3}, -5\right)$. One possibility is:

$$3\left(\tfrac{4}{3}\right) - (-5) = 9 \Longrightarrow 3x - y = 9$$
$$6\left(\tfrac{4}{3}\right) + (-5) = 3 \Longrightarrow 6x + y = 3$$

10. $y = ax^2 + bx + c$

$(0, 6)$: $\quad 6 = c$

$(-2, 2)$: $\quad 2 = 4a - 2b + c$

$\left(3, \frac{9}{2}\right)$: $\quad \frac{9}{2} = 9a + 3b + c$

Solving this system yields: $a = -\frac{1}{2}, b = 1,$ and $c = 6$.

Thus, $y = -\frac{1}{2}x^2 + x + 6$.

11. $2x + y \le 4$
$$2x - y \ge 0$$
$$x \ge 0$$

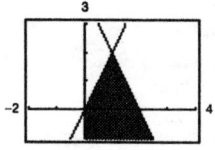

12. $y < -x^4 + x^2 + 4$
$$y > 4x$$

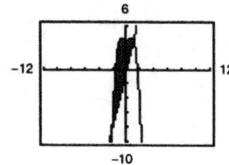

13. Line through (0, 15) and (9, 12): $x + 3y = 45$

Line through (9, 12) and (12, 5): $7x + 3y = 99$

Line through (12, 5) and (12, 0): $x = 12$

Inequalities: $x + 3y \leq 45$

$$7x + 3y \leq 99$$

$$x \leq 12$$

$$x \geq 0$$

$$y \geq 0$$

14. Maximize $z = 20x + 12y$ subject to:

$x \geq 0, \quad y \geq 0$

$x + 4y \leq 32$

$3x + 2y \leq 36$

At (0, 0) we have $z = 0$.

At (0, 8) we have $z = 96$.

At (8, 6) we have $z = 232$.

At (12, 0) we have $z = 240$.

The maximum value, $z = 240$, occurs at (12, 0).

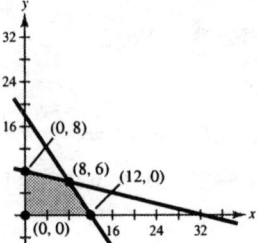

15. Maximize $P = 55x + 75y$ subject to:

$x \geq 0, \quad y \geq 0$

$x + y \leq 300$

$275x + 400y \leq 100,000$

At (0, 0) we have $z = 0$.

At (0, 250) we have $z = 18,750$.

At (160, 140) we have $z = 19,300$.

At (300, 0) we have $z = 16,500$.

The merchant should stock 160 units of the $275 model and 140 units of the $400 model to maximize profit.

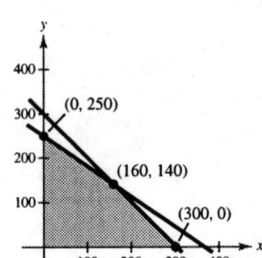

❏ Chapter Test Solutions for Chapter 7

1.
$$\begin{bmatrix} 1 & -1 & 5 \\ 6 & 2 & 3 \\ 5 & 3 & -3 \end{bmatrix}$$

$$\begin{matrix} \\ -6R_1 + R_2 \to \\ -5R_1 + R_3 \to \end{matrix} \begin{bmatrix} 1 & -1 & 5 \\ 0 & 8 & -27 \\ 0 & 8 & -28 \end{bmatrix}$$

$$\begin{matrix} \\ \\ -R_2 + R_3 \to \end{matrix} \begin{bmatrix} 1 & -1 & 5 \\ 0 & 8 & -27 \\ 0 & 0 & -1 \end{bmatrix}$$

$$\begin{matrix} R_2 + R_1 \to \\ \frac{1}{8}R_2 \to \\ -R_3 \to \end{matrix} \begin{bmatrix} 1 & 0 & \frac{13}{8} \\ 0 & 1 & -\frac{27}{8} \\ 0 & 0 & 1 \end{bmatrix}$$

$$\begin{matrix} -\frac{13}{8}R_3 + R_1 \to \\ \frac{27}{8}R_3 + R_2 \to \\ \\ \end{matrix} \begin{bmatrix} 1 & 0 & 0 \\ 0 & 1 & 0 \\ 0 & 0 & 1 \end{bmatrix}$$

2.
$$\begin{bmatrix} 1 & 0 & -1 & 2 \\ -1 & 1 & 1 & -3 \\ 1 & 1 & -1 & 1 \\ 3 & 2 & -3 & 4 \end{bmatrix}$$

$$\begin{matrix} \\ R_1 + R_2 \to \\ -R_1 + R_3 \to \\ -3R_1 + R_4 \to \end{matrix} \begin{bmatrix} 1 & 0 & -1 & 2 \\ 0 & 1 & 0 & -1 \\ 0 & 1 & 0 & -1 \\ 0 & 2 & 0 & -2 \end{bmatrix}$$

$$\begin{matrix} \\ \\ -R_2 + R_3 \to \\ -2R_2 + R_4 \to \end{matrix} \begin{bmatrix} 1 & 0 & -1 & 2 \\ 0 & 1 & 0 & -1 \\ 0 & 0 & 0 & 0 \\ 0 & 0 & 0 & 0 \end{bmatrix}$$

3.
$$\begin{bmatrix} 4 & 3 & -2 & \vdots & 14 \\ -1 & -1 & 2 & \vdots & -5 \\ 3 & 1 & -4 & \vdots & 8 \end{bmatrix} \to \begin{bmatrix} 1 & 0 & 0 & \vdots & 1 \\ 0 & 1 & 0 & \vdots & 3 \\ 0 & 0 & 1 & \vdots & -\frac{1}{2} \end{bmatrix}$$

Solution: $\left(1, 3, -\frac{1}{2}\right)$

4. $y = ax^2 + bx + c$

$(-2, -2)$: $-2 = 4a - 2b + c$

$(2, 2)$: $2 = 4a + 2b + c$

$(4, -2)$: $-2 = 16a + 4b + c$

Solving this system yields $a = -\frac{1}{2}, b = 1,$ and $c = 2$.

Thus, $y = -\frac{1}{2}x^2 + x + 2$.

5. (a) $A - B = \begin{bmatrix} 5 & 4 & 4 \\ -4 & -4 & 0 \end{bmatrix} - \begin{bmatrix} 4 & -1 & 6 \\ -4 & 0 & -3 \end{bmatrix}$

$= \begin{bmatrix} 1 & 5 & -2 \\ 0 & -4 & 3 \end{bmatrix}$

(b) $3A = 3\begin{bmatrix} 5 & 4 & 4 \\ -4 & -4 & 0 \end{bmatrix} = \begin{bmatrix} 15 & 12 & 12 \\ -12 & -12 & 0 \end{bmatrix}$

(c) $3A - 2B = 3\begin{bmatrix} 5 & 4 & 4 \\ -4 & -4 & 0 \end{bmatrix} - 2\begin{bmatrix} 4 & -1 & 6 \\ -4 & 0 & -3 \end{bmatrix}$

$= \begin{bmatrix} 15 & 12 & 12 \\ -12 & -12 & 0 \end{bmatrix} - \begin{bmatrix} 8 & -2 & 12 \\ -8 & 0 & -6 \end{bmatrix}$

$= \begin{bmatrix} 7 & 14 & 0 \\ -4 & -12 & 6 \end{bmatrix}$

6. $\begin{bmatrix} 2 & -2 & 6 \\ 3 & -1 & 7 \\ 2 & 0 & -2 \end{bmatrix}\begin{bmatrix} 4 & 4 \\ 3 & 2 \\ 1 & -2 \end{bmatrix} = \begin{bmatrix} 8 & -8 \\ 16 & -4 \\ 6 & 12 \end{bmatrix}$

7. $\begin{bmatrix} -6 & 4 \\ 10 & -5 \end{bmatrix}^{-1} = \begin{bmatrix} \frac{1}{2} & \frac{2}{5} \\ 1 & \frac{3}{5} \end{bmatrix}$

8. $\begin{bmatrix} \frac{1}{2} & \frac{2}{5} \\ 1 & \frac{3}{5} \end{bmatrix}\begin{bmatrix} 10 \\ 20 \end{bmatrix} = \begin{bmatrix} 13 \\ 22 \end{bmatrix}$

Solution: $(13, 22)$

9. $\begin{vmatrix} 4 & 0 & 3 \\ 1 & -8 & 2 \\ 3 & 2 & 2 \end{vmatrix} = -2$

10. $A = -\frac{1}{2}\begin{vmatrix} -5 & 0 & 1 \\ 4 & 4 & 1 \\ 3 & 2 & 1 \end{vmatrix} = -\frac{1}{2}(-14) = 7$

❏ Cumulative Test Solutions for Chapters 6–8

1. $y = 3 - x^2$

$2(y - 2) = x - 1 \implies 2(3 - x^2 - 2) = x - 1$

$$2(1 - x^2) = x - 1$$

$$2 - 2x^2 = x - 1$$

$$0 = 2x^2 + x - 3$$

$$0 = (2x + 3)(x - 1)$$

$$x = -\tfrac{3}{2} \quad \text{or} \quad x = 1$$

$$y = \tfrac{3}{4} \qquad \qquad y = 2$$

Solutions: $\left(-\frac{3}{2}, \frac{3}{4}\right), (1, 2)$

2. $\begin{aligned} x + 3y &= -1 \implies & 4x + 12y &= -4 \\ 2x + 4y &= 0 \implies & -6x - 12y &= 0 \\ & & -2x &= -4 \end{aligned}$

$$x = 2 \implies y = -1$$

Solution: $(2, -1)$

3. $\begin{aligned} -2x + 4y - z &= 3 \\ x - 2y + 2z &= -6 \\ x - 3y - z &= 1 \end{aligned}$

$$\begin{aligned} -2x + 4y - z &= 3 \\ x - 2y + 2z &= -6 \\ x - 3y - z &= 1 \\ x - 2y + 2z &= -6 \\ 3z &= -9 \quad \text{2 Eq. 1 + Eq. 2} \\ -y - 3z &= 7 \quad \text{−Eq. 1 + Eq. 3} \end{aligned}$$

From Equation 2 we have $z = -3$. Substituting this into Equation 3 yields $y = 2$. Using these in Equation 1 yields $x = 4$.

Solution: $(4, 2, -3)$

4. $x + 3y - 2z = -7$

$-2x + y - z = -5$

$4x + y + z = 3$

$x + 3y - 2z = -7$

$7y - 5z = -19 \qquad 2\,\text{Eq. 1} + \text{Eq. 2}$

$-11y + 9 = 31 \qquad -4\,\text{Eq. 1} + \text{Eq. 3}$

$x + \tfrac{1}{7}z = \tfrac{8}{7} \qquad -3\,\text{Eq. 2} + \text{Eq. 1}$

$y - \tfrac{5}{7}z = -\tfrac{19}{7} \qquad \tfrac{1}{7}\,\text{Eq. 2}$

$\tfrac{8}{7}z = \tfrac{8}{7} \qquad 11\,\text{Eq. 2} + \text{Eq. 3}$

$x = 1 \qquad -\tfrac{1}{7}\,\text{Eq. 3} + \text{Eq. 1}$

$y = -2 \qquad \tfrac{5}{7}\,\text{Eq. 3} + \text{Eq. 2}$

$z = 1 \qquad \tfrac{7}{8}\,\text{Eq. 3}$

Solution: $(1, -2, 1)$

5. Maximize $z = 3x + 2y$.

Subject to: $x + 4y \le 20$

$2x + y \le 12$

$x \ge 0, y \ge 0$

At $(0, 0)$: $z = 0$

At $(0, 5)$: $z = 10$

At $(4, 4)$: $z = 20$

At $(6, 0)$: $z = 18$

Maximum of $z = 20$ at $(4, 4)$

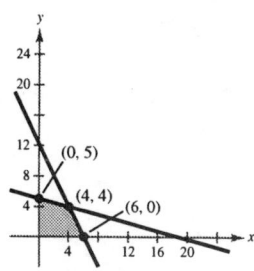

6. $\begin{bmatrix} 4 & -3 \\ 2 & 1 \\ 5 & 0 \end{bmatrix} \begin{bmatrix} 3 & -2 \\ 1 & -3 \end{bmatrix} = \begin{bmatrix} 9 & 1 \\ 7 & -7 \\ 15 & -10 \end{bmatrix}$

7. $\begin{bmatrix} 1 & 2 & -1 \\ 3 & 7 & -10 \\ -5 & -7 & -15 \end{bmatrix}^{-1} = \begin{bmatrix} -175 & 37 & -13 \\ 95 & -20 & 7 \\ 14 & -3 & 1 \end{bmatrix}$

8. $a_n = 4n + 4$

$a_1 = 8, \quad a_{20} = 84$

$S_{20} = \tfrac{20}{2}(8 + 84) = 920$

9. $\displaystyle\sum_{i=0}^{\infty} 3\left(\frac{1}{2}\right)^i = \frac{3}{1 - (1/2)} = 6$

10. $S_1 = 3 = 1(2(1) + 1)$

Assume that $S_k = 3 + 7 + 11 + 15 + \cdots + (4k - 1) = k(2k + 1)$.

Then, $S_{k+1} = 3 + 7 + 11 + 15 + \cdots + (4k - 1) + [4(k + 1) - 1]$

$$= S_k + (4k + 3)$$

$$= k(2k + 1) + (4k + 3)$$

$$= 2k^2 + 5k + 3$$

$$= (k + 1)(2k + 3)$$

$$= (k + 1)[2(k + 1) + 1].$$

Therefore, the formula is valid for all $n \geq 1$.

11. $(z - 3)^4 = z^4 - 4z^3(3) + 6z^2(3)^2 - 4z(3)^3 + (3)^4$

$$= z^4 - 12z^3 + 54z^2 - 108z + 81$$

12. $_{10}P_3 = 720$

13. The first digit is 4 or 5, so the probability of picking it correctly is $\frac{1}{2}$. Then there are two numbers left for the second digit so its probability is also $\frac{1}{2}$. If these two are correct, then the third digit must be the remaining number. The probability of winning is:

$$\left(\tfrac{1}{2}\right)\left(\tfrac{1}{2}\right)(1) = \tfrac{1}{4}$$